Bioenergy and Biofuels

Bioenergy and Biofuels

Edited by
Ozcan Konur

CRC Press is an imprint of the
Taylor & Francis Group, an **informa** business

CRC Press
Taylor & Francis Group
6000 Broken Sound Parkway NW, Suite 300
Boca Raton, FL 33487-2742

First issued in paperback 2021

ISBN 13: 978-1-03-223643-8 (pbk)
ISBN 13: 978-1-138-03281-1 (hbk)

Visit the Taylor & Francis Web site at
http://www.taylorandfrancis.com

and the CRC Press Web site at
http://www.crcpress.com

Contents

Section V: Bioenergy

Preface

I. The societal relevance of this book

As society strives for alternative fuels and energies, biofuels in general and liquid biofuels (biodiesel, bio-oil, bioethanol, biobutanol, and biomethanol), gaseous biofuels (biohydrogen, biogas, and biomethane), solid biofuels (biochars and biomass biofuels), and bioenergy (microbial fuel cells for bioelectricity generation and for bioremediation) in particular have been under close scrutiny at a global scale in recent years.

This book aims to inform readers about the recent developments in bioenergy and biofuels covering the current issues from an interdisciplinary approach, supplemented with scientometric studies of bioenergy and biofuels.

It has an interdisciplinary focus covering energy and fuels, biotechnology, genomics, economics, optimization, chemical engineering, mechanical engineering, nanoscience and nanotechnology, materials science and engineering, microbiology, ecology, agricultural engineering, and algal science, among others.

It has a matrix of biofuel and energy types (biofuels in general, liquid biofuels [biodiesel, bio-oil, bioethanol, and biobutanol], gaseous biofuels [biohydrogen and biogas], solid biofuels [biochars], and bioenergy [microbial fuel cells for bioelectricity generation and bioremediation] and feedstock types [algal biomass, lignocellulosic biomass, other types of biomass, and microbes]). It has 16 book chapters in these areas written by experts in their respective fields.

Because this book adapts an interdisciplinary approach and covers a number of bioenergy and biofuels, the target audience is large including researchers, policy makers, graduate and undergraduate students, engineers and scientists, and researchers in a number of fields including energy and fuels, biotechnology, genomics, economics, optimization, chemical engineering, mechanical engineering, plant sciences, environment engineering, environment sciences, ecology, environmental studies, politics, and algal science. However, this book is also useful for energy and fuel firms, genomic research firms, chemical and analytical instrument makers, and standard-making bodies, among others. Bioenergy and biofuel research firms and organizations, and policy making and executing organizations at a global scale would be interested in this book meeting their information needs.

This book complements many other books in the field of bioenergy and biofuels, providing an integrative approach for those published in the narrow fields of bioenergy and biofuels such as biohydrogen or algal biofuels. This book has significant potential to be used as a textbook and study book for undergraduate, graduate, and professional courses. The typical course names shall be bioenergy and biofuels, energy science and technology, algal biofuels, bioenergy and environment, energy economics, and biofuels and biotechnology.

Because this book has global scope, all countries shall have an interest in this book. As the market leaders in global bioenergy and biofuel research, the United States, China, South Korea, Taiwan, India, Malaysia, Germany, England, Sweden, Brazil, Saudi Arabia, Spain, and Turkey shall be among the key countries interested in this book.

II. This book

Section I: Bioenergy and biofuels in general

This matrix (biofuels in general, liquid biofuels, gaseous biofuels, solid biofuels, and bioenergy) has been based on the scientometric study presented in the introductory chapter by Ozcan Konur in the first section on biofuels in general. He provides for the rationale for the surge of research on bioenergy and biofuels in the 2010s and reasons that this has been due to the public societal concern over energy security, food security, and environmental issues such as global warming, greenhouse gas emission, and climate change. He provides the scientometric overview of more than 106,000 papers on bioenergy and biofuels published between 1980 and April 2017 based on a comprehensive search set of keywords. He then provides an up-to-date and integrated scientometric overview of 50 citation classics and concise summaries for each classical paper, augmented with the tabular and visual data on these papers. He augments the key rules set out for the optimal production of bioenergy and biofuels to address societal concerns in the areas of energy security, food security, and environmental issues by Hill et al. (2006): they should "provide a net energy gain, have environmental benefits, be economically competitive, and be producible in large quantities without risking food security." He urges the research community to focus on the lignocellulosic and algal biomass rather than food crops-based biomass from an interdisciplinary perspective with a special focus on genomics, nanomaterials and nanotechnologies, and microbiology to meet these strict optimal production criteria for bioenergy and biofuels. He also urges for more detailed scientometric studies on the bioenergy and biofuels to meet the crucial information needs of the society at large.

As a second chapter in the first section on biofuels, Koizumi provides an overview of biofuel developments and cooperation among China, Japan, and Malaysia as a case study of the public policy developments and cooperation worldwide. He examines the governmental policies using the optimal production criteria set out by Hill et al. (2006). He highlights the differences between these governmental policies: Energy security is the main incentive for promoting biofuel programs in China, while reducing greenhouse gas emissions is the main incentive for promoting biofuel programs in Japan, and stabilizing the palm oil price is the main incentive for promoting biofuel programs in Malaysia. He concludes that international cooperation is necessary to promote biofuel developments in these countries because promoting the sustainable criteria will become important for them. This study highlights the crucial role adapted by the national governments in providing a key set of incentive structures for the related stakeholders to produce biofuels based on the lignocellulosic or algal biomass instead of food crops-based biomass following the optimal production criteria of Hill et al. (2006).

As a third chapter in the section on biofuels, Danquah and his team focus on the algal biofuels in general section on. They first acknowledge that the algal biomass meets the optimal production criteria set out by Hill et al. (2006). They focus on preparatory technologies for biofuels production, including cultivation and harvesting technologies for microalgal biomass production and algal bioconversion technologies such as gasification, pyrolysis, and transesterification. They also provide an overview of the biochemical characteristics of various algal biomasses, as well as the production of biodiesel, bioethanol, and biogas from algal biomass. They provide a number of recommendations in algal biofuel production technologies such as improving algal biomass productivity, energy usage, water usage, nutrient requirement, production time, and better integration of biorefineries to reduce the production costs and constraints. This chapter highlights the potential of algal biomass to replace the petroleum-based fuels by meeting the optimal production criteria of Hill et al. (2006) in the long run by giving a hint of such future accomplishments based on the academic research as of 2017.

Section II: Liquid biofuels

As the first chapter on liquid biofuels, Moser provides an overview of the current issues in biodiesel ("diesel" or "biodiesel" in some cultures) production and usage in diesel engines as an alternative to petroleum-based diesel fuels. In the first stage, he sets out the basics of biodiesel such as its definition, its chemical structure, the traditional biomass used for biodiesel production (vegetable oils and animal fats), fuel standards, its advantages and disadvantages, the transesterification process used for biodiesel production, the influence of free fatty acids, and catalysts used in the biodiesel production. In the second stage, building on the first stage, he sets out the biodiesel structure–fuel property relationships. In this aspect, he focuses on the low-temperature operability, oxidative stability, kinematic viscosity, cetane number, combustion heat, lubricity, contaminants, and minor components. Building on the first two stages, he finally discusses the alternative biomass for biodiesel production. In this aspect, he focuses on jatropha oil, field pennycress, *Camelina,* and seashore mallow as alternatives to the traditional biodiesel biomass in light of the optimal production criteria set out by Hill et al. (2006). He notes that the key component in the cost of biodiesel has been the cost of biomass (up to 80% of the total cost) and he urges for the use of low-cost biomass such as waste cooking oils instead of virgin cooking oils following a two-step transesterification process. Besides the traditional catalysts such as acids and bases, he focuses on the use of enzymes for biodiesel production. Building on the structure–property relationships of biodiesel covered in this chapter, it is notable that it is possible to produce biodiesel by adjusting the biodiesel composition, usually determined by the national (e.g., the United States) or regional (European) standards. Therefore, the standard-making bodies emerge as key stakeholders in the biodiesel production processes. The key insight from this chapter is that biodiesel has a great potential to replace petroleum-based diesel in the long run, meeting the optimal production criteria set out in Hill et al. (2006) through lowering the biomass cost using low-cost biomass instead of expensive food crops-based oils and fats.

Due to its public importance, the second chapter in the section on liquid biofuels focuses on algal biodiesel ("algae diesel" or "diesel" in some cultures) production and is by Bagchi et al., building on the study on algal biofuels by Chye et al. (Chapter 3) and on the study on biodiesel by Moser (Chapter 4). They first set out the rationale for the production of algal biodiesel as an alternative to petroleum-based diesel fuels. They then focus on algal biomass and the cultivation of algal biomass in photobioreactors and raceway ponds, highlighting the need for optimal production for the biomass with high lipid content to reduce the total cost of the algal biodiesel product. They finally discuss the structure–property relationships of algal biodiesel. They argue that algal biomass is better in relation to the first-generation food crops-based biomass as well as lignocellulosic biomass. They next argue that there is room for the further optimization of the algal cultivation processes to increase the lipid productivity of the algal biomass based on the broad coverage of the literature in this aspect. As in the biofuels from other biomass, the structure and properties of algal biodiesel has been largely regulated by the national or regional standard-making bodies worldwide. Thus, the issue is to produce algal biodiesel complying with a set of standards on biodiesel structure and properties. Supporting the key findings of Moser (Chapter 4) on replacing petroleum-based diesel fuels with lignocellulosic biofuels, Bagchi et al. (Chapter 5) conclude that there is room for the replacement of petroleum-based diesel fuels with algal biodiesel in the long run, meeting the optimal production criteria set out in Hill et al. (2006). However, for this purpose there is a need to lower the production cost of algal biomass with the optimization of the algal biomass production processes, detailed in this chapter. It is likely that the design of algal biomass based on advanced genomic technologies is one of the ways to optimize the algal biomass production in the long run.

The third chapter in the section on liquid biofuels is concerned with bio-oil (in some cultures as in this chapter, "bio-oil," "pyrolysis oil," "bio-crude," or "biocrude") production as a viable alternative to petroleum-based oil and is by Alvarez et al. (Chapter 6). This chapter focuses on the fast pyrolysis of biomass to produce bio-oil for the production of biofuels and biocommodities in a biorefinery through the upgrading processes. The chapter first acknowledges the driving forces for bio-oil production and next provides an overview of fast pyrolysis process for bio-oil production. The chapter next focuses in detail on the technologies used in bio-oil production. As expected, the structure–property relationships are a focus of the chapter: composition and properties, phase separation, and stability. The final focus of the chapter is on the upgrading of bio-oil to produce biofuels and biocommodities. Regarding the discussion on the conversion technologies, it seems that there is ample room for the optimization of these processes on many fronts to optimize bio-oil production. It is pointed out that bio-oil as a microemulsion is not stable and there are ways to increase the stability of bio-oil through a number of measures. The bio-oil upgrading is usually carried out through the steam reforming of bio-oil to produce biohydrogen or through catalytic cracking and deoxygenation of bio-oil to produce biofuels and biocommodities. The chapter concludes that fast pyrolysis of biomass to produce bio-oil is workable as an alternative to petroleum-based oil. The chapter also sets out in detail the parameters of the optimal bio-oil production and upgrading to produce biohydrogen and other biofuels as well as biocommodities. It seems that there is ample room for further optimization of these processes, and thus to replace petroleum-based oil with bio-oil in the long run in the light of Hill et al. (2006).

The fourth chapter in the section on liquid biofuels is concerned with bioethanol production (ethanol or "ethanol fuel," "fuel ethanol," or "bioethanol" in some cultures) and is by Cardona-Alzate et al. (Chapter 7). After setting out the context, the chapter focuses on biomass for bioethanol production in some detail, but with special emphasis on the lignocellulosic biomass and algal biomass as an alternative to food crops-based first-generation biomass. The chapter, as expected, next focuses on the bioconversion technologies from biomass to bioethanol, namely, hydrolysis (saccharification) and fermentation. In this context, the chapter provides the overview of the various bioconversion strategies such as separated hydrolysis and fermentation (SHF), simultaneous saccharification and fermentation (SSF), simultaneous saccharification and cofermentation (SCSF), and the most recent consolidated bioprocessing (CBP). The chapter then discusses bioethanol production in a biorefinery context for optimal production. The chapter finally discusses the strength-weaknesses, opportunities, and threats (SWOT) analysis of bioethanol production. The chapter notes that there are six main groups of lignocellulosic biomass: crop residues, hardwood, softwood, cellulose wastes, herbaceous biomass, and municipal solid wastes. The chapter next highlights the importance of the pretreatment of lignocellulosic biomass for optimal bioethanol production. The chapter provides a detailed SWOT analysis of the bioconversion strategies in which each strategy has its own advantages and disadvantages from the perspective of optimal production. The production of bioethanol in the biorefinery context, as in other biofuels, offers a way for optimal production through lowering the production cost, as the chapter shows through a case study of bioethanol and biodiesel production from oil palm in Colombia. The chapter provides an overview of real-life bioethanol production. The key insight from the chapter is that despite the development of research in bioethanol from lignocellulosic biomass in recent decades, there is still ample room for the optimal production of bioethanol to replace gasoline in the market in the long run in the light of Hill et al. (2006).

The fifth chapter in the section on biofuels is concerned with biobutanol ("butanol" or "biobutanol" in some cultures) production and is by Kumar et al. (Chapter 8). After setting out the context, the chapter discusses the biomass selection for biobutanol production. In this context, lignocellulosic biomass, algal biomass, glycerol (a by-product of biodiesel production), and other biomass are considered for biobutanol production. The chapter next discusses the pretreatment of biomass for biobutanol production: acid, alkaline, enzymatic, and other pretreatments. The chapter notes acetone–butanol–ethanol (ABE) fermentation as the key method for the bioconversion of biomass to biobutanol together with bioacetone and bioethanol. The key insight from the chapter regarding the discussion of pretreatments is that, as in the case of bioethanol, pretreatment of biomass is a crucial step for the optimal production of biobutanol, usually in the combination of more than one method. The chapter next discusses the SWOT analysis of the recovery methods for biobutanol such as distillation. The genomic design of the biomass for the optimal biobutanol production appears as a cost-effective method in the long run. The concluding insight from the chapter is that there is still ample room for the optimal production of biobutanol from lignocellulosic and algal biomass to replace petroleum-based gasoline in the long run, although the ABE process has been in use for a century, in the light of Hill et al. (2006).

Section III: Gaseous biofuels

The first chapter in the section on gaseous biofuels is concerned with the overview of the issues related to biohydrogen ("hydrogen" or "H_2", "molecular hydrogen", "biohydrogen," or "biological hydrogen" in some cultures) and is by Kumari and Das (Chapter 9). It serves as an introduction to the chapters on algal biohydrogen (Tiwari et al., Chapter 11), dark fermentative biohydrogen (Pandey and Srivastava, Chapter 10), and biohydrogen through the microbial electrolysis cells (Kadier et al., Chapter 12). After setting out the context, the chapter discusses the production of biohydrogen in four ways: light-dependent [biophotolysis (water-splitting photosynthesis by algal biomass and hydrogenases) and photofermentation by photosynthetic bacteria and by biomass], and light-independent (dark fermentation by fermentative bacteria and by biomass and through microbial electrolysis cells by electrogenic bacteria and biomass). The chapter next discusses the bacteria involved in biohydrogen production in biophotolysis, photofermentation, and dark fermentation. The chapter then discusses the determinants of dark fermentative biohydrogen production and the modeling of the kinetics of dark fermentative biohydrogen production. The chapter finally discusses the biohythane (biohydrogen and biomethane) production process and the SWOT analysis of biohythane production. The chapter points out first that biohydrogen as a fuel of the future complies well with the optimal production criteria set out in Hill et al. (2006). The chapter considers biomethane production following biohydrogen production through a two-stage dark fermentative process, termed as "biohythane." The chapter also compares the biohythane process with the anaerobic digestion of biomass to produce biogas. The chapter concludes that the dark fermentative biohydrogen method has a favorable future in relation to three other methods. One important benefit of this method is the additional production of biomethane through the biohythane process. The key insight from this chapter is that there is ample room for the optimal production of biohydrogen from biomass to replace the petroleum-based hydrogen in the long run in line with the optimal production criteria set out in Hill et al. (2006). The other important insight from this chapter is the crucial role played by enzymes (hydrogenases and nitrogenases) and bacteria in biohydrogen production.

The second chapter in the section on gaseous biofuels is concerned with dark fermentative biohydrogen production by Pandey and Srivastava (Chapter 10), building on the introductory chapter by Kumari and Das. The chapter first sets out the scene on fermentative hydrogen production. It next deals with the bioreactors used for biohydrogen production. It then discusses the determinants of biohydrogen production and finally focuses on recent advances to improve the productivity of biohydrogen production. The chapter discusses a number of tools such as factorial design and response surface methodology to optimize biohydrogen production. The chapter highlights the role played by the microbial consortia and genomic design in the optimization processes. As in Kumari and Das (Chapter 9), the key insight from this chapter is that there is ample room for the optimal production of biohydrogen from biomass through the dark fermentative processes to replace the petroleum-based hydrogen in the long run in line with the optimal production criteria set out in Hill et al. (2006).

The third chapter in the section on gaseous biofuels is concerned with algal biohydrogen ("algae hydrogen" or "algae H_2" in some cultures) production through water-splitting processes [biophotolysis as classified by in Chapter 9], by Tiwari et al. This chapter (Chapter 11) builds on Chapter 9 by Kumari and Das. After providing background information, the chapter discusses algal biomass and algal biohydrogen production in general terms. The chapter next focuses on the determinants of the algal biohydrogen production [light, temperature, salinity (freshwater or saltwater), carbon source, nitrogen source, molecular nitrogen, oxygen, sulfur, methane, micronutrients, role of uptake hydrogenase, and metabolic potential of microorganisms]. The chapter finally focuses on the challenges faced in algal biohydrogen production: oxygen sensitivity of the hydrogenase enzymes (extending lifetime of hydrogenases), proton supply to the hydrogenases, optimizing light-capture efficiency, and electron supply to the hydrogenases and availability of reduced ferredoxin. The chapter concludes with the discussion of selection and engineering of high-efficiency microalgal biomass, innovative photobioreactors, and microbial electrolysis cells as future prospects. As in Chapters 9 and 10, the key insight from this chapter is that there is ample room for the optimal production of biohydrogen from algal biomass through the water-splitting (biophotolysis) processes to replace the petroleum-based hydrogen in the long run in line with the optimal production criteria set out in Hill et al. (2006).

The fourth chapter in the section on gaseous biofuels is concerned with biohydrogen production together with value-added biocommodities from biomass in microbial electrolysis cells (MECs), by Kadier et al. (Chapter 12), building on Chapter 9 and two chapters on microbial fuel cells (MFCs) (Borole, Chapter 16; Ortiz-Martinez et al., Chapter 15). After providing background information, the chapter discusses operating principles, thermodynamics, and bioelectrochemical evaluation of the MECs as well as electrochemically active bacteria (EAB) and their electron transfer mechanisms. The chapter next focuses in detail on the various MEC reactor designs such as single-chamber, two-chamber, and stacked MECs as well as the operational modes of these MECs such as batch mode, fed-batch mode, and continuous flow mode. The chapter then discusses in detail perhaps the most crucial aspect of MECs, materials used in MECs such as anode and cathode materials and membranes. The chapter finally carries out a SWOT analysis of biohydrogen production in MECs. As in Chapter 9 by Kumari and Das, the key insight from this chapter is that there is ample room for the optimal production of biohydrogen from biomass in microbial electrolysis cells to replace petroleum-based hydrogen in the long run in line with the optimal production criteria set out in Hill et al. (2006). As the chapter notes, biohydrogen production in MECs from biomass is a relatively new technology based on MFCs, and the field of materials science and engineering has a lot to offer for the optimization of anode and cathode materials as well as membranes, especially at the nanoscale through the development of effective nanomaterials and nanotechnologies.

The last chapter in the section on gaseous biofuels, by Wang et al. (Chapter 13), is concerned with biogas ("gas" or "biogas" in some cultures). After providing background information, the chapter discusses the basic processes and principles of biogas production including anaerobic digesters and biomass. The main focus of the chapter is on the optimization of biogas production such as biomass pretreatment, anaerobic codigestion, microbial management, and additives. The chapter finally focuses on the economics and environmental impact of biogas production as well as biogas quality control and biogas upgrading. The chapter notes at the outset that biogas is mainly biomethane and carbon dioxide, produced through the anaerobic digestion of biomass in landfills or anaerobic digesters. The major benefit of the anaerobic digestion of biomass is also bioremediation of biomass, usually low-value or no-value wastes, as an efficient waste treatment option to prevent environmental pollution. For this reason, on-farm biogas plants exist worldwide. Anaerobic digestion of biomass involves four key stages: hydrolysis, acidogenesis, acetogenesis, and methanogenesis. The key determinants of biogas production are given as the solid content of the biomass, C:N ratio, temperature, pH value, and solids and hydraulic retention time. The reviewed biomass for biogas production include algal and aquatic biomass, food crop residues, animal wastes (manure), food wastes, municipal solid wastes, and industrial wastewaters. The chapter pays special attention to the pretreatment of biomass and the use of microbes to optimize the biogas production processes. The chapter also points out the various potential uses of biogas: bioelectricity production, biomethane production, biosyngas production, and biohydrogen production.

The issues discussed in the chapter suggest that biogas production could meet the optimal production criteria set out in Hill et al. (2006); however, there is ample room for the optimal production of biogas from biomass to replace petroleum-based natural gas or syngas in the long run. The other key issue emerging from this chapter, as in the other biofuels, is the crucial role played by the microbes (bacteria) in the anaerobic digestion process.

Section IV: Solid biofuels

Chapter 14 by Nanda et al. is the only chapter in this section and is concerned with biochars ("char," "biochar," "carbon black," or charcoal in some cultures) with a focus on biochar functionality and utility in agronomy. After setting out the context, the chapter starts with biochar production, followed by carbon sequestration and soil amelioration. It continues with the discussion of improvements in crop productivity and adsorption of heavy metals and other soil pollutants. The chapter notes that biochars are the by-product of pyrolysis and gasification processes employed in the biofuel production from biomass. It also highlights the use of biochars in bioenergy production, agronomic practices, biomaterial manufacturing, and greenhouse gas mitigation. It notes that pyrolysis is the most common technology to produce biochar, and pyrolysis operates in the absence of oxygen, resulting in bio-oil, biochar, and biogas. It is argued that biochar production and application as a carbon sequestering agent is an attractive option because it offers the benefits of low cost, soil fertility, and soil amelioration because biochars preserve carbon for centuries and millenniums. Biochar application to soil can significantly improve its quality because it positively alters the soil microflora and composition. Furthermore, biochar amendment to soil results in enhancement of beneficial fungi and nitrogen-fixing microorganisms, which in turn improve crop productivity. The other advantage of biochar application to soil is the reduction of the mobility of heavy metals and other organic soil contaminants such as insecticides. However, the chapter cautions that there is still a gap to understanding the ecological interactions between biochar, soil, plant roots, and soil microorganisms. Nevertheless, the chapter concludes that biochars have a crucial role in maintaining sustainable agricultural activities with reduced greenhouse gas emissions and improved soil quality and plant growth. The key insight from this chapter is that there is ample room for the optimal production of biochars from biomass to replace petroleum-based products in the long run. The other key issue emerging from this chapter is the great positive environmental impact of biochars in light of the optimal production criteria set out in Hill et al. (2006).

Section V: Bioenergy: Microbial fuel cells for bioelectricity generation and bioremediation

The first chapter in the section on bioenergy ("energy," "power," or "bioenergy" in some cultures) is concerned with MFCs in general, with a focus on the materials and MFC design and is Chapter 15 by Ortiz-Martinez et al., complementing Chapter 12 on the biohydrogen production in microbial electrolysis cells (Kadier et al.) as well as Chapter 16 on bioelecrochemical biorefining by Borole. After providing background information, the chapter discusses anode materials, cathode materials, and separators. The chapter then discusses MFC design. The chapter notes that MFCs serve the dual purposes of bioelectricity generation and bioremediation. The chapter notes that carbon-based materials such as carbon paper or graphite plates are most commonly used to prepare MFC anodes and these anodes need pretreatments such as carbon nanotube coatings to facilitate the adhesion of bacteria, electron transfer to an anode electrode, and bioenergy production. Although the anode materials can be used as cathode materials as well, a cathode needs a catalyst for the oxygen-reduction reaction (ORR). The most common cathode assemblies used with catalysts include air cathode and aqueous air cathodes. Biocathodes are usually made of carbon-based materials covered with enzymes or bacteria. The ion exchange membranes are the most commonly used separators. A variety of designs have been tried for MFCs to increase their productivity. The key insights from this chapter are the important role played by the materials science and engineering in MFC design and structure and the important role played by bacteria in the operation of MFCs. Overall, the key insight from this chapter is that there is ample room for the optimal production of bioenergy from biomass in MFCs to replace petroleum-based energy in the long run in line with the optimal production criteria set out in Hill et al. (2006). As the chapter notes, bioenergy production in MFCs from biomass is a relatively new technology and the field of materials science and engineering has a lot to offer for the optimization of the anode and cathode materials as well as membranes, especially at the nanoscale through the development of effective nanomaterials (and especially carbon-based or polymer-based nanomaterials) and nanotechnologies.

The second chapter in the section on bioenergy is concerned with bioelectrochemical biorefining (BER) and is Chapter 16 by Borole et al., complementing Chapter 12 on biohydrogen production in microbial electrolysis cells (Kadier et al.) and Chapter 15 on MFCs by Ortiz-Martinez et al. The chapter first defines the term "bioelectrochemical refining." It next discusses the anode process, followed by thermoconversion-derived substrates. The cathode processes and biorefinery integration are the other topics covered by the chapter. Bioelectrochemical biorefining, as a "marriage between microbial electrochemical technology and biorefineries." As plain marriage between microbial electrochemical technology and biorefineries without double commas as this is merely a statement by the author, Dr. Borole, targets the integration of microbial electrochemical cells into the biorefinery to use low-value or no-value biomass to generate electrons and then use the electrons to produce value-added biofuels and biocommodities. The reactor system in its simplest form includes an anode, a cathode, and potentially a separator. The anode uses the low-value or no-value products of a biorefinery such as wastes or wastewaters. The chapter gives examples of wastes from bioethanol and biobutanol production from the lignocellulosic biomass. Electrochemical utilization of acetate is the most common reaction in the anode; therefore, the rate of utilization of acetate is high. The electrons generated at the anode in a BER cell are used at the cathode of the cell to generate a variety of biofuels and biocommodities through the use of electro-fermentation, a method that uses electrical current to enhance fermentation. The chapter concludes that the demonstration of anodic processes to generate electrons from waste materials as well as the use of electrons provided by the cathode to enhance production of valuable, reduced bioproducts shows that integration of these two half-reactions can enable production of value-added bioproducts using the BER concept. The chapter then recommends that the concept of bioelectrochemical biorefining can be enhanced by genomics to increase economic returns to the BER process. The key insight from this chapter is that there is ample room for the optimal production of biofuels and biocommodities from low-value or no-value biomass, which form the waste in the biorefineries of such as bioethanol or biobutanol in microbial electrolysis cells to replace the petroleum-based fuels and commodities in the long run in line with the optimal production criteria set out in Hill et al. (2006).

Reference

Hill, J., Nelson, E., Tilman, D., Polasky, S., and Tiffany, D. 2006. Environmental, economic, and energetic costs and benefits of biodiesel and ethanol biofuels. *Proceedings of the National Academy of Sciences*, 103(30), 11206–11210.

Acknowledgments

This edited book has been a product of teamwork with a wide range of stakeholders whom the editor would like to thank sincerely: the publisher, CRC Press and Taylor & Francis Group, and the executive editor, T. Michael Slaughter, who provided the opportunity for the publication of this edited book; the outstanding contributors with substantial research experience in their respective fields; the other outstanding research scientists who were not able to contribute a chapter due to the heavy time constraints but who provided collegial support for this book project; the highly cited researchers who have contributed to the development of the interdisciplinary research field of bioenergy and biofuels since the 1980s; and the highly cited researchers who have contributed to the development of the emerging interdisciplinary research field of scientometrics since the 1950s, with special acknowledgement of the late Eugene Garfield, whose studies since the 1950s have provided the foundation for the scientometric field.

Ozcan Konur

Editor

Ozcan Konur as both a materials scientist and social scientist by training, has focused on the bibliometric evaluation of the research in these innovative high-priority research areas at the level of the researchers, journals, institutions, countries, and research areas as well as the social implications of the research in these areas. He has also had extensive research interests in the development of social policies for disadvantaged people on the basis of disability, age, religious beliefs, race, gender, and sexuality at the interface of science and policy. He has published more than 100 journal papers, book chapters, and conference papers. Among them are more than 30 pioneering studies in bioenergy and biofuels. A list of recent publications (after 2015) and a list of earlier publications on bioenergy and biofuels (before 2015) are given below.

Recent publications

Algal materials

1. Konur, O. 2017. The top citation classics in alginates for biomedicine. In: Venkatesan, J., Anil, S., Kim, S. K. (Eds.), *Seaweed Polysaccharides: Isolation, Biological and Biomedical Applications,* 223–250, Elsevier; Amsterdam. [ISBN: 978-0128-09816-5. Publisher link: https://www.elsevier.com/books/seaweed-polysaccharides/venkatesan/978-0-12-809816-5].
2. Konur, O. 2016. Algal Omics: The most-cited papers. In: Kim, S. K. (Ed.) *Marine Omics: Principles and Applications,* CRC Press; Boca Raton, FL, 9–34. [ISBN 978-14822-5820-2, EISBN: 978-1-4822-5821-9. DOI: 10.1201/9781315372303-3. Publisher link: http://www.crcnetbase.com/doi/abs/10.1201/9781315372303-3].
3. Konur, O. 2015. Algal economics and optimization. In: Kim, S. K., Lee, C. G. (Eds.), *Marine Bioenergy: Trends and Developments,* CRC Press; Boca Raton, FL, 691–716. [ISBN: 978-1-4822-2237-1. EISBN:

978-1-4822-2238-8. DOI: 10.1201/b18494-40, Publisher link: http://www.crcnetbase.com/doi/abs/10.1201/b18494-40].

4. Konur, O. 2015. Algal high-value consumer products. In: Kim, S. K., Lee, C. G. (Eds.), *Marine Bioenergy: Trends and Developments*, CRC Press; Boca Raton, FL, 653–682. [ISBN: 978-1-4822-2237-1. EISBN: 978-1-4822-2238-8. DOI: 10.1201/b18494-38. Publisher link: http://www.crcnetbase.com/doi/abs/10.1201/b18494-38].
5. Konur, O. 2015. Algal photobioreactors. In: Kim, S. K., Lee, C. G. (Eds.), *Marine Bioenergy: Trends and Developments*, CRC Press; Boca Raton, FL, 81–108. [ISBN: 978-1-4822-2237-1. EISBN: 978-1-4822-2238-8. DOI: 10.1201/b18494-8. Publisher link: http://www.crcnetbase.com/doi/abs/10.1201/b18494-8].
6. Konur, O. 2015. Algal biosorption of heavy metals from wastes. In: Kim, S. K., Lee, C. G. (Eds.), *Marine Bioenergy: Trends and Developments*, CRC Press; Boca Raton, FL, 597–626. [ISBN: 978-1-4822-2237-1. EISBN: 978-1-4822-2238-8. DOI: 10.1201/b18494-34. Publisher link: http://www.crcnetbase.com/doi/abs/10.1201/b18494-34].
7. Konur, O. 2015. Current state of research on algal bioelectricity and algal microbial fuel cells. In: Kim, S. K., Lee, C. G. (Eds.), *Marine Bioenergy: Trends and Developments*, CRC Press; Boca Raton, FL, 527–556. [ISBN: 978-1-4822-2237-1. EISBN: 978-1-4822-2238-8. DOI: 10.1201/b18494-30. Publisher link: http://www.crcnetbase.com/doi/abs/10.1201/b18494-30].
8. Konur, O. 2015. Current state of research on algal biodiesel. In: Kim, S. K., Lee, C. G. (Eds.), *Marine Bioenergy: Trends and Developments*, CRC Press; Boca Raton, FL, 487–512. [ISBN: 978-1-4822-2237-1. EISBN: 978-1-4822-2238-8. DOI: 10.1201/b18494-27. Publisher link: http://www.crcnetbase.com/doi/abs/10.1201/b18494-27].
9. Konur, O. 2015. Current state of research on algal biohydrogen. In: Kim, S. K., Lee, C. G. (Eds.), *Marine Bioenergy: Trends and Developments*, CRC Press; Boca Raton, FL, 393–422. [ISBN: 978-1-4822-2237-1. EISBN: 978-1-4822-2238-8. DOI: 10.1201/b18494-24. Publisher link: http://www.crcnetbase.com/doi/abs/10.1201/b18494-24].
10. Konur, O. 2015. Current state of research on algal biomethanol. In: Kim, S. K., Lee, C. G. (Eds.), *Marine Bioenergy: Trends and Developments*, CRC Press; Boca Raton, FL, 327–370. [ISBN: 978-1-4822-2237-1. EISBN: 978-1-4822-2238-8. DOI: 10.1201/b18494-22. Publisher link: http://www.crcnetbase.com/doi/abs/10.1201/b18494-22].
11. Konur, O. 2015. Current state of research on algal biomethane. In: Kim, S. K., Lee, C. G. (Eds.), *Marine Bioenergy: Trends and Developments*, CRC Press; Boca Raton, FL, 273–302. [ISBN: 978-1-4822-2237-1. EISBN: 978-1-4822-2238-8. DOI: 10.1201/b18494-20. Publisher link: http://www.crcnetbase.com/doi/abs/10.1201/b18494-20].

12. Konur, O. 2015. Current state of research on algal bioethanol. In: Kim, S. K., Lee, C. G. (Eds.), *Marine Bioenergy: Trends and Developments*, CRC Press; Boca Raton, FL, 217–244. [ISBN: 978-1-4822-2237-1. EISBN: 978-1-4822-2238-8. DOI: 10.1201/b18494-17. Publisher link: http://www.crcnetbase.com/doi/abs/10.1201/b18494-17].
13. Konur, O. 2015. Algal photosynthesis, biosorption, biotechnology, and biofuels. In: Kim, S. K. (Ed.) *Springer Handbook of Marine Biotechnology*, Springer; Berlin, Heidelberg, 1131–1161. [ISBN: 978-3-642-53970-1. EISBN: 978-3-642-53971-8. DOI: 10.1007/978-3-642-53971-8_50. Publisher link: http://link.springer.com/chapter/10.1007%2F978-3-642-53971-8_50].

Nanomedicine

1. Konur, O. 2017. Recent citation classics in antimicrobial nanobiomaterials. In: Ficai, A., Grumezescu, A. (Eds.), *Nanostructures for Antimicrobial Therapy*, Elsevier; Amsterdam, 669–686. [ISBN 978-0-323-46152-8. EISBN 978-0-323-46151-1 June 2017 Publisher link: https://www.elsevier.com/books/nanostructures-for-antimicrobial-therapy/grumezescu/978-0-323-46152-8].
2. Konur, O. 2017. Scientometric overview regarding oral cancer nanomedicine. In: Andronescu, E., Grumezescu, A. M. (Eds.), *Nanostructures for Oral Medicine, Nanostructures for Therapeutic Medicine*, Vol. 5, Elsevier; Amsterdam, 939–962. [ISBN 978-0-323-47720-8. EISBN 978-0-323-47721-5. April 2017. https://www.elsevier.com/books/nanostructures-for-oral-medicine/grumezescu/978-0-323-47720-8].
3. Konur, O. 2016. Scientometric overview regarding the surface chemistry of nanobiomaterials. In: Grumezescu, A. M. (Ed.) *Surface Chemistry of Nanobiomaterials, Applications of Nanobiomaterials*, Vol. 3, Elsevier; Amsterdam, 463–486. [February 2016, ISBN: 978-0-323-42861-3, EISBN: ISBN: 978-0-3234-2884-2, http://dx.doi.org/10.1016/B978-0-323-42861-3.00015-7, Publisher link: http://www.sciencedirect.com/science/article/pii/B9780323428613000157].
4. Konur, O. 2016. Scientometric overview regarding the nanobiomaterials in antimicrobial therapy. In: Grumezescu, A. M. (Ed.) *Nanobiomaterials in Antimicrobial Therapy, Applications of Nanobiomaterials*, Vol. 6, Elsevier; Amsterdam, 511–535. [ISBN: 978-0323-42864-4, EISBN: 978-03234-2887-3, April 2016, http://dx.doi.org/10.1016/B978-0-323-42864-4.00015-4, Publisher link: http://www.sciencedirect.com/science/article/pii/B9780323428644000154].
5. Konur, O. 2016. Scientometric overview regarding the nanobiomaterials in dentistry. In: Grumezescu, A. M. (Ed.) *Nanobiomaterials in Dentistry: Applications of Nanobiomaterials*, Vol. 11, Elsevier;

Amsterdam, 425–453. [ISBN: 978-03-234-2867-5, EISBN: 978-0-323-42890-3, June 2016, http://dx.doi.org/10.1016/B978-0-323-42867-5.00007-2, Publisher link: http://www.sciencedirect.com/science/article/pii/B9780323428675000072].

6. Konur, O. 2016. Scientometric overview in nanobiodrugs. In: Holban, A. M., Grumezescu, A. (Eds.), *Nanoarchitectonics for Smart Delivery and Drug Targeting*, Micro & Nano Technologies Series, Elsevier; Amsterdam, 405–428. [ISBN: 978-0-323-47347-7, EISBN: 978-0-323-47722-2, July 2016, http://dx.doi.org/10.1016/B978-0-323-47347-7.00015-X; Publisher link: http://www.sciencedirect.com/science/article/pii/B978032347347700015X].
7. Konur, O. 2016. The scientometric overview in cancer targeting. In: Holban, A. M., Grumezescu, A. (Eds.), *Nanoarchitectonics for Smart Delivery and Drug Targeting*, Micro & Nano Technologies Series, Amsterdam; Elsevier, 871–895. [ISBN: 978-0-323-47347-7, EISBN: 978-0-323-47722-2, July 2016, http://dx.doi.org/10.1016/B978-0-323-47347-7.00030-6, Publisher link: http://www.sciencedirect.com/science/article/pii/B9780323473477000306].

Nanofood

1. Konur, O. 2017. Scientometric overview in nanopesticides. In: Grumezescu, A. M. (Ed.) *New Pesticides and Soil Sensors, Nanotechnology in the Agri-Food Industry*, Vol. 10, Elsevier; Amsterdam, 719–744. [ISBN: 978-0-12804299-1. eBook ISBN: 978-0-12804370-7. http://dx.doi.org/10.1016/B978-0-12-804299-1.00020-5 February 2017. Publisher link: http://www.sciencedirect.com/science/article/pii/B9780128042991000205].
2. Konur, O. 2017. Scientometric overview regarding water nanopurification. In: Grumezescu, A. M. (Ed.) *Water Purification, Nanotechnology in the Agri-Food Industry*, Vol. 9, Elsevier; Amsterdam, 693–716. [ISBN: 978-0-12-804300-4, EISBN:978-0-12804371-4 January 2017, http://dx.doi.org/10.1016/B978-0-12-804300-4.00020-4. Publisher link: http://www.sciencedirect.com/science/article/pii/B9780128043004000204].
3. Konur, O. 2016. Scientometric overview regarding nanoemulsions used in the food industry. In: Grumezescu, A. M. (Ed.) *Emulsions, Nanotechnology in the Agri-Food Industry*, Vol. 3, Elsevier; Amsterdam, 689–711. [ISBN: 978-0-12-804306-6, July 2016, http://dx.doi.org/10.1016/B978-0-12-804306-6.00020-9. Publisher link: http://www.sciencedirect.com/science/article/pii/B9780128043066000209]. B.EAS.20.
4. Konur, O. 2016. Scientometric overview in food nanopreservation. In: Grumezescu, A. M. (Ed.) *Food Preservation, Nanotechnology in the Agri-Food Industry*, Vol. 6, Elsevier; Amsterdam, 703–729.

[ISBN: 978-0-12-804303-5, EISBN: 978-0-12804374-5, http://dx.doi.org/10.1016/B978-0-12-804303-5.00020-1. Publisher link: http://www.sciencedirect.com/science/article/pii/B9780128043035000201].

Energy and fuels

1. Konur, O. 2015. The scientometric study of the global energy research. In: Prasad, R., Sivakumar, S., Sharma, U. C. (Eds.), *Energy Science and Technology Vol. 1. Opportunities and Challenges*, Studium Press LLC; Houston, TX, 475–489. [ISBN: 978-1-626990-62-3, http://studiumpress.in/indetail.asp?id=289].
2. Konur, O. 2015. The review of citation classics on the global energy research. In: Prasad, R., Sivakumar, S., Sharma, U. C. (Eds.), *Energy Science and Technology Vol. 1. Opportunities and Challenges*, Studium Press LLC; Houston, TX, 490-526. [ISBN: 978-1-626990-62-3, http://studiumpress.in/indetail.asp?id=289].

Glycobiology

1. Konur, O. 2016. Glycoscience: The current state of the research. In: Kim, S. K. (Ed.) *Marine Glycobiology: Principles and Applications*, CRC Press; Boca Raton, FL, 7–21. [ISBN: 978-1-4987-0961-3. EISBN: 978-1-4987-0962-0, October 2016, Publisher book link: https://www.crcpress.com/Marine-Glycobiology-Principles-and-Applications/Kim/9781498709613].

Earlier studies on bioenergy and biofuels

Algal bioenergy

1. Konur, O. 2011. The scientometric evaluation of the research on the algae and bio-energy. *Applied Energy* 88(10), 3532–3540. [DOI: 10.1016/j.apenergy.2010.12.059. [Appl Energ] SCI. SSCI. Publisher full-text link: http://www.sciencedirect.com/science/article/pii/S0306261910005799].

Energy and fuels in general including bioenergy and biofuels

1. Konur, O. 2012. The evaluation of the global energy and fuels research: A scientometric approach. *Energy Education Science and Technology Part A: Energy Science and Research* 30(1), 613–628. SCIE [Ener Educ Sci Tech-A]. [Publisher link: http://www.silascience.com/abstracts/20102012120735.html https://www.researchgate.net/

publication/286050659_The_evaluation_of_the_global_energy_and_fuels_research_A_scientometric_approach].
2. Konur, O. 2012. What have we learned from the citation classics in energy and fuels: A mixed study. *Energy Education Science and Technology Part A: Energy Science and Research* 30 (special issue 1), 255–268. [Ener Educ Sci Tech-A]. Scopus. [Publisher link: http://www.silascience.com/abstracts/31032013112527.html https://www.researchgate.net/publication/287396883_What_have_we_learned_from_the_citation_classics_in_Energy_and_Fuels_A_mixed_study].
3. Konur, O. 2012. 100 citation classics in energy and fuels. *Energy Education Science and Technology Part A: Energy Science and Research* 30 (special issue 1), 319–332. [Ener Educ Sci Tech-A] Scopus. [Publisher link: http://www.silascience.com/journals_detail.aspx?j_id=2&v_no=84https://www.researchgate.net/publication/287396598_100_citation_classics_in_Energy_and_Fuels].

Bioenergy and biofuels

1. Konur, O. 2012. The scientometric evaluation of the research on the production of bioenergy from biomass. *Biomass and Bioenergy* 47, 504–515. SCI. [Biomass Bioenerg]. [DOI: 10.1016/j.biombioe.2012.09.047 Publisher link: http://www.sciencedirect.com/science/article/pii/S096195341200387X].
2. Konur, O. 2012. The evaluation of the biogas research: A scientometric approach. *Energy Education Science and Technology Part A: Energy Science and Research* 29(2), 1277–1292. SCIE. [Ener Educ Sci Tech-A] [Publisher link: http://www.silascience.com/abstracts/08052012185652.html].
3. Konur, O. 2012. The evaluation of the research on the biohydrogen: A scientometric approach. *Energy Education Science and Technology Part A: Energy Science and Research* 29(1), 323–338. SCIE. [Ener Educ Sci Tech-A]. [Publisher link: http://www.silascience.com/abstracts/17022012102309.html https://www.researchgate.net/publication/286054551_The_evaluation_of_the_research_on_the_biohydrogen_A_scientometric_approach].
4. Konur O. 2012. The evaluation of the research on the microbial fuel cells: A scientometric approach. *Energy Education Science and Technology Part A: Energy Science and Research* 29(1), 309–322. SCIE. [Ener Educ Sci Tech-A]. [Publisher link: http://www.silascience.com/abstracts/17022012102116.html https://www.researchgate.net/publication/289212830_The_evaluation_of_the_research_on_the_microbial_fuel_cells_A_scientometric_approach].

5. Konur, O. 2012. The evaluation of the research on the bioethanol: A scientometric approach. *Energy Education Science and Technology Part A: Energy Science and Research* 28(2), 1051–1064. SCIE. [Ener Educ Sci Tech-A]. [Publisher link: http://www.silascience.com/abstracts/14102011184052.html https://www.researchgate.net/publication/289212829_The_evaluation_of_the_research_on_the_bioethanol_A_scientometric_approach].
6. Konur, O. 2012. The evaluation of the research on the biodiesel: A scientometric approach. *Energy Education Science and Technology Part A: Energy Science and Research* 28(2), 1003–1014. SCIE. [Ener Educ Sci Tech-A]. [Publisher link: http://www.silascience.com/abstracts/14102011183828.html https://www.researchgate.net/publication/288305342_The_evaluation_of_the_research_on_the_biodiesel_A_scientometric_approach].
7. Konur, O. 2012. The evaluation of the research on the biofuels: A scientometric approach. *Energy Education Science and Technology Part A: Energy Science and Research* 28(2), 903–916. SCIE. [Ener Educ Sci Tech-A]. [Publisher link: http://www.silascience.com/abstracts/14102011182705.html].
8. Konur, O. 2012. Prof. Dr. Ayhan Demirbas' scientometric biography. *Energy Education Science and Technology Part A: Energy Science and Research* 28(2), 727–738. SCIE. [Ener Educ Sci Tech-A]. [Publisher link: http://www.silascience.com/abstracts/14102011181427.html https://www.researchgate.net/publication/290004479_Prof_Dr_Ayhan_Demirbas%27_scientometric_biography].
9. Konur, O. 2012. The evaluation of the biorefinery research: A scientometric approach. *Energy Education Science and Technology Part A: Energy Science and Research* 30(special issue 1), 347–358. [Ener Educ Sci Tech-A] Scopus. [Publisher link: http://www.silascience.com/abstracts/31032013114013.html https://www.researchgate.net/publication/288771576_The_evaluation_of_the_biorefinery_research_A_scientometric_approach].
10. Konur, O. 2012. The evaluation of the bio-oil research: A scientometric approach. *Energy Education Science and Technology Part A: Energy Science and Research* 30(special issue 1), 379–392. [Ener Educ Sci Tech-A] Scopus. [Publisher link: http://www.silascience.com/abstracts/31032013114134.html http://metrics.stanford.edu/node/evaluation-bio-oil-research-scientometric-approach https://www.researchgate.net/publication/288171481_The_evaluation_of_the_bio-oil_research_A_scientometric_approach].
11. Konur, O. 2011. The scientometric biography of a leading scientist working in the field of bio-energy. In: Karakoc, T. H., Midilli, A., Turan, O., Orman, E., Tunca, F., Yazar, I. (Eds.), *Sixth International Green Energy Conference (IGEC-6) Proceedings*. [ISBN: 978-605-89885-3-8. Eskisehir,

Turkey, Anadolu University in cooperation with International Green Energy Association, 2011].

12. Konur, O. 2011. The scientometric evaluation of the research on the biofuels. In: Karakoc, T. H., Midilli, A., Turan, O., Orman, E., Tunca, F., Yazar, I. (Eds.), *Sixth International Green Energy Conference (IGEC-6) Proceedings*. [ISBN: 978-605-89885-3-8. Eskisehir, Turkey, Anadolu University in cooperation with International Green Energy Association, 2011].

Contributors

Jon Alvarez
Department of Chemical Engineering
University of the Basque Country
Bilbao, Spain

Maider Amutio
Department of Chemical Engineering
University of the Basque Country
Bilbao, Spain

Sourav Kumar Bagchi
Agricultural and Food Engineering Department
Indian Institute of Technology Kharagpur
Kharagpur, India

Tridib Kumar Bhowmick
Department of Bioengineering
National Institute of Technology Agartala
Tripura, India

Javier Bilbao
Department of Chemical Engineering
University of the Basque Country
Bilbao, Spain

Abhijeet P. Borole
Biosciences Division
Oak Ridge National Laboratory
Oak Ridge, Tennessee

and

Department of Chemical and Biomolecular Engineering
and
Bredesen Center for Interdisciplinary Research and Education
University of Tennessee
Knoxville, Tennessee

Carlos Ariel Cardona Alzate
Instituto de Biotecnología y Agroindustria
Departamento de Ingeniería Química
Universidad Nacional de Colombia
Manizales, Colombia

Kuppam Chandrasekhar
School of Applied Biosciences
Kyungpook National University
Dong-Daegu, South Korea

Jonah Teo Teck Chye
Department of Chemical Engineering
Curtin University
Sarawak, Malaysia

Ajay K. Dalai
Department of Chemical and Biological Engineering
University of Saskatchewan
Saskatoon, Saskatchewan, Canada

Michael K. Danquah
Department of Chemical Engineering
Curtin University
Sarawak, Malaysia

Debabrata Das
Department of Biotechnology
Indian Institute of Technology Kharagpur
Kharagpur, India

Carlos Andrés García
Instituto de Biotecnología y Agroindustria
Departamento de Ingeniería Química
Universidad Nacional de Colombia
Manizales, Colombia

Kalyan Gayen
Department of Chemical Engineering
National Institute of Technology Agartala
Tripura, India

Iskender Gökalp
Institut de Combustion
Aérothermique, Réactivité et Environnement
Centre National de la Recherche Scientifique
Orléans, France

F. J. Hernández-Fernández
Chemical and Environmental Engineering Department
Polytechnic University of Cartagena
Cartegena, Spain

Yong Jiang
State Key Joint Laboratory of Environment Simulation and Pollution Control
School of Environment
Tsinghua University
Beijing, P.R. China

Gail Joseph
Department of Energy and Environmental Systems
North Carolina Agricultural and Technical State University
Greensboro, North Carolina

Lau Yien Jun
Department of Chemical Engineering
Curtin University
Sarawak, Malaysia

Abudukeremu Kadier
Department of Chemical and Process Engineering
National University of Malaysia
Selangor, Malaysia

Mohd Sahaid Kalil
Department of Chemical and Process Engineering
National University of Malaysia
Bangi, Malaysia

Thomas Kiran
Department of Biotechnology
Noida International University
Noida, India

Tatsuji Koizumi
Policy Research Institute
Ministry of Agriculture, Forestry, and Fisheries
Tokyo, Japan

Ozcan Konur
Department of Materials Engineering
Ankara Yildirim Beyazit University
Ankara, Turkey

Janusz A. Kozinski
Department of Earth and Space Science and Engineering
York University, Toronto
Ontario, Canada

Manish Kumar
Department of Biology and Biological Engineering
Chalmers University of Technology,
Gothenburg, Sweden

Sinu Kumari
Advanced Technology Development Center
Indian Institute of Technology Kharagpur
Kharagpur, India

Bin Lai
Advanced Water Management Centre
University of Queensland
Brisbane, Australia

Gartzen Lopez
Department of Chemical Engineering
University of the Basque Country
Bilbao, Spain

Nirupama Mallick
Agricultural and Food Engineering Department
Indian Institute of Technology Kharagpur
Kharagpur, India

Azah Mohamed
Department of Electrical Electronic and System Engineering
National University of Malaysia
Bangi, Malaysia

Bryan R. Moser
U.S. Department of Agriculture
Agricultural Research Service
National Center for Agricultural Utilization Research
Peoria, Illinois

Sonil Nanda
Department of Earth and Space Science and Engineering
York University
Toronto, Ontario, Canada

Martin Olazar
Department of Chemical Engineering
University of the Basque Country
Bilbao, Spain

V. M. Ortiz-Martínez
Chemical and Environmental Engineering Department
Polytechnic University of Cartagena
Cartegena, Spain

Sharadwata Pan
Fluid Dynamics of Complex Biosystems
Technical University of Munich
Freising, Germany

Anjana Pandey
Department of Biotechnology
MNNIT Allahabad
Allahabad, India

Kamal K. Pant
Department of Chemical Engineering
Indian Institute of Technology Delhi
New Delhi, India

Reeza Patnaik
Agricultural and Food Engineering Department
Indian Institute of Technology Kharagpur
Kharagpur, India

J. Quesada-Medina
Chemical Engineering Department
University of Murcia
Murcia, Spain

Pankaj Kumar Rai
Department of Biotechnology and Bioinformatics Center
Barkatullah University
Bhopal, India

A. P. de los Ríos
Chemical Engineering Department
University of Murcia
Murcia, Spain

Supreet Saini
Department of Chemical Engineering
Indian Institute of Technology Bombay
Maharashtra, India

M. J. Salar-García
Chemical and Environmental Engineering Department
Polytechnic University of Cartagena
Cartegena, Spain

Sebastián Serna-Loaiza
Instituto de Biotecnología y Agroindustria
Departamento de Ingeniería Química
Universidad Nacional de Colombia
Manizales, Colombia

Saumya Srivastava
Department of Biotechnology
MMNIT Allahabad
Allahadad, India

Archana Tiwari
Department of Biotechnology
Noida International University
Noida, India

F. Tomás-Alonso
Chemical Engineering Department
University of Murcia
Murcia, Spain

Lijun Wang
Department of Natural Resources and Environmental Design
and
Department of Chemical Biological and Bioengineering
North Carolina Agricultural and Technical State University
Greensboro, North Carolina

Lau Sie Yon
Department of Chemical Engineering
Curtin University
Sarawak, Malaysia

Bo Zhang
Department of Natural Resources and Environmental Design
North Carolina Agricultural and Technical State University
Greensboro, North Carolina

section one

Bioenergy and biofuels in general

chapter one

Bioenergy and biofuels science and technology

Scientometric overview and citation classics

Ozcan Konur

Contents

1.1 Introduction

With the ever-increasing human population and the ever-diminishing petroleum-based energy resources, energy security has become a public concern in recent decades (Chester, 2010; Jacobson, 2009; Kruyt et al., 2009; Pimentel, 1991; Winzer, 2012; Yergin, 2006). These concerns require that alternative sustainable energy sources should be developed.

Another public concern in recent decades has been related to global warming (Cox et al., 2000; Hansen, 1998; Held and Soden, 2006; Hughes, 2000; Meinshausen et al., 2009; Vitousek, 1994), greenhouse gas emissions (Johnson et al., 2007; Lashof and Ahuja, 1990; Riahi et al., 2011; Snyder et al., 2009; Woodcock et al., 2009), and climate change (Davidson and Janssens, 2006; Hoegh-Guldberg et al., 2007; Hughes et al., 2003; Lal, 2004; Moss et al., 2010; Oreskes, 2004; Parmesan and Yohe, 2003; Thomas et al., 2004; Vorosmarty et al., 2000; Walther et al., 2002). These concerns have strong implications for the design of environment-friendly energy and fuels.

In the meantime, with the increasing use of the food crops for bioenergy and biofuel production as an alternative to petroleum, food security has become a public concern in recent decades as well (Godfray et al., 2010; Lobell et al., 2008; Maxwell, 1996; Pinstrup-Andersen, 2009; Rosegrant and Cline, 2003; Schmidhuber and Tubiello, 2007; Tscharntke et al., 2012).

As Hill et al. (2006) note, an optimal set of bioenergy and biofuels should "provide a net energy gain, have environmental benefits, be economically competitive, and be producible in large quantities without risking food security." With these insights, there has been ever-increasing investment in the research on bioenergy and biofuels at a global scale in recent decades to produce an optimal set of bioenergy and biofuels from biomass through biological processes.

Besides the research on bioenergy and biofuels, a new academic discipline has emerged in recent decades with the pioneering studies of Garfield starting in the 1950s for the quantitative evaluation of the research in many domains, to complement the qualitative peer review of the research (Garfield, 1972, 1979, 2006, Glanzel and Moed, 2002, MacRoberts and MacRoberts, 1996, Moed et al., 1995; Nederhof, 2006; van Raan, 2005; Weingart, 2005). One publicly important stream of the scientometric research has been the study of the citation classics in respective research fields starting with the pioneering studies of Garfield in this subfield as well (Baltussen and Kindler, 2004a,b; Dubin et al., 1993; Garfield, 1987, 2010; Garfield and Welljams-Dorof, 1992; Paladugu et al., 2002; Ratnatunga and Romano 1997).

Although there have been more than 106,000 papers on bioenergy and biofuels, the scientometric studies in this field only started with the pioneering work of Kajikawa and Takeda (2008) and surged in the mid 2010s after the publication of a large number of pioneering papers by Konur (Konur, 2011, 2012a–m, 2015a–m, 2016, 2017; Chen and Ho, 2015; Coelho and Barbosa, 2014; Costantini et al., 2015; Curci and Ospina, 2016; de Souza et al., 2015; Ferreira et al., 2014; Gomes and Dewes, 2017; Ho, 2017; Hsu and Lin, 2016; Hsu et al., 2014, 2015; Liu et al., 2014; Mao et al., 2015a,b; Mercuri et al., 2016; Mittal, 2013; Mohan, 2015; Qian, 2013; Siegmeier and Moller, 2013; Wang et al., 2013, 2015a,b; Yaoyang and Boeing, 2013; Zhang et al., 2016a,b). It is notable that a number of scientometric approaches, databases, and materials were used in these recent pioneering scientometric studies on bioenergy and biofuels and there is a crucial need for more scientometric studies in this field.

These scientometric studies provide valuable insights for the key stakeholders on the research development in bioenergy and biofuels and provide valuable tools as well to conduct a strength-weakness-opportunities-threats (SWOT) analysis in these areas (Hill and Westbrook, 1997; Jackson et al., 2003; Kim, 2005; Kurttila et al., 2000; Yuksel and Dagdeviren, 2007). Such insights and tools are valuable in the light of the New Institutional Theory developed by North (Denzau and North, 1994; Konur, 2000, 2002a,b, 2004, 2006a,b, 2007a,b, 2013; Milgrom and North, 1990; North, 1986, 1993; North and Weingast, 1989). North's teachings suggest that the key stakeholders should inform themselves timely about the progress of the rules governing the related processes and should revise these rules to design the further efficient incentive structures for the progressive development of the research in these fields.

The studies in bioenergy and biofuels mostly focus on the structure–property relationships of the resulting products as in materials science and engineering (Konur and Matthews, 1989; Li et al., 1990; McCullough et al., 1993; Mishra et al., 2009; Ostuni et al., 2001; Scherf and List, 2002). The starting materials such as biomass and the biological processes to convert the biomass to a set of biofuels and bioenergy are designed to comply with the insights of Hill et al. (2006), resulting in an optimal set of bioenergy and biofuels with a net energy gain, having environmental benefits, being economically competitive, and being producible in large quantities without risking food security. With these insights, there has been an ever-increasing investment in research on the starting materials and the biological conversion processes of bioenergy and biofuels at a global scale in recent decades.

This paper builds on the earlier pioneering scientometric studies in bioenergy and biofuels to present a detailed and up-to-date scientometric study of the research in bioenergy and biofuels as of April 2017. This paper also provides the most up-to-date search strategy in these fields to evaluate a representative population of the research in these fields.

The next section provides information on the methods and materials used in the paper. A scientometric overview of over 106,000 papers in bioenergy and biofuels is provided in Section 1.3. Based on this overview, a scientometric overview of 50 citation classics is provided in Section 1.4. Section 1.5 provides the summaries for each of the 50 citation classics under five broad topical headings (Table 1.1) to inform the key stakeholders about the citation classics significantly impacting the development of the research in bioenergy and biofuels: Biofuels in general, liquid biofuels, gaseous biofuels, solid biofuels, and bioenergy. A list of the key references is provided in the section on references. A figure and five tables (Tables 1.2 through 1.6) complement the text by providing tabular and visual data on the state of the research in these 50 citation classics.

Table 1.1 The types of biofuels and bioenergy considered in the citation classics

Major topics	Minor topics	Number of papers
1. Biofuels in general		19
	1.1. Biofuel production overviews	6
	1.2. Algal biofuels	2
	1.3. Lignocellulosic biofuels	7
	1.4. Biofuels economics optimization	2
	1.5. Biofuels environment	2
2. Liquid biofuels		17
	2.1. Biodiesel	9
	2.2. Bio-oil	2
	2.3. Bioethanol	3
	2.4. Biobutanol	1
	2.5. Other liquid biofuels	2
	2.6. Biomethanol	0
3. Gaseous biofuels		6
	3.1. Biohydrogen	5
	3.2. Biogas	1
	3.3. Biomethane	0
4. Solid biofuels		3
	4.1. Biomass fuels	3
	4.2. Biochars	0
5. Bioenergy		5
	5.1. Microbial fuel cells	5
	5.2. Enzymatic biofuel cells	0

Table 1.2 The citation classics in biofuels in general

Rank	Paper reference	Year	Document type	Affiliation	Country	Number of authors	M/F	Lead authors	Journal	Subject area	Topic	Fuels	Ingredients	Cits.
2	Huber et al.	2006	R	Polytechnic university of Valencia	Spain	3	M	Huber*, Corma*	*Chem. Rev.*	Chem. Mult.	Production of biofuels for transportation from biomass	Biofuels	Biomass	3260
3	Ragauskas et al.	2006	R	Georgia Inst. Technol., Imperial Coll. Sci. Technol. & Med +1	United States, England	14	M	Ragauskas	*Science*	Mult. Sci.	Production of biofuels and biomaterials	Biofuels	Biomass	2664
4	Mosier et al.	2005	R	Purdue Univ., Dartmouth Coll. +4	United States	7	M	Wyman*, Dale, Holtzapple, Ladisch	*Bioresour. Technol.*	Biot. Appl. Microb., Ener. Fuels	Pretreatment of lignocellulosic biomass for the efficient production of biosugars and biofuels	Biofuels	Lignocellulose	2531
6	Corma et al.	2007	R	Polytechnic university of Valencia	Spain	4	M	Corma*	*Chem. Rev.*	Chem. Mult.	Production of biocommodities from biomass	Biocommodities	Biomass	2436

(*Continued*)

Table 1.2 (Continued) The citation classics in biofuels in general

Rank	Paper reference	Year	Document type	Affiliation	Country	Number of authors	M/F	Lead authors	Journal	Subject area	Topic	Fuels	Ingredients	Cits.
9	Lynd et al.	2002	R	Dartmouth Coll., Univ. Stellenbosch +1	United States, South Africa	4	M	Lynd	*Microbiol. Mol. Biol. Rev.*	Microbiol.	Enzymatic production of biofuels and biocommodities from lignocellulosic biomass	Biofuels	Lignocellulosic biomass	2047
11	Searchinger et al.	2008	A	Princeton Univ., Iowa State Univ. +4	United States	9	M	Houghton, Hayes	*Science*	Mult. Sci.	Environmental impact of biofuels due to the land-use change	Biofuels	Biomass	1940
12	Himmel et al.	2007	R	Cornell Univ. +1	United States	7	M	Himmel, Nimlos	*Science*	Mult. Sci.	Biomass recalcitrance of lignocellulosic biomass	Biofuels	Lignocellulosic biomass	1815
14	Fargione et al.	2008	A	Univ. Minnesota +1	United States	5	M	Tilman*, Polasky*	*Science*	Mult. Sci.	Adverse environmental impact of biofuels due to the land-use change	Biofuels	Biomass	1614

(Continued)

Table 1.2 (Continued) The citation classics in biofuels in general

Rank	Paper reference	Year	Document type	Affiliation	Country	Number of authors	M/F	Lead authors	Journal	Subject area	Topic	Fuels	Ingredients	Cits.
16	Yang et al.	2007	A	Huazhong Univ. Sci. & Technol., Nanyang Technol. Univ.	China, Singapore	5	F	Yang	*Fuel*	Ener. Fuel, Eng. Chem	Pyrolysis of lignocellulosic biomass	Biofuels	Lignocellulosic biomass	1430
17	Hu et al.	2008	R	Arizona State Univ., Colorado Sch. Mines +1	United States	7	M	Seibert, Hu, Sommerfeld, Ghirardi F	*Plant J.*	Plant Sci.	Production of biofuels from algae	Biofuels	Algal biomass	1406
18	Farrell et al.	2006	A	Univ. Calif Berkeley	United States	6	M	O'Hare, Kammen	*Science*	Mult. Sci.	Economics and environmental optimization of biofuels	Bioethanol	Lignocellulosic biomass	1405
21	Hendriks and Zeeman	2009	R	Univ. Wageningen & Res. Ctr.	Netherlands	2	M	Zeeman F	*Bioresour. Technol.*	Biot. Appl. Microb., Ener. Fuels	Pretreatment of lignocellulosic biomass for the production of biofuels and biocommodities	Biofuels	Lignocellulosic biomass	1315
22	McKendry	2002	R	Appl. Environm. Res. Ctr. Ltd.	England	1	M	McKendry	*Bioresour. Technol.*	Biot. Appl. Microb., Ener. Fuels	Production of biofuels from biomass	Biofuels	Biomass	1302

(Continued)

Table 1.2 (Continued) The citation classics in biofuels in general

Rank	Paper reference	Year	Document type	Affiliation	Country	Number of authors	M/F	Lead authors	Journal	Subject area	Topic	Fuels	Ingredients	Cits.
27	Hill et al.	2006	A	Univ. Minnesota	United States	5	M	Tilman*, Polasky*	*Proc. Natl. Acad. Sci. U. S. A.*	Mult. Sci.	Techno-economic and environmental optimization of biofuels	Biofuels, biodiesel	Lignocellulosic biomass	1197
28	Brennan and Owende	2010	R	Univ. Coll Dublin +1	Ireland	2	M	Brennan, Owende	*Renew. Sust. Energ. Rev.*	Ener. Fuels., Green. Sust. Sci. Technol.	Production of biofuels from algae	Biofuels	Algal biomass	1193
31	Zakzeski et al.	2010	R	Univ. Utrecht	Netherlands	4	M	Weckhuysen, Bruijnincx	*Chem. Rev.*	Chem. Mult.	Catalytic transformations of lignin for the production of biofuels and biocommodities	Biofuels	Lignocellulosic biomass	1155
32	Chheda et al.	2007	R	Univ. Wisconsin	United States	3	M	Dumesic*, Huber*	*Angew. Chem.-Int. Edit.*	Chem. Mult.	Production of biofuels and biocommodities from biomass	Biofuels	Biomass	1126

(*Continued*)

Table 1.2 (Continued) The citation classics in biofuels in general

Rank	Paper reference	Year	Document type	Affiliation	Country	Number of authors	M/F	Lead authors	Journal	Subject area	Topic	Fuels	Ingredients	Cits.
34	Bozell and Petersen	2010	R	Univ. Tennessee	United States	2	M	Bozell	*Green Chem.*	Green. Sust. Sci. Technol., Chem. Mult.	Production of biofuels and biocommodities from biomass in a biorefinery	Biofuels	Carbohydrates	1086
38	Kumar et al.	2009	R	Univ. Calif Davis	United States	4	F	Stroeve	*Ind. Eng. Chem. Res.*	Eng. Chem.	Pretreatment of lignocellulosic biomass for the efficient production of biofuels	Biofuels	Lignocellulosic biomass	1053

Note: A = article; R = review; M = male; F = female; Cits. = number of citations received in the Web of Science database as of April 2017.

* Top-cited researchers in 2016.

Table 1.3 The citation classics in liquid biofuels

Rank	Paper reference	Year	Document type	Affiliation	Country	Number of authors	M/F	Lead authors	Journal	Subject area	Topic	Fuels	Ingredients	Cits.
1	Chisti	2007	R	Massey Univ.	New Zealand	1	M	Chisti	*Biotechnol. Adv.*	Biot. Appl. Microb.	Production of biodiesel from microalgae	Biodiesel	Algal biomass	3378
5	Ma and Hanna	1999	R	Univ. Nebraska	United States	2	M	Hanna	*Bioresour. Technol.*	Biot. Appl. Microb., Ener. Fuels	Biodiesel production	Biodiesel	Biomass	2522
7	Sun and Cheng	02	R	N. Carolina State Univ.	United States	2	F	Sun	*Bioresour. Technol.*	Biot. Appl. Microb., Ener. Fuels	Production of bioethanol through the enzymatic hydrolysis of lignocellulosic biomass	Bioethanol	Lignocellulose	2333
10	Mohan et al.	2006	R	Mississippi State Univ. +1	United States	3	M	Mohan*, Pitmann	*Energy Fuels*	Ener. Fuels, Eng. Chem.	Production of microalgae for biodiesel production	Bio-oil	Lignocellulosic biomass	2019

(Continued)

Table 1.3 (Continued) The citation classics in liquid biofuels

Rank	Paper reference	Year	Document type	Affiliation	Country	Number of authors	M/F	Lead authors	Journal	Subject area	Topic	Fuels	Ingredients	Cits.
15	Mata et al.	2010	R	Univ. Porto +1	Portugal	4	F	Mata F	*Renew. Sust. Energ. Rev.*	Ener. Fuels., Green. Sust. Sci. Technol.	Microalgae for biodiesel production	Biodiesel	Algal biomass	1501
20	Meher et al.	2006	R	Indian Inst. Technol.	India	4	M	Naik	*Renew. Sust. Energ. Rev.*	Ener. Fuels, Green Sust. Sci. Eng.	Biodiesel production from animal fats and vegetable oils	Biodiesel	Animal fats and vegetable oils	1327
26	Alvira et al.	2010	R	CIEMAT	Spain	4	M	Ballesteros F	*Bioresour. Technol.*	Biot. Appl. Microb., Ener. Fuels +1	Pretreatment technologies for the bioethanol production from the enzymatic hydrolysis of lignocellulosic biomass	Bioethanol	Lignocellulosic biomass	1202

(*Continued*)

Table 1.3 (Continued) The citation classics in liquid biofuels

Rank	Paper reference	Year	Document type	Affiliation	Country	Number of authors	M/F	Lead authors	Journal	Subject area	Topic	Fuels	Ingredients	Cits.
30	Czernik and Bridgwater	2004	A	NREL, Aston Univ.	United States, England	2	M	Bridgwater	*Energy Fuels*	Ener. Fuels, Eng. Chem.	Applications of bio-oil obtained through fast pyrolysis of biomass	Bio-oil	Biomass	1164
35	Zhao et al.	2007	R	Pacific NW Natl. Labs.	United States	4	M	Zhang	*Science*	Mult. Sci.	Production of 5-hydroxy-methylfurfural (HMF) through catalytic conversion	Hydroxy-methyl-furfural	Biomass	1081
36	Palmqvist and Hahn-Hagerdal	2000	R	Univ. Lund	Sweden	2	F	Hahn-Hagerdal F	*Bioresour. Technol.*	Biot. Appl. Microb., Ener. Fuels +1	Production of bioethanol through fermentation of the lignocellulosic biomass	Bioethanol	Lignocellulosic biomass	1079

(Continued)

Table 1.3 (Continued) The citation classics in liquid biofuels

Rank	Paper reference	Year	Document type	Affiliation	Country	Number of authors	M/F	Lead authors	Journal	Subject area	Topic	Fuels	Ingredients	Cits.
37	Jones and Woods	1986	R	Univ. Cape Town	United States	2	M	Jones, Woods	*Microbiol. Rev.*	Microbiol.	Production of biobutanol together with acetone through fermentation of lignocellulosic biomass	Biobutanol	Lignocellulosic biomass	1072
39	Freedman et al.	1984	A	USDA	United States	3	M	Mounts	*J. Am. Oil Chem. Soc.*	Food Sci. Technol., Chem. Appl.	Biodiesel production from vegetable oils	Biodiesel	Vegetable oils	1040
41	Agarwal	2007	R	Indian Inst. Technol.	India	1	M	Agarwal	*Prog. Energy Combust. Sci.*	Ener. Fuels, Therm. +2	Biodiesel for internal combustion engines	Biofuels, bioethanol, biodiesel	Biomass	1023

(*Continued*)

Table 1.3 (Continued) The citation classics in liquid biofuels

Rank	Paper reference	Year	Document type	Affiliation	Country	Number of authors	M/F	Lead authors	Journal	Subject area	Topic	Fuels	Ingredients	Cits.
44	Rodolfi et al.	2009	A	Univ. Florence, CNR	Italy	7	F	Tredici	*Biotechnol. Bioeng.*	Biot. Appl. Microb.	Production of microalgae for biodiesel production	Bio-oil	Algal biomass	999
45	Fukuda et al.	2001	R	Kobe Univ. +1	Japan	3	M	Kondo, Fukuda	*J. Biosci. Bioeng.*	Biot. Appl. Microbiol., Food Sci. Technol.	Biodiesel production from animal fats and vegetable oils	Biodiesel	Animal fats and vegetable oils	978
46	Graboski and McCormick	1998	R	Colorado Sch. Mines	United States	2	M	McCormick	*Prog. Energy Combust. Sci.*	Ener. Fuels, Therm. +2	Combustion of biodiesel from animal fats and vegetable oils	Biodiesel	Vegetable oils, animal fats	967
47	Roman-Leshkov et al	2007	A	Univ. Wisconsin	United States	4	M	Dumesic*	*Nature*	Mult. Sci.	Production of dimethylfuran	Biofuels, dimethylfuran	Carbohydrates	943

Note: A = article; R = review; M = male; F = female; Cits. = number of citations received in the Web of Science database as of April 2017.

* Top-cited researchers in 2016.

Table 1.4 The citation classics in gaseous biofuels

Rank	Paper reference	Year	Document type	Affiliation	Country	Number of authors	M/F	Lead authors	Journal	Subject area	Topic	Fuels	Ingredients	Cits.
24	Peters et al.	1998	A	Utah State Univ.	United States	4	M	Seefeldt, Peters	*Science*	Mult. Sci.	Hydrogenases	Biohydrogen	Hydrogenases	1250
25	Chen et al.	2008	R	N. Carolina State Univ.	United States	3	M	Chen F	*Bioresour. Technol.*	Biot. Appl. Microb., Ener. Fuels +1	Anaerobic digestion of lignocellulosic biomass	Biogas	Lignocellulosic biomass	1212
29	Volbeda et al.	1995	A	CNRS	France	4	F	Fontecilla-Camps	*Nature*	Mult. Sci.	Hydrogenases	Biohydrogen	Hydrogenases	1169
33	Cortright et al.	2002	A	Univ. Wisconsin	United States	3	M	Dumesic*	*Nature*	Mult. Sci.	Production of biohydrogen through catalytic reforming of biofuels	Biohydrogen	Biomass, biofuels	1094

(*Continued*)

Table 1.4 (Continued) The citation classics in gaseous biofuels

Rank	Paper reference	Year	Document type	Affiliation	Country	Number of authors	M/F	Lead authors	Journal	Subject area	Topic	Fuels	Ingredients	Cits.
40	Das and Veziroglu	2001	R	Indian Inst. Technol., Univ. Miami	India, United States	2	M	Veziroglu*, Das	*Int. J. Hydrog. Energy*	Ener. Fuels, Electrochem. +1	Production of biohydrogen	Biohydrogen	Biomass	1039
48	Nicolet et al.	1999	A	CNRS	France	5	M	Fontecilla-Camps, Legrand	*Struct. Fold. Des.*	Bioch. Mol. Biol., Cell Biol. +1	Hydrogenases	Biohydrogen	Hydrogenases	937

Note: A = article; R = review; M = male; F = female; Cits. = number of citations received in the Web of Science database as of April 2017.

* Top-cited researchers in 2016.

Table 1.5 The citation classics in solid biofuels

Rank	Paper reference	Year	Document type	Affiliation	Country	Number of authors	M/F	Lead authors	Journal	Subject area	Topic	Fuels	Ingredients	Cits.
13	Andreae and Merlet	2001	R	Max Planck Inst.	Germany	2	M	Andreae*	*Glob. Biogeochem. Cycle*	Env. Sci., Geosci. Mult. +1	Emissions of biomass fuels	Biomass fuels	Biomass	1673
19	Crutzen and Andreae	1990	R	Max Planck Inst. Chem.	Germany	2	M	Andreae*, Crutzen	*Science*	Mult. Sci.	Emissions for biomass fuels	Biomass fuels	Biomass	1373
50	Bond et al.	2013	R	Univ. Illinois, Univ. Washington +21	United States, England, Germany, Austria, Switzerland +4	29	F	Bond F	*J. Geophys. Res.-Atmos.*	Meteor. Atmosph. Sci.	Impact of the black carbon emissions on climate change	Biofuels	Biomass	920

Note: A = article; R = review; M = male; F = female; Cits. = number of citations received in the Web of Science database as of April 2017.

* Top-cited researchers in 2016.

Table 1.6 The citation classics in bioenergy

Rank	Paper reference	Year	Document type	Affiliation	Country	Number of authors	M/F	Lead authors	Journal	Subject area	Topic	Fuels	Ingredients	Cits.
8	Logan et al.	2006	R	Penn State Univ., Wageningen Univ. +4	United States, Netherlands +3	9	M	Logan*, Verstraete*, Rabaey*	*Environ. Sci. Technol.*	Eng. Env., Env. Sci.	Microbial fuel cells	Microbial fuel cells	Microbes	2072
23	Lovley and Phillips	1988	A	U.S. Geol. Survey	United States	2	M	Lovley*	*Appl. Environ. Microbiol.*	Biot. Appl. Microb., Microb.	Organic carbon oxidation coupled to dissimilatory reduction of iron or manganese	Bioenergy	Microbes	1280
42	Bond and Lovley	2003	A	Univ. Massachusetts	United States	2	M	Lovley*	*Appl. Environ. Microbiol.*	Biot. Appl. Microb., Microbiol.	Bioelectricity production by microbes attached to electrodes	Microbial fuel cells	Microbes	1022
43	Lovley	1991	R	U.S. Geol. Survey	United States	1	M	Lovley*	*Microbiol. Rev.*	Microb.	Dissimilatory Fe(III) and Mn(IV) reduction	Bioenergy	Microbes	1009
49	Reguera et al.	2005	A	Univ. Massachusetts	United States	6	M	Lovley*	*Nature*	Mult. Sci.	Extracellular electron transfer via microbial nanowires	Bioenergy	Microbes	929

Note: A = article; R = review; M = male; F = female; Cits. = number of citations received in the Web of Science database as of April 2017.

* Top-cited researchers in 2016.

1.2 *Methods and materials*

The Web of Science database (v. 5.23.2), comprising Science Citation Index-Expanded (SCI-E), Social Sciences Citation Index (SSCI), and Arts and Humanities Citation Index (A&HCI), was used in April 2017 to locate references on bioenergy and biofuels (Falagas et al., 2008; Hongli et al., 2010; Kulkarni et al., 2009; Meho and Yang, 2007). Because the field of bioenergy and biofuels is an interdisciplinary research field, SSCI and A&HCI should also be searched for the social and humanitarian aspects of the research in bioenergy and biofuels (Braun and Schubert, 2003; Leydesdorff, 2007; Morillo et al., 2001; Rafols and Meyer, 2010; Schummer, 2004).

The keywords for the search of the database were selected based on the experiments carried out in 2017 and on the published pioneering studies on the bioenergy and biofuels by Konur and the other scientists as well. There were two sets of keywords. The first set of keywords was related to biofuels and bioenergy: the keywords for the topical areas and authors and journals. The second set of keywords was related to the biomass and fuels and energy as well as to the subject category of Energy and Fuels. The references found were subjected to excluding terms to clean the data. The number of references for each search stage is also given to show the approximate size of the respective research fields.

I Bioenergy and biofuels: TI = ((("*bio* fuel*" or *biofuel*) or ("*bio* energy*" or *bioenergy*) or ("*bio* ethanol*" or *bioethanol*) or ("*bio* methanol" or *biomethanol*) or ("*bio* gas*" or *biogas*) or ("*bio* hydrogen*" or *biohydrogen* or hydrogenase* or (*ferment* and (*hydrogen* or H-2 or H2)) or (biological* and hydrogen*) or (biophotolysis) or ("*bio* oil*" or *biooil* or "microbial* oil*" or (*pyroly* and *oil*) or *biocrude* or "*bio* crude*") or ("bio* methane" or *biomethane*) or ("bio* butanol*" or *biobutanol* or (*ferment* and *butanol*)) or "bio* pentanol*" or ("*bio* refin*" or *biorefin* or photobioreactor*) or ("*bio* diesel*" or *biodiesel* or "trans* ester*" or *transester* or (*diesel* and *oil*) or glycerol) or "bio* alkane*" or ("*microbial electrolysis cell*" or "*microbial desalination cell*" or "*microbial electrosynthesis*" or "*microbial *electrochem*" or "*enzym* *fuel* cell*" or "*enzymatic fuel cell*" or "*microbial *fuel* cell*" or mfcs or "bio* *fuel* cell*" or "*biofuel cell*" or "glucose fuel cell*" or (*implant* and "fuel cell*") or "*bioelectrochemical system*" or "*bio* electrochemical system*") or ("bio* electric*" or *bioelectric*) or ("*bio* char" or "*bio* chars" or *biochar or biochars or charcoal* or "*char* coal*") or (*biomass* and (*power* or *fuel* or *combust* or *burn*)) or (*fuel* and (*waste* or *wood* or *cellulos*))) 80,463 references.

OR SO = (biofuels* or "journal of biobased*" or "biotechnology for biofuels" or bioenergy*) or AU = ("Logan BE" or "Ragauskas AJ" or "Demirbas A" or "Bridgwater A" or "Huber GW" or "Lehmann J" or "Chisti Y" or "Lovley DR" or "Wyman CE" or "Faaij AP" or "Ghirardi ML" or "Dale B" or "Zeeman G" or "Taherzadeh MJ" or "Himmel ME" or "Lynd LR" or "Mccormick RL" or "Knothe G" or "Van Gerpen J" or "Naik SN" or "Posten C" or "Kruse O" or "Pryde EH" or "Dewil R" or "Weiland R" or "Ptasinski KJ" or "Verstraete W" or "Rabaey K" or "Atanassov P" or "Liu Hong" or "Elliott DC" or "Kammen DM" or "Hahn-Hagerdal B" or "Zacchi G" or "Cardona CA" or "Ahring BK" or "Angelidaki I" or "Ballesteros M" or "Levin DB" or "Fontecilla-Camps JC" or "Hawkes FR" or "Andreae MO" or "Schauer JJ" or "Cass GR") 20,817 references.

98,651 references for search stage I.

II A Biomass: (TI = ((*biomass* or "*bio* mass*") or *microb* or (*alga* or Chlamydomonas or Chlorella or Spirulina or Nannochloropsis or Neochloris or Scenedesmus or Dunaliella or Haematococcus or Botryococcus or Schizochytrium or Synechococcus or Sargassum or *cyanobacter* or *seaweed* or "*sea* weed*" or diatom*) or (*cellulos* or *grass* or *wood* or *jatropha* or "*sugar cane" or *corn* or canola* or linseed* or "cotton stalk*"or "*wheat straw*"or *cassava or *maize or *sugar cane* or "*sugar beet*" or "*energy crop*" or sorghum or miscanthus or "nut shell*" or "rice husk*" or "forestry residue*" or "olive pit*" or cottonseed* or "cotton* seed*" or safflower* or sunflower* or barley or oat* or rapeseed* or "rape* seed*" or "rubber seed*" or "oleaginous microorganism*") or ("*cooking oil*" or "*frying oil*" or "*edible oil*" or "*vegetable oil*" or "animal fat*" or oilseed* or "seed oil*" or "palm oil*" or "oil palm*" or "soybean oil*") or *waste* or *manure* or "sewage sludge" or *carbohydrate* or *lignin* or triacylglycerol* or triglyceride*) OR SO = (biomass* or waste* or algal*)) 934,980 references.

II B Energy and fuels: TI = (((*fuel* near/5 (*production or *generation or transportation or biomass or *cellulos* or crop* or *hydrocarbon* or *waste* or *alga*)) or *fuel* or (*energy* near/5 (*production or generation or biomass or *cellulos* or *waste* or *hydrocarbon* or crop* or *alga*)) or (electricity* near/5 (*production or *generation or biomass or *cellulos* or *waste* or *alga*)) or *electricity or (*power* near/5 (*production or *generation or biomass or *cellulos* or *waste*)) or (*methan* near/5 (*production or *generation or biomass or *cellulos* or waste* or gasification)) or (*ethanol* near/5 (*production or fuel* or clostrid* or energy or biomass or *cellulos* or *waste* or ferment* or *alga*)) or (*diesel* or ("fatty acid*" and "methyl ester*") or fames) or (*gas* near/5 (*production or biomass or *cellulos* or methan* or manure or sludge or *waste*)) or (oil* near/5 (*production or generation or evolution or biomass or *alga* or cyanobacter* or *cellulos*)) or ((*hydrogen* or H-2 or H2) near/5 (*production or generation or evolution or clostrid* or biomass or *alga* or cyanobacter* or *ferment*)) or (*butanol* near/5 (*production or *ferment* or clostrid* or biomass or *cellulos*)) or *butanol* or "*fuel cell*" or *liquefaction* or "*anaerobic digestion" or (*pyroly* and biomass) or *pyroly* or (("pre* treatment*" or pretreatment*) and (biomass or *cellulos* or *waste* or *alga* or cyanobacter*)) or "*trans* ester*" or *transester* or ((*hydrolysis* or hydrolysate*) and (biomass* or *cellulos* or *alga* or cyanobacter* or *waste*)) or (*hydrolysis* or hydrolysate*) or *ferment* or *torrefaction* or (*gasification* and biomass) or *gasification* or dewatering or hydrodeoxygenation* or *hydroxymethylfurfural* or *alkane* or "higher alcohol*" or glycerol or "synthesis gas" or syngas or *dimethylfuran or "black carbon*")) OR WC = (energy*) OR SO = ("Journal of Analytical and Applied Pyrolysis") 682,721 references.

IIA AND IIB: 91,729 references.

I OR (IIA or IIB): 154,451 references.

NOT (TI = ("biomass c" or "biochemical mass*" or algan* or algain* or algaa* or amalgam* or Calgary* or biotin or ligas* or *gast* or *dye* or "water splitting" or "waste heat*" or "bioelectrical impedance" or "hydrogen peroxide" or immun* or biogenic* or transplantation or pharm* or dna or colon* or aquaculture or coronary or *textile* or adsor* or *sorption* or *sorbent* or photovoltaic* or retin* or rat* or gene* or *colour* or human* or photocataly* or disease* or removal* or nuclear or Arabidopsis or (food* and *ferment*)) OR WC = (nutr* or pharm* or immun* or vet* or nuclear* or med*))

The overall number of references found: 124,957 after cleaning the data.

It is notable that the first stage of the search processes focusing on bioenergy and biofuels directly in the titles of the papers produces more than 80,000 references, nearly half of the whole population of references of more than 150,000 references. Therefore, for a representative set of population in bioenergy and biofuels research, there is a crucial need for the second stage of the search processes, which does not directly refer to bioenergy and biofuels in the titles of these papers.

While the "biofuel" and "bioenergy" spellings are searched in the first set of keywords, "fuel" and "energy" spellings are searched in the second set of keywords because in a substantial number of cases, instead of "biofuel" and "bioenergy," "fuel" and "energy" spellings were used in the titles of the papers in the literature. This is more pronounced in bioethanol and biodiesel literature. The other related issue is the presence of "bio-fuel" and "bio-energy" spellings instead of "biofuel" and "bioenergy" spellings, although the recent trend in the bioenergy and biofuels literature has been on the latter set of spellings. Therefore, the keyword sets should be adjusted to cover these kinds of spellings as was done in this paper.

The first most-cited 1900 references were selected for the topical analysis and the key topical areas and keyword sets were located. These references were also used to assess the representativeness of the references of bioenergy and biofuels research areas.

Next, the 50 most-cited papers were selected for this study to locate 50 citation classics in bioenergy and biofuels. The brief summaries were prepared for each classical paper to inform the key stakeholders about the contents of these classical papers and the key data were given in the tables for each topical area. In some seminal papers, the summaries were kept relatively longer with respect to other summaries due to their public importance.

The scientometric data were collected for the whole of set of references for the bioenergy and biofuels and for these citation classics. Based on these data, the scientometric overviews of the research in bioenergy and biofuels in general using a set of over 106,000 papers as well as of these citation classics were provided to inform the key stakeholders about the current state of the research in these research areas as of April 2017. Six tables and one figure were used to complement the textual information on these citation classics.

1.3 Bioenergy and biofuels science and technology: Scientometric overview

As reported in the previous section, nearly 125,000 references were located for bioenergy and biofuels after cleaning the more than 150,000 references found, published between 1980 and April 2017.

There were 105,482 articles, 7359 meeting abstracts, 6251 proceedings papers, 5033 reviews, 2388 news items, 1923 editorial materials, 1168 notes, and other publication items as well. The sample was refined to only 110,515 articles and reviews. The sample was further refined to 106,719 articles and reviews published in English. The other primary languages were German (909), Chinese (656), Russian (438), Portuguese (347), Polish (323), Japanese (303), and Spanish (259) (Van Leeuwen et al., 2001; van Raan et al., 2011).

There were 101,818 articles and 4909 reviews on the final sample, where reviews formed 4.6% of the sample. Additionally, 6175 of these papers were published as proceedings papers as well. It was notable that only 13 papers were retracted (Steen, 2011; Steen et al., 2013; Unger and Couzin, 2006; Van Noorden, 2011).

The most-cited author data show that the 10 most-cited authors were Liu H (2023 papers), Wu QY (823), Lehmann J (704), Verstraete W (519), Demirbas A (443), Zhang Y (385), Andreae MO (364), Wang Y (350), Lovley DR (337), and Logan BE (327). It was observed that the data for Chinese authors (Liu H, Wu QY, Zhang Y, and Wang Y) suggest more than one author with the same last name and initials (D'Angelo et al., 2011; Smalheiser and Torvik, 2009; Strotmann and Zhao, 2012; Tang and Walsh, 2010; Torvik et al., 2005). For example, for "Wu QY," a number of names appear in the literature: Qiu-Yan Wu, Qing Yun Wu, Qian-Yuan Wu, Qiong-You Wu, Qing-Yun Wu, Qian-Yue Wu, Qiu-Yue Wu, and Qin-Yi Wu. Because all these names are consolidated into "Wu QY," the case of author name disambiguation erupts for such authors.

The country data show that the most prolific country was the United States with 23,610 papers, forming 22.1% of the sample. This result was expected due to the heavy research funding by the United States on bioenergy and biofuels in recent decades and due to the public policies providing incentives for the investment in bioenergy and biofuels.

China followed the United States with 18,452 papers, forming 17.2% of the sample. The other most prolific countries were India (5.7%), Germany (5.5%), Japan (5.0%), Spain (4.6%), Canada (4.4%), England (4.1%), France (3.8%), and Brazil (3.7%). In general, 139 countries and territories contributed to the research in bioenergy and biofuels.

It should be noted, however, that China surpassed the United States in the 2010s. For example, China and the United States published 28.8% and 16.9% of 11,814 papers, respectively, in 2016 and it is expected that the race between the United States and China will likely continue in the coming decades, ending the supremacy of the United States in bioenergy and biofuels at least in terms of the number of publications (Fu and Ho, 2013; Hather et al., 2010; Kostoff, 2012; Kristensen, 2015; Leydesdorff and Wagner, 2009).

The research funding data show that only 50.9% of the papers acknowledged research funding received. As expected, the Chinese and the U.S. research funding bodies headed the list of the research funding bodies. The National Natural Science Foundation of China was the most prolific research funding body with 6024 papers funded. In the light of North's New Institutional Theory, funding provides strong incentives for the stakeholders to carry out research in bioenergy and biofuels (Butler, 2003; Johnes and Johnes, 1995; Leydesdorff and Wagner, 2009; Man et al., 2004; Weingart, 2005).

The most prolific national institutions publishing papers in bioenergy and biofuels were the Chinese Academy of Sciences (3339 papers), U.S. Department of Energy (DOE) (2693), Centre National de la Recherche Scientifique (CNRS) (1938), U.S. Department of Agriculture (USDA) (1915), Council of Scientific Industrial Research (CSIR) (1157), Consejo Superior de Investigaciones Cientificas (CSIC) (1102), and Max Planck Society (783 papers).

On the other hand, the most prolific single institutions were the Indian Institute of Technology (1054 papers), University of Chinese Academy of Sciences (914), Ghent University (904), Universidade de Sao Paulo (858), University of Wisconsin–Madison (844), Tsinghua University (824), Technical University of Denmark (820), and Zhejiang University (811 papers). In total, 25,053 national and single institutions contributed to the research in bioenergy and biofuels at a global context.

Figure 1.1 shows the research output over time in bioenergy and biofuels. It is notable that the number of papers increased over time between 1980 and 2016. There were 491 papers published in 1980, whereas 11,468 papers were published in 2016. It was notable that 60.4% of the papers were published in the 2010s as of April 2017. The publication rates were 6.5%, 10.2%, and 22.8% for the 1980s, 1990s, and 2000s, respectively. These findings suggest in light of North's New Institutional Theory, the incentive structures to carry out research in bioenergy and biofuels have been efficiently devised by the stakeholders in the 2010s, attracting more researchers, institutions, journals, and countries for carrying out research in these fields. The research trend seen in the figure suggests that the number of papers would continue to increase in the coming decades with the contribution of a new set of researchers, institutions, and journals.

In total, 5158 journals published papers in bioenergy and biofuels between 1980 and April 2017. The most prolific journal was *Bioresource Technology* with 8810 papers, forming 8.3% of the sample. The other most

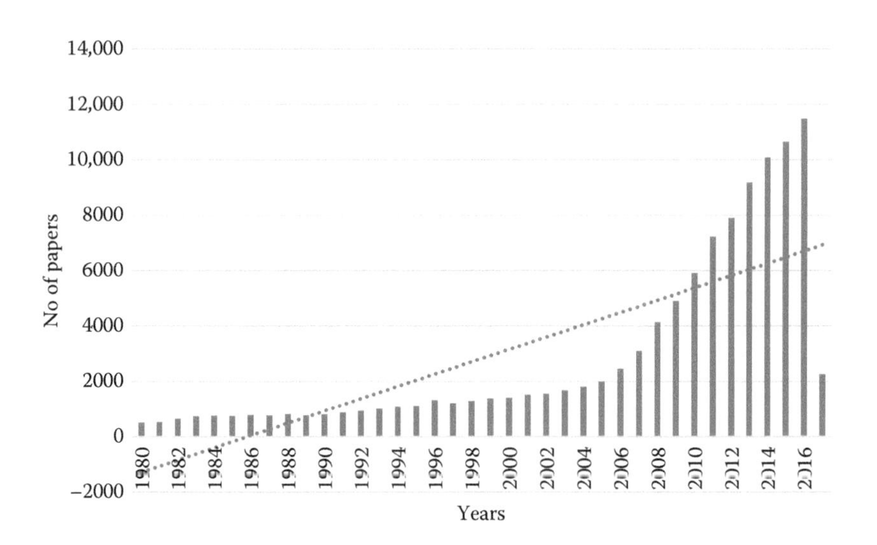

Figure 1.1 The research output over time in bioenergy and biofuels.

prolific journals were *Biomass Bioenergy* (4.0%), *Fuel* (2.6%), *Energy Fuels* (2.2%), *International Journal of Hydrogen Energy* (1.9%), *Journal of Analytical and Applied Pyrolysis* (1.1%), *Fuel Processing Technology* (1.1%), *Renewable Sustainable Energy Reviews* (1.0%), *Energy* (1.0%), *Applied Energy* (1.0), and *Renewable Energy* (1.0%). It is evident from these journal data that *Bioresource Technology* was the major outlet for the bioenergy and biofuels research, followed by *Biomass Bioenergy*. Both journals were indexed under the subject categories of Energy and Fuels as well as Biotechnology and Applied Microbiology.

As expected, the most prolific subject category was Energy and Fuels with 37,295 papers, forming 34.9% of the sample. However, these figures also suggest that most of the research in bioenergy and biofuels (65.1% of the sample) were published in journals indexed under other subject categories. The practical implication of this finding for the search strategy is that the set of keywords should be selected carefully to locate the related research in bioenergy and biofuels outside the subject category of energy and fuels as well because it is customary to limit the search strategies to journals and to subject categories in the scientometric literature.

The other most prolific subject areas were biotechnology applied microbiology (24.1%), engineering chemical (17.1%), agricultural engineering (13.3%), environmental sciences (11.8%), chemistry physical (7.5%), engineering environmental (6.3%), chemistry multidisciplinary (6.1%), and biochemistry molecular biology (4.9%).

These data suggest the importance of subject categories related to biotechnology, applied microbiology, chemical engineering, agricultural engineering, environmental engineering, and biochemistry and molecular biology for the research in bioenergy and biofuels.

It is notable to see the lack of the presence in the top lists of the subject categories related to the materials science and engineering. The major subject categories related to materials were materials science multidisciplinary (2.7%), polymer science (1.9%), materials science paper wood (1.5%), nanoscience nanotechnology (0.8%), and materials science biomaterials (0.8%). As the contribution of nanomaterials and polymeric materials to the research in bioenergy and biofuels increases, it is likely that in the coming decades these subject categories would move up in the category lists.

1.4 *The citation classics in bioenergy and biofuels science and technology: Scientometric overview*

The concise information about the 50 citation classics in bioenergy and biofuels as a text and tables and a figure are given in the following respective sections. These tables (Tables 1.2 through 1.6) inform the readers to locate the key information about these classical papers in one place together with the other related papers.

The first column gives the information about the rank of the paper among these classical papers. The top citation classic was a review paper on algal biodiesel with 3378 citations by Chisti (2007). This paper was closely followed by a review paper on biofuels for transportation from the lead laboratories of Huber and Corma with 3260 citations received (Huber et al., 2006). The following top papers were a review paper on biofuels and biocommodities from Ragauskas' lead laboratory (Ragauskas et al., 2006), a review paper on the pretreatment of lignocellulosic biomass for biofuels from the lead laboratories of Wyman, Dale, Holtzapple, and Ladisch (Mosier et al., 2005), and an early review paper on biodiesel production from the lead laboratory of Hanna (Ma and Hanna, 1999). Thus, these top papers focus on biofuels in general, algal biofuels, lignocellulosic biofuels, and biodiesel. These highlight the public interest in algal and lignocellulosic biofuels as an alternative to food crops-based first-generation biofuels.

The data in the third column on the publication years of the papers show that 68% of these classical papers were published in the 2000s, mostly in 2006 and 2007, with seven and eight papers, respectively. The publication rates for the 1980s, 1990s, and 2010s were 4%, 16%, and 12%, respectively. These suggest that the papers published in the last decade have largely influenced the development of the research in these fields.

The fourth column gives information about the type of papers, where A is article and R is review. The data show that the review papers were overrepresented, forming 68% of the sample, while review papers formed just over 4.6% of the sample for the whole population of papers on bioenergy and biofuels. These data suggest that review papers written by the subject experts significantly influenced the development of the research in their respective fields.

The fifth column gives the data on the affiliation of the authors for each paper. The Indian Institute of Technology, CNRS, Max Planck Institute, Dartmouth College, University of Massachusetts, University of Wisconsin, U.S. Geological Survey, and Valencia Polytechnic University were among the most prolific institutions. It is also notable that 60% of the papers were produced by single institutions, whereas 40% of these papers had more than one university as an affiliation. This suggests that interuniversity collaboration was not much preferred in this field.

The most prolific country was the United States with 58% of the sample. The United States was overrepresented among these classical papers because the United States contributed to only 22.1% of over 106,000 papers in biofuels and bioenergy. Because China followed the United States with 18,452 papers, forming 17.2% of the large sample of over 106,000 papers, it is notable that there was only one citation classic from China. This finding suggests that although China surpassed the United States in recent years in bioenergy and biofuels, it did not show parallel performance among citation classics.

The other most prolific countries with two or more papers were England, France, Germany, India, Netherlands, and Spain. It is also notable that intercountry collaboration was insignificant, with 68% of these classical papers being produced from one country only.

The data on the number of authors for each paper show that the average number of authors per paper was 4.4. These papers were mostly written by two and four authors with 14 and 11 papers, respectively. It is notable that only four papers were written by a single author. These data on the number of authors for papers suggest that these classical papers were produced in general by the research teams led by the subject leaders.

The gender gap in engineering and hard sciences as in some other fields has long been a public concern (Carrell et al., 2010; Chesler and Chesler, 2002; Ding et al., 2006; Jagsi et al., 2006; Kulis et al., 2002; Moss-Racusin et al., 2012). Therefore, the data on the gender of the first author of these papers were given in the table as M for male and F for female. These data show that only 16% of these papers were written by female first authors, providing further evidence of the significant gender gap in bioenergy and biofuels.

Furthermore, the gender gap is more significant when the genders of the leading authors were taken into account because only 10 of these papers were written by female researchers alone or with other leading authors. This finding confirms the presence of a significant gender gap in bioenergy and biofuels.

The data on the leading authors show that the most prolific leading authors were Andrea on biomass fuel emissions, Dumesic on liquid biofuels and biohydrogen, Fontecilla-Camps on hydrogenases, Corma on biofuels and biocommodities, Lovley on microbial fuel cells (MFCs) and bioelectrochemical reduction, and Tilman on environmental effects of biofuel production due to the land-use change.

It is also notable that Andreae, Corma, Dumesic, Huber, Lovley, Mohan, Tilman, Polasky, Veziroglu, and Wyman were considered as the top scientists ("some of the world's most influential minds," [Clarivate Analytics, 2017]) in their respective fields based on the number of citations received for their respective papers using the Web of Science database in 2016 as published online in http://hcr.stateofinnovation.com/ (Abramo et al., 2016; Basu, 2006; Bornmann and Bauer, 2015a,b; Bornmann et al., 2017; Li, 2016; Oravec, 2017; Parker et al., 2013; Podlubny and Kassayova, 2006). These highly cited authors were shown with a* in Tables 1.2 through 1.6. These data suggest that bioenergy and biofuels emerge as one of the most-cited research areas and provide incentives for the expansion of the workforce in these research fields as evidenced by the increasing number of researchers globally engaged in these research fields over time.

The most prolific journal, as expected, was *Bioresource Technology* with 10 papers, forming 20% of the sample. This journal was overrepresented because its contribution to the whole sample was only 8.3%. This top journal was followed closely by *Science* with eight papers, forming 16% of the sample. The other most prolific journals were *Chemical Reviews Nature* and *Renewable Sustainable Energy Reviews* with three papers each, and *Applied and Environmental Microbiology, Energy Fuels, Microbiology Reviews,* and *Progress in Energy and Combustion Science* with two papers each.

It is notable that *Science, Chemical Reviews, Nature, Applied and Environmental Microbiology, Energy Fuels, Microbiology Reviews,* and *Progress in Energy and Combustion Science* were overrepresented as well because they had respectively lower contribution to the whole population of more than 106,000 papers. On the other hand, it is also notable that *Biomass Bioenergy, Fuel, Energy Fuels, International Journal of Hydrogen Energy, Journal of Analytical and Applied Pyrolysis, Fuel Processing Technology, Energy, Applied Energy,* and *Renewable Energy* were underrepresented among these citation classics.

It is also notable that these most prolific journals such as *Bioresource Technology, Science, Nature, Chemical Reviews,* and *Applied and Environmental Microbiology* are also top journals in the Web of Science in terms of a number of citation measures such as H-index. These data suggest that the research on bioenergy and biofuels find a place in these top journals as evidence of the increasing public importance of these research fields as perceived by society at large.

The next column in the tables shows the data for the subject categories with which each of the journals were indexed. The most prolific subject category, as expected, was energy and fuels with 19 papers, forming 38% of the sample, where bioresource technology was the key component. Biotechnology and applied microbiology as well as multidisciplinary sciences were the following most prolific subject categories with 13 papers each. In the former subject category, bioresource technology was the key component, and in the latter one science was the key component. Agricultural engineering was the other most prolific subject category with 10 papers, where bioresource technology was the key component.

The other most prolific subject categories were chemistry multidisciplinary and microbiology with five papers each, green and sustainable science and technology with four papers, engineering chemical with three papers, and food science and technology with two papers.

The most relevant insight emerging from these data is the fact that only 38% of these classical papers were indexed under the subject category of energy and fuels. This finding has strong implications for the search design discussed in Section 1.2: the search should not be limited to the relevant subject categories such as energy and fuels as here, but the whole Web of Science database should be searched for the relevant papers because it is customary to limit the search strategies to journals and subject categories in the scientometric literature.

The data in the column titled "Fuels" give the information of the primary type of biofuels or bioenergy studied for each paper, as summarized in Table 1.1. Because this paper is structured mainly in terms of the types of fuels or energies, this piece of information is highly critical.

Five primary topical areas emerged in these classical papers: Biofuels in general, liquid biofuels, gaseous biofuels, solid biofuels, and bioenergy with 38%, 34%, 12%, 6%, and 10% of the classical paper sample, respectively. A similar structure was also followed in the constitution of this book.

The data in this table show that "biofuels in general" was the most prolific type of fuel with 38% of the sample. Lignocellulosic biofuels and biofuel production overview were the most prolific sub-biofuel categories in this sample with seven and six papers, respectively, following top one. Algal biofuels, biofuels economics and optimization, and biofuels and environment were the other subtopics.

The representation of lignocellulosic biofuels and algal biofuels in this sample suggests that these biofuels emerged as the key types of biofuels over time as an alternative to food crops-based biofuels due to the risks to food security and environmental protection in light of the optimal design rules set out by Hill et al. (2006).

On the other hand, the inclusion of the papers on biofuels economics and optimization, and biofuels and environment in the sample list suggests that society had also been concerned with the economics, optimization, and environmental risks of biofuel production. The price competition with the petroleum-based energy and fuels as well as with other types of fuels and energies such as nuclear power and hydropower would be a crucial part of the fuel and energy selection matrix for many countries in the coming decades. Besides the price competition, the availability in the required amounts for the bioenergy and biofuels would be another criterion to consider. With the increasing public concern about global warming, greenhouse gases, and climate change in recent decades, bioenergy and biofuels, which do not contribute to global warming, greenhouse gases, and climate change, would be sought after by many countries in the coming years.

The studied liquid biofuels in this classical paper sample were biodiesel as an alternative to petroleum-based diesel in diesel engines for transportation, bioethanol as an alternative to petroleum-based gasoline in car engines, bio-oil as an alternative to the petroleum-based oils, biobutanol, and other liquid biofuels as alternatives to gasoline with nine, three, two, and two papers, respectively. However, there was no study on biomethanol in this sample.

The gaseous biofuels studied in this sample included biohydrogen and biogas with five and one papers, respectively. There was no study on the biomethane in this sample. The overrepresentation of biohydrogen (10% of the sample) shows the public interest in biohydrogen as a biofuel of the future because it meets the conditions of the optimal bioenergy and biofuels design as stated in Hill et al. (2006).

Solid biofuels studied in this sample included biomass biofuels, mostly focusing on the gas emissions from the combustion of biomass biofuels. There were no papers on biochars and biocharcoal in this sample. There were also no papers on the combustion of biomass biofuels in the context of residential heating and coal-fired power plants. The overrepresentation of papers in gas emissions from the combustion of biomass biofuels shows the importance of the public concern on the environmental impacts of biomass biofuels.

Bioenergy studied in this sample included only microbial fuel cells for the bioelectricity generation and bioremediation purposes. There were no papers on the enzymatic biofuel cells in this sample. The overrepresentation of microbial fuel cells in this sample shows the importance of public interest in microbial fuel cells for both bioelectricity generation and bioremediation purposes.

The data in the last column of the tables show the ingredients used in the production of bioenergy and biofuels. Lignocellulosic biomass and algal biomass were the most prolific biomass used in the classical papers. Carbohydrates and vegetable oils were the other ingredients studied. In some papers, biomass is considered in the general terms. The focus on the lignocellulosic and algal biofuels shows the importance of public interest in these biomasses because they meet the conditions for the optimal biofuels design as stated in Hill et al. (2006).

1.5 *The citation classics in bioenergy and biofuels science and technology*

In this section, the concise summaries for each of 50 classical papers are provided to inform the stakeholders about the key issues considered in these studies in the order described in Table 1.1: Biofuels in general, liquid biofuels, gaseous biofuels, solid biofuels, and bioenergy.

Additionally, the key bibliographic data are provided for each paper in the topical tables, Tables 1.2 through 1.6: Paper references, publication years, document type (article/review), affiliation of authors (institution and country), the number of authors for each paper, the gender of the first author (male/female), lead authors, journal titles, categorical subject areas, brief topic, type of biofuels and bioenergy, ingredients used in the production of these biofuels and bioenergy, and the number of citations received for each paper in the Web of Science database as of April 2017.

1.5.1 Biofuels in general

In this section, the concise summaries for each of the classical papers are provided to inform the stakeholders about the key issues considered in these studies in the order described in Table 1.1: biofuels production overview, algal biofuels, lignocellulosic biofuels, biofuels economics and optimization, and biofuel and environment (see also Table 1.2).

1.5.1.1 Biofuel production in general

There were six classical papers on the production of biofuels and biocommodities from biomass in the biorefinery context to reduce the production cost of biofuels. The most prolific authors were Corma, Dumesic, and Huber.

Huber et al. (2006) discuss the production of biofuels for transportation from biomass in a comprehensive seminal review paper originating from the lead laboratories of Huber and Corma in Spain with 3260 citations. The key headings for the review were biomass chemistry and growth rates, biomass production (pretreatment, hydrolysis, and hydrogenation), biomass gasification, syngas (biohydrogen, biomethanol, and bioalkanes), utilization, bio-oil production (fast pyrolysis and hydrothermal liquefaction), bio-oil upgrading (hydrodeoxygenation and steam reforming), conversion of sugars into biofuels (bioethanol, biohydrogen, and biomethane), conversion of nonsugar monomers from lignocellulose (lignin, leuvinic acid, and furfural), triglyceride conversion (transesterification, pyrolysis, hydrotreating, microemulsions, and glycerol utilization), and ethical considerations.

Ragauskas et al. (2006) discuss the production of biofuels and biomaterials from biomass in a seminal review paper originating from the lead laboratory of Ragauskas in the United States with 2664 citations. They note that biorefinery is a new manufacturing concept for the sustainable and techno-economic production of these biocommodities from energy crops.

Corma et al. (2007) discuss production of biocommodities from biomass, vegetable oils, animal fats, and terpenes in a comprehensive review paper originating from the lead laboratory of Corma in Spain with 2436 citations.

McKendry (2002) discusses the production of biofuels from biomass in a review paper originating from his laboratory in England with 1302 citations. He notes that the type of biomass required is largely determined by the biofuel conversion process and the form in which the biofuel is required. He focuses on the background to biomass production and biomass properties.

Chheda et al. (2007) discuss the production of biofuels and biocommodities from biomass in a review paper originating from the lead laboratories of Huber and Dumesic in the United States with 1126 citations. They focus on the development of catalytic processes. They note that the key reactions involved in the processing of biomass are hydrolysis, dehydration, isomerization, aldol condensation, reforming, hydrogenation, and oxidation. They then develop strategies for the control of reaction pathways and process conditions to produce H_2/CO_2 or H_2/CO gas mixtures by aqueous-phase reforming, to produce furan compounds by selective dehydration of carbohydrates, and to produce liquid bioalkanes by the combination of aldol condensation and dehydration/hydrogenation processes.

Bozell and Petersen (2010) discuss the production of biofuels and biocommodities from biomass in a biorefinery in a review paper originating from the lead laboratory of Bozell in the United States with 1086 citations. They note that the U.S. DOE developed a selection process for biochemical products of the biorefinery in 2004 and they present an updated evaluation of potential biochemical products using similar selection methodology, and provide an overview of the technology developments that led to the inclusion of a given biocommodity.

1.5.1.2 Algal biofuels

There were two classical papers on the production of algal biofuels from algae as an alternative to the biofuel production from food crops. These papers highlight the potential of these biofuels to replace petroleum-based fuels.

Hu et al. (2008) discuss the production of biofuels from algae in a review paper originating from the lead laboratories of Hu, Sommerfield, Ghirardi, and Seibert in the United States with 1406 citations. They first focus on the oleaginous algae and their fatty acid and triacylglycerol (TAG) biosynthesis and then they discuss algal model systems and genomic approaches to a better understanding of TAG production within the wider historical and technological context.

Brennan and Owende (2010) discuss the production of biofuels from algae in a review paper originating from their laboratories in Ireland with 1193 citations. They focus on the algal biomass production, harvesting, conversion technologies, and the extraction of useful coproducts. They highlight synergistic coupling of microalgal propagation with carbon sequestration and wastewater treatment potential for mitigation of environmental impacts associated with energy conversion and utilization. They argue that although there are outstanding issues related to photosynthetic efficiencies and algal biomass output, algal biofuels could progressively substitute a significant proportion of the fossil fuels.

1.5.1.3 Lignocellulosic biofuels

There were seven classical papers on the production of biofuels from lignocellulosic biomass in general terms.

1.5.1.3.1 Pretreatments Five of these papers were related to the pretreatment of lignocellulosic biomass to make them suitable for the following hydrolysis and fermentation stages of production. Thus, pretreatment of biomass reduces the cost of biofuel production from lignocellulosic biomass. The key insights from these papers are that the pretreatment processes should be optimized for each type of biomass to enhance the efficiency of the conversion of biomass to biofuels. These papers are also most relevant for the bioethanol production from lignocellulosic biomass.

Mosier et al. (2005) discuss the pretreatment of lignocellulosic biomass for the efficient production of biosugars and biofuels in a review paper originating from the lead laboratories of Wyman, Dale, Holtzapple, and Ladisch in the United States with 2531 citations. They note that the goal of any pretreatment technology is to alter or remove structural and compositional impediments to hydrolysis of biomass in order to improve the rate of enzymatic hydrolysis and to increase yields of fermentable biosugars from cellulose or hemicellulose. They argue that experimental investigation of physical changes and chemical reactions that occur during pretreatment is required for the development of effective and mechanistic models that can be used for the rational design of pretreatment processes. Furthermore, they recommend that pretreatment processing conditions must be tailored to the specific chemical and structural composition of the various, and variable, sources of lignocellulosic biomass.

Himmel et al. (2007) discuss biomass recalcitrance of lignocellulosic biomass, the natural resistance of plant cell walls to both microbial and enzymatic deconstruction, in a review paper originating from the lead laboratory of Himmel in the United States with 1,815 citations. They note that biomass recalcitrance of lignocellulosic biomass is largely responsible for the high cost of lignocellulosic conversion to biofuels and biocommodities. They argue that it would be necessary to overcome the chemical and structural properties that have evolved in lignocellulosic biomass to prevent its disassembly through a number of pretreatment technologies.

Hendriks and Zeeman (2009) discuss the pretreatment of lignocellulosic biomass for the production of biofuels and biocommodities in a review paper originating from the lead laboratory of Zeeman in the Netherlands with 1315 citations. They focus on the different effects of several pretreatments on the three main parts of the lignocellulosic biomass—cellulose, hemicellulose, and lignin—to improve its digestibility. They argue that steam pretreatment, lime pretreatment, liquid hot water pretreatments, and ammonia-based pretreatments are pretreatments with high potentials. They note that the main effects of such pretreatments are dissolving hemicellulose and alteration of lignin structure, providing an improved accessibility of the cellulose for hydrolytic enzymes.

Zakzeski et al. (2010) discuss the catalytic transformations of lignin, one of the main part of the lignocellulosic biomass, for the production of biofuels and biocommodities in a comprehensive review paper originating from the lead laboratory of Weckhuysen in the Netherlands with 1155 citations. They focus on lignin structure, pretreatment of lignins, lignin model compounds, biorefineries, lignin dissolution, and catalytic lignin transformations.

Kumar et al. (2009) discuss the pretreatment of lignocellulosic biomass for the efficient production of biofuels in a review paper originating from the lead laboratory of Stroeve in the United States with 1053 citations. They focus on the various pretreatment process methods and the use of these technologies for pretreatment of various lignocellulosic biomasses for the efficient hydrolysis of lignocellulosic biofuels.

1.5.1.3.2 Production There were two papers on the production of biofuels from lignocellulosic biomass. They focus on the enzymatic production of biofuels and biocommodities and pyrolysis of biomass. These papers are also most relevant for the production of bioethanol from lignocellulosic biomass.

Lynd et al. (2002) discuss the enzymatic production of biofuels and biocommodities from lignocellulosic biomass in a review paper originating from the lead laboratory of Lynd in the United States with 2047 citations. They focus on the structure and composition of cellulosic biomass, taxonomic diversity, cellulase enzyme systems, molecular biology of cellulase enzymes, physiology of cellulolytic microorganisms, ecological aspects of cellulase-degrading communities, and rate-limiting factors in nature. They introduce consolidated bioprocessing (CBP) in which the production of cellulolytic enzymes, hydrolysis of cellulosic biomass, and fermentation of resulting biosugars to biofuels and biocommodities take place in one step. They then develop strategies in developing CBP-enabling microorganisms to improve product yield and tolerance in microorganisms with the ability to utilize cellulose, or to express a heterologous system for cellulose hydrolysis and utilization in microorganisms that exhibit high product yield and tolerance.

Yang et al. (2007) discuss pyrolysis of lignocellulosic biomass based on the main components of the biomass, cellulose, hemicellulose, and lignin in an experimental paper originating from the lead laboratory of Yang in China with 1430 citations. In thermal analysis, they observe that the pyrolysis of hemicellulose and cellulose occurred quickly, while lignin was more difficult to decompose. Regarding the energy consumption, they observe that the pyrolysis of the cellulose was endothermic, while that of hemicellulose and lignin was exothermic. The main gas products from pyrolyzing the three components were similar, including CO_2, CO, CH_4, and some organics. They observe that hemicellulose had higher CO_2 yield, cellulose generated higher CO yield, and lignin owned higher H_2 and CH_4 yield.

1.5.1.4 Biofuel economics and optimization

There were two papers on the economics and optimization of biofuels using case studies of bioethanol and biodiesel from food crops such as corn and soybeans. They lay down the rules for the optimal biofuel production: biofuels should provide "a net energy gain, have environmental benefits, be economically competitive, and be producible in large quantities without risking food security" (Hill et al., 2006). The key insights from these studies are that biofuels should not be produced from food crops and other alternatives such as algal biomass and lignocellulosic biomass should be considered for biofuel production. These papers are also the most relevant for bioethanol production.

Farrell et al. (2006) discuss economics and environmental optimization of biofuels with a case study of bioethanol from food crops by evaluating six representative analyses of bioethanol in a paper originating from the lead laboratories of O'Hare and Kammen in the United States with 1405 citations. They argue that studies that reported negative net energy incorrectly ignored coproducts and used some obsolete data. All studies indicated that current corn-based bioethanol technologies are much less petroleum-intensive than gasoline but have greenhouse gas emissions similar to those of gasoline. However, they argue that many important environmental effects of biofuel production are poorly understood and recommend further research on environmental metrics. They recommend the bioethanol production from lignocellulosic biomass rather than food crops such as corn.

Hill et al. (2006) discuss techno-economic and environmental optimization of biofuels in a case study of bioethanol from corn grain and biodiesel from soybeans, both food crops, in a paper originating from the lead laboratories of Tilman and Polasky in the United States with 1197 citations. They note that optimal biofuels should provide "a net energy gain, have environmental benefits, be economically competitive, and be producible in large quantities without risking food security." They use these criteria to evaluate bioethanol and biodiesel compared to gasoline and diesel, through life-cycle assessment. They find that bioethanol yields 25% net energy gain, whereas biodiesel yields 93% more. Compared with bioethanol, biodiesel releases just 1.0%, 8.3%, and 13% of the agricultural nitrogen, phosphorus, and pesticide pollutants, respectively, per net energy gain. Relative to gasoline and diesel, greenhouse gas emissions are reduced 12% by the production and combustion of bioethanol and 41% by biodiesel. Biodiesel also releases less air pollutants per net energy gain than bioethanol. They reason that these advantages of biodiesel over bioethanol emanate from lower agricultural inputs and more efficient conversion of biomass feedstocks to biofuels. However, neither biofuel can replace much petroleum without risking food security. They note that even dedicating all U.S. corn and soybean production to biofuels would meet only 12% of gasoline demand and 6% of diesel demand. Until recent increases in petroleum prices, high production costs made biofuels unprofitable without subsidies. They argue that biodiesel provides sufficient environmental advantages to merit subsidy. They recommend that transportation biofuels should be produced from low-input biomass grown on agriculturally marginal land or from lignocellulosic biomass because they would provide much greater supplies and environmental benefits than food-based biofuels.

1.5.1.5 Biofuels and environment

In addition to two papers on economics and optimization of biofuels in which the environmental impact of biofuel production from energy crops are considered, there were two more papers solely discussing the environmental impact of biofuels due to the land-use change. The key insights from these studies are that due to the land-use change, the production of biofuels from food crops or switchgrass results in adverse environmental impacts in terms of increased greenhouse gas emissions. These findings imply that biofuels should not be produced from food crops and other alternatives such as algal biomass and lignocellulosic biomass should be considered for biofuel production.

Searchinger et al. (2008) discuss the environmental impact of biofuels due to the land-use change using a case study of bioethanol from corn or switchgrass biomass in a paper originating from the lead laboratories of Houghton and Hayes in the United States with 1940 citations. By using a global agricultural model to estimate greenhouse gas emissions from land-use change, they find that corn-based bioethanol, instead of producing a 20% savings, nearly doubles greenhouse gas emissions over 30 years and increases greenhouse gases for 167 years. On the other hand, biofuels from switchgrass, if grown on U.S. corn lands, increase greenhouse gas emissions by 50%. They reason that as farmers respond to higher prices, they convert forest and grassland to new cropland to replace the grain (or crop land) diverted to bioethanol, resulting in increased carbon emissions due to the land-use change. They recommend the production of bioethanol from lignocellulosic biomass rather than from food crops.

Fargione et al. (2008) discuss the adverse environmental impact of biofuels due to the land-use change in a paper originating from the lead laboratories of Tilman and Polasky in the United States with 1614 citations. They note that converting rainforests, peatlands, savannas, or grasslands to produce food crop-based biofuels in Brazil, Southeast Asia, and the United States creates a carbon debt by releasing 17–420 times more CO_2 than the annual greenhouse gas reductions that these biofuels would provide by displacing fossil fuels. In contrast, biofuels made from waste biomass or from biomass grown on degraded and abandoned agricultural lands planted with perennials incur little or no carbon debt and can offer immediate and sustained greenhouse gas advantages.

1.5.2 *Liquid biofuels*

In this section, the concise summaries for each of the classical papers are provided to inform the stakeholders about the key issues considered in these studies in the order described in Table 1.1: biodiesel, bio-oil, bioethanol, biobutanol, and other liquid biofuels (see also Table 1.3). There were no classical papers in biomethanol.

1.5.2.1 Biodiesel

In addition to Hill et al. (2006) in which the environmental impact of biodiesel from corn and soybeans is discussed, there were two papers on biodiesel focusing on the biodiesel engine performance, four papers on biodiesel production from animal fats and vegetable oils, and three papers on algal biodiesel.

1.5.2.1.1 Biodiesel engine performance Agarwal (2007) discusses biodiesel for internal combustion engines in a review paper originating from his laboratory in India with 1023 citations. He focuses on performance and emission of biodiesel in internal combustion engines, combustion analysis, wear performance on long-term engine usage, and economic feasibility.

Graboski and McCormick (1998) discuss the combustion of biodiesel from animal fats and vegetable oils in a review paper originating from the lead laboratory of McCormick in the United States with 967 citations. They note that the price of the biomass is the major factor determining biodiesel price. The use of biodiesel in neat or blended form has no effect on the energy-based engine fuel economy. The lubricity of biodiesel is superior to conventional diesel at levels above 20 vol%. Emissions of particulate matter can be reduced dramatically through use of biodiesel in engines that are not high lube oil emitters. Emissions of NO_x increase significantly for both neat and blended fuels in both two- and four-stroke engines. They recommend that biodiesel of well-known composition and purity should be used with detailed analyses. The purity levels necessary for achieving adequate engine endurance, compatibility with coatings and elastomers, cold flow properties, stability, and emissions performance must be better defined.

1.5.2.1.2 Biodiesel production Ma and Hanna (1999) discuss biodiesel production in a review paper originating from the lead laboratory of Hanna in the United States with 2522 citations. They note that the waste cooking oils as raw material, adaption of continuous transesterification process, and recovery of glycerol from biodiesel by-product are primary routes to lower the cost of biodiesel. They note that although there are four primary ways to make biodiesel, the most commonly used method is transesterification of vegetable oils and animal fats. They further note that the determinants of the transesterification reaction are molar ratio of glycerides to alcohol, catalysts, reaction temperature, reaction time, and free fatty acids and water content of oils or fats.

Meher et al. (2006) discuss the biodiesel production from animal fats and vegetable oils by transesterification in a review paper originating from the lead laboratory of Naik in India with 1327 citations. They note that the determinants of the transesterification are the mode of reaction condition, molar ratio of alcohol to oil, type of alcohol, type and amount of catalysts, reaction time, and temperature and purity of reactants. They focus on the various methods of production of biodiesel with different combinations of oil and catalysts.

Freedman et al. (1984) discuss the biodiesel production from vegetable oils such as cotton seed, peanut, soybean, and sunflower oils by transesterification in an early experimental paper originating from the lead laboratory of Mounts in the United States with 1040 citations. They find that transesterification reaction variables that affect yield and purity of the product esters include molar ratio of alcohol to vegetable oil, type of catalyst (alkaline versus acidic), temperature, and degree of refinement of the vegetable oil. They observe that at moderate temperatures (32°C), vegetable oils were 99% transesterified in approximately 4 h with an alkaline catalyst, while transesterification by acid catalysis was much slower than by alkali catalysis.

Fukuda et al. (2001) discuss the biodiesel production from animal fats and vegetable oils by enzymatic transesterification using lipases in a review paper originating from the lead laboratories of Fukuda and Kondo in Japan with 978 citations. They note that enzymatic transesterification using lipase has become more attractive for biodiesel fuel production because the glycerol produced as a by-product can easily be recovered and the purification of fatty methyl esters is simple to accomplish. However, the main hurdle to the commercialization of this system is the cost of lipase production. As a means of reducing the cost, they argue that the use of whole-cell biocatalysts immobilized within biomass support particles is significantly advantageous because immobilization can be achieved spontaneously during batch cultivation, and in addition no purification is necessary. They further note that the lipase production cost can be further lowered using genetic engineering technology, such as by developing lipases with high levels of expression and/or stability toward methanol.

1.5.2.1.3 Algal biodiesel There were three classical papers on algal biodiesel highlighting the public importance of these biofuels as an alternative to food crops-based biofuels. These studies imply that the cultivation, harvesting, and processing of algal biomass should be optimized to reduce the production costs to be competitive with biofuels from lignocellulosic biomass, food crops, and petroleum.

Chisti (2007) discusses production of biodiesel from microalgae in a seminal review paper originating from his laboratory in New Zealand with 3378 citations. He argues that microalgae is the only source of renewable biodiesel that is capable of meeting the global demand for transportation fuels compared to biodiesel from food crops and lignocellulosic biomass. He notes that like plants, microalgae use sunlight to produce oils, but they do so more efficiently than crop plants, and oil productivity of many microalgae greatly exceeds the oil productivity of the best producing oil crops. He focuses on the approaches for making microalgal biodiesel economically competitive with petroleum-based diesel.

Mata et al. (2010) discuss microalgae for biodiesel production in a review paper originating from the lead laboratory of Mata in Portugal with 1501 citations. They focus on the use of microalgae for biodiesel production, including their cultivation, harvesting, and processing. They give an overview of the development of algal cultivation systems such as photobioreactors and open ponds. They finally discuss its use in the biosequestration of CO_2, in wastewater treatment, in human health as food additive, and for aquaculture.

Rodolfi et al. (2009) discuss the production of microalgae for biodiesel production in an experimental paper originating from the lead laboratory of Tredici in Italy with 999 citations. They screen 30 microalgal strains for their biomass productivity and lipid content, select four strains, and cultivate them under nitrogen deprivation. Only the two marine microalgae accumulated lipid under such conditions. One of them, the *Nannochloropsis* sp., attained 60% lipid content after nitrogen starvation. They find that fatty acid content increased with high irradiances following both nitrogen and phosphorus deprivation. They further find that lipid productivity increased in nutrient sufficient media with a 32% lipid content to more than 60% final lipid content in nitrogen-deprived media. They obtain an increase of both lipid content and areal lipid productivity attained through nutrient deprivation in an outdoor algal culture: 20 tons of lipid per hectare and more than 30 tons of lipid per hectare in sunny tropical areas.

1.5.2.2 Bio-oil

There were two classical papers on the production of bio-oil through the fast pyrolysis of lignocellulosic biomass as well as their applications as biofuels and biocommodities through a number of upgrading processes.

Mohan et al. (2006) discuss the production of bio-oil through pyrolysis of lignocellulosic wood biomass in a review paper originating from the lead laboratories of Mohan and Pittman in the United States with 2019 citations. They focus on the determinants of the properties of bio-oil such as the effect of the wood composition and structure, heating rate, and residence time during pyrolysis on the overall reaction rate and the yield of the volatiles. They note that very fast and very slow pyrolysis of biomass produce markedly different products and the variety of heating rates, temperatures, residence times, and feedstock varieties make generalizations difficult to define.

Czernik and Bridgwater (2004) discuss the applications of bio-oil obtained through fast pyrolysis of biomass in a review paper originating from the lead laboratory of Bridgwater in England with 1164 citations. They highlight the applications of bio-oil as biofuels and biocommodities.

1.5.2.3 Bioethanol

In addition to Farrell et al. (2006), Hill et al. (2006), and Searchinger et al. (2008) in which the economic and environmental impact of bioethanol production from food crops are discussed, there were three more papers on bioethanol production from lignocellulosic biomass as advised by Farrell et al. (2006), Hill et al. (2006), and Searchinger et al. (2008).

These studies also build on Mosier et al. (2005), Himmel et al. (2007), Hendriks and Zeeman (2009), Zakzeski et al. (2010), and Kumar et al. (2009) in which the pretreatment of lignocellulosic biomass for the efficient biological conversion of biomass to biofuels is discussed. These studies show that the pretreatment of lignocellulosic pretreatment is a necessary step for the cost-effective production of bioethanol from lignocellulosic biomass. These studies additionally highlight the use of enzymes during the conversion processes of hydrolysis and fermentation.

Sun and Cheng (2002) discuss the production of bioethanol through the enzymatic hydrolysis of lignocellulosic biomass in a review paper originating from the lead laboratory of Sun in the United States with 2333 citations. They note that there are mainly two processes involved in the conversion of biomass to bioethanol: hydrolysis of cellulose in the lignocellulosic biomass to produce reducing biosugars, and fermentation of the biosugars to bioethanol. Because the cost of bioethanol production from lignocellulosic materials is relatively high based on technologies available in the early 2000s, they argue that the main challenges are the low yield and high cost of the hydrolysis process. They further argue that pretreatment of lignocellulosic materials to remove lignin and hemicellulose can significantly enhance the hydrolysis of cellulose. Optimization of the cellulase enzymes and the enzyme loading can also improve the hydrolysis. Simultaneous saccharification and fermentation effectively removes glucose, which is an inhibitor to cellulase activity, thus increasing the yield and rate of cellulose hydrolysis.

Alvira et al. (2010) discuss the pretreatment technologies for the bioethanol production from the enzymatic hydrolysis of lignocellulosic biomass in a review paper originating from the lead laboratory of Ballesteros in Spain with 1202 citations. They note that because the physical and chemical barriers caused by the close association of the main components of lignocellulosic biomass hinder the hydrolysis of cellulose and hemicellulose to fermentable sugars, the main goal of pretreatment is to increase the enzyme accessibility, improving digestibility of cellulose. Because each pretreatment has a specific effect on the cellulose, hemicellulose, and lignin fraction, they recommend that different pretreatment methods and conditions should be chosen according to the process configuration selected for the subsequent hydrolysis and fermentation steps. They point out several key properties that should be targeted for low-cost and advanced pretreatment processes.

Palmqvist and Hahn-Hagerdal (2000) discuss the production of bioethanol through fermentation of the lignocellulosic biomass in a review paper originating from the lead laboratory of Hahn-Hagerdal in Sweden with 1079 citations. They focus on the generation of inhibitors to microorganisms (weak acids, furan derivatives, and phenolic compounds) during degradation of lignocellulosic materials, and the effect of these on fermentation yield and productivity discussing inhibiting mechanisms of individual compounds present in the hydrolysates and their interaction effects.

1.5.2.4 Biobutanol

There was only one classical paper on the biobutanol production from lignocellulosic biomass.

Jones and Woods (1986) discuss the production of biobutanol together with acetone through fermentation of lignocellulosic biomass in a review paper originating from their laboratories in South Africa with 1072 citations. They focus on the history, fermentation, fermentation substrates (lignocellulosic, nonlignocellulosic), biochemistry and physiology, regulation of electron flow, triggering of solventogenesis, events with solventogenesis, solvent toxicity, genetics and strain improvement, and process development.

1.5.2.5 Other liquid biofuels

There were two more classical papers on the production of other liquid biofuels, 5-hydroxymethylfurfural (HMF) and dimethylfuran, which are the intermediate products for biofuel production, besides biodiesel, bioethanol, bio-oil, and biobutanol.

Zhao et al. (2007) discuss the production of 5-hydroxymethylfurfural (HMF), an intermediate for the biofuel production, through catalytic conversion in an experimental paper originating from the lead laboratory of Zhang in the United States with 1081 citations. They find that chromium(II) chloride is an efficient catalyst with 70% glucose to HMF yield. A wide range of metal halides catalyze the conversion of fructose to HMF.

Roman-Leshkov et al. (2007) discuss the production of dimethylfuran in a paper originating from the lead laboratory of Dumesic in the United States with 943 citations. They present a catalytic strategy for the production of 2,5-dimethylfuran from fructose as an alternative to bioethanol. Compared to bioethanol, they find that 2,5-dimethylfuran has a higher energy density, a higher boiling point, and is not soluble in water.

1.5.3 Gaseous biofuels

In this section, the concise summaries for each of the classical papers are provided to inform the stakeholders about the key issues considered in these studies in the order described in Table 1.1: biohydrogen and biogas (see also Table 1.4). However, it is notable that there were no classical papers in biomethane.

1.5.3.1 Biohydrogen

There were five citation classics in biohydrogen. Usually, there are five research streams in biohydrogen research: general biohydrogen research, dark fermentative biohydrogen research, biohydrogen through the catalytic reforming of biofuels, algal photobiological hydrogen production, and biohydrogen production in microbial fuel cells. There was also a strong research stream on the hydrogenases, enzymes facilitating biohydrogen production.

However, among citation classics there was one paper each on general biohydrogen research and biohydrogen through the catalytic reforming of biofuels, as well as three papers on hydrogenases. There were no classical papers in dark fermentative biohydrogen research, algal photobiological hydrogen production, and hydrogen production in microbial fuel cells.

1.5.3.1.1 Overview Das and Veziroglu (2001) discuss the production of biohydrogen in a seminal review paper originating from their laboratories in India and the United States, respectively, with 1039 citations. They present the microorganisms and biochemical pathways involved in biohydrogen production processes. They find an immobilized system suitable for the continuous biohydrogen production. And approximately 28% of energy can be recovered in the form of biohydrogen using sucrose as a substrate because fermentative biohydrogen production processes have some edge over the other biological processes.

1.5.3.1.2 Biohydrogen through the catalytic reforming of biofuels Cortright et al. (2002) discuss the production of biohydrogen through catalytic reforming of biofuels in an experimental paper originating from the lead laboratory of Dumesic with 1094 citations. They show that biohydrogen can be produced from biosugars and bioalcohols at temperatures near 500 K in a single-reactor aqueous-phase reforming process using a platinum-based catalyst. They convert glucose—which makes up the major energy reserves in plants and animals—to biohydrogen and gaseous bioalkanes, with biohydrogen constituting 50% of the products. They find that the selectivity for biohydrogen production increases when they use molecules that are more reduced than sugars, with ethylene glycol and methanol being almost completely converted into biohydrogen and carbon dioxide. These findings suggest that catalytic aqueous-phase reforming might prove useful for the generation of hydrogen-rich fuel gas from carbohydrates extracted from renewable biomass and biomass waste streams.

1.5.3.1.3 Hydrogenases, enzymes facilitating biohydrogen production Peters et al. (1998) discuss hydrogenases to provide insights into the mechanism of biohydrogen activation in an experimental paper originating from the lead laboratories of Peters and Seefeldt in the United States with 1250 citations. They determine a three-dimensional structure for the monomeric iron-containing hydrogenase (CpI) from *Clostridium pasteurianum* to 1.8-Å resolution by x-ray crystallography. They find that CpI contains 20-gram atoms of iron per mole of protein, arranged into five distinct [Fe-S] clusters. The probable active-site cluster, previously termed the H-cluster, was an unexpected arrangement of six iron atoms existing as a [4Fe-4S] cubane subcluster covalently bridged by a cysteinate thiol to a [2Fe] subcluster. The iron atoms of the [2Fe] subcluster both exist with an octahedral coordination geometry and are bridged to each other by three nonprotein atoms, assigned as two sulfide atoms and one carbonyl or cyanide molecule.

Volbeda et al. (1995) discuss hydrogenases to provide insights into the electron and proton transfer pathways in an experimental paper originating from the lead laboratory of Fontecilla-Camps in France with 1169 citations. They determine the x-ray structure of the heterodimeric

Ni–Fe hydrogenase from *Desulfovibrio gigas,* the enzyme responsible for the metabolism of molecular hydrogen, at 2.85-Å resolution. They find that the active site, containing, besides nickel, a second metal ion, is buried in the 60K subunit. The 28K subunit, which coordinates one [3Fe–4S] and two [4Fe–4S] clusters, contains an amino-terminal domain with similarities to the redox protein flavodoxin.

Nicolet et al. (1999) discuss the hydrogenases in an experimental paper originating from the lead laboratory of Fontecilla-Camps and Legrand in France with 937 citations. They report the structure of the heterodimeric Fe-only hydrogenase from Desulfovibrio desulfuricans. They find that with the exception of a ferredoxin-like domain, the structure represents a novel protein fold. The so-called H-cluster of the enzyme is composed of a typical [4Fe–4S] cubane bridged to a binuclear active-site Fe center containing putative CO and CN ligands and one bridging 1,3-propanedithiol molecule. The conformation of the subunits can be explained by the evolutionary changes that have transformed monomeric cytoplasmic enzymes into dimeric periplasmic enzymes. They identify plausible electron- and proton-transfer pathways and a putative channel for the access of hydrogen to the active site. They conclude that the unrelated active sites of Ni–Fe and Fe-only hydrogenases have several common features: coordination of diatomic ligands to a Fe ion, a vacant coordination site on one of the metal ions representing a possible substrate-binding site, a thiolate-bridged binuclear center, and plausible proton- and electron-transfer pathways and substrate channels. The diatomic coordination to Fe ions makes them low spin and favors low redox states, which may be required for catalysis. Complex electron paramagnetic resonance signals typical of Fe-only hydrogenases arise from magnetic interactions between the [4Fe–4S] cluster and the active-site binuclear center.

1.5.3.2 *Biogas*

There was only one classical paper in biogas. Chen et al. (2008) discuss anaerobic digestion of lignocellulosic biomass such as wastes for biogas production and biowaste treatment in a review paper originating from the lead laboratory of Chen in the United States with 1212 citations. They focus on the inhibition of anaerobic processes. The inhibitors commonly present in anaerobic digesters include ammonia, sulfide, light metal ions, heavy metals, and organics. Due to the difference in anaerobic inocula, waste composition, and experimental methods and conditions, literature results on inhibition caused by specific toxicants vary widely. They note that codigestion with other waste, adaptation of microorganisms to inhibitory substances, and incorporation of methods to remove or counteract toxicants before anaerobic digestion can significantly improve the waste treatment efficiency.

1.5.4 *Solid biofuels*

In this section, the concise summaries for each of the classical papers are provided to inform the stakeholders about the key issues considered in these studies in the order described in Table 1.1: biomass emissions from biomass biofuels (see also Table 1.5). There were no classical papers in biochars, usually a by-product of biofuel production.

1.5.4.1 *Biomass fuels*

There were three papers on the gas emissions from combustion of biomass biofuels in residential heating or in coal-fired power plants as well due to the land-use change building on Searchinger et al. (2008) and Fargione et al. (2008). These classical papers highlight the adverse impact of these biomass biofuels. There were no separate classical papers on the combustion of biomass biofuels. The summaries for some of these papers were kept long due to the data they provide on the adverse effects of combustion of biomass biofuels complementing the issues raised in Searchinger et al. (2008) and Fargione et al. (2008).

1.5.4.1.1 Biomass gas emissions Andreae and Merlet (2001) discuss emissions of biomass fuels in a seminal review paper originating from the lead laboratory of Andrea in Germany with 1673 citations. They present a set of emission factors for a large variety of species emitted from biomass fires. They derive global estimates of pyrogenic emissions for important species emitted by the various types of biomass burning and compare them with results from inverse modeling studies.

Crutzen and Andreae (1990) discuss emissions for biomass fuels in a review paper originating from their laboratories in Germany with 1373 citations. They note that biomass burning is widespread, especially in the tropics. It serves to clear land for shifting cultivation, to convert forests to agricultural and pastoral lands, and to remove dry vegetation in order to promote agricultural productivity and the growth of higher yield grasses. Furthermore, much agricultural waste and fuel wood is being combusted, particularly in developing countries. Biomass containing 2 to 5 petagrams of carbon is burned annually (1 petagram = 10^{15} grams), producing large amounts of trace gases and aerosol particles that play important roles in atmospheric chemistry and climate. Emissions of carbon monoxide and methane by biomass burning affect the oxidation efficiency of the atmosphere by reacting with hydroxyl radicals, and emissions of nitric oxide and hydrocarbons lead to high ozone concentrations in the tropics during the dry season. Large quantities of smoke particles are produced as well, and these can serve as cloud condensation nuclei. These particles may thus substantially influence cloud microphysical and optical properties, an effect that could have repercussions for the radiation budget and the hydrological cycle in the tropics. Widespread burning may also disturb biogeochemical cycles, especially that of nitrogen. Approximately 50% of the nitrogen in the biomass fuel can be released as molecular nitrogen. This pyrdenitrification process causes a sizable loss of fixed nitrogen in tropical ecosystems, in the range of 10–20 teragrams per year (1 teragram = 10^{12} grams).

Bond et al. (2013) discuss the impact of black carbon emissions on climate change in a review paper originating from the lead laboratory of Bond in the United States with 920 citations. They evaluate black carbon climate forcing with the main forcing terms: direct solar absorption; influence on liquid, mixed phase, and ice clouds; and deposition on snow and ice. They determine that the predominant sources are combustion related: fossil fuels for transportation, solid fuels for industrial and residential uses, and open burning of biomass. Total global emissions of black carbon are 7500 Gg $year^{-1}$ in the year 2000 with an uncertainty range of 2000–29,000. The best estimate for the industrial-era (1750–2005) direct radiative forcing of atmospheric black carbon is +0.71 W m^{-2} with 90% uncertainty bounds of (+0.08, +1.27) W m^{-2}. Total direct forcing by all black carbon sources is estimated as +0.88 (+0.17, +1.48) W m^{-2}. The best estimate of industrial-era climate forcing of black carbon through all forcing mechanisms, including clouds and cryosphere forcing, is +1.1 W m^{-2} with 90% uncertainty bounds of +0.17 to +2.1 W m^{-2}. Thus, there is a very high probability that black carbon emissions, independent of coemitted species, have a positive forcing and warm the climate. They estimate that black carbon, with a total climate forcing of +1.1 W m^{-2}, is the second most important human emission in terms of its climate forcing in the present-day atmosphere; only carbon dioxide is estimated to have a greater forcing. When the principal effects of short-lived coemissions, including cooling agents such as sulfur dioxide, are included in net forcing, energy-related sources (fossil fuel and biofuel) have an industrial-era climate forcing of +0.22 (−0.50 to +1.08) W m^{-2} during the first year after emission. For a few of these sources, such as diesel engines and possibly residential biofuels, warming is strong enough that eliminating all short-lived emissions from these sources would reduce net climate forcing (i.e., produce cooling). When open burning emissions, which emit high levels of organic matter, are included in the total, the best estimate of net industrial-era climate forcing by all short-lived species from black-carbon-rich sources becomes slightly negative (−0.06 W m^{-2} with 90% uncertainty bounds of −1.45 to +1.29 W m^{-2}). The uncertainties in net climate forcing from black-carbon-rich sources are substantial, largely due to a lack of knowledge about cloud interactions with both black carbon and coemitted organic carbon. In prioritizing potential black-carbon mitigation actions, nonscience factors, such as technical feasibility, costs, policy design, and implementation feasibility, play important roles. The major sources of black carbon are presently in different stages with regard to the feasibility for near-term mitigation.

1.5.5 *Bioenergy: Microbial fuel cells*

In this section, the concise summaries for each of the classical papers are provided to inform the stakeholders about the key issues considered in these studies in the order described in Table 1.1: microbial fuel cells (see also Table 1.6). However, there were no classical papers in enzymatic biofuel cells. The key researchers in this field were Lovley and Logan.

1.5.5.1 *Overview*

Logan et al. (2006) discuss MFCs in a seminal review paper originating from the lead laboratories of Logan, Schroeder, Verstraete, and Rabaey in the United States, Germany, and the Netherlands, respectively, with 2072 citations. They provide an overview of the different materials and methods used to construct MFCs, techniques used to analyze system performance, and recommendations on what information to include in MFC studies and the most useful ways to present results.

1.5.5.2 *Bioremediation and electricity generation through microbial fuel cells*

Lovley and Phillips (1988) discuss organic carbon oxidation coupled to dissimilatory reduction of iron or manganese in an experimental paper originating from the lead laboratory of Lovley with 1280 citations. They isolate a dissimilatory Fe(III)- and Mn(IV)-reducing microorganism, GS-15, from freshwater sediments of a river. They show that microorganisms can completely oxidize organic compounds with Fe(III) or Mn(IV) as the sole electron acceptor and that oxidation of organic matter coupled to dissimilatory Fe(III) or Mn(IV) reduction can yield energy for microbial growth. GS-15 provides a model for how enzymatically catalyzed reactions can be quantitatively significant mechanisms for the reduction of iron and manganese in anaerobic environment.

Bond and Lovley (2003) discuss bioelectricity production by *Geobacter sulfurreducens* attached to electrodes in an experimental paper originating from the lead laboratory of Lovley with 1022 citations. They inoculate *G. sulfurreducens* into chambers in which a graphite electrode served as the sole electron acceptor and acetate or hydrogen was the electron donor. When a small inoculum of *G. sulfurreducens* was introduced into electrode-containing chambers, electrical current production was dependent upon oxidation of acetate to carbon dioxide and increased exponentially, indicating that electrode reduction supported the growth of this organism. They show that the effectiveness of microbial fuel cells can be increased with organisms such as *G. sulfurreducens* that can attach to electrodes and remain viable for long periods of time while completely oxidizing organic substrates with quantitative transfer of electrons to an electrode.

Lovley (1991) discusses dissimilatory Fe(III) and Mn(IV) reduction in a review paper originating from his laboratory in the United States with 1009 citations. He notes that microorganisms that can effectively couple the oxidation of organic compounds to the reduction of Fe(III) or Mn(IV) have recently been discovered. With Fe(III) or Mn(IV) as the sole electron acceptor, these organisms can completely oxidize fatty acids, hydrogen, or a variety of monoaromatic compounds. This metabolism provides energy to support growth. Sugars and amino acids can be completely oxidized by the cooperative activity of fermentative microorganisms and hydrogen- and fatty-acid-oxidizing Fe(III) and Mn(IV) reducers. This provides a microbial mechanism for the oxidation of the complex assemblage of sedimentary organic matter in Fe(III)- or Mn(IV)-reducing environments. The available evidence indicates that this enzymatic reduction of Fe(III) or Mn(IV) accounts for most of the oxidation of organic matter coupled to reduction of Fe(III) and Mn(IV) in sedimentary environments.

Reguera et al. (2005) discuss extracellular electron transfer via microbial nanowires in an experimental paper originating from the lead laboratory of Lovley in the United States with 929 citations. They report that a pilus-deficient mutant of *G. sulfurreducens* could not reduce Fe(III) oxides but could attach to them. They confirm that the pili were highly conductive. They reason that the pili of *G. sulfurreducens* might serve as biological nanowires, transferring electrons from the cell surface to the surface of Fe(III) oxides. They speculate that electron transfer through pili indicates possibilities for other unique cell-surface and cell–cell interactions, and for bioengineering of novel conductive materials.

1.6 Conclusion

This paper built upon the pioneering scientometric studies in bioenergy and biofuels to provide the data highlighting the current state of the research in this field. The common theme in the research on the bioenergy and biofuels has been to design a set of bioenergy and biofuels to meet the criteria set down in Hill et al. (2006): bioenergy and biofuels should "provide a net energy gain, have environmental benefits, be economically competitive, and be producible in large quantities without risking food security."

It is notable that these criteria imply that this field has an interdisciplinary character spanning many academic disciplines in addition to the subject category of energy and fuels. Therefore, it is important to obtain a representative sample of papers using a well-designed keyword search set beyond the search relying solely on this subject category. For this purpose, this paper introduced a detailed keyword set for the optimal search of the literature on the bioenergy and biofuels.

This study highlighted the key players such as authors, funding bodies, institutions, and journals as well as the key issues and papers shaping the development of the research in bioenergy and biofuels over time since 1980.

This paper also introduced a topical classification based on the topical analysis of the research in bioenergy and biofuels along five primary headings, which are also adopted in the structuring of the book: biofuels in general, liquid biofuels, gaseous biofuels, solid biofuels, and bioenergy (microbial fuel cells for bioremediation and bioelectricity generation).

It is recommended that further research should be carried out, building on the pioneering scientometric studies in bioenergy and biofuels along the research streams identified in this paper.

Acknowledgments

This paper acknowledges 50 citation classics in bioenergy and biofuels published between 1980 and April 2017. Many of these classical papers were written by the research teams led by the highly cited researchers. Therefore, this paper acknowledges the valuable contribution of these highly cited researchers as well.

References

Abramo, G., D'Angelo, C. A., and Soldatenkova, A. 2016. The ratio of top scientists to the academic staff as an indicator of the competitive strength of universities. *Journal of Informetrics*, 10(2), 596–605.

Agarwal, A. K. 2007. Biofuels (alcohols and biodiesel) applications as fuels for internal combustion engines. *Progress in Energy and Combustion Science*, 33(3), 233–271.

Alvira, P., Tomas-Pejo, E., Ballesteros, M., and Negro, M. J. 2010. Pretreatment technologies for an efficient bioethanol production process based on enzymatic hydrolysis: A review. *Bioresource Technology*, 101(13), 4851–4861.

Andreae, M. O. and Merlet, P. 2001. Emission of trace gases and aerosols from biomass burning. *Global Biogeochemical Cycles*, 15(4), 955–966.

Baltussen, A. and Kindler, C. H. 2004a. Citation classics in anesthetic journals. *Anesthesia & Analgesia*, 98(2), 443–451.

Baltussen, A. and Kindler, C. H. 2004b. Citation classics in critical care medicine. *Intensive Care Medicine*, 30(5), 902–910.

Basu, A. 2006. Using ISI's 'Highly Cited Researchers' to obtain a country level indicator of citation excellence. *Scientometrics*, 68(3), 361–375.

Bond, T. C., Doherty, S. J., Fahey, D. W. et al. 2013. Bounding the role of black carbon in the climate system: A scientific assessment. *Journal of Geophysical Research: Atmospheres*, 118(11), 5380–5552.

Bond, D. R. and Lovley, D. R. 2003. Electricity production by *Geobacter sulfurreducens* attached to electrodes. *Applied and Environmental Microbiology*, 69(3), 1548–1555.

Bornmann, L. and Bauer, J. 2015a. Evaluation of the highly-cited researchers' database for a country: Proposals for meaningful analyses on the example of Germany. *Scientometrics*, 105(3), 1997–2003.

Bornmann, L. and Bauer, J. 2015b. Which of the world's institutions employ the most highly cited researchers? An analysis of the data from highlycited.com. *Journal of the Association for Information Science and Technology*, 66(10), 2146–2148.

Bornmann, L., Bauer, J., and Schlagberger, E. M. 2017. Characteristics of highly cited researchers 2015 in Germany. *Scientometrics*, 111(1), 543–545.

Bozell, J. J. and Petersen, G. R. 2010. Technology development for the production of biobased products from biorefinery carbohydrates—The US Department of Energy's "Top 10" revisited. *Green Chemistry*, 12(4), 539–554.

Braun, T. and Schubert, A. 2003. A quantitative view on the coming of age of interdisciplinarity in the sciences 1980–1999. *Scientometrics*, 58(1), 183–189.

Brennan, L. and Owende, P. 2010. Biofuels from microalgae—A review of technologies for production, processing, and extractions of biofuels and co-products. *Renewable and Sustainable Energy Reviews*, 14(2), 557–577.

Butler, L. 2003. Explaining Australia's increased share of ISI publications—The effects of a funding formula based on publication counts. *Research Policy*, 32(1), 143–155.

Carrell, S. E., Page, M. E., and West, J. E. 2010. Sex and science: How professor gender perpetuates the gender gap. *Quarterly Journal of Economics*, 125(3), 1101–1144.

Chen, Y., Cheng, J. J., and Creamer, K. S. 2008. Inhibition of anaerobic digestion process: A review. *Bioresource Technology*, 99(10), 4044–4064.

Chen, H. and Ho, Y. S. 2015. Highly cited articles in biomass research: A bibliometric analysis. *Renewable and Sustainable Energy Reviews*, 49, 12–20.

Chesler, N. C. and Chesler, M. A. 2002. Gender-informed mentoring strategies for women engineering scholars: On establishing a caring community. *Journal of Engineering Education*, 91(1), 49–55.

Chester, L. 2010. Conceptualising energy security and making explicit its polysemic nature. *Energy Policy*, 38(2), 887–895.

Chheda, J. N., Huber, G. W., and Dumesic, J. A. 2007. Liquid-phase catalytic processing of biomass-derived oxygenated hydrocarbons to fuels and chemicals. *Angewandte Chemie International Edition*, 46(38), 7164–7183.

Chisti, Y. 2007. Biodiesel from microalgae. *Biotechnology Advances*, 25(3), 294–306.

Clarivate Analytics. 2017. Highly Cited Researchers. http://hcr.stateofinnovation.com/ (Accessed on 05.04.2017).

Coelho, M. S. and Barbosa, F. G. 2014. The scientometric research on macroalgal biomass as a source of biofuel feedstock. *Algal Research*, 6, 132–138.

Corma, A., Iborra, S., and Velty, A. 2007. Chemical routes for the transformation of biomass into chemicals. *Chemical Reviews*, 107(6), 2411–2502.

Cortright, R. D., Davda, R. R., and Dumesic, J. A. 2002. Hydrogen from catalytic reforming of biomass-derived hydrocarbons in liquid water. *Nature*, 418(6901), 964–967.

Costantini, V., Crespi, F., and Curci, Y. 2015. A keyword selection method for mapping technological knowledge in specific sectors through patent data: The case of biofuels sector. *Economics of Innovation and New Technology*, 24(4), 282–308.

Cox, P. M., Betts, R. A., Jones, C. D., Spall, S. A., and Totterdell, I. J. 2000. Acceleration of global warming due to carbon-cycle feedbacks in a coupled climate model. *Nature*, 408(6809), 184–187.
Crutzen, P. J. and Andreae, M. O. 1990. Biomass burning in the tropics: Impact on atmospheric chemistry and biogeochemical cycles. *Science*, 250(4988), 1669–1679.
Curci, Y. and Ospina, C. A. M. 2016. Investigating biofuels through network analysis. *Energy Policy*, 97, 60–72.
Czernik, S. and Bridgwater, A. V. 2004. Overview of applications of biomass fast pyrolysis oil. *Energy & Fuels*, 18(2), 590–598.
D'Angelo, C. A., Giuffrida, C., and Abramo, G. 2011. A heuristic approach to author name disambiguation in bibliometrics databases for large-scale research assessments. *Journal of the American Society for Information Science and Technology*, 62(2), 257–269.
Das, D. and Veziroglu, T. N. 2001. Hydrogen production by biological processes: A survey of literature. *International Journal of Hydrogen Energy*, 26(1), 13–28.
Davidson, E. A. and Janssens, I. A. 2006. Temperature sensitivity of soil carbon decomposition and feedbacks to climate change. *Nature*, 440(7081), 165–173.
Denzau, A. T. and North, D. C. 1994. Shared mental models: Ideologies and institutions. *Kyklos*, 47(1), 3–31.
de Souza, L. G. A., de Moraes, M. A. F. D., Dal Poz, M. E. S., and da Silveira, J. M. F. J. 2015. Collaborative networks as a measure of the innovation systems in second-generation ethanol. *Scientometrics*, 103(2), 355–372.
Ding, W. W., Murray, F., and Stuart, T. E. 2006. Gender differences in patenting in the academic life sciences. *Science*, 313(5787), 665–667.
Dubin, D., Hafner, A. W., and Arndt, K. A. 1993. Citation classics in clinical dermatologic journals: Citation analysis, biomedical journals, and landmark articles, 1945-1990. *Archives of Dermatology*, 129(9), 1121–1129.
Falagas, M. E., Pitsouni, E. I., Malietzis, G. A., and Pappas, G. 2008. Comparison of PubMed, Scopus, Web of Science, and Google Scholar: Strengths and weaknesses. *FASEB Journal*, 22(2), 338–342.
Fargione, J., Hill, J., Tilman, D., Polasky, S., and Hawthorne, P. 2008. Land clearing and the biofuel carbon debt. *Science*, 319(5867), 1235–1238.
Farrell, A. E., Plevin, R. J., Turner, B. T. et al. 2006. Ethanol can contribute to energy and environmental goals. *Science*, 311(5760), 506–508.
Ferreira, R. B., Neto, A. C. B., Nabout, J. C. et al. 2014. Trends in global scientific literature about biodiesel: A scientometrics analysis. *Bioscience Journal*, 30(5), 547–554.
Freedman, B., Pryde, E. H., and Mounts, T. L. 1984. Variables affecting the yields of fatty esters from transesterified vegetable oils. *Journal of the American Oil Chemists Society*, 61(10), 1638–1643.
Fu, H. Z. and Ho, Y. S. 2013. Independent research of China in Science Citation Index Expanded during 1980–2011. *Journal of Informetrics*, 7(1), 210–222.
Fukuda, H., Kondo, A., and Noda, H. 2001. Biodiesel fuel production by transesterification of oils. *Journal of Bioscience and Bioengineering*, 92(5), 405–416.
Garfield, E. 1972. Citation analysis as a tool in journal evaluation. *Science*, 178(4060), 471–479.
Garfield, E. 1979. Is citation analysis a legitimate evaluation tool? *Scientometrics*, 1(4), 359–375.

Garfield, E. 1980. Most-cited articles of the 1960s. 4. Clinical research. *Current Contents*, (6), 5–14.
Garfield, E. 1987. 100 citation classics from the *Journal of the American Medical Association. JAMA*, 257(1), 52–59.
Garfield, E. 2006. The history and meaning of the journal impact factor. *JAMA*, 295(1), 90–93.
Garfield, E. and Welljams-Dorof, A. 1992. Of nobel class: A citation perspective on high impact research authors. *Theoretical Medicine and Bioethics*, 13(2), 117–135.
Glanzel, W. and Moed, H. 2002. Journal impact measures in bibliometric research. *Scientometrics*, 53(2), 171–193.
Godfray, H. C. J., Beddington, J. R., Crute, I. R. et al. 2010. Food security: The challenge of feeding 9 billion people. *Science*, 327(5967), 812–818.
Gomes, J. and Dewes, H. 2017. Disciplinary dimensions and social relevance in the scientific communications on biofuels. *Scientometrics*, 110(3), 1173–1189.
Graboski, M. S. and McCormick, R. L. 1998. Combustion of fat and vegetable oil derived fuels in diesel engines. *Progress in Energy and Combustion Science*, 24(2), 125–164.
Hansen, J. E. 1998. Sir John Houghton: Global warming: The complete briefing. *Journal of Atmospheric Chemistry*, 30(3), 409–412.
Hather, G. J., Haynes, W., Higdon, R. et al. 2010. The United States of America and scientific research. *PLoS One*, 5(8), e12203.
Held, I. M. and Soden, B. J. 2006. Robust responses of the hydrological cycle to global warming. *Journal of Climate*, 19(21), 5686–5699.
Hendriks, A. T. W. M. and Zeeman, G. 2009. Pretreatments to enhance the digestibility of lignocellulosic biomass. *Bioresource Technology*, 100(1), 10–18.
Hill, J., Nelson, E., Tilman, D., Polasky, S., and Tiffany, D. 2006. Environmental, economic, and energetic costs and benefits of biodiesel and ethanol biofuels. *Proceedings of the National Academy of Sciences*, 103(30), 11206–11210.
Hill, T. and Westbrook, R. 1997. SWOT analysis: It's time for a product recall. *Long Range Planning*, 30(1), 46–52.
Himmel, M. E., Ding, S. Y., Johnson, D. K. et al. 2007. Biomass recalcitrance: Engineering plants and enzymes for biofuels production. *Science*, 315(5813), 804–807.
Ho, Y. S. 2017. Comments on "Past, current and future of biomass energy research: A bibliometric analysis" by Mao et al. (2015). *Renewable and Sustainable Energy Reviews*. doi: 10.1016/j.rser.2017.04.120.
Hoegh-Guldberg, O., Mumby, P. J., Hooten, A. J. et al. 2007. Coral reefs under rapid climate change and ocean acidification. *Science*, 318(5857), 1737–1742.
Hongli, D., Zidong, W., Huijun, G. et al. 2010. Web of Science®. *IEEE Transactions on Signal Processing*, 58(4), 1957–1966.
Hsu, C. W., Chang, P. L., Hsiung, C. M., and Lin, C. Y. 2014. Commercial application scenario using patent analysis: Fermentative hydrogen production from biomass. *International Journal of Hydrogen Energy*, 39(33), 19277–19284.
Hsu, C. W., Chang, P. L., Hsiung, C. M., and Wu, C. C. 2015. Charting the evolution of biohydrogen production technology through a patent analysis. *Biomass and Bioenergy*, 76, 1–10.
Hsu, C. W. and Lin, C. Y. 2016. Using social network analysis to examine the technological evolution of fermentative hydrogen production from biomass. *International Journal of Hydrogen Energy*, 41(46), 21573–21582.

Hu, Q., Sommerfeld, M., Jarvis, E. et al. 2008. Microalgal triacylglycerols as feedstocks for biofuel production: Perspectives and advances. *Plant Journal*, 54(4), 621–639.
Huber, G. W., Iborra, S., and Corma, A. 2006. Synthesis of transportation fuels from biomass: Chemistry, catalysts, and engineering. *Chemical Reviews*, 106(9), 4044–4098.
Hughes, L. 2000. Biological consequences of global warming: Is the signal already apparent? *Trends in Ecology & Evolution*, 15(2), 56–61.
Hughes, T. P., Baird, A. H., Bellwood, D. R. et al. 2003. Climate change, human impacts, and the resilience of coral reefs. *Science*, 301(5635), 929–933.
Jackson, S. E., Joshi, A., and Erhardt, N. L. 2003. Recent research on team and organizational diversity: SWOT analysis and implications. *Journal of Management*, 29(6), 801–830.
Jacobson, M. Z. 2009. Review of solutions to global warming, air pollution, and energy security. *Energy & Environmental Science*, 2(2), 148–173.
Jagsi, R., Guancial, E. A., Worobey, C. C. et al. 2006. The "gender gap" in authorship of academic medical literature—A 35-year perspective. *New England Journal of Medicine*, 355(3), 281–287.
Johnes, J. and Johnes, G. 1995. Research funding and performance in UK university departments of economics: A frontier analysis. *Economics of Education Review*, 14(3), 301–314.
Johnson, J. M. F., Franzluebbers, A. J., Weyers, S. L., and Reicosky, D. C. 2007. Agricultural opportunities to mitigate greenhouse gas emissions. *Environmental Pollution*, 150(1), 107–124.
Jones, D. T. and Woods, D. R. 1986. Acetone-butanol fermentation revisited. *Microbiological Reviews*, 50(4), 484–524.
Kajikawa, Y. and Takeda, Y. 2008. Structure of research on biomass and bio-fuels: A citation-based approach. *Technological Forecasting and Social Change*, 75(9), 1349–1359.
Kim, G. J. 2005. A SWOT analysis of the field of virtual reality rehabilitation and therapy. *Presence: Teleoperators and Virtual Environments*, 14(2), 119–146.
Konur, O. 2000. Creating enforceable civil rights for disabled students in higher education: An institutional theory perspective. *Disability & Society*, 15(7), 1041–1063.
Konur, O. 2002a. Access to nursing education by disabled students: Rights and duties of nursing programs. *Nurse Education Today*, 22(5), 364–374.
Konur, O. 2002b. Assessment of disabled students in higher education: Current public policy issues. *Assessment and Evaluation in Higher Education*, 27(2), 131–152.
Konur, O. 2004. Disability and racial discrimination in employment in higher education. In: Law, I., Phillips, D., Turney, L. (Eds.), *Institutional Racism in Higher Education*, Trentham Books Ltd., Stoke-on-Trent, U.K., 83–92.
Konur, O. 2006a. Participation of children with dyslexia in compulsory education: Current public policy issues. *Dyslexia*, 12(1), 51–67.
Konur, O. 2006b. Teaching disabled students in higher education. *Teaching in Higher Education*, 11(3), 351–363.
Konur, O. 2007a. Computer-assisted teaching and assessment of disabled students in Higher Education: The interface between academic standards and disability rights. *Journal of Computer Assisted Learning*, 23(3), 207–219.
Konur, O. 2007b. A judicial outcome analysis of the Disability Discrimination Act: A windfall for the employers? *Disability & Society*, 22(2), 187–204.

Konur, O. 2011. The scientometric evaluation of the research on the algae and bioenergy. *Applied Energy*, 88(10), 3532–3540.
Konur, O. 2012a. 100 citation classics in Energy and Fuels. *Energy Education Science and Technology Part A: Energy Science and Research*, 30(special issue 1), 319–332.
Konur, O. 2012b. The evaluation of the biogas research: A scientometric approach. *Energy Education Science and Technology Part A: Energy Science and Research*, 29(2), 1277–1292.
Konur, O. 2012c. The evaluation of the bio-oil research: A scientometric approach. *Energy Education Science and Technology Part A: Energy Science and Research*, 30(special issue 1), 379–392.
Konur, O. 2012d. The evaluation of the biorefinery research: A scientometric approach. *Energy Education Science and Technology Part A: Energy Science and Research*, 30(special issue 1), 347–358.
Konur, O. 2012e. The evaluation of the global energy and fuels research: A scientometric approach. *Energy Education Science and Technology Part A: Energy Science and Research*, 30(1), 613–628.
Konur, O. 2012f. The evaluation of the research on the biodiesel: A scientometric approach. *Energy Education Science and Technology Part A: Energy Science and Research*, 28(2), 1003–1014.
Konur, O. 2012g. The evaluation of the research on the bioethanol: A scientometric approach. *Energy Education Science and Technology Part A: Energy Science and Research*, 28(2), 1051–1064.
Konur, O. 2012h. The evaluation of the research on the biofuels: A scientometric approach. *Energy Education Science and Technology Part A: Energy Science and Research*, 28(2), 903–916.
Konur, O. 2012i. The evaluation of the research on the biohydrogen: A scientometric approach. *Energy Education Science and Technology Part A: Energy Science and Research*, 29(1), 323–338.
Konur, O. 2012j. The evaluation of the research on the microbial fuel cells: A scientometric approach. *Energy Education Science and Technology Part A: Energy Science and Research*, 29(1), 309–322.
Konur, O. 2012k. Prof. Dr. Ayhan Demirbas' scientometric biography. *Energy Education Science and Technology Part A: Energy Science and Research*, 28(2), 727–738.
Konur, O. 2012l. The scientometric evaluation of the research on the production of bioenergy from biomass. *Biomass and Bioenergy*, 47, 504–515.
Konur, O. 2012m. What have we learned from the citation classics in Energy and Fuels?: A mixed study. *Energy Education Science and Technology Part A: Energy Science and Research*, 30(special issue 1), 255–268.
Konur, O. 2013. The policies and practices for the academic assessment of blind students in higher education and professions. *Energy Education Science and Technology Part B: Social and Educational Studies*, 5(1), 463–472.
Konur, O. 2015a. Algal biosorption of heavy metals from wastes. In: Kim, S. K., Lee, C. G. (Eds.), *Marine Bioenergy: Trends and Developments*, CRC Press, Boca Raton, FL, 597–626.
Konur, O. 2015b. Algal economics and optimization. In: Kim, S. K., Lee, C. G. (Eds.), *Marine Bioenergy: Trends and Developments*, CRC Press, Boca Raton, FL, 691–716.
Konur, O. 2015c. Algal high-value consumer products. In: Kim, S. K., Lee, C. G. (Eds.), *Marine Bioenergy: Trends and Developments*, CRC Press, Boca Raton, FL, 653–682.

Konur, O. 2015d. Algal photobioreactors. In: Kim, S. K., Lee, C. G. (Eds.), *Marine Bioenergy: Trends and Developments*, CRC Press, Boca Raton, FL, 81–108.
Konur, O. 2015e. Algal photosynthesis, biosorption, biotechnology, and biofuels. In: Kim, S. K. (Ed.), *Springer Handbook of Marine Biotechnology*, Springer, Berlin, Heidelberg, 1131–1161.
Konur, O. 2015f. Current state of research on algal biodiesel. In: Kim, S. K., Lee, C. G. (Eds.), *Marine Bioenergy: Trends and Developments*, CRC Press, Boca Raton, FL, 487–512.
Konur, O. 2015g. Current state of research on algal bioelectricity and algal microbial fuel cells. In: Kim, S. K., Lee, C. G. (Eds.), *Marine Bioenergy: Trends and Developments*, CRC Press, Boca Raton, FL, 527–556.
Konur, O. 2015h. Current state of research on algal bioethanol. In: Kim, S. K., Lee, C. G. (Eds.), *Marine Bioenergy: Trends and Developments*, CRC Press, Boca Raton, FL, 217–244.
Konur, O. 2015i. Current state of research on algal biohydrogen. In: Kim, S. K., Lee, C. G. (Eds.), *Marine Bioenergy: Trends and Developments*, CRC Press, Boca Raton, FL, 393–422.
Konur, O. 2015j. Current state of research on algal biomethane. In: Kim, S. K., Lee, C. G. (Eds.), *Marine Bioenergy: Trends and Developments*, CRC Press, Boca Raton, FL, 273–302.
Konur, O. 2015k. Current state of research on algal biomethanol. In: Kim, S. K., Lee, C. G. (Eds.), *Marine Bioenergy: Trends and Developments*, CRC Press, Boca Raton, FL, 327–370.
Konur, O. 2015l. The review of citation classics on the global energy research. In: Prasad, R., Sivakumar, S., Sharma, U. C. (Eds.), *Energy Science and Technology. V. 1. Opportunities and Challenges*, Studium Press LLC, Houston, TX, 490–526.
Konur, O. 2015m. The scientometric study of the global energy research. In: Prasad, R., Sivakumar, S., Sharma, U. C. (Eds.), *Energy Science and Technology. V. 1. Opportunities and Challenges*, Studium Press LLC, Houston, TX, 475–489.
Konur, O. 2016. Algal OMICS: The most-cited papers. In: Kim, S. K. (Ed.), *Marine OMICS: Principles and Applications*, CRC Press, Boca Raton, FL, 9–34.
Konur, O. 2017. The top citation classics in alginates for biomedicine. In: Venkatesan, J., Anil, S., Kim, S. K. (Eds.), *Seaweed Polysaccharides: Isolation, Biological and Biomedical Applications*, Elsevier, Amsterdam, 223–250.
Konur, O. and Matthews, F. L. 1989. Effect of the properties of the constituents on the fatigue performance of composites: A review. *Composites*, 20(4), 317–328.
Kostoff, R. N. 2012. China/USA nanotechnology research output comparison—2011 update. *Technological Forecasting and Social Change*, 79(5), 986–990.
Kristensen, P. M. 2015. Revisiting the "American social science"—Mapping the geography of international relations. *International Studies Perspectives*, 16(3), 246–269.
Kruyt, B., van Vuuren, D. P., de Vries, H. J. M., and Groenenberg, H. 2009. Indicators for energy security. *Energy Policy*, 37(6), 2166–2181.
Kulis, S., Sicotte, D., and Collins, S. 2002. More than a pipeline problem: Labor supply constraints and gender stratification across academic science disciplines. *Research in Higher Education*, 43(6), 657–691.
Kulkarni, A. V., Aziz, B., Shams, I., and Busse, J. W. 2009. Comparisons of citations in Web of Science, Scopus, and Google Scholar for articles published in general medical journals. *JAMA*, 302(10), 1092–1096.

Kumar, P., Barrett, D. M., Delwiche, M. J., and Stroeve, P. 2009. Methods for pretreatment of lignocellulosic biomass for efficient hydrolysis and biofuel production. *Industrial & Engineering Chemistry Research*, 48(8), 3713–3729.

Kurttila, M., Pesonen, M., Kangas, J., and Kajanus, M. 2000. Utilizing the analytic hierarchy process (AHP) in SWOT analysis—A hybrid method and its application to a forest-certification case. *Forest Policy and Economics*, 1(1), 41–52.

Lal, R. 2004. Soil carbon sequestration impacts on global climate change and food security. *Science*, 304(5677), 1623–1627.

Lashof, D. A. and Ahuja, D. R. 1990. Relative contributions of greenhouse gas emissions to global warming. *Nature*, 344(6266), 529–531.

Leydesdorff, L. 2007. Mapping interdisciplinarity at the interfaces between the Science Citation Index and the Social Science Citation Index. *Scientometrics*, 71(3), 391–405.

Leydesdorff, L. and Wagner, C. 2008. Is the United States losing ground in science? A global perspective on the world science system. *Scientometrics*, 78(1), 23–36.

Leydesdorff, L. and Wagner, C. 2009. Macro-level indicators of the relations between research funding and research output. *Journal of Informetrics*, 3(4), 353–362.

Li, J. T. 2016. What we learn from the shifts in highly cited data from 2001 to 2014? *Scientometrics*, 108(1), 57–82.

Li, S., Garreau, H., and Vert, M. 1990. Structure–property relationships in the case of the degradation of massive poly (α-hydroxy acids) in aqueous media. *Journal of Materials Science: Materials in Medicine*, 1(4), 198–206.

Liu, W., Gu, M., Hu, G., et al. 2014. Profile of developments in biomass-based bioenergy research: A 20-year perspective. *Scientometrics*, 99(2), 507–521.

Lobell, D. B., Burke, M. B., Tebaldi, C. et al. 2008. Prioritizing climate change adaptation needs for food security in 2030. *Science*, 319(5863), 607–610.

Logan, B. E., Hamelers, B., Rozendal, R. et al. 2006. Microbial fuel cells: Methodology and technology. *Environmental Science & Technology*, 40(17), 5181–5192.

Lovley, D. R. 1991. Dissimilatory Fe(III) and Mn(IV) reduction. *Microbiological Reviews*, 55(2), 259–287.

Lovley, D. R. and Phillips, E. J. P. 1988. Novel mode of microbial energy metabolism: Organic carbon oxidation coupled to dissimilatory reduction of iron or manganese. *Applied and Environmental Microbiology*, 54(6), 1472–1480.

Lynd, L. R., Weimer, P. J., van Zyl, W. H., and Pretorius, I. S. 2002. Microbial cellulose utilization: Fundamentals and biotechnology. *Microbiology and Molecular Biology Reviews*, 66(3), 506–577.

Ma, F. and Hanna, M. A. 1999. Biodiesel production: A review. *Bioresource Technology*, 70(1), 1–15.

MacRoberts, M. and MacRoberts, B. 1996. Problems of citation analysis. *Scientometrics*, 36(3), 435–444.

Man, J. P., Weinkauf, J. G., Tsang, M., and Sin, J. H. D. D. 2004. Why do some countries publish more than others? An international comparison of research funding, English proficiency and publication output in highly ranked general medical journals. *European Journal of Epidemiology*, 19(8), 811–817.

Mao, G., Liu, X., Du, H., Zuo, J., and Wang, L. 2015a. Way forward for alternative energy research: A bibliometric analysis during 1994–2013. *Renewable and Sustainable Energy Reviews*, 48, 276–286.

Mao, G., Zou, H., Chen, G., Du, H., and Zuo, J. 2015b. Past, current and future of biomass energy research: A bibliometric analysis. *Renewable and Sustainable Energy Reviews*, 52, 1823–1833.
Mata, T. M., Martins, A. A., and Caetano, N. S. 2010. Microalgae for biodiesel production and other applications: A review. *Renewable and Sustainable Energy Reviews*, 14(1), 217–232.
Maxwell, S. 1996. Food security: A post-modern perspective. *Food Policy*, 21(2), 155–170.
McCullough, R. D., Tristram-Nagle, S., Williams, S. P., Lowe, R. D., and Jayaraman, M. 1993. Self-orienting head-to-tail poly (3-alkylthiophenes): New insights on structure–property relationships in conducting polymers. *Journal of the American Chemical Society*, 115(11), 4910–4911.
McKendry, P. 2002. Energy production from biomass (Part 1): Overview of biomass. *Bioresource Technology*, 83(1), 37–46.
Meher, L. C., Sagar, D. V., and Naik, S. N. 2006. Technical aspects of biodiesel production by transesterification—A review. *Renewable and Sustainable Energy Reviews*, 10(3), 248–268.
Meho, L. I. and Yang, K. 2007. Impact of data sources on citation counts and rankings of LIS faculty: Web of Science versus Scopus and Google Scholar. *Journal of the American Society for Information Science and Technology*, 58(13), 2105–2125.
Meinshausen, M., Meinshausen, N., Hare, W. et al. 2009. Greenhouse-gas emission targets for limiting global warming to 2°C. *Nature*, 458(7242), 1158–1162.
Mercuri, E. G. F., Kumata, A. Y. J., Amaral, E. B., and Vitule, J. R. S. 2016. Energy by microbial fuel cells: Scientometric global synthesis and challenges. *Renewable and Sustainable Energy Reviews*, 65, 832–840.
Milgrom, P. R. and North, D. C. 1990. The role of institutions in the revival of trade: The law merchant, private judges, and the champagne fairs. *Economics & Politics*, 2(1), 1–23.
Mishra, A., Fischer, M. K., and Bauerle, P. 2009. Metal-free organic dyes for dye-sensitized solar cells: From structure: Property relationships to design rules. *Angewandte Chemie International Edition*, 48(14), 2474–2499.
Mittal, R. 2013. Biofuel research and data mining. *Performance Measurement and Metrics*, 14(1), 71–92.
Moed, H., De Bruin, R., and Van Leeuwen, T. H. 1995. New bibliometric tools for the assessment of national research performance: Database description, overview of indicators and first applications. *Scientometrics*, 33(3), 381–422.
Mohan, S. R. 2015. Structure and growth of research on biohydrogen generation using wastewater. *International Journal of Hydrogen Energy*, 40(46), 16056–16069.
Mohan, D., Pittman, C. U., and Steele, P. H. 2006. Pyrolysis of wood/biomass for bio-oil: A critical review. *Energy & Fuels*, 20(3), 848–889.
Morillo, F., Bordons, M., and Gomez, I. 2001. An approach to interdisciplinarity through bibliometric indicators. *Scientometrics*, 51(1), 203–222.
Mosier, N., Wyman, C., Dale, B. et al. 2005. Features of promising technologies for pretreatment of lignocellulosic biomass. *Bioresource Technology*, 96(6), 673–686.
Moss, R. H., Edmonds, J. A., Hibbard, K. A. et al. 2010. The next generation of scenarios for climate change research and assessment. *Nature*, 463(7282), 747–756.

Moss-Racusin, C. A., Dovidio, J. F., Brescoll, V. L., Graham, M. J., and Handelsman, J. 2012. Science faculty's subtle gender biases favor male students. *Proceedings of the National Academy of Sciences*, 109(41), 16474–16479.

Nederhof, A. J. 2006. Bibliometric monitoring of research performance in the social sciences and the humanities: A review. *Scientometrics*, 66(1), 81–100.

Nicolet, Y., Piras, C., Legrand, P., Hatchikian, C. E., and Fontecilla-Camps, J. C. 1999. *Desulfovibrio desulfuricans* iron hydrogenase: The structure shows unusual coordination to an active site Fe binuclear center. *Structure*, 7(1), 13–23.

North, D. C. 1986. The new institutional economics. *Journal of Institutional and Theoretical Economics (JITE)/Zeitschrift fur die Gesamte Staatswissenschaft*, 142(1), 230–237.

North, D. C. 1993. Institutions and credible commitment. *Journal of Institutional and Theoretical Economics (JITE)/Zeitschrift fur die Gesamte Staatswissenschaft*, 149(1), 11–23.

North, D. C. and Weingast, B. R. 1989. Constitutions and commitment: The evolution of institutions governing public choice in seventeenth-century England. *Journal of Economic History*, 49(04), 803–832.

Oravec, J. A. 2017. The manipulation of scholarly rating and measurement systems: Constructing excellence in an era of academic stardom. *Teaching in Higher Education*, 22(4), 423–436.

Oreskes, N. 2004. The scientific consensus on climate change. *Science*, 306(5702), 1686–1686.

Ostuni, E., Chapman, R. G., Holmlin, R. E., Takayama, S., and Whitesides, G. M. 2001. A survey of structure–property relationships of surfaces that resist the adsorption of protein. *Langmuir*, 17(18), 5605–5620.

Paladugu, R., Schein, M., Gardezi, S., and Wise, L. 2002. One hundred citation classics in general surgical journals. *World Journal of Surgery*, 26(9), 1099–1105.

Palmqvist, E. and Hahn-Hagerdal, B. 2000. Fermentation of lignocellulosic hydrolysates. II: Inhibitors and mechanisms of inhibition. *Bioresource Technology*, 74(1), 25–33.

Parker, J. N., Allesina, S., and Lortie, C. J. 2013. Characterizing a scientific elite (B): Publication and citation patterns of the most highly cited scientists in environmental science and ecology. *Scientometrics*, 94(2), 469–480.

Parmesan, C. and Yohe, G. 2003. A globally coherent fingerprint of climate change impacts across natural systems. *Nature*, 421(6918), 37–42.

Peters, J. W., Lanzilotta, W. N., Lemon, B. J., and Seefeldt, L. C. 1998. X-ray crystal structure of the Fe-only hydrogenase (CpI) from *Clostridium pasteurianum* to 1.8 angstrom resolution. *Science*, 282(5395), 1853–1858.

Pimentel, D. 1991. Ethanol fuels: Energy security, economics, and the environment. *Journal of Agricultural and Environmental Ethics*, 4(1), 1–13.

Pinstrup-Andersen, P. 2009. Food security: Definition and measurement. *Food Security*, 1(1), 5–7.

Podlubny, I. and Kassayova, K. 2006. Towards a better list of citation superstars: Compiling a multidisciplinary list of highly cited researchers. *Research Evaluation*, 15(3), 154–162.

Qian, G. 2013. Scientometrics analysis on the intellectual structure of the research field of bioenergy. *Journal of Biobased Materials and Bioenergy*, 7(2), 305–308.

Rafols, I. and Meyer, M. 2010. Diversity and network coherence as indicators of interdisciplinarity: Case studies in bionanoscience. *Scientometrics*, 82(2), 263–287.

Ragauskas, A. J., Williams, C. K., Davison, B. H. et al. 2006. The path forward for biofuels and biomaterials. *Science*, 311(5760), 484–489.
Ratnatunga, J. and Romano, C. 1997. A "citation classics" analysis of articles in contemporary small enterprise research. *Journal of Business Venturing*, 12(3), 197–212.
Reguera, G., McCarthy, K. D., Mehta, T. et al. 2005. Extracellular electron transfer via microbial nanowires. *Nature*, 435(7045), 1098–1101.
Riahi, K., Rao, S., Krey, V. et al. 2011. RCP 8.5—A scenario of comparatively high greenhouse gas emissions. *Climatic Change*, 109(1–2), 33–57.
Rodolfi, L., Chini Zittelli, G. C., Bassi, N. et al. 2009. Microalgae for oil: Strain selection, induction of lipid synthesis and outdoor mass cultivation in a low-cost photobioreactor. *Biotechnology and Bioengineering*, 102(1), 100–112.
Roman-Leshkov, Y., Barrett, C. J., Liu, Z. Y., and Dumesic, J. A. 2007. Production of dimethylfuran for liquid fuels from biomass-derived carbohydrates. *Nature*, 447(7147), 982–985.
Rosegrant, M. W. and Cline, S. A. 2003. Global food security: Challenges and policies. *Science*, 302(5652), 1917–1919.
Scherf, U. and List, E. J. 2002. Semiconducting polyfluorenes—Towards reliable structure–property relationships. *Advanced Materials*, 14(7), 477–487.
Schmidhuber, J. and Tubiello, F. N. 2007. Global food security under climate change. *Proceedings of the National Academy of Sciences*, 104(50), 19703–19708.
Schummer, J. 2004. Multidisciplinarity, interdisciplinarity, and patterns of research collaboration in nanoscience and nanotechnology. *Scientometrics*, 59(3), 425–465.
Searchinger, T., Heimlich, R., Houghton, R. A. et al. 2008. Use of US croplands for biofuels increases greenhouse gases through emissions from land-use change. *Science*, 319(5867), 1238–1240.
Siegmeier, T. and Moller, D. 2013. Mapping research at the intersection of organic farming and bioenergy—A scientometric review. *Renewable and Sustainable Energy Reviews*, 25, 197–204.
Smalheiser, N. R. and Torvik, V. I. 2009. Author name disambiguation. *Annual Review of Information Science and Technology*, 43(1), 1–43.
Snyder, C. S., Bruulsema, T. W., Jensen, T. L., and Fixen, P. E. 2009. Review of greenhouse gas emissions from crop production systems and fertilizer management effects. *Agriculture, Ecosystems & Environment*, 133(3), 247–266.
Steen, R. G. 2011. Retractions in the scientific literature: Is the incidence of research fraud increasing? *Journal of Medical Ethics*, 37(4), 249–253.
Steen, R. G., Casadevall, A., and Fang, F. C. 2013. Why has the number of scientific retractions increased? *PloS One*, 8(7), e68397.
Strotmann, A. and Zhao, D. 2012. Author name disambiguation: What difference does it make in author-based citation analysis? *Journal of the American Society for Information Science and Technology*, 63(9), 1820–1833.
Sun, Y. and Cheng, J. Y. 2002. Hydrolysis of lignocellulosic materials for ethanol production: A review. *Bioresource Technology*, 83(1), 1–11.
Tang, L. and Walsh, J. P. 2010. Bibliometric fingerprints: Name disambiguation based on approximate structure equivalence of cognitive maps. *Scientometrics*, 84(3), 763–784.
Thomas, C. D., Cameron, A., Green, R. E. et al. 2004. Extinction risk from climate change. *Nature*, 427(6970), 145–148.

Torvik, V. I., Weeber, M., Swanson, D. R., and Smalheiser, N. R. 2005. A probabilistic similarity metric for Medline records: A model for author name disambiguation. *Journal of the American Society for Information Science and Technology*, 56(2), 140–158.

Tscharntke, T., Clough, Y., Wanger, T. C. et al. 2012. Global food security, biodiversity conservation and the future of agricultural intensification. *Biological Conservation*, 151(1), 53–59.

Unger, K. and Couzin, J. 2006. Even retracted papers endure. *Science*, 312(5770), 40–41.

Van Leeuwen, T., Moed, H., Tijssen, R., Visser, M., and van Raan, A. 2001. Language biases in the coverage of the Science Citation Index and its consequences for international comparisons of national research performance. *Scientometrics*, 51(1), 335–346.

Van Noorden, R. 2011. The trouble with retractions. *Nature*, 478(7367), 26–28.

van Raan, A. F. 2005. Fatal attraction: Conceptual and methodological problems in the ranking of universities by bibliometric methods. *Scientometrics*, 62(1), 133–143.

van Raan, A. F., van Leeuwen, T. N., and Visser, M. S. 2011. Severe language effect in university rankings: Particularly Germany and France are wronged in citation-based rankings. *Scientometrics*, 88(2), 495–498.

Vitousek, P. M. 1994. Beyond global warming: Ecology and global change. *Ecology*, 75(7), 1861–1876.

Volbeda, A., Charon, M. H., Piras, C. et al. 1995. Crystal structure of the nickel-iron hydrogenase from *Desulfovibrio gigas*. *Nature*, 373(6515), 580–587.

Vorosmarty, C. J., Green, P., Salisbury, J., and Lammers, R. B. 2000. Global water resources: Vulnerability from climate change and population growth. *Science*, 289(5477), 284–288.

Walther, G. R., Post, E., Convey, P. et al. 2002. Ecological responses to recent climate change. *Nature*, 416(6879), 389–395.

Wang, M. Y., Fang, S. C., and Chang, Y. H. 2015a. Exploring technological opportunities by mining the gaps between science and technology: Microalgal biofuels. *Technological Forecasting and Social Change*, 92, 182–195.

Wang, L. H., Wang, Q., Zhang, X., Cai, W., and Sun, X. 2013. A bibliometric analysis of anaerobic digestion for methane research during the period 1994–2011. *Journal of Material Cycles and Waste Management*, 15(1), 1–8.

Wang, J., Zheng, T., Wang, Q., Xu, B., and Wang, L. 2015b. A bibliometric review of research trends on bioelectrochemical systems. *Current Science*, 109(12), 2204–2211.

Weingart, P. 2005. Impact of bibliometrics upon the science system: Inadvertent consequences? *Scientometrics*, 62(1), 117–131.

Winzer, C. 2012. Conceptualizing energy security. *Energy Policy*, 46, 36–48.

Woodcock, J., Edwards, P., Tonne, C. et al. 2009. Public health benefits of strategies to reduce greenhouse-gas emissions: urban land transport. *Lancet*, 374(9705), 1930–1943.

Yang, H. P, Yan, R., Chen, H., Lee, D. H., and Zheng, C. 2007. Characteristics of hemicellulose, cellulose and lignin pyrolysis. *Fuel*, 86(12), 1781–1788.

Yaoyang, X. Y. and Boeing, W. J. 2013. Mapping biofuel field: A bibliometric evaluation of research output. *Renewable and Sustainable Energy Reviews*, 28, 82–91.

Yergin, D. 2006. Ensuring energy security. *Foreign Affairs*, 85(2), 69–82.

Yuksel, I. and Dagdeviren, M. 2007. Using the analytic network process (ANP) in a SWOT analysis—A case study for a textile firm. *Information Sciences*, 177(16), 3364–3382.

Zakzeski, J., Bruijnincx, P. C., Jongerius, A. L., and Weckhuysen, B. M. 2010. The catalytic valorization of lignin for the production of renewable chemicals. *Chemical Reviews*, 110(6), 3552–3599.

Zhang, M., Gao, Z., Zheng, T. et al. 2016a. A bibliometric analysis of biodiesel research during 1991–2015. *Journal of Material Cycles and Waste Management*, doi: 10.1007/s10163-016-0575-z.

Zhang, Y., Kou, M., Chen, K., Guan, J., and Li, Y. 2016b. Modelling the Basic Research Competitiveness Index (BR-CI) with an application to the biomass energy field. *Scientometrics*, 108(3), 1221–1241.

Zhao, H., Holladay, J. E., Brown, H., and Zhang, Z. C. 2007. Metal chlorides in ionic liquid solvents convert sugars to 5-hydroxymethylfurfural. *Science*, 316(5831), 1597–1600.

chapter two

Biofuel developments and cooperation among China, Japan, and Malaysia

Tatsuji Koizumi

Contents

2.1 Introduction

The governments of Asian countries are promoting biofuel programs to address energy security and environmental problems and to contribute to agricultural and rural development. In their efforts, they have produced biofuels from various feedstock. This chapter covers biofuel developments and cooperation among China, Japan, and Malaysia. China produced 2750 million liters of bioethanol and was the largest bioethanol-producing country in Asia in 2016 (F.O. Licht, 2016b). Malaysia produced 802 million tons of biodiesel and ranked as the second-largest biodiesel-producing country after Indonesia in 2015 (F.O. Licht, 2016a). Japanese biofuel production is much lower than that of China and Malaysia, but its program and approach to increasing biofuel production differs from that of other countries.

Several studies have addressed East Asian biofuel programs. For example, Koizumi and Ohga (2007) conducted an economic analysis of the availability of supplies of biofuel production and its feedstock in Asian countries, Wang et al. (2009) examined the distribution and development of biofuel crops and bioenergy industry in China, Qiu et al. (2012) reviewed the Chinese bioethanol program, Koizumi (2013) reviewed the developments and perspectives of the Japanese biofuel program, and Lim and Lee (2012) discussed international cooperation between Malaysia and Japan to implement the biofuel target in Malaysia.

This chapter reviews biofuel programs and developments for China, Malaysia, and Japan and examines their cooperation in developing their technologies and sustainable criteria for biofuels. The next section further discusses biofuel programs and developments in China. The third section focuses on biofuel development in Japan. The fourth section reviews biofuel programs in Malaysia. In the fifth section, biofuel cooperation for technical issues and sustainability among the three countries are discussed, and the last section provides the conclusion.

2.2 *Chinese biofuel program*

2.2.1 *Background of Chinese biofuel program and production*

In China, petroleum consumption is increasing rapidly, and imports of crude oil are rising. The increase in petroleum consumption is causing a serious air pollution problem. In addition, excessive stock of grain, especially corn, led to crucial problems from 1996 to 2000. To deal with energy security, air pollution, and excessive grain stocks, the Chinese government strongly promoted the national bioethanol program.

In China, the concept of alternative energy was expressly stated in the Five-Year Plan of 1982. Furthermore, the Five-Year Plan for the period 2001–2005 included the promotion of biomass energy. In June 2002, the Chinese government started to mandate the use of bioethanol blend gasoline in five cities in Heilongjiang and Henan. In October 2004, the government introduced the compulsory use of a 10% blend of bioethanol to gasoline (E10) in all areas within Heilongjiang, Jilin, Liaoning, Henan, and Anhui. The government expanded the E10 program in 27 cities within Shandong, Jiangsu, Hebei, and Hubei starting in 2006.

Corn and wheat comprise a major part of the feedstock for bioethanol. Bioethanol is produced from corn in Heilongjiang, Jilin, Anhui, and Henan. It is also produced from wheat in Henan. Cassava is used to produce bioethanol in Guangxi and Henan. In addition, it is produced from sweet sorghum stalks in Inner Mongolia. Currently, seven bioethanol production facilities in China have operating licenses from the government (Table 2.1).

China also produces biodiesel for fuel use. There are four major facilities in Fujiang, Jiangsu, Hebei, and Beijing. However, China's production capacity has been estimated at 1.14 billion liters in 2015 because of a lack of feedstock availability. The main feedstock for biodiesel is used cooking oil. Although Chinese mills prefer to produce biodiesel from vegetable oil, securing vegetable oil for biodiesel use can be difficult because China is a net importer of oilseed and vegetable oil. Securing feedstock is a crucial problem for expanding biodiesel production in China, further exacerbated by the fact that biofuels are sold to only two state-owned companies, China Petroleum and Chemical Corporation (Sinopec) and China National Petroleum Corporation (CNPC), for blending with gasoline (Zhou and Thomson, 2009).

In China, the cost of corn-based bioethanol was US$1.56/L, and the feedstock cost of corn was US$0.64/L (Table 2.2). The feedstock cost of cassava ranged from US$0.53/L to US$0.87/L, and the total cost of cassava-based bioethanol ranged from US$0.94/L to US$1.27/L. The feedstock cost of sweet sorghum ranged from US$0.36/L to US$0.72/L, and the total cost is from US$0.66/L to US$1.01/L. The feedstock cost of sweet potato ranged from US$0.51/L to US$0.68/L, and the total cost was from US$0.91/L to US$1.08/L. The feedstock cost of cellulosic bioethanol was US$0.29/L, the nonfeedstock cost was US$1.17/L, and the total cost was US$1.46/L. The feedstock cost of waste oil-based biodiesel ranged from US$0.66/L to US$1.04/L, and the total cost was from US$0.93/L to US$1.31/L. The

Table 2.1 Chinese bioethanol production

Location	Producer	Main feeds tock	Capacity (1000 t)
Heilongjiang	COFCO Zhaodong	Corn, wheat	450.0
Jilin	Jilin Fuel Ethanol Co.	Corn, wheat	750.0
Henan	Henan Tianguan	Corn, wheat, tapioca, cassava	750.0
Anhui	COFCO Anhui	Corn, cassava	700.0
Guangxi	COFCO Guangxi	Cassava, tapioca	400.0
Shangdong	Shandong Longlive	Corncob	80.0
Inner Mongolia	ZTE Energy	Sweet sorghum stalks	80.0
Total			3210.0

Source: U.S. Department of Agriculture, Foreign Agricultural Service (USDA-FAS). 2015. *China-Peoples Republic of Biofuels Annual 2015.*

https://gain.fas.usda.gov/Recent%20GAIN%20Publications/Biofuels%20Annual_Beijing_China%20-%20Peoples%20Republic%20of_9-3-2015.pdf (accessed 01/28/17).

Table 2.2 Biofuel production cost in China

Feedstock	Feedstock cost	Nonfeedstock cost	Total producing cost
Corn	0.64	0.920	1.56
Cassava	0.53–0.87	0.400	0.94–1.27
Sweet sorghum	0.36–0.72	0.300	0.66–1.01
Sweet potato	0.51–0.68	0.400	0.91–1.08
Cellulosic ethanol	0.29	1.170	1.46
Waste oil	0.66–1.04	0.270	0.93–1.31
Jatropha curcas	0.56–1.13	0.370	0.94–1.50

Source: Chang, S., L. Zhao., Timilsina G. R., and X. Zhang. 2012. *Energy Policy* 51: 64–79 and Song, A., G. Pei., F. Wang., D. Wan., and C. Feng. 2008. *Academic Report of Agricultural Process* 24(3): 302–307.

Note: Biofuel production is converted from tons to liters (0.88 kg/L); values given in USD/L, with US$1 = 6.0597 CNY (November 2013).

feedstock cost of *Jatropha curcas* for biodiesel ranged from US$0.56/L to US$1.13/L, and the total cost was from US$0.94/L to US$1.50/L. The government introduced a US$112.3* per ton subsidy for nonfood grain feedstock such as cassava and sweet sorghum. Cellulosic bioethanol production receives US$119.8 per ton (USDA-FAS, 2015).

2.2.2 *Feedstock for bioethanol production from nonfood resources*

Corn is the main feedstock for bioethanol production in China. Corn accounted for 60.9% of fuel bioethanol production in 2013, and another 18.9% came from wheat. A total of 9.1% of biofuel production is produced using cassava, and less than 6.5% is produced from sweet sorghum. An additional 4.7% of biofuel production is cellulosic bioethanol derived from corn cobs (USDA-FAS, 2015).

Chinese corn consumption for feed and starch use has increased since 1990, and the domestic corn price has also increased since December 2004. Chinese corn ending stocks were estimated to decrease dramatically from 123.8 million tons in 1999–2000 to 36.6 million tons in 2006–2007. When the government started to expand the corn-based bioethanol program, corn ending stocks were abundant, and the government tried to eliminate them. In China, the domestic corn wholesale price increased by 30.0% from February 2005 to September 2006† due to the tight Chinese corn supply-and-demand situation. Corn consumption for bioethanol competed with corn consumption for feed, food, and other industries. In this regard, the Chinese government started to regulate corn-based bioethanol expansion

* US$1 = 6.67 CNY (September 15, 2016).

† This was derived from the Institute of Agricultural Economics, Chinese Academy of Agricultural Science (October 2007).

on December 21, 2006. This regulation facilitated the current bioethanol production level in Heilongjiang and Jilin, but limited further expansion of corn-based bioethanol production. This regulation applies to wheat-based bioethanol production as well.

Instead of expanding corn-based bioethanol production, the government now seeks to diversify bioethanol production, especially that from cassava and nonfood resources. The first cassava bioethanol plant was built in Guangxi in 2007. Total cassava production in China was 4.6 million tons in 2014, which was much smaller than cassava production in Thailand (30.0 million tons in 2014). Although Guangxi is trying to increase its cassava production, it is assumed that China has difficulty securing enough cassava to meet domestic consumption for bioethanol production. China will have to rely on cassava imports from Thailand. China has mastered cassava-based bioethanol technology by constructing a demonstration project in Guangxi, but in the areas of liquefaction, scarification, fermentation, separation process, and sterilization devices, it still lags behind advanced international levels (Wang, 2011). A key to success for developing cassava-based bioethanol production in China is technical innovation for mass production.

In 2015, 3.381 million tons of corn were used for bioethanol production (Table 2.3). The corn-use ratio of domestic corn consumption was 1.6% in 2015. The wheat-use ratio in domestic wheat consumption was 0.9% in 2015. A total of 504,000 tons of cassava were used for bioethanol production in 2015. The cassava use ratio for domestic cassava production was 10.8% in 2014. The use ratio of cassava is much higher than that of corn and wheat. Cassava production now competes with feed use and for agricultural resources, such as land and water use.

To encourage the development of noncereal-based bioethanol production, this policy announcement also makes it clear that any new bioethanol production based on cereal crops will not be supported or subsidized (Qiu et al., 2012). This regulation can be evaluated as an adequate policy measure to regulate bioethanol production from corn and wheat. Without this regulation, the international corn price could have been much higher than the real prevailing price from 2006 to 2008 and food security for China and food-importing developing countries could have been damaged.

A sweet sorghum bioethanol plant was produced in Inner Mongolia in 2012. In June 2014, the plant started to supply fuel to three cities in the country's provinces (USDA-FAS, 2014). Sweet sorghum can grow under dry conditions in saline alkaline soil. Although a number of provinces are trying to increase their sweet sorghum production, its production is much lower than corn.* Corn cobs were used for cellulosic bioethanol produc-

* In 2012–2013, sorghum production was 2.5 million tons and corn production was 205.6 million tons (USDA-FAS, 2017).

Table 2.3 Biofuel feedstock use ratio in China

Feedstock	Unit	2006	2007	2008	2009	2010	2011	2012	2013	2014	2015
Biofuel production	Million liters	1664	1731	2257	2466	2479	2566	2858	2934	2951	3078
Corn-use ratio	%	2.2	2.3	2.4	2.2	2.0	1.9	1.8	1.7	1.7	1.6
Corn use for bioethanol	1000 tons	3233	3385	3608	3614	3641	3641	3645	3563	3389	3381
Domestic corn consumption	1000 tons	145,000	150,000	153,000	165,000	180,000	188,000	200,000	208,000	202,000	217,500
Wheat-use ratio	%	1.0	1.0	1.0	1.0	1.0	0.9	0.8	0.9	0.9	0.9
Wheat use for bioethanol	1000 tons	1050	1050	1050	1050	1050	1050	1050	1050	1050	1050
Domestic wheat consumption	1000 tons	102,000	106,000	105,500	107,000	110,500	122,500	125,000	116,500	116,500	112,000
Cassava-use ratio	%	0.0	0.0	8.3	10.4	8.6	7.4	7.3	8.5	10.8	–
Casssava use for bioethanol	1000 tons	0	0	364	467	392	336	336	392	504	504
Domestic cassava production[a]	1000 tons	4313	4362	4409	4506	4565	4514	4574	4600	4680	–
Sweet sorghum	%	–	–	–	–	–	–	–	–	0.7	3.5
Sweat sorghum use for bioethanol	1000 tons	0	0	0	0	0	0	0	0	90	360
Domestic sweet sorghum production	1000 tons	2050	2000	2000	1900	2200	2200	3200	6800	12,900	10,300

Source: U.S. Department of Agriculture, Foreign Agricultural Service (USDA-FAS). 2015. *China-Peoples Republic of Biofuels Annual 2015.*
*https://*gain.fas.usda.gov/Recent%20GAIN%20Publications/Biofuels%20Annual_Beijing_China%20-%20Peoples%20Republic%20of_9-3-2015.pdf (accessed 01/28/17) and Food and Agricultural Organization of the United Nations (FAO) *FAOSTAT*. http://faostat.fao.org/ (accessed 01/20/17).

[a] Domestic cassava production is derived from the Food and Agriculture Organization of the United Nations. Domestic cassava production are alternative data for cassava consumption.

tion. In 2015, 80 thousand tons of corn cobs were used for cellulosic bioethanol (USDA-FAS, 2015). However, production of this type of fuel is in its experimental stage. In addition to these crops, bioethanol production from crop stalks and straw, sugarcane, sweet potatoes, sugar beets, woody biomass, and others are also the experimental stage.

2.2.3 *Developments of the Chinese biofuel program*

The utilization and development of renewable energy in China is a crucial national program that not only contributes to energy security and improves environmental problems, but also develops rural areas, promoting new industries and technical innovation. In 2006, the government enacted the Renewable Energy Law to promote renewable energy utilization and production. The government promotes a biomass energy program, which is divided into four categories: biofuels, rural biomass, biogas, and bioelectricity. The national bioethanol program was started in 2001, and the government strongly promoted the bioethanol program to provide an alternative fuel to gasoline. The government will promote the bioethanol program in the future because of the increasing gravity of the energy security problem and the air pollution problem.

In September 2007, the National Development and Reform Commission (NDRC) provided a midterm to long-term plan for renewable energy. The plan indicated that bioethanol from nonfood-grade crops would reach 11.4 million liters in 2020. The plan also indicated that biodiesel production would increase to 2.2 million liters in 2020.* In 2014, the Chinese government approved a National Climate Change Plan (NCCP) that set out emission and clean energy targets for 2020. The NCCP set a target of 130 billion cubic meters of biofuel production by 2020 (USDA-FAS, 2015).

However, it is uncertain whether China can meet these goals. China will have to diversify feedstock for biofuel production. China has switched from grain-based biofuels to nonfood-grade biofuel production, using feedstock such as sweet sorghum and cassava. However, biofuel production levels from these nonfood resources are still being determined in a pilot-scale project at present, and it is difficult to expand bioethanol production from cassava and sweet sorghum because of the difficulty in procuring feedstock.

The Chinese commercial cellulosic bioethanol plant, which uses corn cobs, reached an annual capacity of 63.4 million liters. The plant has supplied bioethanol for seven cities in the Shandong provinces since 2012. Other bioethanol plants are reportedly preparing to build demonstration-scale cellulosic bioethanol projects using a variety of feedstock, such as

* Biofuel production has been converted from tons to liters (0.88 kg/L).

corn stover and wheat straw. Chinese research and development (R&D) for second-generation biofuel production has just begun. Further governmental support is necessary to expand second-generation biofuel production.

2.3 Japanese biofuel program

2.3.1 Development of the Japanese biofuel program

Japan's biofuel resources were developed as an emergency alternative fuel for gasoline and diesel during World War II. The quality and production costs of biofuels were not suitable for commercial use after World War II, when most of these technologies were abandoned.* After World War II, Japan continued to produce bioethanol from imported molasses. However, the two oil crises in the 1970s shifted the focus of Japan's energy policy to energy savings and to reducing the country's reliance on oil†, with the result that the adoption of biofuels was not considered until 2002. However, under the Kyoto Protocol, Japan became committed to cutting greenhouse gas (GHG) emissions by 6% from the 1990 levels before the end of the first commitment period (2008–2012). The decision to promote the recycling of various types of resources, including biomass, was made, resulting in the creation of the Basic Law for Promoting the Formation of a Recycling-Oriented Society in 2001. The first time the government announced a plan to promote biofuel production and utilization of biofuels was in the Biomass Nippon Strategy,‡ which the cabinet adopted in December 2002.

The cabinet adopted the Kyoto Protocol Target Achievement Plan in April 2005, with the target of achieving a reduction in CO_2 equivalent to 500 million liters of crude oil.§ When the Kyoto Protocol came into force in April 2005, Japan determined that, to meet its targets, it would be necessary to convert biomass energy into useful forms of energy, such as transportation fuels, and to draw a road map for the adoption of domestically produced biomass as transportation fuel. In March 2006, the cabinet adopted the revised Biomass Nippon Strategy, the most striking feature of which was that biofuels became the main force among various biomass products.¶

The Japanese government has been promoting bioethanol production and its use in automobiles since 2003. Verification tests and large-scale

* For more detailed information, please refer to Koizumi (2013).

† Japan relied on oil for 77.4% of energy consumption in 1973, and 71.5% in 1979, but this dropped to 49.4% in 2001 (Ministry of Economy, Trade and Industry, 2009).

‡ *Nippon* means Japan in Japanese.

§ 500 million liters of crude oil is equivalent to 800 million liters of bioethanol.

¶ As for Japanese biofuel production and programs, it depends on Koizumi (2013).

projects for bioethanol production were launched at 10 locations in Japan. Demonstration projects include large-scale projects that began in 2007 to collect data for domestic transportation biofuels and to support a model project for the local utilization of biomass. In 2015, the bioethanol production level in Japan was at 20 million liters.* The municipal government and nongovernmental organizations are promoting the production of biodiesel from used cooking oil blended with diesel used for public buses, official cars, and municipal garbage trucks. Twenty biodiesel fuel projects have started since 2007. The Ministry of Economy, Trade and Industry (MITI) has supported biodiesel production, transportation, and utilization since 2012. Biodiesel production resulted in 15.0 million liters in 2015.†

MITI established sustainable criteria for biofuels that covered two major issues: First, domestic and imported biofuels should reduce GHG to 50%, compared to gasoline and diesel. Second, oil refiners should ensure sufficient food availability. At present, only Brazilian sugarcane-based bioethanol production can meet the criteria. MITI also set up the target of bioethanol utilization from 216 million liters in 2010 to 500 million liters in 2017. The bioethanol is used for ethyl tert-butyl ether (ETBE), not a direct blend of gasoline. This is a mandate for domestic oil refiners. Japanese oil refiners can meet this target to utilize imported bioethanol from Brazil and ETBE.

The domestic costs of bioethanol were much higher than those of gasoline and imported bioethanol because of the expensive land usage. The production cost of bioethanol from nonfood-grade wheat was US$1.53/L (Table 2.4). This type of bioethanol was used in Hokkaido for ETBE production. The price of bioethanol for ETBE use was based on the imported Brazilian bioethanol price, determined by the Petroleum Association of Japan (PAJ). The total price of bioethanol from Brazil was US$1.29/L, and the price difference between that of nonfood-grade wheat and the Brazilian bioethanol price was US$0.24/L. Food-based biofuels are not produced in Japan, so these biofuel production costs are theoretical figures (Table 2.4). It is not realistic to produce bioethanol from food use grains in Japan because the production costs (US$5.70/L) are much higher than what other biofuels and gasoline cost.

These price differences present crucial challenges to expand biofuel production in Japan. Bioethanol producers bear the price deficiencies using subsidies. No bioethanol producers can operate their production facilities without subsidies from the Ministry of Agriculture, Forestry and Fisheries (MAFF) and the Ministry of the Environment (MOE). The tax reduction was established for the portion of bioethanol out of bioethanol-blended gasoline in February 2009. In the case of the 3% bioethanol

* Data from Ministry of Agriculture, Forestry and Fisheries (June 2016).

† Data from Ministry of Agriculture, Forestry and Fisheries (June 2016).

Table 2.4 Japanese bioethanol producing cost (Based on MAFF Japan 2013)

Feedstock	Feedstock cost	Processing cost	Gasoline tax	Total cost
Molasses	0.07	0.85	0.53	1.45
Wheat (nonfood use)	0.53	0.47	0.53	1.53
Wheat (food use)	2.88	0.47	0.53	3.88
Rice (bioethanol use)	0.46	0.50	0.53	1.49
Rice (food use)	4.66	0.50	0.53	5.69
	Wholesale/cost, insurance, and freight (CIF) price	**Import tariff**	**Gasoline tax**	**Total cost**
Domestic gasoline	0.61[a]	–	0.55	1.15
Imported bioethanol from Brazil	0.67[b]	0.09	0.53	0.62

Note: Production cost includes capital cost and variable cost. Retail price includes transportation cost and consumption tax. These data are based on Ministry of Agriculture, Forestry and Fisheries of Japan (2013); values given in USD/L, with US$1 = 98.19 JPY (November 2013).

[a] The wholesale price of gasoline was the average March 2010 price from the Oil Information Center of Japan.

[b] The Brazilian bioethanol CIF price was the average March 2010 price from trade statistics. The custom tariff was 13.4%.

blended in gasoline (E3), US$0.02/L is tax exempt. However, this is not good enough to promote biodiesel utilization. Reducing the cost of producing bioethanol is the key to increasing its domestic production, but it will be difficult to reduce the domestic bioethanol cost to the price level of gasoline and imported bioethanol.

In 2014, 10 bioethanol production projects were under operation. It was difficult for most of these facilities to increase their production levels due to limited feedstock. Furthermore, there was a strong critical opinion that food-based biofuels would damage domestic and world food availability because Japan is one of the largest food-importing countries in the world. The Japanese government became concerned about the impact of biofuels on food security. In order to increase the volume of domestically produced bioethanol in Japan, it was necessary to produce biofuels from agricultural residuals and other nonfood resources.

MAFF promoted the soft cellulose-based bioethanol project from 2008 to 2012 to use rice straw and wheat straw to produce bioethanol. Japanese research institutes, universities, and private companies are researching ways to increase the efficiency of cellulose-based bioethanol production. MAFF also conducted a research project with research institutes, universities, and private companies from 2010 to 2015. The Universities of Tokyo, Tsukuba, and Kobe, and several private companies [Euglena, Denso,

Yamaha, Ishikawajima-Harima Heavy Industries (IHI), and others] participated in this project. The research is designed to help companies extract oil produced by algae and to develop mass production technology for jet fuel use.

2.3.2 *Prospects of Japanese biofuel program*

After the earthquake and tsunami disasters in March 2011, one of the priority policy tasks in Japan was to revitalize the economies of disaster-stricken areas in the northeast region (the Fukushima, Iwate, and Miyagi prefectures). The severe effects of the radiation leakage accident caused by the disaster are still being felt by the Japanese economy and society. Following the disaster, there was a crucial debate on whether Japan should abandon nuclear energy or not and whether Japan should increase its renewable energy supply ratio. Therefore, energy security became a national concern after the disaster in Japan, and renewable energy is expected to mitigate energy security problems and stimulate rural development, especially in the disaster area. The feed-in tariff (FIT) system started to promote renewable energy in heat and power plants in 2012. However, biofuels for power plants were excluded from this system.

MAFF supported three bioethanol facilities in Hokkaido and Niigata. Hokkaido bioethanol and Oenon Holdings started bioethanol production in 2009, and Japan Agricultural Cooperation (JA) Nigata started in 2008. Hokkaido Bioethanol Co. Ltd in Hokkaido produces bioethanol from surplus sugar beets and substandard wheat. Oenon Holdings in Hokkaido produced bioethanol from nonfood rice. JA Agricultural Cooperatives in Niigata City, in Niigata Prefecture, produced bioethanol from nonfood rice.

In December 2013, a chair of Administrative Reform Promotion Headquarters of the Liberal Democratic Party requested that MAFF review the biofuel program. MAFF organized a review committee for bioethanol projects in February 2014. The committee published the report in May 2014, which pointed out that the continuation of their business would be very difficult without governmental support. The committee suggested that MAFF should decide whether to continue providing assistance with the bioethanol project or not. Based on the report from the committee, MAFF decided to abolish subsidies for these facilities after Fiscal Year 2015 because these plants had difficulty running independent of subsidies. Hokkaido Bioethanol Co. Ltd. abandoned its bioethanol production after this decision in July 2014 and declared bankruptcy in June 2015. Two other facilities decreased their bioethanol production after this decision.

MOE continues to support a bioethanol facility on Miyakojima Island, Okinawa. This facility produces bioethanol from molasses and continues to supply E3 for gas stations on the island. However, it produced 17,000 L in

2015. Thus, it is estimated that Japanese bioethanol production decreased after 2014 due to the abolishment of subsidies from MAFF.* MITI supported biodiesel production, transportation, and utilization. However, its subsidy expired at the end of Fiscal Year 2015. It is difficult to increase biodiesel demand and production without subsidies. Thus, subsidies to promote biofuel production from MITI and MAFF are expired at present. Only MOE supports the small size of bioethanol production in Okinawa.

2.4 *Malaysian biofuel program*

The main incentives for promoting biofuel programs in Malaysia are stable palm oil price and energy security. Malaysia is the second-largest palm oil producer after Indonesia. The National Biofuel Policy was released in 2006 and the Biofuel Industry Act was enacted in 2007. The government of Malaysia decided to start a 5% biodiesel blend (B5) mandate to diesel oil from 2007. The biodiesel production was 186 million liters in 2008 and increased to 247 million liters in 2009 (Table 2.1). The main feedstock for biodiesel production is palm oil. A total of 195 thousand tons of palm oil was used for biodiesel production in 2008 (USDA-FAS, 2016b).

When the government decided to start the national biodiesel program in 2006, world crude oil price was high and palm oil price was low. It was ideal economic condition to produce biodiesel from palm oil. However, the palm oil price increased from 2006 to 2008 (Figure 2.1). It was correlated to soybean oil, other vegetable oil, and crude oil price hikes. The feedstock cost accounts for 86% of total biodiesel production cost in 2007 (Koizumi, 2009). Thus, the economic condition of producing biodiesel deteriorated from 2006 to 2012.

High production costs inhibited its production. The biodiesel price was 4.2 Malaysian ringgit (MYR)/L (3.6 MYR/L, with the refinery cost was 0.6 MYR/L) in 2007. Contrary to this, diesel was sold at 1.6 MYR/L including fuel subsidy (0.58 MYR/L) in 2007 (Koizumi, 2009). The fossil retail price was much lower than the international price. Relatively higher producing cost than diesel price was the main obstacle to increasing biodiesel production in Malaysia. At that time, the government gave preference treatment for stabilizing fossil fuel price rather than the biodiesel program. Therefore, domestic biodiesel did not have any economic advantage comparing with fossil fuel.

The European Union is the largest biodiesel market. It sets bioenergy mandate,† mainly from biofuel, but EU production cannot satisfy the demand derived from the mandate. Therefore, the European Union

* There is no official production data in 2014.

† The detailed information will be introduced in the next section.

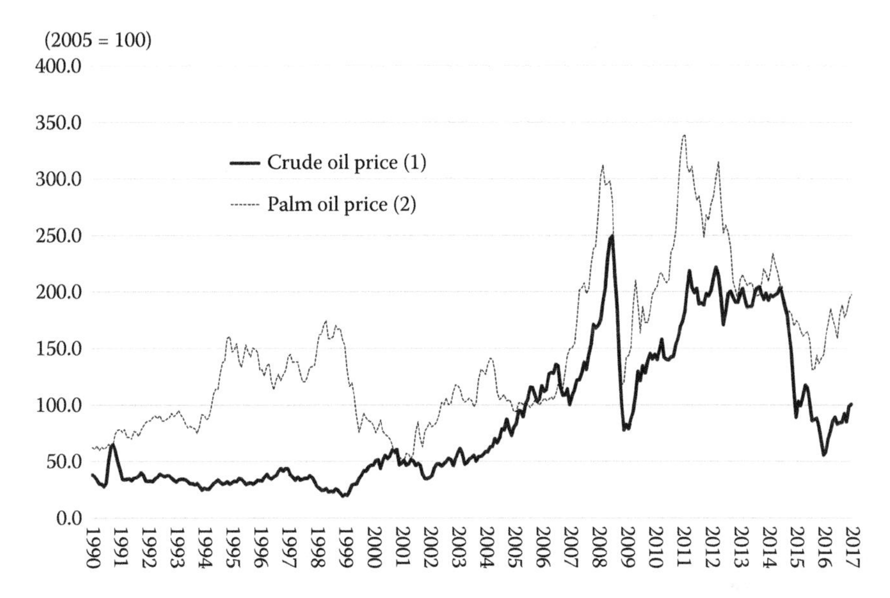

Figure 2.1 International crude oil price and palm oil price (Based on International Monetary Fund). (1) Crude oil (petroleum): simple average of three spot prices; Dated Brent, West Texas Intermediate, and the Dubai Fateh; (2) palm oil: Malaysia palm oil Futures (first contract forward) 4%–5% FFA.

imports biofuels and their feedstock from Indonesia, Malaysia, Argentina, and other countries.

The biodiesel price in the European Union was US$1.57/L in 2007.* The Malaysian biodiesel price was equivalent to US$1.32/L, while the Indonesian biodiesel price was equivalent to US$1.10/L (Koizumi, 2009). However, relatively higher feedstock cost than Indonesia was the main obstacle to increasing biodiesel exports in Malaysia.

Biodiesel production was 247 million liters in 2009 and had stagnated from 2010 to 2012 (Table 2.1). Its exports reached 258.5 million liters in 2009 and decreased to 33.0 million liters in 2012. The B5 program was planned to start in 2007, but it was heavily delayed because the domestic biodiesel production did not satisfy the demand for B5. The production increased gradually from 2011 and the B5 program was implemented gradually. The B5 program was implemented in the Central Region in June 2011, in the Southern Region in July 2013, in the Northern Region in March 2014, and nationwide in March 2014 (Table 2.2). B7 was planned to start in January 2015, but it was implemented nationwide in November 2014. The implementation of B10, originally scheduled for October 1, 2015, had been delayed due to resistance from key diesel vehicle manufactures,

* It was produced from rapeseed oil. Data are derived from Tan (2007).

claiming the 10% blend would have adverse effects on the engine and lubrication systems (USDA-FAS, 2016b).

The government of Malaysia terminated fuel subsidies on December 1, 2014. The diesel and gasoline prices are decided by the average price of crude oil during the previous month and the prices are elastic with crude oil price. This termination changed the domestic biodiesel production condition dramatically. The palm oil price was decreasing since 2012 and the biodiesel production increased from 271 million tons in 2012 to 833 million tons in 2016. The export increased to 210 million liters in 2016 as a result of increasing domestic production. Malaysia exported 85.1% of biodiesel to the European Union in 2015 (USDA-FAS, 2016b) (Table 2.5).

The government of Malaysia is promoting biodiesel production from *J. curcas*, but it is still in the experimental stage. The development of the second-generation biodiesel program is not active in Malaysia. The palm kernel shells (PKSs) were used for the heat source of the boiler as a cogeneration. The PKSs are exported to Thailand and South Korea and sold at 120–150 MYR/ton.* Small farmers own 10% of total palm field and large enterprise owns more than 60% of total palm field in Malaysia (Koizumi, 2009). Large enterprises dominate most of the palm plantation. Palm oil refining companies own biodiesel facilities as a part of their production process. Accordingly, there are no companies specialized for biodiesel production. Some leading companies, such as Sime Darby Plantation, established the Malaysian Biodiesel Association. It has strong political power to decide the national biodiesel program. The companies increase biodiesel production when their production is profitable. It means that they decrease production when the operation margin is not profitable. The feedstock cost accounts for more than 80% of total biodiesel producing cost. Thus, biodiesel production in Malaysia can be elastic with palm oil price. The government plans to implement the B10 program and to start the B15 program, but that depends on domestic production and palm oil prices (Table 2.6).

2.5 Discussion

2.5.1 Bioenergy feedstock trade in China, Japan, and Malaysia

India and the European Union are the main palm oil trade partners for Malaysia. Malaysia exported 21.1% of global palm oil exports, including crude palm oil (CPO) and refined palm oil (RPO), to India, 13.9% to the European Union, 13.6% to China, and 3.2% to Japan.† Palm oil is used for

* From interview with Ministry of Plantation Industries and Commodities, Malaysia (December 2016).

† Palm oil trade data are derived from MPOB (2015).

Table 2.5 Biodiesel market in Malaysia

Feedstock	2008	2009	2010	2011	2012	2013	2014	2015	2016
				Biodiesel					
Production	186	247	127	188	271	513	451	550	833
Export	198	247	97	54	32	190	95	195	210
Consumption	0	0	0	26	125	188	370	525	530
Ending stocks	3	3	33	141	255	390	376	206	299
				Palm oil					
Production	17,259	17,763	18,211	18,202	19,321	20,161	19,879	17,700	20,000
Exports	15,990	16,610	17,151	17,586	18,524	17,344	17,378	16,621	17,500
Consumption	2688	2301	2204	2150	2451	2869	2941	2990	3170
Biodiesel use	195	222	95	96	103	329	263	361	341

Source: Based on U.S. Department of Agriculture, Foreign Agricultural Service (USDA-FAS). 2016b. *Malaysia Biofuels Annual.* *https://*gain.fas.usda.gov/Recent%20GAIN%20Publications/Biofuels%20Annual_Kuala%20Lumpur_Malaysia_7-27-2016.pdf (accessed 01/15/17).

Note: Values given in millions of liters.

Table 2.6 National biodiesel program: Original plan and implementation

Biodiesel	Original Plan	Implementation
B5	2008	June 2011 (Central Region) July 2013 (Southern Region) March 2014 (Northern Region) March 2014 (Nationwide)
B7	January 2015	November 2014 (Nationwide)
B10	October 2015	None

Source: Interview from Malaysian Palm Oil Board and Malaysian Biodiesel Association (December 2016).

making edible oils, margarine, shortening, soaps, and other products in China and Japan; however, it is not used for biodiesel feedstock in China and Japan. The palm oil trade relationship from Malaysia to China is very active, but Malaysia exported 1.5% of the biodiesel to Japan and 0.9% to China.* Thus, the trade relation is not from Malaysia to Japan. As mentioned previously, most Malaysian biodiesel was exported to the European Union and biodiesel from Malaysia to China and Japan is not active.

The government of Japan started a FIT system for electricity from renewable energy sources such as solar and wind power in 2012. Under the FIT system, power generation companies have to purchase electricity at fixed rates for fixed periods (Table 2.7). After the nuclear accident in Fukushima in 2011, power generation companies began to use biomass such as wood pellets as a source for thermal power generation. However,

Table 2.7 FIT purchase rates for fiscal year 2016

	FIT rate (JPY)	FIT rate (USD)
Solar power	24–33 JPY per kWh	0.21–0.29 USD per kWh
Wind power	22–55 JPY per kWh	0.19–0.48 USD per kWh
Hydro power	14–34 JPY per kWh	0.12–0.30 USD per kWh
Geothermal power	26–40 JPY per kWh	0.230.35 USD per kWh
	Biomass-based power	
Wood materials	13–40 JPY per kWh	0.11–0.35 USD per kWh
Waste materials	17 JPY per kWh	0.15 USD per kWh
Biogas derived from methane fermentation	39 JPY per kWh	0.34 USD per kWh

Source: Ministry of Economy, Trade and Industry. 2016. Feed in tariffs (FIT) scheme for fiscal year 2016. Tokyo, http://www.enecho.meti.go.jp/category/saving_and_new/saiene/kaitori/kakaku.html (accessed 01/23/17).

* Export data are derived from USDA-FAS (2016b).

domestic production cannot satisfy the demand from the power generation companies. Therefore, imported wood pellets are used as prices decrease compared to those of domestically produced products. Japanese wood pellet production and import are increasing. In 2015, Japanese imports of wood pellets increased 140% from the previous year to 232 thousand tons (USDA-FAS, 2016a). Some companies utilized PKSs for thermal power generation as a substitute for wood pellet use because the price for PKSs is lower than that of domestic and imported wood pellets in Japan. It is difficult to grow palm botanically in Japan, and PKSs are imported from Indonesia and Malaysia. Accordingly, the government of Japan expects that PKS imports will increase in the future. Under the FIT system, PKS trade from Malaysia to Japan can be active in the future.

2.5.2 Technical cooperation from Japan to China and Malaysia

The governments of China and Japan are promoting biofuel programs to deal with energy security and environmental problems and to enhance agricultural and rural development. These countries are promoting biofuels from agricultural residuals and other nonfood resources. Rice and wheat straw and crop stalk are used for biofuel production in China and Japan. However, high enzyme cost and collection costs are the main obstacles to expanding their production. Rice and wheat straw are categorized as second-generation biofuels. The Chinese government has tried to produce biofuels from nonfood feedstock such as cassava, sweet potato, and other crops. Japanese research institutes, universities, and private companies have developed R&D technologies to reduce the production cost for nonfood resource-based biofuels. Therefore, the technical transfer from Japan to China and other Asian countries is a crucial factor to promote biofuel production from nonfood resources in China and other Asian countries.

Development of biodiesel production from nonfood resources is inactive in Malaysia. There seems to be no room for technical cooperation from Japan to Malaysia. However, the Japan Automobile Manufacturers Association (JAMA)* is entrusted by the government of Malaysia and inspected diesel engine durability performance to blend 5% and 7% of biodiesel in Malaysia. Its inspection result can enact a crucial judgment to implement a biodiesel blend program in Malaysia. JAMA is inspecting the engine performance in the case of B10. JAMA plays a crucial role in judgment to implement the Malaysian biodiesel program, and it is a crucial technical cooperation to implement the biodiesel blend target in Malaysia.

* JAMA is not a governmental organization in Japan.

2.5.3 Sustainable criteria

Japan is importing biofuel and feedstock for biofuels from other countries. To ensure the sustainability of biofuel not only in their countries, but also on a global scale, they have to take care of the environment, food availability, and social consequences of their trading partner countries. Thus, establishing sustainable criteria is a crucial factor in promoting biofuel utilization and production in these countries. The Japanese government decided on the sustainable criteria for biofuel production and utilization. The Act on Sophisticated Methods of Energy Supply Structures, enacted in 2009, required oil refiners (petroleum and gas enterprises) to use biofuel and biogas. To decide on sustainable criteria for the use of biofuel in Japan, the government organized a study panel to discuss the introduction of the criteria in 2009, and in 2010 the criteria were finally stipulated in Notification No. 242 of the Ministry of Economy, Trade and Industry.

The criteria covered several issues: First, the biofuel should eliminate 50% of GHGs compared to gasoline or diesel. Second, oil refiners should pay attention to ensure food availability and not impair such availability in the course of promoting biofuel utilization.* Third, oil refiners should recognize the impact of biofuel production on biodiversity and obey domestic laws and regulations related to these areas. Fourth, oil refiners should promote cellulose-based and algae-based biofuel R&D and utilization.

These sustainable criteria took into account not only domestic biofuel production, but also imported biofuel. At present, most of the Japanese biofuel production does not satisfy the criteria (50% GHG reduction), with the exception of waste woods and sugar beet for bioethanol use.

The government of Japan decided on mandatory sustainable criteria for biofuel production and utilization. The criteria cover the limitation of GHG emissions, while paying attention to biodiversity and food availability. However, the criteria do not cover social consequences and other environmental issues, such as air quality, water availability, and others. On this account, Japan has been contributing to discussions in the Global Bioenergy Partnership (GBEP) to establish international guidelines for sustainable criteria for biofuel with the Food and Agricultural Organization of the United Nations (FAO) and other countries since 2007. The category of proposed sustainable criteria in the GBEP is much wider than those of Japan. GBEP brings together public, private, and civil society stakeholders in a joint commitment to promote bioenergy for sustainable development. The proposed criteria cover environmental (GHG emissions, productivity capacity of the land and ecosystems, air

* If they are concerned that bioethanol production of the trading partner country will dramatically decrease, oil refiners should report their situation to the Japanese government.

quality, water availability, use efficiency and quality, biological diversity, and direct and indirect land-use change), social, economic, and energy-related security.

The European Union is a crucial palm oil and biodiesel trade partner for Malaysia. The European Union imports 166 million liters of biodiesel from Malaysia. However, the European Union prefers importing palm oil to biodiesel. EU oil refiners can produce edible oil, industrial oil, and biodiesel with flexibility when they import CPO; however, they cannot produce palm oil products with flexibility when they import biodiesel. The European Union utilized 1.7 million tons of crude palm oil for biodiesel use, and CPO accounts for 14.3% of the total feedstock for biodiesel production in the European Union* (USDA-FAS, 2016b). Reducing GHG emissions, diversifying energy sources, and developing the rural economy are the main incentives to promote biofuel programs in the European Union. The Renewable Energy Directive (RED) came into force in June 2009. The RED set several goals for 2020, including a 20% share of renewable energy in the EU total energy mix, and part of this 20% share is a 10% minimum mandate for renewable energy consumed in transport use.† The latest RED amended a 7% cap on renewable energy in the transporting sector coming from food or feed crop.‡ The mandated biofuels have to meet the sustainable criteria, reducing GHG emissions by at least 35% compared to fossil fuels. From 2017, the reduction must reflect a 50% savings and at least a 60% savings for new installations. Malaysian palm oil is used for biodiesel production, and biodiesel must satisfy the sustainable criteria in the European Union.

The International Sustainability and Carbon Certification (ISCC) is one of the leading certification systems for sustainability and greenhouse gas emissions. The European Commission recognizes ISCC as one of the certifications for compliance with RED requirements. ISCC certification can be applied to meet the legal requirements in the bioenergy markets as well as to demonstrate the sustainability and traceability of feedstock in the food, feed, and chemical industries§ (ISCC). Most Malaysian producers obtained ISCC certification for biodiesel and palm oil, and therefore Malaysian palm oil and biodiesel can be exported to and utilized in the European Union. As stated previously, the government of Japan

* This includes CPO from Indonesia.

† The package had some goals for 2020: (1) 20% share of renewable energy in the EU total energy mix; part of this 20% share is a 10% minimum mandate for renewable energy consumed in transport use; (2) 20% reduction in GHG emissions compared to 1990; (3) 20% improvement in energy efficiency compared to 2020.

‡ It was amended in 2015.

§ ISCC ensures that (1) greenhouse gas emissions are reduced; (2) biomass is not produced on land with high biodiversity and high carbon stock; (3) good agricultural practices and the protection of soil, water, and air are applied; and (4) human, labor, and land rights are respected (ISCC).

introduced its own sustainable criteria for bioenergy, and the Malaysian palm oil refiners obtained the certification to satisfy the sustainable criteria for the European Union. However, it does not appear that the government of China has introduced its own or other sustainability criteria for bioenergy.

2.6 Conclusion

The governments of China, Japan, and Malaysia are promoting biofuel programs to address energy security and environmental problems and to contribute to agricultural and rural development. Energy security is the main incentive for promoting biofuel programs in China. Reducing GHG emissions is the main incentive for promoting biofuel programs in Japan. In Malaysia, stabilizing the palm oil price is the main incentive for promoting biofuel programs.

Malaysia is exporting palm oil to China and Japan. However, palm oil is not used for biodiesel feedstock in China and Japan, so the trade relation for biodiesel among the three countries is not active. On the other hand, under the Japanese FIT system, PKS trade from Malaysia to Japan can be active in the future. The governments of China and Japan are promoting biofuels from nonfood resources. Japanese research institutes, universities, and private companies have accumulated R&D technologies to reduce the production cost for nonfood resource-based biofuels. Therefore, the technical transfer from Japan to China is a crucial factor in the promotion of biofuel production from nonfood resources in China and other Asian countries. Development of biodiesel production from nonfood resources is not active in Malaysia, and there seems to be no room for technical cooperation from Japan to Malaysia. However, Japanese nongovernmental organizations play a crucial role in judgment to implement the Malaysian biodiesel target, and it is crucial that technical cooperation be exercised in order to implement biodiesel programs in Malaysia.

The government of Japan introduced its own sustainable criteria for bioenergy, but the government of Malaysia did not. The European Union is a crucial palm oil and biodiesel trade partner for Malaysia. Therefore, it is reasonable for Malaysian palm oil refiners, including biodiesel producers, to obtain sustainable criteria from the European Union. However, it is not cleared for the government of China to introduce its own or comply with other sustainability criteria for bioenergy. Internationally, the government will be required to introduce its own sustainable criteria or comply with other criteria. Therefore, the Japanese criteria and Malaysian compliance can serve as a good reference for the government of China and other Asian countries.

Trade, technical, and sustainability criteria cooperation are necessary to promote biofuel developments in China, Japan, and Malaysia. All Asian countries have turned their focus toward expanding biofuel production; however, promoting the sustainable criteria will become important for Asian countries. Therefore, international cooperation for sustainable criteria from Japan and Malaysia can be crucial not only for China, but also for other Asian countries.

References

Chang, S., L. Zhao., Timilsina G. R., and X. Zhang. 2012. Biofuel development in China: Technology options and policies needed to meet the 2020 target. *Energy Policy* 51: 64–79.

Food and Agricultural Organization of the United Nations (FAO). *FAOSTAT.* http://faostat.fao.org/ (accessed 01/20/17).

F.O. Licht. 2016a. 2017 world biodiesel demand may reach record level. London.

F.O. Licht. 2016b. World ethanol production to remain stable in 2016. London.

International Sustainability and Carbon Certification (ISCC). ISCC System. https://www.iscc-system.org/

Koizumi, T. 2009. *Biofuel and World Food Markets* (in Japanese). Tokyo, Nourin Toukei Kyokai.

Koizumi, T. 2013. Biofuels and food security in China and Japan. *Renewable and Sustainable Energy Reviews* 21: 102–109.

Koizumi, T. and K. Ohga. 2007. Biofuels policies in Asian countries: Impact of the expanded biofuels programs on world agricultural markets. *Journal of Agricultural & Food Industrial Organization* 5(2): 1190–1190.

Lim, S. and K. T. Lee. 2012. Implementation of biofuels in Malaysian transportation sector towards sustainable development: A case study of international cooperation between Malaysia and Japan. *Renewable and Sustainable Energy Reviews* 16: 1790–1800.

Malaysian Palm Oil Board (MPOB). 2015. *Malaysian Oil Palm Statistics 2015,* Selangor, Malaysia.

Ministry of Economy, Trade and Industry. 2009. General Energy Statistics 2008FY. Tokyo, http://www.enecho.meti.go.jp/info/statistics/jukyu/index.htm (accessed 01/20/17).

Ministry of Economy, Trade and Industry. 2016. Feed in tariffs (FIT) scheme for Fiscal Year 2016. Tokyo, http://www.enecho.meti.go.jp/category/saving_and_new/saiene/kaitori/kakaku.html (accessed 01/23/17).

Qiu, H., L. Sun, Huang, J., and S. Rozelle. 2012. Liquid biofuels in China: Current status, government policies, and future opportunities and challenges. *Renewable and Sustainable Energy Reviews* 16: 3095–3104.

Song, A., G. Pei., F. Wang., Wan, D., and C. Feng. 2008. Survey for fuel biofuel feedstock multiple production. *Academic Report of Agricultural Process* 24(3): 302–307.

Tan, C. 2007. *Palm biodiesel in Southern Asia.* Rabobank International, Singapore.

U.S. Department of Agriculture, Foreign Agricultural Service (USDA-FAS). 2014. *China-Peoples Republic of Biofuels Annual 2014.* http://gain.fas.usda.gov/Recent%20GAIN%20Publications/Biofuels%20Annual_Beijing_China%20-%20People%20Republic%20of_11-42014.pdf (accessed 01/25/17).

U.S. Department of Agriculture, Foreign Agricultural Service (USDA-FAS). 2015. *China-Peoples Republic of Biofuels Annual 2015.* https://gain.fas.usda.gov/Recent%20GAIN%20Publications/Biofuels%20Annual_Beijing_China%20-%20Peoples%20Republic%20of_9-3-2015.pdf (accessed 01/28/17).
U.S. Department of Agriculture, Foreign Agricultural Service (USDA-FAS). 2016a. *Japan Biofuels Annual.* https://gain.fas.usda.gov/Recent%20GAIN%20Publications/Biofuels%20Annual_Tokyo_Japan_8-26-2016.pdf (accessed 01/12/17).
U.S. Department of Agriculture, Foreign Agricultural Service (USDA-FAS). 2016b. *Malaysia Biofuels Annual.* https://gain.fas.usda.gov/Recent%20GAIN%20Publications/Biofuels%20Annual_Kuala%20Lumpur_Malaysia_7-27-2016.pdf (accessed 01/15/17).
U.S. Department of Agriculture, Foreign Agricultural Service (USDA-FAS) Production, Supply and Distribution Online. 2017. http://www.fas.usda.gov/psdonline/psdQuery.aspx (accessed 01/23/17).
Wang, Q. A. 2011. Time for commercializing non-food biofuels in China. *Renewable & Sustainable Energy Reviews* 15(1): 621–629.
Wang, F., Xing, X. R., and C. Z. Liu. 2009. Biofuel in China: Opportunities and challenges. *In Vitro Cellular & Developmental Biology Plant* 5(3): 342–349.
Zhou, A. and E. Thomson. 2009. The development of biofuels in Asia. *Applied Energy* 86: S11–S20.

chapter three

Biofuel production from algal biomass

Jonah Teo Teck Chye, Lau Yien Jun, Lau Sie Yon, Sharadwata Pan, and Michael K. Danquah

Contents

3.1 Introduction

Today, approximately 88% of worldwide energy consumption is a consequence of the combustion of fossil fuels. This may be because of its dominant exploitation in the transportation sector, which makes up for 58% of the total fossil fuel usage (Brennan and Owende, 2010; Gaurav et al., 2017). Nevertheless, exhaustive consumption of fossil fuels leads to adverse impacts on human health and triggers environmental concerns. In particular, there is an imminent threat of global warming due to extraordinary emission of greenhouse gases (GHGs), environmental pollutants, and devastation of flora and fauna habitats due to oil resources exploration and subsequent utilization. During past decades, biofuels have been widely used as a viable alternative to fossil fuel resources, thanks to the crisis of gradual oil resources depletion, as well as due to an increase in worldwide fuel demand and consumption. The term "biofuel" refers to a fuel that is formed directly or indirectly from the biomass feedstock. Biomass is defined as biological materials that are either energy crops, wastes and by-products from agricultural and forestry, manure or municipal solid wastes, or any other material containing microbial biomass. Charcoal, fuelwood, bioethanol, biomethane, biohydrogen, and biodiesel are some examples of biofuels that are found in continuous production (Saladini et al., 2016). Figure 3.1 illustrates the feedstock sources and products of biofuels.

The first generation of biofuels, such as bioethanol, biobutanol, and biodiesels, is usually produced using edible, conventional crops (Ullah et al., 2015). They may be produced either by utilizing feedstock via starch fermentation of crops such as wheat, barley, potato, corn, sugar beet, and sugarcane (mostly used), or chemically by using rapeseed, sunflower, soybeans, palm, coconut, and animal fats as feedstocks (Lee and Lavoie,

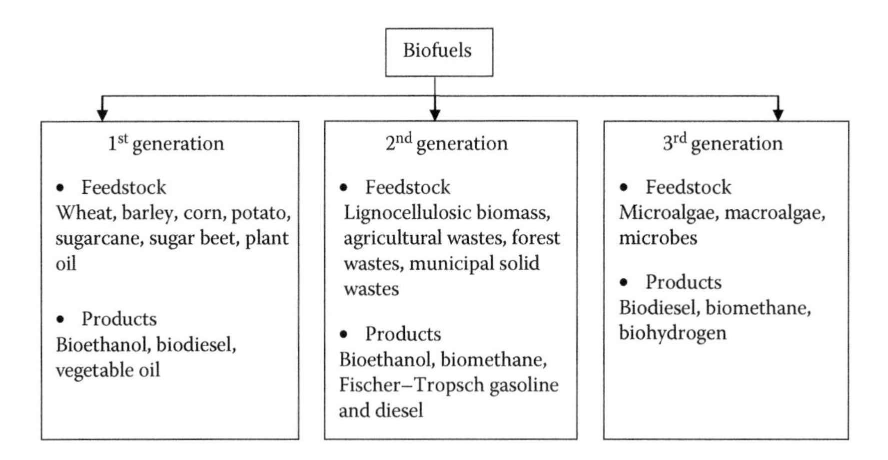

Figure 3.1 Classification of biofuel production: the three generations of biofuels.

2013; Maity et al., 2014). The important characteristics of first-generation biofuels include their ability to be blended with petroleum-based fuels and their efficiency in internal combustion engines, as well as the compatibility with flexible fuel vehicles (FFVs). However, the principal disadvantage of using food-based crops as feedstock is the concomitant increase in food prices due to the concealed crisis of food shortage (Naik et al., 2010). Additionally, the crop feedstock requires large agricultural areas to produce sufficient quantities of biomass, which invites competition between food and biofuel production. In terms of environmental issues, this increased agriculture yield and subsequent harvesting may lead to enhanced land clearing, loss in biodiversity due to habitat destruction, water depletion, and air pollution (Brennan and Owende, 2010).

The second generation of biofuels is produced mainly from nonedible lignocellulosic biomass. The chief feedstocks for this kind of biofuels include agricultural and forest wastes, municipal solid wastes, and manures. For example, second-generation bioethanol and biodiesel have been produced from *Jatropha*, cassava, and *Miscanthus* (Maity et al., 2014). According to Lavoie (2016), the second-generation biofuels have an advantage over their first-generation counterparts in terms of possessing significantly inexpensive supplementary feedstocks. Additionally, second-generation biofuel sources do not have direct rivalry with food production because the utilized plants are specifically grown for bioenergy production, are inedible, and ideally make use of infertile lands that are otherwise detrimental for food crop production (Aro, 2016). Nevertheless, the second-generation biofuels too face steep challenges in the form of technical difficulties during the pretreatment process and inefficient conversion of lignocellulosic materials due to their complex structures (Lee and Lavoie, 2013).

In a search for viable and cost-effective alternatives to fossil fuels, past studies have reported superior capabilities of algae-derived biomass for the production of an improved version: the third-generation biofuels. Gaurav et al. (2017) have shown that algae have the ability to produce crude oil, which may then be easily processed to manufacture diesel and gasoline. The particular algae species, which may be categorized as either microalgae or macroalgae, may as well be genetically modified so that the carbon metabolic pathway facilitates the production of important end products like ethanol. While macroalgae such as red and brown algae typically include seaweeds from marine areas, microalgae are mostly blue-green algae, such as dinoflagellates and bacillariophyta. It should be considered that compared to other feedstock, seaweed has high cellulose-to-lignin content ratio. Carneiro et al. (2017) have highlighted the benefits of the production of third-generation over first-generation biofuels including efficient land utilization and the ability to thwart competition with food crops for the latter case. Algae have the ability to produce more energy per acre of land as compared with other conventional feedstock crops such as sugarcane bagasse and corn. For these reasons, microalgae are touted to be the biomass with the maximum potential to act as a substitute for petroleum-derived transport diesel without adversely affecting the food supply and other crop products (Chisti 2007, 2008).

3.2 *Biofuel production from microalgae*

Typically, microalgae are unicellular photosynthetic microorganisms with the capability to store CO_2 and convert it into energy-intensive compounds such as fatty acids and starch (Gambelli et al., 2017). Compared to other traditional biofuel crops, it is a better source for oil extraction (Gaurav et al., 2017), and may be utilized to synthesize a multitude of diverse biofuels such as biomethane (via anaerobic digestion) and biodiesel (using microalgal oil), as well as biohydrogen (via photobiological synthesis) (Chisti, 2007). A wide array of benefits are associated with microalgae including rapid growth rate, enhanced ability to fix greenhouse gases, ability to greatly reduce overall emission in the life-cycle analysis, and increased lipid synthesis capacity (Alam et al., 2012). Microalgae can duplicate their own biomass within a day, are able to achieve a complete growth cycle within a few days (Chisti, 2008), and are deemed to be 10–20 times more productive than typical biofuel crops, such as soybean and palm oil (Narala et al., 2016). While the lipid content in unit dry biomass weight of microalgae may vary for different species, the species volumetric productivity also needs to be considered to assess feasibility and for selecting the best microalgae for biofuel production (Mata et al., 2010).

Table 3.1 shows the comparison of microalgae with other biodiesel feedstocks on their oil content in dry weight, oil yield per hectare per

Table 3.1 Comparison of several biodiesel feedstock sources

Raw material	Oil content (% in biomass dry weight)	Oil yield (L/ha/year)	Land used (m^2/year/kg biodiesel)	Biodiesel productivity (kg biodiesel/ ha/year)
Soybeans	18	636	18	562
Rapeseed	41	974	12	862
Sunflower	40	1070	11	946
Palm oil	36	5366	2	4747
Castor	48	1307	9	1156
Microalgae	70	136,900	0.1	121,104

Source: Mata, T. M., A. A. Martins, and N. S. Caetano. 2010. Microalgae for biodiesel production and other applications: A review. *Renewable and Sustainable Energy Reviews* 14 (1): 217–232.

year, land utilized, water footprint, and biodiesel production efficiencies. The data adequately demonstrate the superiority of microalgae as the most promising alternative over other conventional feedstock resources. According to Ullah et al. (2015), algae have the highest growth rate, which makes it a viable source of biomass. Moreover, because it mostly contains lipid oil, it possesses the highest oil yield and annual biodiesel productivity per year in comparison with other feedstocks. Although it carries out photosynthesis in a similar way as other plants, it typically possesses more solar energy converters due to several factors: its comparatively simpler cellular structure; ability to grow in aqueous suspensions; easy access to rudimentary nutrients, water, and CO_2; and ability to register optimal growth between 20°C and 30°C (Alam et al., 2012).

Unlike second-generation biofuel feedstocks, microalgae may adapt to thrive in diverse environmental conditions depending on the species variant (Mata et al., 2010). Thus, cultivation of algal biofuels does not necessitate fertile land because it can be grown anywhere with access to reasonable sunlight and water. Subsequently, it can help in minimizing land-use change because it may effectively utilize deteriorated lands that had once been previously used for intensive farming. It can also indirectly diminish habitat losses for native species (Correa et al., 2017); can be cultivated in wastewaters, thereby saving fresh water resources; and does not constrain pesticides and fertilizers for microalgae cultivation. Correa et al. (2017) have reported that microalgal cultivation systems are less likely to cause eutrophication because fertilizer runoff is easily controlled via wastewater recycling in the system. This is in contrast to probable soil erosion or fertilizer runoff into water sources as noticed in case of first- and second-generation biofuels. Production of biofuel from microalgae includes several steps: culturing the algae, recovery or harvesting, and downstream processing to extract the metabolites from the biomass (Molina Grima et al., 2003).

3.3 Microalgal cultivation technologies

In addition to nutrients such as phosphorus, nitrogen, potassium, zinc, and calcium, microalgal cultivation requires water, carbon dioxide, and sunlight to produce biomass through photosynthesis, by transforming solar energy into organic chemical energy stored in the microalgal cells. Three major types of microalgae cultivation techniques have been established based on their growth conditions: photoautotropic, heterotrophic, and mixotrophic cultures. Photoautotrophic cultivation utilizes light as the sole energy source that is converted to chemical energy through photosynthetic reactions. Conversely, heterotrophic production requires organic carbon materials as carbon and energy source to simulate growth. In mixotrophic cultivation, the organisms are able to thrive either autotrophically or heterotrophically, depending on the concentration of organic compounds and available light intensity. Among the aforementioned methods, photoautotrophic production is most widely used because it is economically feasible and suitable for large-scale algal biomass production. The choice of microalgal cultivation system from the most commonly available ones, whether open, closed, or hybrid, is made based on the products and the strains to be cultivated (Gambelli et al., 2017; Narala et al., 2016).

3.3.1 Open-air systems

Open-air systems, also known as open raceway ponds systems, are made of closed-loop recirculation channels, with a typical depth range of 10–50 cm. The ponds are kept shallow specifically to allow adequate solar penetration for efficient photosynthesis of the algal medium. These ponds incorporate paddle wheels, with daylong continuous operation to prevent settling, for effective gas circulation and liquid mixing. It is open to the surroundings, which enables liquid evaporation and temperature regulation of the process (Jorquera et al., 2010). The culture is fed continuously from the front part of the paddlewheel where the flow begins. Inorganic elements such as nitrogen, phosphorus, iron, and silicon constitute some of the bioalgal nutrients. The open-air systems are advantageous because they are relatively cheaper than the photobioreactors systems because they possess lower construction and operational costs, as well as a low energy requirement for mixing. However, these systems face significant challenges too: requirement of larger areas; risk of contamination from outside sources like animals, other microalgae, bacteria, algal grazers, or the weather (Chisti, 2007; Dragone et al., 2010); and control of growth affecting operational conditions like temperature and water evaporation (Narala et al., 2016) (Figure 3.2).

3.3.2 Closed photobioreactor systems

A photobioreactor (PBR) is a system in which microalgal cultivation is carried out in a recycling reactor. Dragone et al. (2010) have detailed different types of PBRs: tubular PBRs, flat PBRs, and bubble column PBRs. A tubular PBR has an array of straight transparent tubes made out of either plastic or glass that are arranged in parallel, and aims to capture sunlight for photosynthesis of the algal medium in the stream. During daytime, PBR will be operational, circulating microalgal mixture from the degassing column through the solar arrays, and while a portion will be circulated back into the degassing column, the other part is harvested. At night, the flow is stopped, the mixture in the degassing column is constantly mixed to prevent settling, and approximately

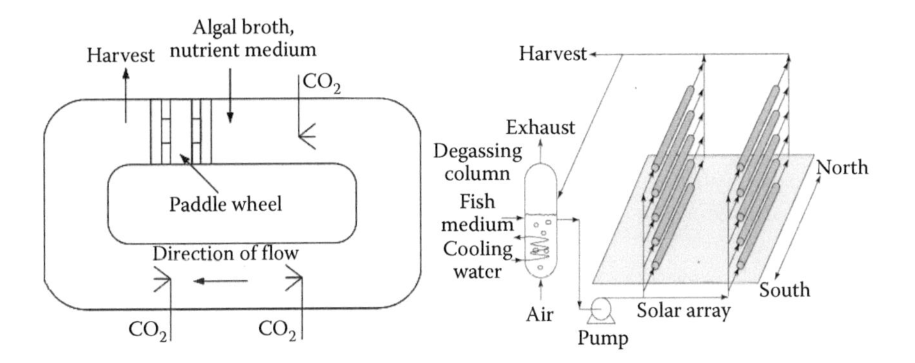

Figure 3.2 Open raceway pond system (left) and tubular photobioreactor system (right). (From Chisti, Y. 2008. *Trends in Biotechnology* 26: 126–131; Jorquera, O., A. Kiperstok, E A. Sales, M. Embirucu, and M. L. Ghirardi. 2010. *Bioresource Technology* 101: 1406–1413.)

25% of the biomass produced during the day is consumed to sustain the cell culture until the next day. Variation in the algal growth is a consequence of the light levels during daytime, growth temperature, culture pH, dissolved oxygen levels in the medium, and the night temperature (Chisti, 2008). Because photosynthesis will be inhibited at 400% of the air saturation value of the dissolved oxygen in the medium, degassing column is necessary for oxygen removal (Molina et al., 2001). According to Jorquera et al. (2010), PBRs have higher volumetric productivity compared to the open-air ponds because they are more efficient in capturing solar energy and require less area. Besides, the PBR system permits controlled cultivation conditions, and the system protects the microalgae from probable contamination. Nonetheless, they are more expensive in terms of capital and operational costs owing to elevated levels of energy consumption and suffer from limited sunlight penetration due to algal attachment. Zhou et al. (2012) established a multilayered PBR structure that can reduce the occupying space, as well as mitigate the sunlight penetration issue.

3.3.3 Two-stage hybrid system

The two-stage hybrid cultivation system combines the distinct growth stages of both open-pond and PBR systems, and cashes in on the advantages of both systems. Beal et al. (2015) reported hybrid systems in which large PBRs continuously provide microalgae for open ponds. Additionally, Zhou et al. (2012) developed an effective hetero-photoautotrophic two-stage microalgae culture system, with dual purposes of low-cost wastewater treatment and production of renewable biofuels from microalgal biomass. The model also displayed improved wastewater nutrient removal efficiency, enhanced lipid yield, lower algal harvesting cost, and carbon dioxide fixation. Ji et al. (2013) studied biodiesel production from a microalgal system coupled with wastewater treatment and showed that the hybrid system can achieve high lipid productivity with an extremely high nutrient excretion rate (up to 99%) from wastewater.

3.3.4 *Lipid induction technique*

According to Chiaramonti et al. (2017), to produce diesel-like biofuels based on microalgal lipids, special cultivation techniques such as the deprivation of nitrogen and phosphorous may improve oil yield and efficacy depending on the microalgal strain. The removal of nutrients inhibits cell division and creates a stressful environment, which enables the microalgae to accumulate lipids and cause an increase in cell size. Strains of microalgae known to exhibit this are *Chlorella vulgaris* and *Nannochloropsis* sp. According to Miao and Wu (2004), *Chlorella protothecoides* grown under heterotrophic culture conditions, by addition of glucose and reduction of nitrogen, causes accumulation of lipids and has yields 3.4 times more as compared to autotrophic cultivation.

3.4 *Microalgal harvesting technologies*

After the microalgal cultivation process, the biomass undergoes a harvesting process in order to separate the biomass from the bioreactor effluent. Harvesting of the microalgal products contribute significantly (20%–30%) to the total cost of biomass production (Li et al., 2008). Thus, it is essential to establish a cost-effective harvesting technology for mass production of biofuels from microalgae. Majorly used techniques for harvesting microalgae include filtration, centrifugation, sedimentation, flocculation, and flotation. The selection of the particular harvesting method depends on the microalgal characteristic, for instance size, cell density, and the desired product value. Dragone et al. (2010) have reported that tiny algal cells, with approximately 2–20 μm diameters and coupled with low concentrations, are responsible for the higher costs associated with harvesting methodologies. The two-stage microalgal harvesting process includes bulk harvesting and thickening. For bulk harvesting, the biomass is separated from the bulk suspension through techniques such as sedimentation, flocculation, and flotation. On the other hand, the objective of the thickening process is to achieve a concentrated form of the slurry using techniques like centrifugation and filtration. Table 3.2 lists the advantages and disadvantages of a variety of harvesting methods.

Table 3.2 Assets and liabilities of different harvesting methods

Harvesting method	Assets	Liabilities
Sedimentation	• Easy to set up and operate • Cheap and energy efficient	• Time-consuming • Inefficient for small microalgal species
Flocculation	• Efficient in biomass harvest • High volumetric capacity and ability • Easy to scale-up	• Flocculants might be expensive • Toxic materials and high dosage will affect water recycling ability and the quality of final extracted products
Flotation	• Applicable on a large scale • Fast and efficient	• Might be expensive due to the added agents • Consumes energy because fine bubbles need to be generated • Toxic materials and high dosage will affect water reuse and the quality of the final extracted products
Filtration	• Applicable on lab scale • Suitable for the processing of larger algal species, such as *Spirulina platensis*	• Long processing time • Scale-up is not suitable due to frequent clogging • Limited capacity and ability to harvest biomass
Centrifugation	• Easy to set up and operate • Efficient and fast because microalgae can achieve up to 20% of total solid matters in a short time	• Scale-up is unsuitable due to increased energy consumption; invites higher costs

Source: Zhu, L., Y. K. Nugroho, S. R. Shakeel et al. 2017. *Renewable and Sustainable Energy Reviews* 78: 391–400.

3.5 Technologies for biomass to biofuel conversion technologies

Available processes that carry out transformation of microalgal biomass to different energy sources are classified as biochemical conversion, thermochemical conversion, chemical reactions, and direct combustion, as shown in Figure 3.3.

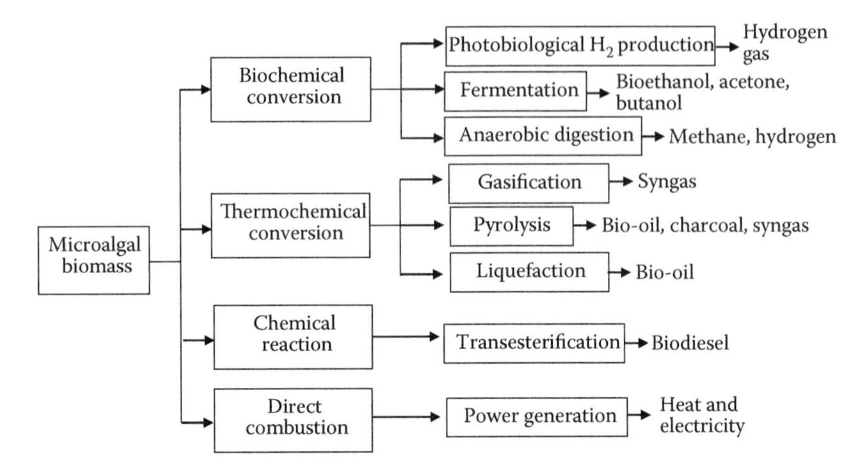

Figure 3.3 Microalgal biomass conversion processes. (Adapted from Dragone, G., B. D. Fernandes, A. Vicente, and J. A. Teixeira. 2010. *Current Research, Technology and Education Topics in Applied Microbiology and Microbial Biotechnology,* ed. A. Mendez-Vilas, 1355–1366. FORMATEX; and Zinnai, A., C. Sanmartin, I. Taglieri, G. Andrich, and F. Venturi. 2016. *Journal of Supercritical Fluids* 116: 126–131.)

3.5.1 *Thermochemical conversion*

Thermochemical processes such as gasification, pyrolysis, hydrothermal liquefaction (HTL), and supercritical fluid extraction may be used to convert microalgal biomass into biofuels. These processes have the potential to convert not only lipids or carbohydrates, but the whole algal biomass into biofuels. Special cultivation conditions to maximize lipid content, such as performing nitrogen deprivation, is not mandatory if the selected processing pathway involves thermochemical conversion (Hallenbeck et al., 2016).

3.5.2 *Gasification*

Gasification is achieved by the reaction of microalgal biomass in a gasifier in the absence of any form of combustion under partial oxidation of air, oxygen, or steam. The conventional gasification process is usually accompanied by other downstream processes such as drying, pyrolysis, combustion, and reduction, and produces hydrogen, methane, and carbon monoxide (collectively known as syngas), alongside certain undesired tar, ash, and solid by-products (Raheem et al., 2015). Gasification may encompass two broad domains: conventional gasification and supercritical water gasification (SCWG). Conventional gasification is achieved at high temperatures of 800–1000°C either in a fluidized or fixed bed. SCWG has an operational temperature range of 375–550°C at 22.1–36 MPa (water critical point), and may be aided by catalysts. Unlike conventional gasification, SCWG does not require an energy-intensive drying process because it directly converts microalgae into the gas product beyond the water critical point (Chen et al., 2015). According to Ni et al. (2006), production of syngas from supercritical water gasification operating at moderately low temperatures has the potential to be more cost-effective as compared to electrolysis or conventional gasification. Algal biomass with moisture content beyond 40% will significantly reduce the gasifier efficiency and produce syngas of lower energy content (Raheem et al., 2015). The SCWG process typically possesses a residence time range from seconds to minutes, allowing for a more flexible reactor volume design based on the required production rate (Chen et al., 2015). Gas yield, carbon yield, and energy recovery for SCWG may be maximized by operating at higher temperatures, placing longer holding times, employing higher water-to-microalgae load ratio, and applying appropriate catalysts (Chen et al., 2015). Gas yield and composition of formed gas as a consequence of gasification also depends on the composition of feedstock. Schumacher et al. (2011) reported that seaweed samples with high carbohydrate and low protein content and large amounts of inorganic salts can produce high amounts of gas rich in hydrogen and methane. They also reported an excessive low carbon monoxide content in the produced gas, i.e., less than 1% (v/v). According to Onwudili et al. (2013), the macroalga *Saccharina* sp. outperformed *Spirulina* sp. and *Chlorella* sp. in SCWG process at 500°C and 36 MPa in an Inconel batch reactor, with the addition of NaOH and nickel-alumina catalyst. The addition of nickel catalyst to the SCWG process reduced the production of hydrogen while increasing methane production as compared to that with no catalyst because it aided in degrading the microalgae and improving methanation reaction (Krishnan et al., 2016). The addition of alkali such as NaOH, KOH, and $Ca(OH)_2$ promoted water-gas shift reaction during gasification. The mechanism is that the alkali reacts with the CO_2 formed as a consequence of gasification, which alters the water-gas shift reaction equilibrium so that it always proceeds in the forward direction, producing more H_2. The addition of alkali also suppressed tar and char formation (Muangrat et al., 2010).

3.5.3 *Pyrolysis*

Microalgae are thermally decomposed by heating in the absence of oxygen or air. Operating conditions normally proceed at atmospheric pressure, with a temperature range of 400–600°C for conventional pyrolysis, up to 800°C for microwave pyrolysis and a minimum of 300°C for catalytic pyrolysis (Chen et al., 2015). Fast or flash pyrolysis is known to have highly intensive heating rates and a short residence time, and maximizes the crude bio-oil yield (Chiaramonti et al., 2017). Conventional or slow pyrolysis produces biochar and maximizes the production of pyrolytic gases such as methane and CO_2 (Suali and Sarbatly, 2012). Feeding a pyrolysis unit strictly requires a drying unit for the biomass in the inlet because it has no moisture tolerance. It has been shown that fast pyrolysis at 500°C (at a heating rate of 600°C s^{-1}) with 2–3 s of residence time yielded 18% and 24% of liquid product yield based on *C. protothecoides* and *Microcystis aeruginosa,* respectively. The biofuel heating value was 29 MJ/kg, an increase of 1.4 times as compared to the case where wood was employed as feedstock (Miao et al., 2004). Miao and Wu (2004) stated that for *C. protothecoides,* heterotrophic growth and fast pyrolysis at 450°C attained a very high bio-oil yield of 57.9%. Wang et al. (2017) performed thermogravimetric analysis (TGA) and detailed a wide range of diverse pyrolysis reaction pathways for lipids, proteins, and carbohydrates. Compared to conventional pyrolysis, microwave-enhanced pyrolysis (MEP) has advantages in terms of fast, targeted, and uniform heating, and easy control. The enhancement in the rate of maximum temperature and pyrolysis temperature is governed by the magnitude of the microwave strength. Hu et al. (2012) observed the effects of microwave on *C. vulgaris* and determined the optimal microwave power to be 1500 and 2250 W attaining a maximum bio-oil yield and biofuel yield of 35.83% and 74.93%, respectively. According to Li et al. (2013), an increment of microwave power from 750 to 2250 W shows an increase of gas yield by 39.45% and decrease in solid residues by 22.05%. They also reported that a microwave power of 1500 W is optimal for pyrolysis with the addition of MgO as catalyst for maximum effect on the product mass per unit of power consumed.

3.5.4 *Liquefaction*

In HTL of algae, the physical and chemical conversion is performed at high temperatures of approximately 250–500°C under high autothermal water pressures of 5–20 bar with or without catalysts, with the microalgae making up to 5%–50% mass fraction of the slurry feed (Chen et al., 2015; Raheem et al., 2015). The products of liquefaction include bio-oil and additional gaseous coproducts of methane. The pressure is kept high to maintain the water in liquid phase, and the residence time is approximately 5–60 min. Chen et al. (2015) reported production of different strains of microalgae based on diverse operating conditions as well as the bio-oil yield and efficacy in terms of higher heating values (HHVs). Similar to SCWG, this process does not require drying of the wet algal biomass, making it an energy-efficient and more economically viable process. The advantage of HTL over SCWG is that it is suitable for wet biomass with high moisture contents. The physical properties of water under these conditions promote the decomposition of macromolecules in the biomass as well as polymerization to form bio-oil (Guo et al., 2015). Microalgal biomass is a better feedstock for HTL than lignocellulosic biomass because thermal decomposition of lignin may form solid residues due to condensation or repolymerization, leading to lower yields and a longer residence time (Demirbas, 2000). The water phase effluent of the HTL process concentrates trace amounts of nitrogen, phosphorus, and potassium (NPK) as well as secondary micronutrients, which may be integrated for microalgal cultivation (Jena et al., 2011). Microalgal liquefaction studies have shown that HTL may only achieve yields of up to 54% in case of protein-rich *Spirulina* sp., without catalyst, which is lower than that of pyrolysis and gasification (Matsui et al., 1997; Suali and Sarbatly, 2012).

3.5.5 Biochemical conversion

Biochemical conversion of microalgae involves the utilization of microorganisms and/or enzymes to break down algae into liquid fuels. These are normally slower and less energy-intensive as compared to the thermochemical conversion ones. The biochemical transformation procedure includes anaerobic digestion, fermentation, and photobiological H_2 production, and often requires pretreatment of the biomass, especially for anaerobic digestion and fermentation.

3.5.6 Anaerobic digestion

Anaerobic digestion is a process through which organic wastes, specifically algal biomass, are converted into biomethane and carbon dioxide with traces of hydrogen sulfide via the breakdown of organic matter by enzymes and/or microorganisms. It is suitable for high-moisture-containing organic matter, including algae (Brennan and Owende, 2010). In general, the process depends on the digester design (batch or continuous), hydraulic retention time (HRT), system microorganism colony, and temperature modes in specific cases of either mesophilic (30–38°C) or thermophilic microbes (49–57°C) (Fasahati et al., 2017). The process may be attributed to three sequential steps: hydrolysis, acetogenesis, and methanogenesis. Because hydrolysis is the rate-limiting step for anaerobic digestion, the vulnerability of the algae to enzymatic attack is an important factor and may be achieved by pretreatment. Although the usual mechanical pretreatment is the general norm for microalgae, they are more suited for thermal pretreatment (Rodriguez et al., 2015). The biomethane production potential of algae highly depends on its composition so that the constituent minerals fulfill the nutrient deficits of the anaerobic microorganisms (Sialve et al., 2009). Ehimen et al. (2013) expressed doubts by touting the anaerobic digestion of single-celled species, such as *Chlorella* sp., as unreasonable and extravagant due to the exorbitant costs associated with the cultivation and harvesting stages. It was reported that filamentous algae such as *Rhizoclonium*, after experimentation, proved to be unproductive with only 62–97 mL CH_4/g VS as compared to *Dunaliella* sp. and *Chlorella* sp. with 440–450 mL CH_4/g VS and 310–370 mL CH_4/g VS, respectively. El Asri et al. (2017) studied three macroalgae: *Caulerpa prolifera, Gracilaria bursa-pastoris,* and *Colpomenia sinuosa* originating from the Marchica Lagoon in Morroco. The results showed that *G. bursa-pastoris* and *C. sinuosa* are potential substrates owing to their capability to produce 86.35 and 74.68 mL/g VS, respectively. In addition, anaerobic bacteria involved in the digestion of *G. bursa-pastoris* seem to be thriving, growing from 0.19×10^6 to 7×10^6 colony-forming units (CFU)/g. Harvesting costs for macroalgae are significantly lowered due to the microalgal size difference, and may possess better economical potentials for industrial-scale anaerobic digestion.

3.5.7 Fermentation

In the fermentation process, the constituent carbohydrates such as sugar, starch, or cellulose are converted to bioethanol. Microalgae with high starch contents are traditionally suitable for fermentation. This enzymatic process involves the hydrolysis of starch to simple sugars, followed by yeast fermentation, producing bioethanol (Milano et al., 2016). Similar to anaerobic digestion, fermentation requires pretreatment to release carbohydrates from the algal cell wall by either ultrasonication or hydrolytic enzymatic conversion to allow for the biomass fermentation (Suali and Sarbatly, 2012; Naveena et al., 2015). The pectinase enzyme group has a high potential for excretion of fermentable sugars present in microalgae. For instance, it has been reported that pectinase from *Aspergillus aculeatus* may achieve a saccharification yield of 79% during hydrolysis of *C. vulgaris* at 50°C after 3 days (Kim et al., 2014). Ethanol production via fermentation involves four principal stages. The first step is glycolysis, characterized by formation of two molecules of pyruvate ($CHCOCOO^-$) and water and hydrogen ions (H^+) as by-products from glucose breakdown. During glycolysis, coenzymes adenosine diphosphate (ADP) and nicotinamide adenine dinucleotide (NAD^+) are reduced to adenosine triphosphate (ATP) and NADH, respectively. Following glycolysis, the second step is the production of acetaldehyde, CO_2, and H^+ from pyruvate, catalyzed by the enzyme pyruvate decarboxylase. The third step involves conversion of acetaldehyde into ethanol anion, aided by coenzyme NADH, which is synthesized during glycolysis. The last step involves the protonation of ethanol anion by hydrogen ions forming ethanol (Suali and Sarbatly, 2012). It is a fact that higher yeast concentrations may improve ethanol production from microalgae. Based on the yeast fermentation studies by Harun et al. (2010) on microalgae *Chlorococum* sp., the highest ethanol concentration of 3.58 g L^{-1} could be obtained when the yeast concentration was at its peak at 14.25 g L^{-1}. Harun et al. also concluded that lipid-extracted microalgae has higher ethanol content than dried or pure microalgae. Ji et al. (2016) studied the use of thermophilic bacterium *Defluviitalea phaphyphila* Alg1 for bioethanol fermentation and concluded that high ethanol yields of 0.44 g/g glucose, 0.47 g/g mannitol, and 0.3 g/g alginate were attained at 60°C.

3.5.8 *Transesterification*

Transesterification is characterized by the conversion of microalgal biomass to biodiesel and is subcategorized as chemical changeover of algal biomass. Triglyceride or lipid reacts with a monoalcohol with the additive of catalysts in the form of acid, alkali, or enzymes to produce fatty acid methyl ester (FAME) and glycerol (Milano et al., 2016). The process has garnered a lot of attention in recent times for biodiesel production, but its economic feasibility is still limited due to the high enzyme cost. Razzak et al. (2013) reported that alkali-catalyzed transesterification is the most commonly used method for biodiesel production. If the raw materials are rich in free fatty acids, they may react with the alkaline catalysts and undergo saponification, which is undesirable. High water content may hydrolyze the triglycerides to diglycerides, which in turn generates more free fatty acids, demanding pretreatment to eliminate the acidic materials. The transesterification process may either be conventional or *in situ* (direct). While conventional transesterification demands drying, lipid extraction before the actual process, and a subsequent purification, *in situ* transesterification omits the lipid-extraction stage (Milano et al., 2016). *In situ* transesterification is a simpler and cheaper form of algal biodiesel manufacturing, compared to the conventional method. *In situ* studies of transesterification of *C. vulgaris* by Velasquez-Orta et al. (2012) showed that for a methanol-to-lipid molar ratio of 600:1 and catalyst-to-lipid ratio of 0.15:1, an alkaline catalyst (NaOH) could outperform an acid catalyst (H_2SO_4), attaining higher conversion into FAME (77.6 wt%) for low residence time of 75 min at room temperature. This was the case when acid catalyst-to-lipid ratio of 0.35:1 (≈96.8 wt% conversion) was achieved for longer reaction times up to 20 h. In a similar study, *in situ* transesterification of *Chlorella* sp. with 20 wt% sulfuric acid showed a 96%–98% yield of FAME within 4 h (Viegas et al., 2015). The study showed that higher weight compositions of sulfuric acid produced higher FAME yields, but it is economically unfeasible because sulfuric acid is expensive. Teo et al. (2016) investigated a nanocatalyst of calcium methoxide, $Ca(OCH_3)_2$, with 3 wt% of catalyst loading, a methanol-to-lipid molar ratio of 30:1, reaction time of 3 h, at hydrothermal temperature of 80°C, and managed to achieve 96%–99% of FAME yield.

3.5.9 Direct combustion

For direct combustion, algal biomass is burnt in the presence of air in a furnace, boiler, or steam turbine to convert the chemical energy of the algal biomass into heat or electricity. Direct combustion of microalgae is only feasible with moisture content of less than 50% of its dry weight (Goyal et al., 2008). However, the process suffers from a limitation because it requires pretreatment of algae, for instance drying, which may lead to a somewhat diminished energy efficiency. Instead, cofiring of algal biomass and coal is an effective way to generate electricity. During cofiring of algal biomass, the algae is cultivated and fed with the CO_2 recycle stream from the power plant, and the cultivated algae is fed into the power plant along with coal for combustion (Milano et al., 2016).

3.5.10 Ultrasonication to aid biofuel yields

The ultrasonification process necessitates the incorporation of acoustic energy with frequencies of 10 kHz to 20 MHz, which is applied to form constantly expanding and shrinking microbubbles. The microbubbles destabilize as the acoustic energy goes beyond a certain threshold. When it resonates with the natural frequency of the microbubbles, it collapses, causing high-velocity microjets of more than 100 ms^{-1}, which carry more than 100 MPa of pressure toward the biomass surface. This causes cavitation and the process enhances biochemical reactions and can aid in maximizing lipid yield during lipid extraction of microalgae, or cause rupture of the cell wall to aid the hydrolysis step for fermentation to bioethanol or anaerobic digestion (Naveena et al., 2015). According to Suali and Sarbatly (2012), ultrasonication is able to reduce lipid extraction time by one-tenth of that of conventional methods, and achieve 50%–500% more yield as well.

3.6 Biochemical characterization of different algal species

Scott et al. (2010) have reported the existence of nearly 300,000 species of algae, with varied oil contents and growth rates. Microalgal biomass is mainly constituted of lipids, carbohydrates, and proteins. Different microalgal species have significantly different compositions, where the average lipid oil content typically varies between 8% and 31%. This lipid oil can be extracted and converted into biodiesel. Mendoza et al. (2015) reported that microalgae have a high variability of lipid extractability, which is related to the resistance of the cell walls to physical and chemical extraction treatments. Del Río et al. (2017) reported that *Chlorococcum oleofaciens* and *Pseudokirchneriella subcapitata* have daily maximal production of 110 mg fatty acids L^{-1} and 160 mg fatty acids L^{-1}, respectively. It was also reported that the strains react differently in terms of nitrate availability, and the species *P. subcapacitata* has a more acceptable fatty acid profile for biodiesel production because it has a low nitrogen stress threshold that activates fatty acid accumulation. Quantification of lipid content in microalgae may be carried out by gravimetric methods, which is, however, a time-consuming process. Balduyck et al. (2015) achieved lipid staining with Nile red followed by lipid quantification using a spectrofluorometric device. However, this method is argued to be species-dependent and requires a species-based optimization. Table 3.3 illustrates the biochemical compositions of various types of microalgae, including their lipid, protein, and carbohydrate contents on the dry mass basis (Alam et al., 2012; Dragone et al., 2010).

3.7 *Biodiesel from algal biomass*

Biodiesel, an alternative to petroleum-based diesel, is a low-emission fuel manufactured from renewable biomass and waste lipids. Lipid content in the algal biomass drives the quality and yield of biodiesel produced, and plays an important role in determining its potency to substitute petroleum-based diesel (Milano et al., 2016). Algae-based biodiesel may also perform as petroleum diesel because it does not contain sulfur and has reduced emissions of particulate matter, CO, hydrocarbons, and SO_x, but may have an increase in NO_x depending on the engine type (Mata et al., 2010). However, biodiesel has lower oxidative stability as compared to petroleum-based diesel along with cold-weather performance issues, which impede its ability to act as a potent and suitable jet fuel. Algal biodiesel may cause potential contamination of the esters with metals, toxins, chlorophyll, or catalyst poisons, and has a difference in performance criteria based on different compositions as compared to petroleum-based diesel, with end-product variability for different strains of microalgae (U.S. Department of Energy, 2016). Milano et al. (2016) suggested an improvement in the oxidative stability of biodiesel by using microalgal strains with higher oleic acid contents in the microalgal fatty acids, which includes *Chlorella* sp., *Saccharomyces cerevisiae, Botryococcus* sp., *Scenedesmus* sp., *Picochlorum* sp., and *Nannochloropsis oculata.*

Lipid, or more specifically triacylglyceride (TAG), composition, is a crucial factor in the selection of microalgal species for biodiesel production (Chernova and Kiseleva, 2017). Low lipid contents of algal biomass; high cost of nutrient fertilizers; high energy demands from biomass harvesting, drying, and conversion processes; and quality of biodiesel

Table 3.3 Biochemical composition of microalgae expressed in % dry weight

Microalgal species	Lipid	Protein	Carbohydrates
Anabaena cylindrica	4–7	43–56	25–30
Botryococcus braunii	33	–	–
Chaetoceros muelleri	34	–	–
Chaetoceros calcitrans	15–40	–	–
Chlamydomonas rheinhardii	21	48	17
Chlorella emersonii	25–63	–	–
Chlorella protothecoides	15–58	–	–
Chlorella pyrenoidosa	2	57	26
Chlorella sorokiniana	19–22	–	–
Chlorella vulgaris	14–22	51–58	12–17
	5–58	–	–
Chlorococcum sp.	20	–	–
Dunaliella bioculata	8	49	4
Dunaliella primolecta	23	–	–
Dunaliella tertiolecta	11	–	–
	18–71	–	–
Ellipsoidion sp.	14–20	39–61	14–18
Euglena gracilis	14–20	–	–
Haematococcus pluvialis	25	–	–
Isochrysis galbana	7–40	–	–
Isochrysis sp.	7–33	–	–
Nannochloropsis oculata	29.7	–	–
	23–30	–	–
Nannochloris sp.	20–56	–	–
Neochloris oleoabundans	29–65	–	–
Pavlova salina	31	–	–
Pavlova lutheri	36	–	–
Phaeodactylum tricornutum	18–57	–	–
Porphyridium cruentum	9–14	28–39	40–57
Pyrmnesium parvum	22–39	28–45	25–33
Scenedesmus dimorphus	16–40	8–18	21–52
Scenedesmus obliquus	12–14	50–56	10–17
	11–55	–	–
Scenedesmus sp.	20–21	–	–
Spirogyra sp.	11–21	6–20	33–64
Spirulina maxima	6–7	60–71	13–16

(*Continued*)

Table 3.3 (Continued) Biochemical composition of microalgae expressed in % dry weight

Microalgal species	Lipid	Protein	Carbohydrates
Spirulina platensis	4–9	46–63	8–14
	4–17	–	–
Synechoccus sp.	11	63	15
Tetraselmis maculate	3	52	15

produced are some of the pertinent challenges that mar biodiesel production (Mehrabadi et al., 2016). Past studies have shown that lipid induction may be achieved by applying stressors, such as either by limiting the nitrogen and/or phosphorus content or by removing it entirely, which results in lipid accumulation in certain strains of microalgae. As for diatomic microalgae, silicium is a potent stressor for lipid growth. Other potential stressors include irradiation, low light intensity, dark exposure, and shortened light period. Ultraviolet irradiation and combined effects were also reported to induce TAG agglomeration (Chernova and Kiseleva, 2017). Biodiesel quality is determined based on the cetane number (CN), calorific value, melting point, and viscosity, all of which significantly impact the fuel performance in a diesel engine. It should be noted that the values of the aforementioned quantities are proportional to the chain length. The higher the degree of saturation of fatty acids in the biodiesel, the greater is its oxidation stability, viscosity, and CN (Chung et al., 2017). Pure biodiesel and biodiesel blends have lower CN and heating values compared to petroleum-based diesel, which causes a drop in the engine performance. Engine tests on pure biodiesel (B100) have shown that engine power output was 6% lower than that of conventional diesel, but with a significant drop in CO emissions (Piloto-Rodríguez et al., 2017; Tuccar and Aydin, 2013).

Currently, the increased cost factor for biodiesel production from algae is still an issue and widescale commercialization is yet to be achieved. Furthermore, biodiesel manufacturing generates the by-product glycerin, which has already reached the market saturation point, causing depreciation of its market value and affecting the commercialization potential of biodiesel (Galan et al., 2009; Patil et al., 2017). Patil et al. (2017) have studied a glycerin-free noncatalytic transesterification process using supercritical methyl acetate, and based on initial outcomes have commented on its high efficiency in producing relatively stable algal biodiesel. Chung et al. (2017) proposed multiple strategies to improve microalgal cultures, lipid accumulation, and improvement of biodiesel quality, which includes bioengineering of microalgal strains as well as other molecular strategies.

3.8 Bioethanol from algal biomass

Algae, in general, are rich in lipids and carbohydrates. Algal cell walls are constituted of cellulose, mannans, xylans, and sulfated glycans, which can be broken down either chemically or through enzymatic action to produce simple sugars which are then converted to bioethanol, using bacteria or yeast, under anaerobic conditions (Sirajunnisa and Surendhiran, 2016). Pure ethanol is a favorable fuel alternative that is comparable to gasoline, having approximately 66% of the energy of gasoline by volume, but with a higher octane number (Nigam and Singh, 2011). A higher octane number of ethanol indicates that it may burn at higher compression ratios with a faster combustion time, which suits the engine better. The heat of vaporization of ethanol is significantly higher than gasoline, boasting a higher volumetric efficiency and power output (Zabed et al., 2017). Gasoline has the tendency to explode and burn accidentally, depositing gum (on storage surfaces) and carbon (in the combustion chamber), and eject air pollutants, thereby making it ecologically unsafe. Ethanol burns cleaner, producing less CO, hydrocarbons, and nitrogen oxides, and is even hypothetically compatible with compression-ignition (CI) engines as well (Agarwal, 2007).

Production of algal bioethanol covers several broad mechanisms: algal selection and culturing, pretreatment processes, liquefaction, saccharification, fermentation, and purification by distillation (Lee and Lee, 2016). Production of bioethanol from microalgae is reported to have almost double the productivity of that of sugarcane and nearly five times that of corn (Zhang et al., 2014). Bioethanol manufacturing has seen a massive enhancement in the past few years and production capacity rose from 1 billion liters (in 1975) to 86 billion liters (in 2010), with a projection of 160 billion liters by 2020 (Khambhaty et al., 2012). Conversion of algal biomass to bioethanol is especially gaining more attention due to the disadvantages and conflicts between previous generations of biofuels with respect to sustainability, usage of food crops as energy crops, and land-usage modifications. Brown algae has generated a lot of interest and has been progressively and collectively studied for industrial-scale applications (Fasahati et al., 2015; Fasahati and Liu 2012, 2015). It contains approximately 30–67 dry wt% of carbohydrates, with principal constituents being alginate (up to 60% of total sugars), laminaran, and mannitol (Lee and Lee, 2016; Song et al., 2015). According to Zhang et al. (2014), while the latter two components are easily consumed by microbes, the same cannot be ascertained about alginate. This motivated them to study different inoculum strains that can degrade alginate and produce higher bioethanol yields through fermentation. It has been reported that metabolic engineering of alginate-metabolizing *Sphingomonas* sp. A1 cells could potentially produce bioethanol from alginate (Takeda et al., 2011). Lee and Lee (2016) extensively reported advances in bioethanol production from brown algae including synthetic yeast platform and a wide range of studies of metabolic alterations focusing on a variety of microbial strains to better convert and aid bioethanol fermentation of brown algae.

3.9 Biogas and biohydrogen from algal biomass

Biogas with methane and carbon dioxide as chief constituents is produced mostly by anaerobic digestion and as a coproduct in anaerobic fermentation. In its current state, biogas production from algae has more limitations than benefits. Biomass algae from water resources that are exposed to high levels of eutrophication also pose technical difficulties in the biogas production systems. This may be because it is difficult to assess different substrates and microalgal species for characterization and process optimization purposes. Limitations associated with the utilization of algae for biogas production include production of substances or compounds that are toxic to anaerobic microbes, and an undesirable C:N ratio in the microalgae biomass, which may vary depending on the species studied (Dębowski et al., 2013). Dębowski et al. (2013) reported that the C:N ratio in anaerobic digestion tanks is optimal at the range of 20:1 to 30:1, and will experience lower efficiencies at ratios below 10:1. They also suggested an improvement by adding a secondary biomass with a high organic carbon concentration, commonly referred to as anaerobic codigestion. Anaerobic codigestion has been carried out by Yen and Brune (2007), who have improved the C:N ratio of 6:1 of an algal slurry to 20–25:1. In addition, Yen and Brune have demonstrated betterment for the methane production procedure. Akunna and Hierholtzer (2016) studied codigestion of brown marine macroalgae *Laminaria digitata* with green pea and reported that the introduction of the algae into the digester affected methanogenesis and that there was excessive volatile acid accumulation causing inhibition of methanogens and process destabilization. The study concluded that certain seaweed compounds may inhibit anaerobic microbes, even at trace concentrations, thereby limiting the ratio of seaweed to cobiomass to achieve process stability. Despite these limitations, algal biomass still remains relevant for biogas production, which can be attributed to several factors: high contents of polysaccharides and lipids in most marine microalgae, absence of lignin with low cellulose content, simplicity of transformation, and higher growth rates and lower land usage as compared to the lignocellulosic biomass. As an added benefit, the solid by-product of anaerobic digestion may also be used as a soil additive (Dębowski et al., 2013; Vergara-Fernandez et al., 2008).

Biohydrogen is a promising fuel source because of its high energy density (energy content of 2.2 lb. of hydrogen is the same as 6.2 lb. of gasoline) (Nagarajan et al., 2017). Among all fuels, hydrogen has the highest gravimetric energy density of 142 MJ/kg and a high conversion efficiency of approximately 94% in a fuel cell, as compared to 60% efficiency of the internal combustion engines (Hallenbeck et al., 2016). Certain microalgae possess the enzyme hydrogenase, which reduces protons to hydrogen. This may be achieved by cultivating the microalgae in the dark, thereby rendering it anaerobic, and then exposing the culture to light. However, this method of hydrogen production is challenging due to the short life of the reaction because oxygen generated due to photosynthesis deactivates the sensitive hydrogenase enzyme almost immediately (Hallenbeck, 2011; Hallenbeck et al., 2016). Strategies to mitigate or alleviate the oxygen problem enable nutrient deprivation of sulfur, which leads to prolonged periods of hydrogen production, because it limits photosynthesis. However, this may prove to be counterproductive because photosynthetic efficiency is equally important for algal growth. A better strategy is to improve the tolerance of the hydrogenase toward oxygen, which requires further technological advances in metabolic engineering (Hallenbeck et al., 2016). Past studies have utilized microbial process methodologies such as dark fermentation, microbial electrolysis, biophotolysis, photofermentation, and the biological water-gas shift reaction to produce hydrogen (Hallenbeck, 2011; Ni et al., 2006).

3.10 Integrated algal biomass production

A biorefinery concept represents an integration of different industrial units for mutual production benefits, with a vision on superior sustainability and economic benefits. This integration may include interactions such as utilization of flue gas effluents for CO_2 sequestration and simultaneous microalgal growth, as well as recycling of nutrients from wastewater streams or anaerobic digestion effluents.

Li et al. (2008) suggested that integration and utilization of a flue gas or flaring gas streams containing high CO_2 contents can be fed to a microalgal cultivation system allowing for CO_2 mitigation because certain species have high tolerance to CO_2 and demonstrate high CO_2-capturing capability during photosynthesis. In November 2006, U.S. Green Energy Technology Company in cooperation with Arizona Public Service Company commenced a microalgal production system in which flue gas generated from a 1040 MW capacity power plant was fed to the microalgal cultivation that had produced approximately 5000–10,000 gallons per acre per year (Sirajunnisa and Surendhiran, 2016). Direct combustion of microalgae to generate heat or electricity also produces flue gas, which may also be integrated into the cultivation system.

There are significant lacunae between laboratory-scale studies of lipid yield and achievable industrial-scale yield, mainly due to the nutrient requirements for the growth of algal biomass. The necessary nutrients, mainly phosphorus and nitrogen, are expensive and industrial exploitation of fertilizers for algal cultivation may induce competition in the food production chain. Generally, microalgae are known to thrive in wastewaters, especially in effluents from sewage treatment plants. Therefore, an integrated biomass production in conjunction with wastewater treatment plants may alleviate the need for fertilizer use (Hallenbeck et al., 2016). Wastewater streams that are integrated with the microalgal cultivation system are able to remove nitrogen and phosphorus from effluents to prevent algal blooms and eutrophication (Li et al., 2008). However, the same cannot be ascertained for cases when the effluent is purged to the environment. In the case of anaerobic digestion of algal biomass, plant nutrients are the leftover residues. These nutrients can be subsequently recycled and utilized for the algal cultivation, while simultaneously reducing their water footprint as well as eutrophication (Moreira and Pires, 2016). Guldhe et al. (2017) studied the applicability of aquaculture wastewater for cultivation of algal biomass of *Chlorella sorokiniana* using heterotrophic cultivation. They concluded that the wastewater was highly suitable and generated high lipid, carbohydrate, and protein yields in the microalgae. For the HTL process, the liquid effluent may be integrated for nutrient recycling as well because it has high nitrogen, phosphorus, and potassium contents (Figure 3.4).

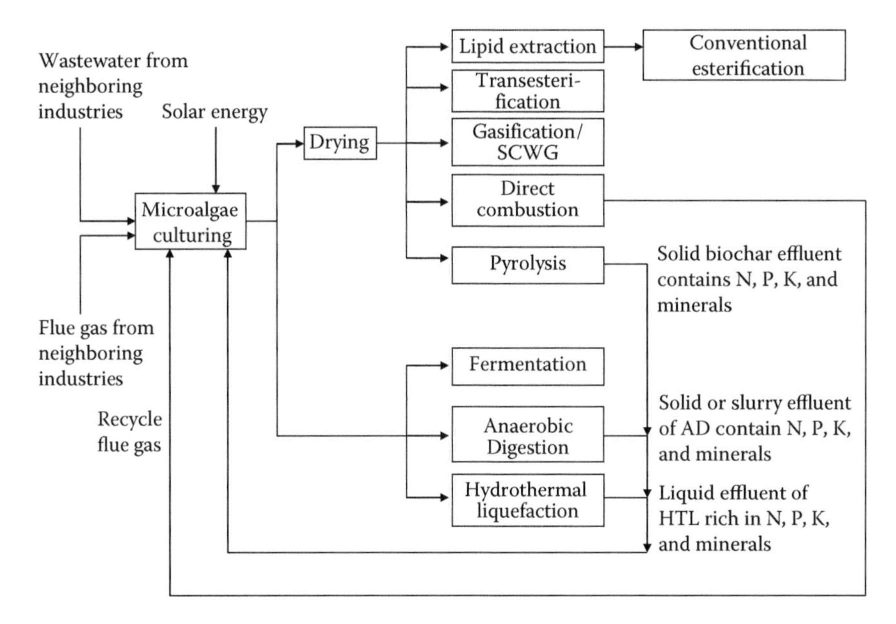

Figure 3.4 Potential process integration options for different methods of biofuel production.

3.11 *Future outlook and commercial prospects of algal biofuels*

Currently, a plethora of imminent challenges is looming over the actualization of commercial feasibility of algal biofuel production. These include species selection based on their characterization that best suits the production of biofuel, extraction of coproducts, attainment of higher photosynthetic efficiencies for cultivation systems, reduction of evaporative losses and CO_2 diffusion losses (for open ponds), and ensuring a positive net energy balance after incorporating a life-cycle analysis of the product, as well as lack of data due to insufficient algal biofuel production plants (Brennan and Owende, 2010). Chlorophyceae have been reported to be the most productive strains for algal cultivation, which grow well in open pond systems, have a high tolerance to CO_2, and can accumulate lipids under low temperatures of 10°C (Neofotis et al., 2016; Su et al., 2017).

The idea of integration of algal cultivation in a biorefinery for supplementary usage may be beneficial not only in terms of carbon sequestration, but also reduction in carbon taxation costs. Nutrient recycling may also diminish fertilizer as well as wastewater treatment costs. According to Judd et al. (2017), a cost reduction of 35%–86% could be realized from integrated CO_2 and nutrient recycling from waste streams. Fasahati et al. (2017) modeled and studied the economics of heat and power production based on anaerobic digestion of brown algae *Laminaria japonica* with a plant production scale of 400,000 dry tons/year. Based on sensitivity studies, Fasahati et al. outlined that loading of solids, anaerobic digestion yield, and retention time have the highest impact on the commercial viability of this process.

Collective studies have determined that microalgal biofuel production is yet to achieve industrial commercialization due to significant production costs. Based on the current state of the art, the most significant bottleneck is the exorbitant costs associated with algal cultivation and harvesting (Laurens et al., 2017). Doshi et al. (2016) claimed that microalgal biofuel is not yet competitive with fossil fuels, although they may have better market prospects as aviation fuels. Future works and research concerning the domain of algal biofuel production should aim to improve algal biomass productivity, energy usage, water usage, nutrient requirement, production time, and better integration of biorefineries to reduce the production costs and constraints, all of which may go a long way in realizing full commercialization of algal biofuels. Dutta et al. (2016) emphasized the need to attain higher-value coproducts in addition to the biofuel to make the process economically viable.

3.12 Conclusion

In this chapter, the first, second, and third generations of biofuels have been reviewed and compared, and the benefits of algae as a crucial feedstock for biofuel production have been detailed. A sequential review has been done for the entire process of biofuel production, starting from cultivation, harvesting, pretreatments, and the conversion processes. Algal cultivation technologies have been discussed including their advantages and disadvantages. A wide range of harvesting technologies have been reviewed based on their respective benefits and pitfalls as an aid for technology selection. Conversion technologies of the harvested algal biomass have been outlined, with several optimization recommendations for each process. Characterization of algal biomass based on their lipid, carbohydrate, and protein contents have been reviewed, and characterization profiles of different species have been compiled based on previous studies. This chapter also discusses the suitability and applicability of biodiesel, bioethanol, biogas, and biohydrogen as biofuel products, comparing their performance with fossil fuels. Furthermore, there is a focus on the biorefinery concept whereby integration of several industrial units or integration within a biomass production plant is possible by recycling of nutrients and CO_2. Future outlooks and commercial prospects of algal biofuel have been reviewed and a few recommendations, based on the bottlenecks of the current state of the art, have been stated.

References

Agarwal, A. K. 2007. Biofuels (alcohols and biodiesel) applications as fuels for internal combustion engines. *Progress in Energy and Combustion Science* 33 (3): 233–271.

Akunna, J. C. and A. Hierholtzer. 2016. Co-digestion of terrestrial plant biomass with marine macro-algae for biogas production. *Biomass and Bioenergy* 93: 137–143.

Alam, F., A. Date, R. Rasjidin et al. 2012. Biofuel from algae- Is it a viable alternative? *Procedia Engineering* 49: 221–227.

Aro, E.-M. 2016. From first generation biofuels to advanced solar biofuels. *Ambio* 45: 24–31.

Balduyck, L., C. Veryser, K. Goiris et al. 2015. Optimization of a Nile red method for rapid lipid determination in autotrophic, marine microalgae is species dependent. *Journal of Microbiological Methods* 118: 152–158.

Beal, C. M, L. N. Gerber, D. L Sills et al. 2015. Algal biofuel production for fuels and feed in a 100-Ha facility: A comprehensive techno-economic analysis and life cycle assessment. *Algal Research* 10: 266–279.

Brennan, L. and P. Owende. 2010. Biofuels from microalgae—A review of technologies for production, processing, and extractions of biofuels and co-products. *Renewable and Sustainable Energy Reviews* 14: 557–577.

Carneiro, M. L. N. M., F. Pradelle, S. L. Braga et al. 2017. Potential of biofuels from algae: Comparison with fossil fuels, ethanol and biodiesel in Europe and Brazil through life cycle assessment (LCA). *Renewable and Sustainable Energy Reviews* 73: 632–653.

Chen, W.-H., B.-J. Lin, M.-Y. Huang, and J.-S. Chang. 2015. Thermochemical conversion of microalgal biomass into biofuels: A review. *Bioresource Technology* 184: 314–327.

Chernova, N. I. and S. V. Kiseleva. 2017. Microalgae biofuels: Induction of lipid synthesis for biodiesel production and biomass residues into hydrogen conversion. *International Journal of Hydrogen Energy* 42: 2861–2867.

Chiaramonti, D., M. Prussi, M. Buffi, A. M. Rizzo, and L. Pari. 2017. Review and experimental study on pyrolysis and hydrothermal liquefaction of microalgae for biofuel production. *Applied Energy* 185: 963–972.

Chisti, Y. 2007. Biodiesel from microalgae. *Biotechnology Advances* 25 (3): 294–306.

Chisti, Y. 2008. Biodiesel from microalgae beats bioethanol. *Trends in Biotechnology* 26: 126–131.

Chung, Y.-S., J.-W. Lee, and C.-H. Chung. 2017. Molecular challenges in microalgae towards cost-effective production of quality biodiesel. *Renewable and Sustainable Energy Reviews* 74: 139–144.

Correa, D. F., H. L. Beyer, H. P. Possingham et al. 2017. Biodiversity impacts of bioenergy production: Microalgae vs. first generation biofuels. *Renewable and Sustainable Energy Reviews* 74: 1131–1146.

Dębowski, M., M. Zielinski, A. Grala, and M. Dudek. 2013. Algae biomass as an alternative substrate in biogas production technologies—Review. *Renewable and Sustainable Energy Reviews* 27: 596–604.

Del Río, E., E. Garcia-Gomez, J. Moreno, M. G. Guerrero, and M. Garcia-Gonzalez. 2017. Microalgae for oil. Assessment of fatty acid productivity in continuous culture by two high-yield strains, *Chlorococcum oleofaciens* and *Pseudokirchneriella subcapitata*. *Algal Research* 23: 37–42.

Demirbas, A. 2000. Mechanisms of liquefaction and pyrolysis reactions of biomass. *Energy Conversion and Management* 41 (6): 633–646.

Doshi, A., S. Pascoe, L. Coglan, and T. J. Rainey. 2016. Economic and policy issues in the production of algae-based biofuels: A review. *Renewable and Sustainable Energy Reviews* 64: 329–337.

Dragone, G., B. D. Fernandes, A. Vicente, and J. A. Teixeira. 2010. Third generation biofuels from microalgae. In *Current Research, Technology and Education Topics in Applied Microbiology and Microbial Biotechnology*, ed. A. Mendez-Vilas, 1355–1366. Formatex, Portugal.

Dutta, S., F. Neto, and M. C. Coelho. 2016. Microalgae biofuels: A comparative study on techno-economic analysis & life-cycle assessment. *Algal Research* 20: 44–52.

Ehimen, E. A., J. B. Holm-Nielsen, M. Poulsen, and J. E. Boelsmand. 2013. Influence of different pre-treatment routes on the anaerobic digestion of a filamentous algae. *Renewable Energy* 50: 476–480.

El Asri, O., M. Ramdani, L. Latrach et al. 2017. Comparison of energy recovery after anaerobic digestion of three Marchica Lagoon algae (*Caulerpa Prolifera, Colpomenia Sinuosa, Gracilaria Bursa-Pastoris*). *Sustainable Materials and Technologies* 11: 47–52.

Fasahati, P. and J. J. Liu. 2012. Process simulation of bioethanol production from brown algae. *IFAC Proceedings Volumes* 45: 597–602.

Fasahati, P. and J. J. Liu. 2015. Economic, energy, and environmental impacts of alcohol dehydration technology on biofuel production from brown algae. *Energy* 93: 2321–2336.

Fasahati, P., C. M. Saffron, H. C. Woo, and J. J. Liu. 2017. Potential of brown algae for sustainable electricity production through anaerobic digestion. *Energy Conversion and Management* 135: 297–307.

Fasahati, P., H. C. Woo, and J. J. Liu. 2015. Industrial-scale bioethanol production from brown algae: Effects of pretreatment processes on plant economics. *Applied Energy* 139: 175–187.

Galan, M.-I., J. Bonet, R. Sire, J.-M. Reneaume, and A. E. Plesu. 2009. From residual to useful oil: Revalorization of glycerine from the biodiesel synthesis. *Bioresource Technology* 100: 3775–3778.

Gambelli, D., F. Alberti, F. Solfanelli, D. Vairo, and R. Zanoli. 2017. Third generation algae biofuels in Italy by 2030: A scenario analysis using Bayesian networks. *Energy Policy* 103: 165–178.

Gaurav, N., S. Sivasankari, G. S. Kiran, A. Ninawe, and J. Selvin. 2017. Utilization of bioresources for sustainable biofuels: A review. *Renewable and Sustainable Energy Reviews* 73: 205–214.

Goyal, H. B., D. Seal, and R. C. Saxena. 2008. Bio-fuels from thermochemical conversion of renewable resources: A review. *Renewable and Sustainable Energy Reviews* 12: 504–517.

Guldhe, A., F. A. Ansari, P. Singh, and F. Bux. 2017. Heterotrophic cultivation of microalgae using aquaculture wastewater: A biorefinery concept for biomass production and nutrient remediation. *Ecological Engineering* 99: 47–53.

Guo, Y., T. Yeh, W. Song, D. Xu, and S. Wang. 2015. A review of bio-oil production from hydrothermal liquefaction of algae. *Renewable and Sustainable Energy Reviews* 48: 776–790.

Hallenbeck, P. C. 2011. Microbial paths to renewable hydrogen production. *Biofuels* 2: 285–302.

Hallenbeck, P. C., M. Grogger, M. Mraz, and D. Veverka. 2016. Solar biofuels production with microalgae. *Applied Energy* 179: 136–145.

Harun, R., M. K. Danquah, and G. M. Forde. 2010. Microalgal biomass as a fermentation feedstock for bioethanol production. *Journal of Chemical Technology & Biotechnology* 85: 199–203.

Hu, Z., X. Ma, and C. Chen. 2012. A study on experimental characteristic of microwave-assisted pyrolysis of microalgae. *Bioresource Technology* 107: 487–493.

Jena, U., N. Vaidyanathan, S. Chinnasamy, and K. C. Das. 2011. Evaluation of microalgae cultivation using recovered aqueous co-product from thermochemical liquefaction of algal biomass. *Bioresource Technology* 102: 3380–3387.

Ji, M.-K., R. A. I. Abou-Shanab, S.-H. Kim et al. 2013. Cultivation of microalgae species in tertiary municipal wastewater supplemented with CO_2 for nutrient removal and biomass production. *Ecological Engineering* 58: 142–148.

Ji, S.-Q., B. Wang, M. Lu, and F.-L. Li. 2016. Direct bioconversion of brown algae into ethanol by thermophilic bacterium *Defluviitalea phaphyphila*. *Biotechnology for Biofuels* 9: 81.

Jorquera, O., A. Kiperstok, E A. Sales, M. Embirucu, and M. L. Ghirardi. 2010. Comparative energy life-cycle analyses of microalgal biomass production in open ponds and photobioreactors. *Bioresource Technology* 101: 1406–1413.

Judd, S. J., F. A. O. Al Momani, H. Znad, and A. M. D. Al Ketife. 2017. The cost benefit of algal technology for combined CO_2 mitigation and nutrient abatement. *Renewable and Sustainable Energy Reviews* 71: 379–387.

Khambhaty, Y., K. Mody, M. R. Gandhi et al. 2012. *Kappaphycus Alvarezii* as a source of bioethanol. *Bioresource Technology* 103: 180–185.

Kim, K. H., I. S. Choi, H. M. Kim, S. G. Wi, and H.-J. Bae. 2014. Bioethanol production from the nutrient stress-induced microalga *Chlorella vulgaris* by enzymatic hydrolysis and immobilized yeast fermentation. *Bioresource Technology* 153: 47–54.

Krishnan, V., S. Thachanan, Y. Matsumura, and Y. Uemura. 2016. Supercritical water gasification on three types of microalgae in the presence and absence of catalyst and salt. *Procedia Engineering* 148: 594–599.

Laurens, L. M. L., M. Chen-Glasser, and J. D. McMillan. 2017. A perspective on renewable bioenergy from photosynthetic algae as feedstock for biofuels and bioproducts. *Algal Research* 24: 261–264.

Lavoie, J.-M. 2016. Implementing 2nd generation liquid biofuels in a fossil fuel-dominated market: Making the right choices. *Current Opinion in Green and Sustainable Chemistry* 2: 45–47.

Lee, R. A. and J.-M. Lavoie. 2013. From first- to third-generation biofuels: Challenges of producing a commodity from a biomass of increasing complexity. *Animal Frontiers* 3 (2): 6–11.

Lee, O. K. and E. Y. Lee. 2016. Sustainable production of bioethanol from renewable brown algae biomass. *Biomass and Bioenergy* 92: 70–75.

Li, Y., M. Horsman, N. Wu, C. Q. Lan, and N. Dubois-Calero. 2008. Biofuels from microalgae. *Biotechnology Progress* 24 (4): 815–820.

Li, L., X. Ma, Q. Xu, and Z. Hu. 2013. Influence of microwave power, metal oxides and metal salts on the pyrolysis of algae. *Bioresource Technology* 142: 469–474.

Maity, J. P., Bundschuh, J., Chen, C.-Y., and P. Bhattacharya. 2014. Microalgae for third generation biofuel production, mitigation of greenhouse gas emissions and wastewater treatment: Present and future perspectives—A mini review. *Energy* 78: 104–113.

Mata, T. M., A. A. Martins, and N. S. Caetano. 2010. Microalgae for biodiesel production and other applications: A review. *Renewable and Sustainable Energy Reviews* 14 (1): 217–232.

Matsui, T.-O., A Nishihara, C. Ueda et al. 1997. Liquefaction of micro-algae with iron catalyst. *Fuel* 76: 1043–1048.

Mehrabadi, A., R. Craggs, and M. M. Farid. 2016. Biodiesel production potential of wastewater treatment high rate algal pond biomass. *Bioresource Technology* 221: 222–233.

Mendoza, H., L. Carmona, P. Assuncao et al. 2015. Variation in lipid extractability by solvent in microalgae. Additional criterion for selecting species and strains for biofuel production from microalgae. *Bioresource Technology* 197: 369–374.

Miao, X. and Q. Wu. 2004. High yield bio-oil production from fast pyrolysis by metabolic controlling of *Chlorella protothecoides*. *Journal of Biotechnology* 110: 85–93.

Miao, X., Q. Wu, and C. Yang. 2004. Fast pyrolysis of microalgae to produce renewable fuels. *Journal of Analytical and Applied Pyrolysis* 71: 855–863.

Milano, J., H. C. Ong, H. H. Masjuki et al. 2016. Microalgae biofuels as an alternative to fossil fuel for power generation. *Renewable and Sustainable Energy Reviews* 58: 180–197.

Molina, E., J. Fernandez, F. G. Acien, and Y. Chisti. 2001. Tubular photobioreactor design for algal cultures. *Journal of Biotechnology* 92: 113–131.

Molina Grima, E., E. H. Belarbi, F. G. Acien Fernandez, A. Robles Medina, and Y. Chisti. 2003. Recovery of microalgal biomass and metabolites: Process options and economics. *Biotechnology Advances* 20: 491–515.

Moreira, D. and J. C. M. Pires. 2016. Atmospheric CO_2 capture by algae: Negative carbon dioxide emission path. *Bioresource Technology* 215: 371–379.

Muangrat, R., J. A. Onwudili, and P. T. Williams. 2010. Influence of alkali catalysts on the production of hydrogen-rich gas from the hydrothermal gasification of food processing waste. *Applied Catalysis B: Environmental* 100: 440–449.

Nagarajan, D., D.-J. Lee, A. Kondo, and J.-S. Chang. 2017. Recent insights into biohydrogen production by microalgae—From biophotolysis to dark fermentation. *Bioresource Technology* 227: 373–387.

Naik, S. N., V. V. Goud, P. K. Rout, and A. K. Dalai. 2010. Production of first and second generation biofuels: A comprehensive review. *Renewable and Sustainable Energy Reviews* 14: 578–597.

Narala, R. R., S. Garg, K. Sharma, and P. M. Schenk. 2016. Comparison of microalgae cultivation in photobioreactor, open raceway pond, and a two-stage hybrid system. *Frontiers in Energy Research* 4 (29): 1–10.

Naveena, B., P. Armshaw, and J. T. Pembroke. 2015. Ultrasonic intensification as a tool for enhanced microbial biofuel yields. *Biotechnology for Biofuels* 8: 140.

Neofotis, P., A. Huang, K. Sury, W. Chang et al. 2016. Characterization and classification of highly productive microalgae strains discovered for biofuel and bioproduct generation. *Algal Research* 15: 164–178.

Ni, M., D. Y. C. Leung, M. K. H. Leung, and K. Sumathy. 2006. An overview of hydrogen production from biomass. *Fuel Processing Technology* 87: 461–472.

Nigam, P. S. and A. Singh. 2011. Production of liquid biofuels from renewable resources. *Progress in Energy and Combustion Science* 37: 52–68.

Onwudili, J. A., A. R. Lea-Langton, A. B. Ross, and P. T. Williams. 2013. Catalytic hydrothermal gasification of algae for hydrogen production: Composition of reaction products and potential for nutrient recycling. *Bioresource Technology* 127: 72–80.

Patil, P. D., H. Reddy, T. Muppaneni, and S. Deng. 2017. Biodiesel fuel production from algal lipids using supercritical methyl acetate (glycerin-free) technology. *Fuel* 195: 201–207.

Piloto-Rodríguez, R., Y. Sanchez-Borroto, E. A. Melo-Espinosa, and S. Verhelst. 2017. Assessment of diesel engine performance when fueled with biodiesel from algae and microalgae: An overview. *Renewable and Sustainable Energy Reviews* 69: 833–842.

Raheem, A., W. A. K. G. W. Azlina, Y. H. T. Yap, M. K. Danquah, and R. Harun. 2015. Thermochemical conversion of microalgal biomass for biofuel production. *Renewable and Sustainable Energy Reviews* 49: 990–999.

Razzak, S. A., M. M. Hossain, R. A. Lucky, A. S. Bassi, and H. de Lasa. 2013. Integrated CO_2 capture, wastewater treatment and biofuel production by microalgae culturing—A review. *Renewable and Sustainable Energy Reviews* 27: 622–653.

Rodriguez, C., A. Alaswad, J. Mooney, T. Prescott, and A. G. Olabi. 2015. Pretreatment techniques used for anaerobic digestion of algae. *Fuel Processing Technology* 138: 765–779.

Saladini, F., N. Patrizi, F. M. Pulselli, N. Marchettini, and S. Bastianoni. 2016. Guidelines for emergy evaluation of first, second and third generation biofuels. *Renewable and Sustainable Energy Reviews* 66: 221–227.

Schumacher, M., J. Yanik, A. Sinag, and A. Kruse. 2011. Hydrothermal conversion of seaweeds in a batch autoclave. *Journal of Supercritical Fluids* 58: 131–135.

Scott, S. A., M. P. Davey, J. S. Dennis et al. 2010. Biodiesel from algae: Challenges and prospects. *Current Opinion in Biotechnology* 21: 277–286.

Sialve, B., N. Bernet, and O. Bernard. 2009. Anaerobic digestion of microalgae as a necessary step to make microalgal biodiesel sustainable. *Biotechnology Advances* 27: 409–416.

Sirajunnisa, A. R. and D. Surendhiran. 2016. Algae—A quintessential and positive resource of bioethanol production: A comprehensive review. *Renewable and Sustainable Energy Reviews* 66: 248–267.

Song, M., H. D. Pham, J. Seon, and H. C. Woo. 2015. Marine brown algae: A conundrum answer for sustainable biofuels production. *Renewable and Sustainable Energy Reviews* 50: 782–792.

Su, Y., K. Song, P. Zhang et al. 2017. Progress of microalgae biofuel's commercialization. *Renewable and Sustainable Energy Reviews* 74: 402–411.

Suali, E. and R. Sarbatly. 2012. Conversion of microalgae to biofuel. *Renewable and Sustainable Energy Reviews* 16 (6): 4316–4342.

Takeda, H., F. Yoneyama, S. Kawai, W. Hashimoto, and K. Murata. 2011. Bioethanol production from marine biomass alginate by metabolically engineered bacteria. *Energy & Environmental Science* 4: 2575–2581.

Teo, S. W., A. Islam, and Y. H. Taufiq-Yap. 2016. Algae derived biodiesel using nanocatalytic transesterification process. *Chemical Engineering Research and Design* 111: 362–370.

Tuccar, G. and K. Aydin. 2013. Evaluation of methyl ester of microalgae oil as fuel in a diesel engine. *Fuel* 112: 203–207.

Ullah, K., A. Ahmad, V. K. Sharma et al. 2015. Assessing the potential of algal biomass opportunities for bioenergy industry: A review. *Fuel* 143: 414–423.

U.S. Department of Energy. 2016. *National Algal Biofuels Technology Review.* Energy Efficiency and Renewable Energy, U.S. Department of Energy, Bioenergy Technologies Office, Washington, DC.

Velasquez-Orta, S. B., J. G. M. Lee, and A. Harvey. 2012. Alkaline in situ transesterification of *Chlorella vulgaris*. *Fuel* 94: 544–550.

Vergara-Fernandez, A., G. Vargas, N. Alarcon, and A. Velasco. 2008. Evaluation of marine algae as a source of biogas in a two-stage anaerobic reactor system. *Biomass and Bioenergy* 32: 338–344.

Viegas, C. V., I. Hachemi, S. P. Freitas et al. 2015. A route to produce renewable diesel from algae: Synthesis and characterization of biodiesel via in situ transesterification of *Chlorella* alga and its catalytic deoxygenation to renewable diesel. *Fuel* 155: 144–154.

Wang, X., L. Sheng, and X. Yang. 2017. Pyrolysis characteristics and pathways of protein, lipid and carbohydrate isolated from microalgae *Nannochloropsis* sp. *Bioresource Technology* 229: 119–125.

Yen, H.-W. and D. E. Brune. 2007. Anaerobic co-digestion of algal sludge and waste paper to produce methane. *Bioresource Technology* 98: 130–134.

Zabed, H., J. N. Sahu, A. Suely, A. N. Boyce, and G. Faruq. 2017. Bioethanol production from renewable sources: Current perspectives and technological progress. *Renewable and Sustainable Energy Reviews* 71: 475–501.

Zhang, W., J. Zhang, and H. Cui. 2014. The isolation and performance studies of an alginate degrading and ethanol producing strain. *Chem Biochem Eng Q* 28: 391–398.

Zhou, W., M. Min, Y. Li et al. 2012. A hetero-photoautotrophic two-stage cultivation process to improve wastewater nutrient removal and enhance algal lipid accumulation. *Bioresource Technology* 110: 448–455.

Zhu, L., Y. K. Nugroho, S. R. Shakeel et al. 2017. Using microalgae to produce liquid transportation biodiesel: What is next? *Renewable and Sustainable Energy Reviews* 78: 391–400.

Zinnai, A., C. Sanmartin, I. Taglieri, G. Andrich, and F. Venturi. 2016. Supercritical fluid extraction from microalgae with high content of LC-PUFAs. A case of study: SC-CO_2 oil extraction from *Schizochytrium* sp. *Journal of Supercritical Fluids* 116: 126–131.

section two

Liquid biofuels

chapter four

Biodiesel*

Bryan R. Moser

Contents

4.1 Definition

Biodiesel is defined by ASTM International as a fuel comprised of mono-alkyl esters of long-chain fatty acids (FAs) derived from vegetable oils or

* Mention of trade names or commercial products in this publication is solely for the purpose of providing specific information and does not imply recommendation or endorsement by the U.S. Department of Agriculture. USDA is an equal opportunity provider and employer.

animal fats, designated B100, and meeting the requirements of ASTM D6751. The European biodiesel standard, EN 14214, de facto defines biodiesel as FA methyl esters (FAMEs). Consequently, alternative fuels such as neat vegetable oils, hydrocarbons prepared therefrom via hydrotreatment, or Fischer–Tropsch diesel produced by gasification of biomass do not technically qualify as biodiesel, despite their occasional reference as such. As the name implies, biodiesel is a substitute or blend component for middle distillate fuels (petrodiesel) for combustion in diesel engines and generators and is thus not suited for spark-ignition (gasoline) engines.

The primary reason lipids are converted to biodiesel as opposed to their direct use as fuel is their excessive kinematic viscosity (KV), which is approximately an order of magnitude higher than petrodiesel. For example, KVs (40°C) of soybean oil, soybean oil methyl esters (SMEs) and ultra-low-sulfur [<15 ppm S] diesel (ULSD) fuel are 31.49, 4.12, and 2.30 mm^2/s, respectively (Moser and Vaughn, 2010). High KV results in poor fuel atomization in the combustion chamber, which causes operational problems such as engine deposits (Knothe, 2005).

4.2 Chemical structure

Plant oils, animal fats, and other lipids from which biodiesel is prepared are composed of triacylglycerols (TAGs) consisting of long-chain FAs chemically bound to a glycerol backbone. Table 4.1 summarizes the most

Table 4.1 Systematic, common, and shorthand names of fatty acids

Systematic	Common	Shorthand
Octanoic	Caprylic	C8:0
Decanoic	Capric	C10:0
Dodecanoic	Lauric	C12:0
Tetradecanoic	Myristic	C14:0
Hexadecanoic	Palmitic	C16:0
9Z-hexadecenoic	Palmitoleic	C16:1 9c
Octadecanoic	Stearic	C18:0
9Z-Octadecenoic	Oleic	C18:1 9c
9Z,12Z-Octadecadienoic	Linoleic	C18:2 9c, 12c
9Z,12Z,15Z-Octadecatrienoic	Linolenic	C18:3 9c, 12c, 15c
Eicosanoic	Arachidic	C20:0
11Z-Eicosenoic	Gadoleic	C20:1 11c
Docosanoic	Behenic	C22:0
13Z-Docosenoic	Erucic	C22:1 13c
Tetracosanoic	Lignoceric	C24:0
15Z-Tetracosenoic	Nervonic	C24:1 15c

common FAs encountered in lipids. Of these, palmitic, stearic, oleic, linoleic, and linolenic acids are the most prevalent (Knothe, 2005; Moser, 2009a). The FA profile of biodiesel as determined by gas chromatography flame ionization detector (GC-FID) is equivalent to the TAG from which it was prepared so long as derivatization to FAMEs is quantitative, inert systems for GC analysis are used, and correct data acquisition methods are utilized (Craske and Bannon, 1987).

4.3 Traditional feedstocks

It is generally cost-prohibitive to import or otherwise transport over long distances lipids for biodiesel production due to the low cost of competing petrodiesel. As a consequence, biodiesel is typically prepared from regionally available commodity lipids. Feedstocks for biodiesel production therefore vary according to geography and climate. For instance, sunflower and especially rapeseed (referred to as canola in North America) oils are predominately used in Europe for biodiesel production; soybean oil, animal fats, and mixtures thereof are principally used in the United States; palm and coconut oils are used in tropical countries; canola oil is utilized in Canada; and soybean and palm oils are among those utilized in South America. Table 4.2 summarizes the FA profiles of the most common feedstocks used for production of biodiesel.

Table 4.2 Typical fatty acid compositions (area %) of the leading vegetable oils produced worldwide

Fatty acid	COC	COT	OLI	PAL	PAK	PEA	RAP	SOY	SUN
C8:0	7.3				3.3				
C10:0	6.6				3.5				
C12:0	47.8			0.2	47.8				
C14:0	18.1			1.1	16.3				
C16:0	8.9	23	10.5	44.1	8.5	10.8	3.6	11.0	6.4
C18:0	2.7		2.6	4.4	2.4	2.9	1.6	4.0	4.7
C20:0	0.1			0.2		1.8			
C22:0						1.9			
C18:1	6.4	17	76.9	39.0	15.4	53.1	61.6	23.4	21.0
C18:2	1.6	56	7.5	10.6	2.4	27.3	21.7	53.2	67.7
C18:3				0.2			9.6	7.8	
C20:1						1.0	1.4		
Other	0.5	4.0	2.5	0.2	0.4	1.2	0.5	0.6	0.2

Source: Data from Gunstone, F. D., Harwood, J. L., and A. J. Dijkstra (eds.). 2007. *The Lipid Handbook*, 3rd Ed., Boca Raton: CRC Press.

Note: COC = coconut oil; COT = cottonseed oil; OLI = olive oil; PAL = palm oil; PAK = palm kernel oil; PEA = peanut oil; RAP = rapeseed oil (low erucic acid variety); SOY = soybean oil; SUN = sunflower oil.

4.4 Fuel standards

The two most important biodiesel fuel standards, ASTM D6751 and EN 14214, are summarized in Tables 4.3 and 4.4, respectively. An additional biodiesel standard, ASTM D7467, covers blends of biodiesel from 6–20 vol% in petrodiesel. The petrodiesel standards ASTM D975 and EN 590 allow for inclusion of up to 5 and 7 vol% biodiesel (B5 and B7), respectively. In addition, the U.S. heating oil standard, ASTM D396, specifies that up to B5 is permissible. In Europe, biodiesel (B100; 100% biodiesel) may be directly

Table 4.3 ASTM D6751 biodiesel fuel standard

Property	Test method	Limits	Units
Flash point (closed cup)	ASTM D93	93 min.	°C
Alcohol control:			
One of the following must be met:			
1. Methanol content	EN 14110	0.2 max.	% volume
2. Flash point	ASTM D93	130.0 min.	°C
Water and sediment	ASTM D2709	0.050 max.	% volume
Kinematic viscosity, 40°C	ASTM D445	1.9–6.0	mm^2/s
Sulfated ash	ASTM D874	0.020 max.	% mass
Sulfur[a]	ASTM D5453	0.0015 max. (S15) 0.05 max. (S500)	% mass (ppm)
Copper strip corrosion	ASTM D130	No. 3 max.	
Cetane number	ASTM D613	47 min.	
Cloud point	ASTM D2500	Report	°C
Cold soak filterability	ASTM D7501	360 max.[b]	seconds
Carbon residue	ASTM D4530	0.050 max.	% mass
Acid value	ASTM D664	0.50 max.	mg KOH/g
Free glycerin	ASTM D6584	0.020	% mass
Total glycerin	ASTM D6584	0.240	% mass
Oxidation stability	EN 14112	3.0 min.	hours
Phosphorous content	ASTM D4951	0.001 max.	% mass
Sodium and potassium, combined	EN 14538	5 max.	ppm
Calcium and magnesium, combined	EN 14538	5 max.	ppm
Distillation temperature, Atmospheric equivalent temperature, 90% recovered	ASTM D1160	360 max.	°C

[a] The limits are for Grade S15 and Grade S500 biodiesel, with S15 and S500 referring to maximum allowable sulfur content (ppm).

[b] B100 intended for blending into petrodiesel that is expected to give satisfactory performance at fuel temperatures at or below −12°C shall comply with a maximum cold soak filterability limit of 200 s.

used as a heating oil and is covered by a separate standard, EN 14213. In the cases of ASTM D7467, D975, and D396, the biodiesel component must satisfy the requirements of ASTM D6751 before inclusion in the respective fuels. Correspondingly, biodiesel must be satisfactory according to EN 14214 in Europe before inclusion in petrodiesel, as mandated by EN 590.

4.5 *Advantages and disadvantages*

Biodiesel is miscible with petrodiesel in any proportion and possesses several technical advantages over ULSD, such as inherently good lubricity, low toxicity, derivation from renewable and domestic feedstocks, superior biodegradability, higher flash point (safer handling), negligible sulfur content, positive energy balance, and lower exhaust emissions. Disadvantages include high feedstock cost, inferior storage and oxidative stability, lower volumetric energy content, susceptibility to hydrolysis as well as microbial degradation, and inferior cold flow properties (Knothe, 2005; Moser, 2009a). Many of these deficiencies are mitigated by additives, blending with petrodiesel and reducing storage time (Dunn, 2008a; Knothe, 2007; Moser, 2014; Moser et al. 2008). Additional methods to enhance cold flow properties include crystallization fractionation and transesterification with branched-chain alcohols (Dunn, 2008b, 2009).

Feedstock acquisition accounts for up to 80% of biodiesel production expenses (Haas et al., 2006). A solution to the problem of high feedstock cost is employment of alternatives of lesser value. Such alternatives may include soapstocks, acid oils, tall oils, used cooking oils and waste restaurant greases, various animal fats, nonfood plant oils, and microalgae. However, many of these alternatives contain high levels of undesirable free fatty acids (FFAs), water, and/or insoluble matter, which negatively impact transesterification if not removed beforehand.

4.6 *Transesterification*

Transesterification involves reaction of TAGs with a short-chain monohydric alcohol in the presence of a catalyst at elevated temperature to produce FA alkyl esters (FAAEs) and glycerol (Figure 4.1). Methanol is typically used as the alcohol because it is the least expensive of the short-chain monohydric alcohols. However, ethanol is utilized in regions such as Brazil where it is less expensive to produce FA ethyl esters (FAEEs). Transesterification conducted in the presence of methanol is often referred to as methanolysis, whereas ethanolysis is the common name for ethanol-mediated transesterification. Ethanol has the added benefit of being bio-derived, whereas methanol is produced from natural gas. Consequently, FAEEs are completely derived from biological feedstocks, whereas FAMEs contain a hydrocarbon-derived component.

Table 4.4 EN 14214 biodiesel fuel standard

Property	Test method(s)	Limits	Units
Ester content	EN 14103	96.5 min.	% (mol/mol)
Density, 15°C	ISO 3675, ISO 12185	860–900	kg/m^3
Kinematic viscosity, 40°C	ISO 3104	3.50–5.00	mm^2/s
Flash point	ISO 2719, ISO 3679	101 min.	°C
Sulfur content	ISO 20846, ISO 20884	10.0 max.	mg/kg
Carbon residue (10% distillation residue)	ISO 10370	0.30 max.	% (mol/mol)
Cetane number	ISO 5165	51.0 min.	
Sulfated ash content	ISO 3987	0.02 max.	% (mol/mol)
Water content	ISO 12937	500 max.	mg/kg
Total contamination	EN 12662	24 max.	mg/kg
Copper strip corrosion (3 h, 50°C)	ISO 2160	Class 1	degree of corrosion
Oxidation stability, 110°C	EN 14112, EN 15751	8.0 min.	hours
Cold filter plugging point	EN 116	Variable	°C
Acid value	EN 14104	0.50 max.	mg KOH/g
Iodine value	EN 14111	120 max.	g I_2/100 g
Linolenic acid methyl ester content	EN 14103	12.0 max.	% (mol/mol)
Polyunsaturated ($\geq$ 4 double bonds) methyl esters	EN 14103	1 max.	% (mol/mol)
Methanol content	EN 14110	0.20 max.	% (mol/mol)
MAG content	EN 14105	0.80 max.	% (mol/mol)
DAG content	EN 14105	0.20 max.	% (mol/mol)
TAG content	EN 14105	0.20 max.	% (mol/mol)
Free glycerol	EN 14105 EN 14106	0.02 max.	% (mol/mol)
Total glycerol	EN 14105	0.25 max.	% (mol/mol)
Group I metals (Na, K)	EN 14108 EN 14109 EN 14538	5.0 max.	mg/kg
Group II metals (Ca, Mg)	EN 14538	5.0 max.	mg/kg
Phosphorous content	EN 14107	4.0 max.	mg/kg

Industrially, the lipid, alcohol, and catalyst are mixed in continuously stirred tank reactors (CSTRs) until the reaction has reached equilibrium. Excess alcohol drives the equilibrium toward biodiesel and glycerol. Often, two CSTRs are utilized in series (with glycerol removal in between) to ensure high conversion. The catalyst is dissolved in alcohol prior to addition to the CSTR containing the lipid to minimize the possibility that a concentrated region of catalyst might come into contact with TAGs and promote unwanted soap formation. Separation of glycerol from biodiesel is accomplished with a settling tank or with a disk centrifuge. The glycerol stream is then acidified to neutralize residual catalyst and to decompose soaps into FFAs and salts. Methanol entrained in the glycerol phase is recycled through an evaporative process. The fraction containing biodiesel is also neutralized with acid and methanol is recovered, followed by washing with water and drying to provide finished biodiesel.

The two most significant parameters that determine compatibility of feedstock oils or fats with alkaline-catalyzed transesterification depicted in Figure 4.1 are the contents of FFAs and water (Figure 4.2). Typically, refined vegetable oils or fats contain very low percentages of FFA (<0.05 wt%) and water (<0.1 wt%), so their direct conversion to biodiesel is readily

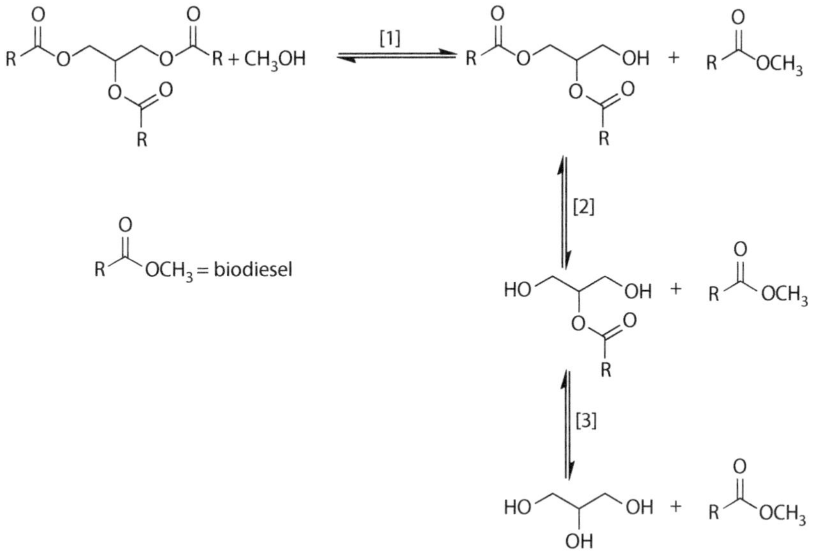

Figure 4.1 General scheme for transesterification of lipids to biodiesel and glycerol. Reaction [1] depicts conversion of TAG to DAG and FAME, with reaction [2] showing conversion of DAG to MAG and FAME and reaction [3] depicting conversion of MAG to FAME and glycerol.

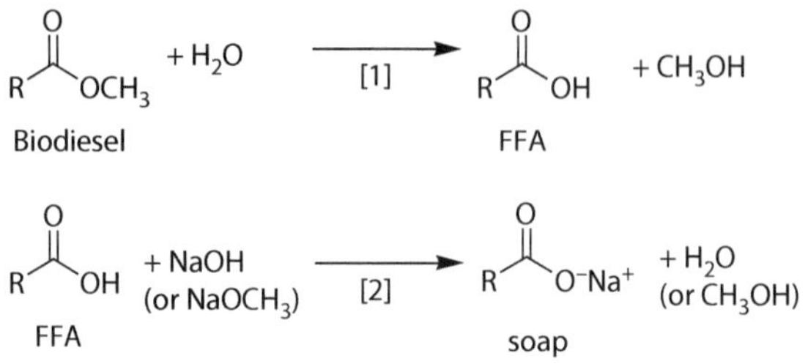

$$\underset{\text{Biodiesel}}{RC(=O)OCH_3} + H_2O \xrightarrow{[1]} \underset{\text{FFA}}{RC(=O)OH} + CH_3OH$$

$$\underset{\text{FFA}}{RC(=O)OH} + NaOH\ (\text{or } NaOCH_3) \xrightarrow{[2]} \underset{\text{soap}}{RC(=O)O^-Na^+} + H_2O\ (\text{or } CH_3OH)$$

Figure 4.2 Hydrolysis of biodiesel (reaction [1]) to yield FFAs and methanol. Formation of soap upon reaction of FFA with alkaline catalyst (reaction [2]).

accomplished. Crude lipids can have considerably higher percentages of these contaminants, so purification is required prior to alkaline-catalyzed transesterification. Water and FFAs can be removed with physical refining (O'Brien, 2009). Additionally, FFAs can be esterified to biodiesel prior to or during methanolysis utilizing acid catalysts (Lotero et al., 2005).

The classic alkaline-catalyzed conditions for methanolysis of anhydrous, low-FFA (<0.50 wt%) TAGs are a 6:1 molar ratio of methanol to TAG, 0.5 wt% catalyst (with respect to TAG), 600 + rpm agitation intensity, 60°C, and 1 h of reaction (Freedman et al., 1984). Alkaline-catalyzed transesterification consists of a sequence of three reversible reactions whereby TAG is converted to biodiesel and glycerol via diacylglycerol (DAG) and monoacylglycerol (MAG) intermediates (Figure 4.1). Stoichiometrically, 3 mol of biodiesel and 1 mol of glycerol are produced for every mole of TAG undergoing complete conversion.

4.7 *Influence of free fatty acids*

If a feedstock contains a significant percentage of FFAs (>2 wt%), alkaline catalysts such as metal hydroxides or methoxides will not be effective as a result of a deleterious side reaction (Figure 4.2) in which the catalyst reacts with FFAs to form soap (metal salt of FA) and water (or methanol in the case of methoxide), thus irreversibly quenching the catalyst and reducing yield (Haas, 2005; Lotero et al., 2005; Van Gerpen, 2005). A further complicating factor of high FFA content is the production of water upon reaction with hydroxide catalysts (Figure 4.2). Water is particularly problematic because in the presence of any remaining catalyst it can hydrolyze biodiesel to produce additional FFAs and methanol (Figure 4.2).

When FFA content is prohibitively high for direct conversion to FAME, a two-step procedure is utilized in which acid pretreatment (to lower FFA content) is followed by alkaline-catalyzed transesterification to produce FAME. Acid-catalyzed esterification of FFAs to FAME is conducted in the presence of heat, excess methanol, and catalytic mineral acid

(Lotero et al., 2005; Moser and Vaughn, 2010; Naik et al., 2008). This two-step procedure accommodates high-FFA-containing, low-cost feedstocks for preparation of biodiesel. For example, the FFA content of crude jatropha (*Jatropha curcas* L.) oil was lowered from 14 wt% to less than 1 wt% upon treatment with H_2SO_4 catalyst and excess methanol. Subsequent transesterification yielded FAME in essentially quantitative yield (Tiwari et al., 2007). Despite the added capital costs, the integrated two-step procedure is being increasingly applied industrially to low-cost but high-FFA-containing feedstocks (Lotero et al., 2005). Other strategies to lower FFA content include refining, bleaching, and/or deodorization (O'Brien, 2009). Alternatively, catalysts that can simultaneously transesterify TAG and esterify FFAs may be used.

4.8 Catalysts

Homogeneous and heterogeneous acids and bases, lipases, and sugars, as well as ion exchange resins and zeolites with acidic or basic functionalities catalyze transesterification (Akoh et al., 2007; Di Serio et al., 2008; Narasimharao et al., 2007; Van Gerpen, 2005). Industrial production of biodiesel most commonly utilizes sodium hydroxide, potassium hydroxide, or sodium methoxide, with the latter preferred because it cannot form water upon reaction with alcohol such as with hydroxides. Transesterification with bases is 4000 times faster and more complete than with mineral acid catalysts (Di Serio et al., 2008). Furthermore, alkaline catalysis is performed at lower temperatures and pressures and is less corrosive to industrial equipment than acid catalysis (Freedman et al., 1984; Lotero et al., 2005; Van Gerpen, 2005). However, alkaline catalysts require anhydrous feedstocks with low FFAs, otherwise hydrolysis, soap formation, and catalyst deactivation become problematic (Figure 4.2). In fact, acids can simultaneously catalyze esterification of FFAs and transesterification of TAGs (Haas et al., 2003; Lotero et al., 2005).

Similar to acid catalysis, enzymatic catalysts can produce biodiesel from feedstocks with high FFA content because enzymes catalyze both esterification and transesterification (Akoh et al., 2007; Talukder et al., 2009). Other advantages include high selectivity, less energy consumption (reactions are performed using mild conditions), ease of catalyst recovery, production of fewer side products, less wastewater production, and tolerance of water (Akoh et al., 2007). Because enzymatic catalysts are heterogeneous, they are removed from biodiesel by simple filtration, thus eliminating the need for water washing of biodiesel and resulting in a cleaner glycerol coproduct stream. Finally, recovered enzymes can be reused multiple times. Disadvantages include enzyme deactivation in the presence of methanol, significantly longer reaction times, higher catalyst requirements, and high cost. The problem of enzyme deactivation is avoided by stepwise addition of methanol or the use of cosolvents such as tert-Butanol (Akoh et al., 2007).

The most prominent example of an enzyme catalyst is *Candida antarctica*, which is referred to as Novozym 435 when immobilized on solid support (acrylic resin). Representative conditions include 4.0 wt% catalyst, 50°C, 13 wt% methanol, and 2 h of reaction to provide FAME from high-FFA (>93 wt%) palm fatty acid distillate in 90.2% yield. The addition of cosolvents such as isooctane or hexane with otherwise similar conditions improves the yield to 95.2% and 94.5%, respectively. In the case of isooctane-mediated conditions, the yield of FAME after 15 cycles of catalyst recovery remained constant at 94.9% (Talukder et al., 2009).

4.9 Influence of biodiesel composition on fuel properties

FA composition, along with the presence of contaminants and minor components, dictates fuel properties of biodiesel. Properties affected by FA composition include cold flow, oxidative stability, KV, acid value (AV), cetane number (CN), lubricity, and energy content. Minor components are defined as naturally occurring (endogenous) constituents found in lipids and may include tocopherols, phospholipids, steryl glucosides, chlorophyll, fat-soluble vitamins, and hydrocarbons such as alkanes, squalene, carotenes, and polycyclic aromatic hydrocarbons (Gunstone, 2004). Contaminants are defined as incomplete or unwanted reaction products or reagents, such as FFAs, soaps, TAG, DAG, MAG, alcohol, catalyst, glycerol, metals, and water.

4.9.1 Low-temperature operability

Low-temperature operability is characterized by cloud point (CP) (ASTM D2500 or D5773), pour point (PP) (ASTM D97 or D5949), cold filter plugging point (CFPP) (ASTM D6371), and cold soak filtration test (CSFT) (ASTM D7501). CP is defined as the temperature at which crystal growth is large enough (diameter $\geq$0.5 μm) to be visible. Crystal agglomerations prevent fluid flow at temperatures below CP. The lowest temperature at which the fuel will pour is defined as the PP. The CFPP is the lowest temperature at which a given volume of biodiesel completely flows under vacuum through a filter within 60 s. CFPP is a more reliable indicator of cold flow behavior than CP or PP because the fuel will contain solids of sufficient size to render the engine inoperable due to fuel filter plugging once the CFPP is reached. CSFT is defined as the time required for a biodiesel sample to pass through a filter under vacuum. However, prior to filtration, the sample is cooled to 4.5°C for 16 h, and then equilibrated to room temperature. Under the conditions of the test, any high melting constituents will precipitate during the 16-h cooling period and remain solid at room temperature. Such solids may result in filtration times of greater than 360 s, thus causing failure of the test. Solids formation is particularly problematic in SMEs and palm oil methyl esters (PMEs) as a result of relatively high levels of steryl glucosides present in these fuels (Bondioli et al., 2008; Moreau et al., 2008).

Structural features such as chain length, degree of unsaturation, and double-bond orientation affect cold flow behavior of FAMEs. As the chain length of saturated FAMEs increases from 12 [laurate, melting point (mp) 4.3°C] to 18 (stearate, mp 37.7°C) carbons, a corresponding increase in mp is observed (Knothe and Dunn, 2009). Compounds of similar chain length but increasing levels of unsaturation display lower mp, as evidenced by methyl esters of stearic (mp 37.7°C), oleic (mp −20.1°C), linoleic (mp −43.1°C), and linolenic (mp −52°C) acids (Knothe and Dunn, 2009). Double-bond orientation also affects cold flow behavior. Nearly all naturally occurring unsaturated FAs contain *cis*-oriented double bonds. However, *trans* isomers may be chemically introduced through catalytic partial hydrogenation or other means. Constituents that contain *trans* double bonds exhibit higher mp than *cis* isomers. For example, the mp of methyl elaidate (methyl 9*E*-octadecenoate) is 9.9°C (Knothe and Dunn, 2009), which is considerably higher than the *cis* isomer (methyl oleate, mp −20.1°C).

Applying the above trends to biodiesel, one can see the influence of saturated FAME on low-temperature operability of SME and PME, which have PP values of 0 and 18°C, respectively, and CFPP values of −4 and 12°C, respectively (Moser, 2008). As seen in Table 4.2, PME has considerably more saturated FAMEs than SME. The influence of double-bond orientation is seen by comparing the CP and PP values of SME (0 and −2°C) to *trans*-containing partially hydrogenated SME (3 and 0°C) (Moser et al., 2007).

4.9.2 *Oxidative stability*

Oxidative stability is determined by induction period (IP) following EN 15751. The units for IP are expressed in hours and fuels with longer IPs are more stable to oxidation. Oxidation initiates at methylene carbons allylic to sites of unsaturation (Frankel, 2005). Autoxidation in the presence of initiators such as light, heat, metals, peroxides, and hydroperoxides produces free radicals by hydrogen abstraction. These free radicals react with oxygen to produce peroxides, which in turn react with unoxidized biodiesel to produce additional free radicals (Frankel, 2005). EN 15751 utilizes elevated temperature (110°C) to accelerate the initial rate-limiting hydrogen abstraction step. Products that ultimately form by oxidation of biodiesel include aldehydes, shorter chain FAs, other oxygenated species (such as ketones), and polymers (Frankel, 2005).

The rate of autoxidation is dependent on the number and location of methylene-interrupted double bonds in biodiesel. FAMEs with more methylene carbons allylic to sites of unsaturation are more vulnerable to autoxidation, as evidenced by the relative rates of oxidation (1, 41, and 98) for ethyl esters of oleic, linoleic, and linolenic acids (Holman and Elmer, 1947). This trend is confirmed by the IPs of methyl esters of stearic (>40 h), oleic (2.5 h), linoleic (1.0 h), and linolenic (0.2 h) acids (Moser, 2009b). Superior oxidative stability is achieved by preparing biodiesel from feedstocks that have high saturated FA content or low polyunsaturated FA content. For example, PME is more stable to oxidation than SME (Moser, 2008). Not only will oxidized biodiesel fail oxidative stability specifications, but oxidative degradation negatively impacts AV and KV (Moser, 2011). Corrosion of engine components, along with damaged fuel lines and fuel filter plugging, also result from oxidatively degraded biodiesel.

Structural features such as degree of unsaturation and chain length affect IP. Double-bond orientation (*cis* versus *trans*) also impacts oxidative stability, as seen by the contrasting IP values of methyl esters of oleic (2.5 h) and elaidic (7.7 h) acids (Moser, 2009b). Biodiesel fuels that contain at least some *trans* FAMEs exhibit enhanced oxidative stability than those that contain a similar number of entirely *cis* double bonds. For instance, partially hydrogenated SME (7.7% *trans* FAME, 16.4% saturated FAME, 44.7% polyunsaturated FAME) has an IP of 6.2 hours versus 2.3 hours for SME (0% *trans*, 16.9% saturated, 59.8% polyunsaturated) (Moser et al., 2007). Finally, oxidative stability and low-temperature operability are inversely related because structural factors that improve oxidative stability negatively affect cold flow properties and vice versa (Moser, 2009a).

4.9.3 *Kinematic viscosity*

KV is determined following ASTM D445 at 40°C. Chain length, degree of unsaturation, and double-bond orientation influence KV. Longer chain length results in higher KV, as seen by comparison of methyl esters of lauric, myristic, palmitic, and stearic acids, which have KVs of 2.43, 3.30, 4.38, and 5.85mm^2/s, respectively. Increasing the level of unsaturation results in lower KVs, as evidenced by methyl esters of stearic (5.85 mm^2/s), oleic (4.51 mm^2/s), linoleic (3.65 mm^2/s), and linolenic (3.14 mm^2/s) acids. The impact of double-bond orientation is seen by comparison of methyl elaidate (5.86 mm^2/s) and methyl linoelaidate (5.33 mm^2/s) to the corresponding *cis* isomers (methyl oleate and linoleate) (Knothe and Steidley, 2005a).

4.9.4 *Cetane number*

CN is determined following ASTM D613 and is related to ignition delay after injection into the combustion chamber. Shorter ignition delay results in higher CN and vice versa. Hexadecane, also known as cetane, is the high-quality reference with a short ignition delay and an assigned CN of 100. The low-quality reference is 2,2,4,4,6,8,8-heptamethylnonane with a long ignition delay and an assigned CN of 15. The end points of the CN scale indicate that branching reduces CN. Chain length also influences CN, with higher CNs observed for longer-chain compounds, as seen for methyl laurate (67), palmitate (86), and stearate (101). Increasing levels of unsaturation lower CN, as evidenced by the methyl stearate (101), oleate (59), linoleate (38), and linolenate (23) (Knothe et al., 2003). The influences of chain length and unsaturation on CN is evident by comparison of PME (62) to SME (54) (DeOliveira et al., 2006).

4.9.5 *Heat of combustion*

Heat of combustion is determined according to ASTM D240 or ASTM D4809, but is not specified in ASTM D6741 or EN 14214. Heat of combustion refers to the thermal energy that is liberated during combustion, and is thus referred to as energy content. The gross heats of combustion of diesel fuel, B20 SME, and SME are 46.7, 43.8, and 38.1 MJ/kg, respectively (DeOliveira et al., 2006). As the biodiesel component in the fuel increases, a decrease in energy content is observed (Moser et al., 2016a).

Factors influencing energy content are chain length and degree of unsaturation. Energy content increases with increasing chain length, as seen by comparison of methyl stearate (40.07 MJ/kg) to methyl laurate (37.97 MJ/kg). Energy content decreases with unsaturation, with methyl oleate (40.09 MJ/kg) possessing higher energy content than methyl linoleate (39.70 MJ/kg), which in turn has more energy than methyl linolenate (39.34 MJ/kg) (Knothe, 2008; Moser et al., 2009a). Energy content and KV are loosely related in that structural features that improve one also improve the other and vice versa.

4.9.6 *Lubricity*

Lubricity is measured according to ASTM D6079. During the experiment, a ball and disk are submerged in biodiesel and rubbed at 60°C against each other for 75 min at a rate of 50 Hz to generate a wear scar. The maximum length of the scar is then measured and this value represents lubricity, with shorter scars indicating better lubricity. Lubricity is not prescribed in ASTM D6751 or EN 14214. However, ASTM D975 and D7467 and EN 590 specify maximum wear scars of 520, 520, and 460 μm, respectively. Biodiesel possesses good lubricity, especially when compared to petrodiesel (Knothe and Steidley, 2005b; Moser et al., 2008). For instance, the lubricity of ULSD (without additives) is 571 μm, whereas B2, B5, B10, B20, and B100 blends of SME display progressively shorter wear scars of 271, 198, 154, 143, and 137 μm (Moser et al., 2009a). Generally, ULSD (without additives) does not possess satisfactory lubricity. The reason for the poor lubricity of ULSD is removal of polar compounds containing heteroatoms such as sulfur, oxygen, and nitrogen during hydrotreatment (Knothe and Steidley, 2005b). Biodiesel has excellent lubricity primarily as a result of its oxygen-containing ester functionality.

Structural features that influence lubricity aside from the presence of heteroatoms include chain length and unsaturation. Compounds of increasing chain length generally display superior lubricity, although the relationship is not perfect. For instance, the lubricities of methyl laurate, myristate, palmitate, and stearate are 416, 353, 357, and 246 μm, respectively. Increasing unsaturation also gives better lubricity, as evidenced by methyl stearate (322 μm), oleate (290 μm), linoleate (236 μm), and linolenate (183 μm). Lubricity of biodiesel is generally in the range of 120–200 μm (Moser, 2009a; Moser and Vaughn, 2010; Moser et al., 2008).

4.9.7 Contaminants

Contaminants may include methanol, water, catalyst, glycerol, FFAs, soaps, metals, MAG, DAG, and TAG. Methanol contamination, arising from insufficient purification, is indirectly measured by flash point because contaminated biodiesel will fail to meet the minimum specified flash point. Corrosion of engine components, along with damage to fuel lines and rubber seals, can result from methanol-contaminated biodiesel.

Water is a significant source of fuel contamination. While fuel leaving a production facility may be free of water, once it enters the distribution and storage network it will come into contact with water as a result of environmental humidity. Water causes corrosion of engine fuel system components and damage to fuel lines and promotes microbial growth and hydrolysis of FAME. Consequently, both ASTM D6751 and EN 14214 contain maximum limits on water. Water may be present in either the dissolved or free form. Dissolved water, which is measured by EN ISO 12937, is water that is soluble (i.e., dissolved) in biodiesel. Free water, which is measured by a centrifugation method (ASTM D2709) as specified in ASTM D6751, arises after biodiesel becomes saturated with dissolved water, thereby resulting in a separate water phase.

Residual catalyst may be present due to insufficient purification after transesterification. The catalyst is detected by inductively coupled plasma spectroscopy (EN 14538) through measurement of combined sodium and potassium. Calcium and magnesium, also determined by EN 14538, may be introduced during purification by washing with hard water or through the use of drying agents. Both ASTM D6751 and EN 14214 contain upper limits on combined sodium and potassium and combined calcium and magnesium. Metal contamination causes elevated ash production during combustion along with poisoning of aftertreatment systems.

Glycerol, which is quantified by GC using ASTM D6584 or EN 14105, is limited in ASTM D6751 and EN 14214. Bound glycerol (MAG + DAG + TAG) is limited in ASTM D6751 by the total glycerol (total = free + bound glycerol) specification and in EN 14214 by individual MAG, DAG, and TAG limits, as well as a total glycerol limit. Bound glycerol results from incomplete conversion and may cause carbon deposits on fuel-injector tips and piston rings of diesel engines due to incomplete combustion. Elevated particulate matter and hydrocarbon emissions may also result, along with toxic acrolein. Bound glycerol influences low temperature operability, lubricity and KV. As in the case of FFAs, bound glycerol has a beneficial effect on lubricity, which is apparent from the wear scars of MAG (139 μm), DAG (186 μm), and TAG (143 μm) versus methyl oleate (290 μm) (Knothe and Steidley, 2005b). However, saturated bound glycerol (saturated FAs bound to glycerol) has a deleterious effect on cold flow properties due to its low solubility in FAME. As a result, high temperatures are required to keep saturated MAG from crystallizing. Finally, bound glycerol increases KV, as evidenced by comparison of triolein (32.94 mm^2/s; Knothe and Steidley, 2005a) to methyl oleate (4.51 mm^2/s).

FFAs may be present in biodiesel prepared from a feedstock with high FFA content or by hydrolysis (Figure 4.2) in the presence of water and catalyst. FFAs are detected by titration following ASTM D664 are reported as AV. Upper limits on AV are specified in ASTM D6751 and EN 14214. FFAs also impact low-temperature performance, lubricity, and KV, and act as pro-oxidants. Damaged fuel lines and corrosion of engine components are also caused by FFAs. FFAs have higher KVs (40°C) and mp than FAME, as evidenced by comparison of oleic acid (19.91 mm^2/s; 12°C) to methyl oleate (4.51 mm^2/s; –20°C) (Gunstone et al., 2007, Knothe and Steidley, 2005a). FFAs have excellent lubricity, as seen by comparison of oleic acid (0 μm) to methyl oleate (290 μm) (Knothe and Steidley, 2005b).

4.9.8 Minor components

Minor components may include tocopherols, phospholipids, steryl glucosides, waxes, chlorophyll, fat-soluble vitamins, and hydrocarbons (Gunstone, 2004). The quantities of these endogenous components depend on feedstock, purification methods, and degree of preprocessing (refining, bleaching, deodorization, degumming, etc.) prior to transesterification. Many minor components, such as tocopherols, serve as beneficial natural antioxidants (Frankel, 2005). Chlorophylls, on the other hand, act as sensitizers for photooxidation (Gunstone, 2004).

Steryl glucosides form solids at temperatures above the CP, which cause engine failure due to fuel filter plugging. Many steryl glucosides have mp in excess of 240°C and are insoluble in biodiesel. This problem is especially prevalent in SME and PME because the parent oils contain high endogenous steryl glucoside levels. Steryl glucosides are not limited in ASTM D6751 or EN 14214, but the CSFT specification in ASTM D6751 is designed to detect solids that form above the CP. Either GC (Bondioli et al., 2008) or high-performance liquid chromatography (HPLC) (Moreau et al., 2008) is used to detect and quantify steryl glucosides. Removal of steryl glucosides is accomplished via cooling of biodiesel to cause precipitation and then filtration to remove the solids (Danzer et al., 2015).

4.10 Alternative feedstocks

An increase in biodiesel production capacity and governmental mandates and incentives for alternative fuels worldwide has necessitated the development of alternative feedstocks because it does not appear possible to meet the increased production capacity and mandated demand with traditional commodity sources of biodiesel (soybean, rapeseed/canola, palm, and coconut oils, and animal fats). For example, if all U.S. soybean production were dedicated to biodiesel, then only 6% of diesel demand would be satisfied (Hill et al., 2006). Furthermore, as discussed previously, commodity oils and fats account for up to 80% of biodiesel production expenses, which renders exploration of economical alternatives especially important.

Feedstocks that have much lower life-cycle greenhouse gas emissions than fossil fuels and do not compete with food production hold the most promise as sources for biofuels. Examples include perennial plants grown on fallow (unproductive) lands, crop residues, wood and forest residues, double crops and mixed cropping systems, and municipal and industrial wastes (Tilman et al., 2009). There are four major biodiesel feedstock categories: microalgae, oilseeds, animal fats, and low-value wastes such as used cooking oils, greases, and soapstocks. Microalgae is reviewed in greater depth by Chisti (2007, 2013).

Desirable characteristics of alternative oilseed feedstocks include adaptability to local growing conditions (rainfall, soil type, latitude), regional availability, high oil content, favorable FA composition, compatibility with existing farm infrastructure, low agricultural inputs (water, fertilizer, pesticides), definable growth season, uniform seed maturation rates, potential markets for agricultural by-products, compatibility with agriculturally undesirable lands, and/or off-season rotation with commodity crops (Moser, 2009a; Moser and Vaughn, 2012). Biodiesel prepared from feedstocks that meet the above criteria hold the most promise as alternatives to petrodiesel. Below are selected examples of alternative oilseed crops that meet some or most of these criteria.

4.10.1 *Jatropha*

Nonedible jatropha (*J. curcas* L.) oil has attracted interest as a feedstock for biodiesel production in India and other climatically similar regions. The jatropha tree is a drought-resistant shrub (up to 5 m) belonging to the *Euphorbiaceae* family whose seeds contain 30–40 wt% oil. Found in tropical and subtropical regions such as Central America, Africa, the Indian subcontinent, and other countries in Asia, jatropha is tolerant of saline soils and produces seed for 50 years (Koh and Ghazi, 2011). Crude jatropha oil contains 15% FFAs, which requires a two-step procedure to obtain FAMEs: esterification of FFAs followed by transesterification of TAGs (Berchmans and Hirata, 2008). In addition, the presence of toxic phorbol esters in jatropha oil prevent its use in edible applications (Devappa et al., 2012). Because jatropha oil contains a high percentage of saturated FAs (20%; Table 4.5), the corresponding biodiesel exhibits relatively poor low-temperature operability, as evidenced by a PP of 2°C (Tiwari et al., 2007).

4.10.2 *Field pennycress*

Field pennycress (*Thlaspi arvense* L.) is a winter annual belonging to the *Brassicaceae* family that is also known as stinkweed or French weed.

Table 4.5 Fatty acid profiles in area % of alternative oilseed feedstocks

Fatty acid	Jatropha	Field pennycress	Camelina	Seashore mallow
C14:0	–	–	0.1	0.2
C16:0	14.2	2.4	6.8	20.7
C16:1 9c	1.4	–	–	0.5
C18:0	6.9	0.2	2.7	1.8
C18:1 9c	43.1	11.0	18.6	13.7
C18:2 9c, 12c	34.4	19.5	19.6	53.1
C18:3 9c, 12c, 15c	–	8.9	32.6	0.6
C20:0	–	2.2	0.2	0.5
C20:1 11c	–	10.2	12.4	–
C20:2 11c, 14c	–	1.6	1.3	–
C22:0	–	0.2	0.2	0.3
C22:1 13c	–	36.2	2.3	–
C24:1 15c	–	3.6	–	–
Others (sum)	0	4.0	3.2	8.6

Note: Data are from Koh, M. Y. and T. I. M. Ghazi. 2011. *Renew. Sustain. Energ. Rev.* 15: 2240–51 (jatropha); Moser, B. R., Evangelista, R. L., and T. A. Isbell. 2016a. *Energ. Fuels* 30: 473–9 (field pennycress); Moser, B., R. and S. F. Vaughn. 2010. *Bioresour. Technol.* 101: 646–53 (camelina); and Moser, B. R., Seliskar, D. M., and J. L. Gallagher. 2016b. *Ind. Crops Prod.* 87: 20–6 (seashore mallow).

Members of this plant family are attractive as alternative feedstocks because they evolved to thrive in and originated mostly within the northern temperate climate zone (Moser, 2012). Many members also produce seeds with high oil content and include commercially important species such as *Brassica napus* (rapeseed). Native to Eurasia but with an extensive distribution throughout the world, field pennycress is highly adapted to a wide variety of climatic conditions. Considered an agricultural pest (weed), field pennycress may serve in a summer–winter rotational cycle with commodity crops (such as soybean), thus not displacing existing agricultural production. Field pennycress is tolerant of fallow lands, requires minimal agricultural inputs, is not part of the food chain, has high oil content (20–36 wt%), and is compatible with existing farm infrastructure (Isbell, 2009). Oil is readily expelled from seeds using a conventional heavy-duty screw press to yield both liquid (oil) and solid (press cake) fractions (Evangelista et al., 2012). The press cake inhibits seedling germination and emergence of agricultural pests, thus suggesting biofumigation properties (Vaughn et al., 2005). Erucic acid is the principal component of field pennycress oil (36.2%; Table 4.5). The remaining FA distribution consists mostly of oleic, linoleic, and linolenic acids.

The content of FFAs in crude field pennycress oil is sufficiently low for direct conversion to FAMEs following alkaline-catalyzed transesterification, thereby eliminating the need for acid pretreatment (Moser et al., 2009b). The cold flow properties of field pennycress oil methyl esters (FPMEs) (CFPP −17°C) are more favorable than commodity oil-derived FAMEs such as SMEs (Moser et al., 2016a). This has important implications during winter months in temperate climates where ambient temperatures regularly reach 0°C or lower. However, the high content of methyl erucate (C22:1 13c) results in a higher KV (5.24 mm^2/s) than typical for commodity-derived FAMEs (4.10–4.50 mm^2/s). Methods to lower KV include reduction of methyl erucate content through genetic modification or distillation, breeding techniques, or simply through blending with less-viscous FAMEs (Isbell et al., 2015; Moser, 2016). The presence of longer-chain FAMEs such as methyl erucate also results in a CN of 60 that is significantly higher than commodity oil-derived FAMEs. In contrast to other biodiesel fuels, cold flow properties of blends with petrodiesel are minimally impacted by blend ratio (Moser, 2014; Moser et al., 2016a). Therefore, field pennycress possesses several positive agronomic characteristics and its biodiesel has several performance advantages over biodiesel prepared from commodity lipids.

4.10.3 Camelina

Camelina [*Camelina sativa* (L.) Crantz], also known as false flax or gold-of-pleasure, is a broadleaf flowering plant of the *Brassicaceae* family that grows optimally in temperate climates. Camelina can be cultivated in a variety of climatic and soil conditions as a spring or summer annual or as a biannual winter crop. Cultivated in Europe since the Bronze Age, camelina has several beneficial agronomic attributes: short growing season (85–100 days), drought tolerance, compatibility with existing farm practices, tolerance of cold weather and semiarid conditions, and compatibility with low-quality soils. Camelina also has lower water, pesticides, and fertilizer requirements than commodity crops. Moreover, camelina is well adapted to the more northerly regions of North America, Europe, and Asia. As such, it may serve in a rotational cycle with winter wheat, thus facilitating disruption of undesirable weed and pest cycles (Moser, 2012).

The lipid content of individual seeds ranges from 28%–40%. Linolenic acid is the most abundant FA in camelina oil at 32.6% (Table 4.5), with most of the remaining content distributed among unsaturated FAs such as linoleic (19.6%), oleic (18.6%), and 11Z-eicosenoic (12.4%) acids. Similar to field pennycress oil, crude camelina oil can be directly converted to FAMEs with alkaline transesterification due to its low content of FFAs (Moser and Vaughn, 2010). The fuel properties of camelina oil methyl esters are similar to those of SME, thereby indicating its acceptability for use as biodiesel. It should be noted that, based on the high amount of unsaturated FAMEs contained in camelina oil-derived biodiesel, antioxidant additives are needed to meet the IP limits prescribed in ASTM D6751 and EN 14214. However, both ASTM D6751 and EN 14214 recommend that antioxidants be added to all biodiesel fuels immediately after their production, regardless of feedstock from which they are prepared.

4.10.4 Seashore mallow

Kosteletzkya pentacarpos (L.) Ledebour, known as seashore mallow, is a noninvasive, herbaceous perennial halophyte native to the Gulf and Atlantic coasts of the southeastern and eastern United States. Tolerant of saline soil and brackish water, seashore mallow is attractive for production of biofuels in coastal areas that are not otherwise suited for traditional agriculture (Moser et al., 2013). The stem core fibers are highly absorbent and have value for hydroseeding and animal bedding (Vaughn et al., 2013). Desirable agronomic characteristics include compatibility with existing farm infrastructure, tolerance of waterlogging, and resistance to disease and seed shattering. Finally, seashore mallow may be cultivated on saline or dry soil and irrigated with brackish estuarine water, thus liberating freshwater and high-quality soil for traditional agriculture while utilizing otherwise unproductive land (Moser et al., 2016b).

The oil content of seashore mallow seeds is around 19%, which is rich in linoleic acid (53.1%; Table 4.5). Most of the remaining content is distributed among palmitic (20.7%) and oleic (13.7%) acids. Cyclopropyl FAs such as malvalic (5.2%) and dihydrosterculic (1.2%) acids are also in the oil (Moser et al., 2016a), which renders it inedible due to negative effects on human and animal health (O'Brien, 2009). Similar to jatropha oil, pretreatment is required prior to alkaline-catalyzed transesterification, which lowers the AV from 2.72 to 0.08 mg KOH/g (Moser et al., 2013). Fuel properties of the FAME are within the specifications listed in ASTM D6751 and EN 14214 with the exception of IP. Similar to camelina, antioxidants are needed to meet IP specifications. Of note are high CN (60) and low iodine value (111 g I_2/100 g), both of which represent advantages over SME. Finally, properties of blends with petrodiesel are within the ranges specified in the petrodiesel standards after addition of antioxidants (Moser et al., 2013).

References

Akoh, C. C., Chang, S. W., Lee, G. C., and J. F. Shaw. 2007. Enzymatic approach to biodiesel production. *J. Agric. Food Chem.* 55: 8995–9005.

Berchmans, H. J. and S. Hirata. 2008. Biodiesel production from crude *Jatropha curcas* L. seed oil with a high content of free fatty acids. *Bioresour. Technol.* 99: 1716–21.

Bondioli, P., Cortesi, N., and C. Mariani 2008. Identification and quantification of steryl glucosides in biodiesel. *Eur. J. Lipid Sci. Technol.* 110: 120–6.

Chisti, Y. 2007. Biodiesel from microalgae. *Biotechnol. Adv.* 25: 294–306.

Chisti, Y. 2013. Constraints to commercialization of algal fuels. *J. Biotechnol.* 167: 201–14.

Craske, J. D. and C. D. Bannon. 1987. Gas liquid chromatography analysis of the fatty acid composition of fats and oils: A total system for high accuracy. *J. Am. Oil Chem. Soc.* 64: 1413–7.

Danzer, M. F., Ely, T. L., Kingery, S. A. et al. 2015. Biodiesel cold filtration process. U.S. Patent 9,109,170.

DeOliveira, E., Quirino, R. L., Suarez, P. A. Z., and A. G. S. Prado. 2006. Heats of combustion of biofuels obtained by pyrolysis and by transesterification and of biofuel/diesel blends. *Thermochim. Acta* 450: 87–90.

Devappa, R. K., Rajesh, S. K., Kumar, V., Makkar, H. P. S., and K. Becker. 2012. Activities of *Jatropha curcas* phorbol esters in various bioassays. *Ecotoxicol. Environ. Saf.* 78: 57–62.

Di Serio, M., Tesser, R., Pengmei, L., and E. Santacesaria. 2008. Heterogeneous catalysts for biodiesel production. *Energ. Fuels* 22: 207–17.

Dunn, R. O. 2008a. Antioxidants for improving storage stability of biodiesel. *Biofuels, Bioprod. Bioref.* 2: 304–18.

Dunn, R. O. 2008b. Crystallization behavior of fatty acid methyl esters. *J. Am. Oil Chem. Soc.* 85: 961–72.

Dunn, R. O. 2009. Cold-flow properties of soybean oil fatty acid monoalkyl ester admixtures. *Energ. Fuels* 23: 4082–91.

Evangelista, R. L., Isbell, T. A., and S. C. Cermak. 2012. Extraction of pennycress (*Thlaspi arvense*) seed oil by full pressing. *Ind. Crops Prod.* 37: 76–81.

Frankel, E. N. 2005. *Lipid Oxidation*, 2nd Ed. Bridgewater: The Oily Press.

Freedman, B., Pryde, E. H., and T. L. Mounts. 1984. Variables affecting the yields of fatty esters from transesterified vegetable oils. *J. Am. Oil Chem. Soc.* 61: 1638–43.

Gunstone, F. D. 2004. *The Chemistry of Oils and Fats. Sources, Composition, Properties and Uses*. Boca Raton: CRC Press.

Gunstone, F. D., Harwood, J. L., and A. J. Dijkstra. (eds.). 2007. Dictionary. In: *The Lipid Handbook*, 3rd Ed., 444–5. Boca Raton: CRC Press.

Haas, M. J. 2005. Improving the economics of biodiesel production through the use of low value lipids as feedstocks: Vegetable oil soapstock. *Fuel Process. Technol.* 86: 1087–96.

Haas, M. J., McAloon, A. J., Yee, W. C., and T. A. Foglia. 2006. A process model to estimate biodiesel production costs. *Bioresour. Technol.* 97: 671–8.

Haas, M. J., Michalski, P. J., Runyon, S., Nunez, A., and K. M. Scott. 2003. Production of FAME from acid oil, a byproduct of vegetable oil refining. *J. Am. Oil Chem. Soc.* 80: 97–102.

Hill, J., Nelson, E., Tilman, D., Polasky, S., and D. Tiffany. 2006. Environmental, economic, and energetic costs and benefits of biodiesel and ethanol biofuels. *Proc. Natl. Acad. Sci.* 103: 11206–10.

Holman, R. T. and O. C. Elmer. 1947. The rates of oxidation of unsaturated fatty acids and esters. *J. Am. Oil Chem. Soc.* 24: 127–9.

Isbell, T. A. 2009. US effort in the development of new crops (lesquerella, pennycress, coriander and cuphea). *Ol., Corps Gras, Lipides* 16: 205–10.

Isbell, T. A., Evangelista, R., Glenn, S. E. et al. 2015. Enrichment of erucic acid from pennycress (*Thlaspi arvense* L.) seed oil. *Ind. Crops Prod.* 66: 188–93.

Knothe, G. 2005. Dependence of biodiesel fuel properties on the structure of fatty acid alkyl esters. *Fuel Process. Technol.* 86: 1059–70.

Knothe, G. 2007. Some aspects of biodiesel oxidative stability. *Fuel Process. Technol.* 88: 669–77.

Knothe, G. 2008. "Designer" biodiesel: Optimizing fatty ester composition to improve fuel properties. *Energ. Fuels* 22: 1358–64.

Knothe, G. and R. O. Dunn. 2009. A comprehensive evaluation of the melting points of fatty acids and esters determined by differential scanning calorimetry. *J. Am. Oil Chem. Soc.* 86: 843–56.

Knothe, G. and K. R. Steidley. 2005a. Kinematic viscosity of biodiesel fuel components and related compounds. Influence of compound structure and comparison to petrodiesel fuel components. *Fuel* 84: 1059–65.

Knothe, G. and K. R. Steidley. 2005b. Lubricity of components of biodiesel and petrodiesel. The origin of biodiesel lubricity. *Energ. Fuels* 19: 1192–200.

Knothe, G., Matheaus, A. C., and T. W. Ryan III. 2003. Cetane numbers of branched and straight-chain fatty esters determined in an ignition quality tester. *Fuel* 82: 971–5.

Koh, M. Y. and T. I. M. Ghazi. 2011. A review of biodiesel production from *Jatropha curcas* L. oil. *Renew. Sustain. Energ. Rev.* 15: 2240–51.

Lotero, E., Liu, Y., Lopez, D. E. et al. 2005. Synthesis of biodiesel via acid catalysis. *Ind. Eng. Chem. Res.* 44: 5353–63.

Moreau, R. A., Scott, K. M., and M. J. Haas. 2008. The identification and quantification of steryl glucosides in precipitates from commercial biodiesel. *J. Am. Oil Chem. Soc.* 85: 761–70.

Moser, B. R. 2008. Influence of blending canola, palm, soybean, and sunflower oil methyl esters on fuel properties of biodiesel. *Energ. Fuels* 22: 4301–6.
Moser, B. R. 2009a. Biodiesel production, properties, and feedstocks. *In Vitro Cell. Dev. Biol.-Plant* 45: 229–66.
Moser, B. R. 2009b. Comparative oxidative stability of fatty acid alkyl esters by accelerated methods. *J. Am. Oil Chem. Soc.* 86: 699–706.
Moser, B. R. 2011. Influence of extended storage on fuel properties of methyl esters prepared from canola, palm, soybean and sunflower oils. *Renew. Energ.* 36: 1221–6.
Moser, B. R. 2012. Biodiesel from alternative oilseed feedstocks: Camelina and field pennycress. *Biofuels* 3: 193–209.
Moser, B. R. 2014. Impact of fatty ester composition on low temperature properties of biodiesel-petroleum diesel blends. *Fuel* 115: 500–6.
Moser, B. R. 2016. Fuel property enhancement of biodiesel fuels from common and alternative feedstocks via complementary blending. *Renew. Energ.* 85: 819–25.
Moser, B. R., Cermak, S. C., and T. A. Isbell. 2008. Evaluation of castor and lesquerella oil derivatives as additives in biodiesel and ultralow sulfur diesel fuels. *Energ. Fuels* 22: 1349–52.
Moser, B. R., Dien, B. S., Seliskar, D. M., and J. L. Gallagher. 2013. Seashore mallow (*Kosteletzkya pentacarpos*) as a salt-tolerant feedstock for production of biodiesel and ethanol. *Renew. Energ.* 50: 833–9.
Moser, B. R., Evangelista, R. L., and T. A. Isbell. 2016a. Preparation and fuel properties of field pennycress (*Thlaspi arvense*) seed oil ethyl esters and blends with ultralow-sulfur diesel fuel. *Energ. Fuels* 30: 473–9.
Moser, B. R., Haas, M. J., Winkler, J. K. et al. 2007. Evaluation of partially hydrogenated methyl esters of soybean oil as biodiesel. *Eur. J. Lipid Sci. Technol.* 109: 17–24.
Moser, B. R., Knothe, G., Vaughn, S. F., and T. A. Isbell. 2009b. Production and evaluation of biodiesel from field pennycress (*Thlaspi arvense*) oil. *Energ. Fuels* 23: 4149–55.
Moser, B. R. Seliskar, D. M., and J. L. Gallagher. 2016b. Fatty acid composition of fourteen seashore mallow (*Kosteletzkya pentacarpos*) seed oil accessions collected from the Atlantic and Gulf coasts of the United States. *Ind. Crops Prod.* 87: 20–6.
Moser, B. R. and S. F. Vaughn. 2010. Evaluation of alkyl esters from *Camelina sativa* oil as biodiesel and as blend components in ultra-low sulfur diesel fuel. *Bioresour. Technol.* 101: 646–53.
Moser, B. R. and S. F. Vaughn. 2012. Efficacy of fatty acid profile as a tool for screening feedstocks for biodiesel production. *Biomass Bioenerg.* 37: 31–41.
Moser, B. R., Williams, A., Haas, M. J., and R. L. McCormick. 2009a. Exhaust emissions and fuel properties of partially hydrogenated soybean oil methyl esters blended with ultralow sulfur diesel fuel. *Fuel Process. Technol.* 90: 1122–8.
Naik, M., Meher, L. C., Naik, S. N., and L. M. Das. 2008. Production of biodiesel from high free fatty acid Karanja (*Pongamia pinnata*) oil. *Biomass Bioenerg.* 32: 354–7.
Narasimharao, K., Lee, A., and K. Wilson. 2007. Catalysts in production of biodiesel: A review. *J. Biobased Mat. Bioenerg.* 1: 19–30.
O'Brien, R. D. 2009. *Fats and Oils. Formulating and Processing for Applications*, 3rd Ed. Boca Raton: CRC Press.

Talukder, M. M. R., Wu, J. C., Lau, S. K. et al. 2009. Comparison of Novozym 435 and Amberlyst 15 as heterogeneous catalyst for production of biodiesel from palm fatty acid distillate. *Energ. Fuels* 23: 1–4.
Tilman, D., Socolow, R., Foley, J. A., et al. 2009. Beneficial biofuels – The food, energy, and environment trilemma. *Science* 325: 270–1.
Tiwari, A.K., Kumar, A., and H. Raheman. 2007. Biodiesel production from jatropha oil (*Jatropha curcas*) with high free fatty acids: An optimized process. *Biomass Bioenerg.* 31: 569–75.
Van Gerpen, J. 2005. Biodiesel processing and production. *Fuel Process. Technol.* 86: 1097–107.
Vaughn, S. F., Isbell, T. A., Weisleder, D., and M. A. Berhow. 2005. Biofumigation compounds released by field pennycress (*Thlaspi arvense*) seedmeal. *J. Chem. Ecol.* 31: 167–77.
Vaughn, S. F., Moser, B. R., Dien, B. S. et al. 2013. Seashore mallow (*Kosteletzkya pentacarpos*) stems as feedstock for biodegradable absorbants. *Biomass Bioenerg.* 59: 300–5.

chapter five

Algal biodiesel production

An overview

Sourav Kumar Bagchi, Reeza Patnaik, and Nirupama Mallick

Contents

5.1 Background and driving forces

5.1.1 Biofuels

The last decade has witnessed a tremendous impetus on biofuel research due to the irreversible depletion of fossil fuels and the escalating emissions of greenhouse gases into the atmosphere. Biofuels are subdivided into two major liquid fuels, namely, bioethanol and biodiesel. The lignocellulosic crops like sugarcane and corn are fermented to produce bioethanol, whereas biodiesel, the other transportation fuel, is produced from various lipid-rich sources starting from many food materials ranging from vegetable oils to tiny microorganisms. Bioethanol is generally used in combination with gasoline and biodiesel is blended with diesel to open options for alternative fuel utilization in the transport sector. However, the main concern for the suitability of a biofuel product currently depends on two major factors, that is, the land use and appropriate biomass to fuel energy conversion (Kosinkova et al., 2015). Global biofuel production has been rapidly growing for the last few years, from 15 billion liters in 2000 to 110 billion liters in 2013, and has captured a 3.5% share of the total transportation fuel in the world (REN21, 2015). We find that biofuels meet nearly one-fourth of the road transport fuel requirements in Brazil. Global biofuel production has been projected to be approximately 140 billion liters by the end of 2018.

5.1.2 Biofuel production scenario in India

India, the second most populous country in the world, now has a large expectation and hope to conserve its national energy security by the development of renewable biofuels. At this time, India is the fourth largest energy consumer and net importer of crude oil and petroleum products in the world after the United States, China, and Japan (EIA, 2014). The oil production and consumption scenario in India is very unique. In the year of 2014, India produced around 767,600 barrel (bbl)/day crude oil. However, this developing country has invested a lot of its budget to import crude and refined oils from other countries. A report showing that in the year of 2013, India imported around 3.812 million bbl/day of crude oil and 312,000 bbl/day of refined oil to fulfill its gross energy consumption (CIA, 2015). India has a substantial amount of biomass resources, but their use in development toward sustainability was significantly low compared to many countries like the United States, China, Brazil, and the United Kingdom.

It is amazing to see the transportation fuel requirements in India, which are quite unique in the world. India consumes almost five times less gasoline or petrol fuel than diesel fuels. In all other countries like the United States, this trend is totally reverse, with gasoline consumption being much higher than diesel fuel (Khan et al., 2009). In continuation of this report, the Petroleum Planning and Analysis Cell (PPAC), Government of India, showed that out of the gross consumption of all petroleum products in 2012–2013, the consumption of diesel oil accounted for 69.08 million tons (43.98% shares), whereas the total petrol consumption was only 15.71 million tons (10.02% shares). Apart from its wide use in the transport sector, diesel is extensively used in the sweetmeat and bakery industries in urban areas of India (PPAC, 2013). The preference of diesel engines over engines run by other forms of fuel in India gives the development of biodiesel technology a greater potential for this nation. Bioethanol, which is used in blends with gasoline, has low cetane numbers and, therefore, is unsuitable for diesel engines. On the other hand, biodiesel can be used to run conventional diesel engines with only minor adjustments (Rao and Gopalkrishnan, 1991). Considering the above facts, the primary need for Indian transport sectors is to produce biodiesel as an alternative to diesel fuel.

5.1.3 Biodiesel

Biodiesel has become a salient part of the field of renewable energy production. Biodiesel is a renewable, biodegradable, sustainable, and environment-friendly fuel that is obtained chemically by reacting lipids resulting from animal or plant sources with an alcohol, preferably methanol due to its low cost, to create fatty acid methyl esters (FAMEs). This chemical reaction is termed "transesterification" and requires a strong acid (sulfuric acid or hydrochloric acid) or a strong base (caustic potash or caustic soda) or enzyme lipase as catalyst (Meher et al., 2006). Transesterification is a reversible reaction and proceeds essentially by mixing the reactants. A by-product of the transesterification reaction is the production of glycerol. Generally, this glycerol by-product also has a high market value, which has made this transesterification process an economically feasible venture. Currently, research is being performed around the globe to produce some chemical building blocks (intermediate chemicals) from glycerol to make this by-product more commercially viable (http://www.bioenergywiki.net/Websites).

In the year of 1850, two scientists, E. Duffy and J. Patrick, designed the first diesel engine. Some years later, on August 10, 1893, another brilliant scientist named Sir Rudolf Diesel made a leading new form of diesel engine with a 3-m iron cylinder and a flywheel in its base point. The new engine was powered with vegetable peanut oil for its first test on August 10, 1893, in Augsburg, Germany. Diesel predicted that biodiesel would replace fossil fuels in the future to fuel different mechanical engines. To mark Sir Diesel's amazing achievements, August 10 is celebrated as International Biodiesel Day every year. In 2015, India also celebrated this day with a great enthusiasm. At this event, the Union Petroleum Ministry, Government of India, launched four retail outlets situated in Delhi, Haldia, Vijaywada, and Visakhapatnam for biodiesel blended with conventional diesel fuel (https://currentaffairs.gktoday.in/10-august-world-bio-fuel-day-08201525506.html). A recent report published by the Global Agricultural Information Network (GAIN) clearly demands that 5% biodiesel blended with diesel (B5) be commercially available from these four retail sale outlets and that private companies also be allowed to sell biodiesel directly if they will produce biodiesel according to the specified official standards (GAIN, 2016).

Biodiesel is the only substitute with energy densities (with the consideration of volume and weight) rivaling liquid petroleum fuels. The report of Worldwatch Institute (2011) stated that the leading biodiesel producer is the European Union with 53% of the total biodiesel production in the world. Another report claimed that the International Energy Agency (IEA) has a target for biodiesel to reach more than 25% of world demand for transportation fuels by the year 2050 to reduce reliance on coal and petroleum fuels (Tanaka, 2011). The major advantages of using biodiesel are primarily its nontoxic and biodegradable nature (Krawczyk, 1996). It burns in conventional diesel engines with minor or without any modification while reducing pollution in comparison to the conventional diesel fuel (McMillen et al., 2005).

5.1.4 Generations of biodiesel production

There are three generations of biodiesel production from various plant sources. First-generation biodiesel was sourced from the edible plant sources like soya, mustard, jojoba, flax, sunflower, palm, coconut, and safflower. However, there are two foremost concerns over the sustainability of first-generation biodiesel production. First, these crops would be in a race for arable lands with conventional food crops. Second, energy savings and greenhouse gas controls over the life-cycle analysis may be less than the predicted values. As an example, the energy input needed for the life cycle was more than 50% of the energy contained in the fuel for the biodiesel produced from soya and rape seeds (Nigam and Singh, 2011). The second generation of biodiesel was produced from nonedible plant sources like jatropha (*Jatropha curcas*), karanja (*Pongamia pinnata*), mahua (*Madhuca longifolia*), and field pennycress (*Thlaspi arvense*) (Dorado et al., 2002; Ghadge and Raheman, 2006). Nevertheless, the cost of biodiesel production is still a major obstacle for large-scale commercial exploitation, mainly due to the large land area requirement and long growth periods for second-generation plants (Chisti, 2008). It is also essential to mention that oil production reduces significantly when jatropha plants are grown on marginal land. Therefore, interest in jatropha cultivation has decreased considerably in recent years (http://biofuel.org.uk). Currently, attention is focused on microalgal biodiesel production technology, which is becoming the third generation of biodiesel. There are a number of advantages for microalgal biodiesel production over the first and second generations' biodiesel:

The photosynthesis in microalgal cellular system is like to that of higher plants; however, the miniature organisms have the capability to capture 10–50 times more solar energy than that of the terrestrial plants (Li et al., 2008).

Algae can be cultivated in freshwater, wastewater, or saline water on nonarable land, thereby reducing competition for arable land, limited freshwater, and nutrients used for conventional agriculture. Moreover, algae cultivation does not need herbicides or pesticides (de Godos et al., 2009). High-quality agricultural land is not required at all to grow the microalgal biomass (Scott et al., 2010).

The doubling time of microalgae is only a few hours during the exponential phase. The yield of microalgal oil has been predicted to be 58,700–136,900 L ha^{-1} $year^{-1}$ with the variation of oil content of algae (Chisti, 2007; Moazami et al., 2011).

Though reports showed that microalgae production costs more per unit mass, there are claims showing the yield could be 100 times more fuel per unit area than the second-generation biofuel crops (Greenwell et al., 2010).

It is also noteworthy that microalgae can biomitigate CO_2 from flue gases emitted from fossil fuel-fired power plants and other sources, thereby reducing emissions of a major greenhouse gas (Watanabe and Saiki, 1997).

After lipid extraction, the spent algal biomass can also be used to produce bioethanol and many other industrially important value-added bioproducts like carotenoids, beta-carotene, and some colorants (Shukla and Dhar, 2014).

In view of the above considerations, microalgae are considered to be viable feedstocks for third-generation advanced biofuel production, minimizing the major drawbacks associated with the first- and second-generation feedstocks.

5.2 *Microalgae: The green gold*

Microalgae, also called phytoplanktons or microphytes, are very small plantlike organisms that do not have roots or leaves and are generally a few micrometers in diameter. Microalgae are typically found in aquatic environments both in freshwater and saline water and can survive in both the water column and sediments (Thurman, 1997). They are unicellular or multicellular species that can exist either individually or in groups or chains. They are capable of performing photosynthesis and are able to generate half of the oxygen required for the survival of other living organisms on earth. There are nearly 200,000–800,000 different microalgal species available on Earth, but only a few hundred have been described in the literature until now (Thrush et al., 2006). Algae may be prokaryotic in nature, namely, cyanobacteria (Chloroxybacteria), or may be eukaryotic in nature like red algae (Rhodophyta), brown algae (Phaeophyta), diatoms (Bacillariophyta), green algae (Chlorophyta), and much more. The most recent widely accepted algae classification of inland waters was reported by Krienitz (2009) based on phylogenetic analysis, morphological characterization, and advanced RNA sequencing. According to that study, algae are mainly divided into two groups, Prokaryota and Eukaryota. Prokaryota has two major divisions, Cyanophyta (Cyanobacteria) and Prochlorophyta. Eukaryota is classified into eight major divisions, namely, Rhodophyta, Chlorophyta, Heterokontophyta, Haptophyta, Chrysophyta, Dinophyta, Euglenophyta, and Charophyta. These divisions are also classified into several other classes and subclasses (Krienitz, 2009). Hence, there are immense possibilities to explore many new and indigenous isolated microalgal species which may be useful for various kinds of human benefits starting from food and healthcare to advanced biofuel production (Singh and Gu, 2010).

Since the last decade, microalgae, and more specifically, green microalgal feedstocks, have been becoming more important in the field of biodiesel production due to their rapid growth rate coupled with elevated lipid contents. However, biodiesel production from microalgae is primarily dependent upon the biomass yield and cellular lipid accumulation potential of the selected species. Moreover, microalgae can be cultivated in wastewater or seawater and an elevated oil content of 50%–60% of dry cell weight (dcw) can be achieved as compared with some best terrestrial oil crops of only 10%–15% oil content (Li et al., 2008).

It is noteworthy that microalgae can bioremediate wastewater by removing $NH4^+$, $NO3^-$, and $PO4^{3-}$ from a variety of wastewater sources. Researchers reported that microalgae are able to grow in various wastewaters and the wastewater sources proved to be a potential source of low-cost lipids for biodiesel production (Woertz et al., 2009). In another report, Mandal and Mallick (2011) showed that the green microalga *Scenedesmus obliquus* was grown successfully in three kinds of waste discharges, namely, the secondary settling tank discharge of municipal wastewater, a fish pond, and poultry litter discharges. The lipid accumulation reached up to 1.0 g L^{-1} with an increased level of saturated fatty acid content.

However, there remain many hurdles in each and every step for microalgal biodiesel production, starting from strain selection and improvement of lipid accumulation and yield to algae mass cultivation to oil extraction and biodiesel conversion (transesterification). Isolation and screening of oleaginous microalgae are two pivotal important upstream factors that should be addressed according to the need of freshwater or marine algae with a consideration that wild-type indigenous isolate can be the best suited for laboratory to large-scale exploitation. There is currently a lot of literature available on microalgal biodiesel production, but only a very few illustrate a detailed stepwise description with the pros and cons of the processes of biodiesel production from microalgae.

5.2.1 Nutrient starvation/deficiency and other stresses can trigger lipid accumulation in microalgae

In the microalgal biodiesel production process, the foremost target is to choose such a microalgal strain that is capable of high lipid productivity because it largely influences the production cost (Sheehan et al., 1998). Lipid yield primarily depends on the microalgal growth rate and the lipid content of the algal strain. Each species of microalga produces different ratios of lipids, carbohydrates, and proteins. Nevertheless, these tiny organisms have the ability to change their metabolism through simple manipulations of the chemical composition of the culture medium (Behrens and Kyle, 1996), thus high lipid productivity can be achieved. Physiological stresses such as nutrient starvation and deficiency, high light intensity, and salt stress have been employed for directing metabolic fluxes to lipid biosynthesis of microalgae. Reports are available in which attempts have been made to raise the lipid pool of microalgal species. Among all the nutrients, nitrogen was found to be the vital nutrient affecting the cellular lipid accumulation in microalgae as was reported by various researchers in several studies (Mandal and Mallick, 2009; Ren et al., 2013; Fan et al., 2014; Shen et al., 2016; Wang et al., 2016).

Many researchers also reported that this nutritional stress led to a boost in the lipid content higher than 50% (dcw) in various microalgal species (Table 5.1). Mallick et al. (2012) found that lipid content of *Chlorella vulgaris* was significantly boosted up to 57% (9% under control culture condition) when cultured under nitrogen, iron, and phosphorus limitations. In another study, the green microalga *Scenedesmus* sp. showed lipid accumulation of 53% (dcw) in nitrogen-deficient medium (Ren et al., 2013). The high light intensity with nitrogen-depleted conditions was also found to raise the lipid content up to 54% (dcw) compared with 11% under control culture conditions in *Nannochloropsis oceanica* IMET1 (Xiao et al., 2015). Fakhry and Maghraby (2015) also reported that limited nitrogen addition (25% of control) triggered the lipid accumulation up to 59% (dcw) in *N. salina*.

An increased lipid content of 51% (dcw) was reported under limited nitrogen addition in the medium for the microalga *C. sorokiniana*, whereas

Table 5.1 Higher lipid accumulation in some microalgal species under various stress conditions

Name of the microalga	Stress condition	Lipid content (% dcw)[a]	Reference
S. obliquus	Nitrogen and phosphorus limitations in the presence of thiosulphate	58 (13)	Mandal and Mallick (2009)
C. vulgaris	Nitrogen, iron, and phosphorus limitations	57 (09)	Mallick et al. (2012)
B. braunii	Phosphorous limitation	54 (17)	Ruangsomboon (2012)
Scenedesmus sp. strain R-16	Nitrogen deficiency	53 (14)	Ren et al. (2013)
C. vulgaris	Ca starvation	40 (12)	Gorain et al. (2013)
S. obliquus		37 (11)	
C. vulgaris	NaCl stress	40 (12)	
S. obliquus		39 (11)	
Scenedesmus sp.	Low Ca^{2+} concentration	47 (10)	Ren et al. (2014)
N. oceanica IMET1	High light intensity and nitrogen-depleted culture conditions	54 (11)	Xiao et al. (2015)
N. salina	Limited nitrogen addition	59 (35)	Fakhry and Maghraby (2015)
C. sorokiniana	Nitrogen limitation	51 (25)	Li et al. (2015)
Chlorella sp.	Light intensity of 10,000 lux	32 (18)	Han et al. (2015)
C. pyrenoidosa	NaCl stress	44 (10)	Bajwa and Bishnoi (2015)
C. vulgaris NIES-227	Nitrogen-starved heterotrophic culture	56 (15)	Shen et al. (2016)
Chlamydomonas reinhardtii CC125	Nitrogen starvation with 200 mM NaCl stress	54 (25)	Kwak et al. (2016)
C. pyrenoidosa	Nitrogen deficiency	52 (34)	Wang et al. (2016)
A. obliquus	Optimum metal stress with EDTA	54 (26)	Singh et al. (2016)
P. kessleri	Sulfur deficiency	51 (15)	Ota et al. (2016)

[a] Values in parentheses represent lipid content under control culture condition.

the value was only 25% (dcw) under the control culture condition (Li et al., 2015). Shen et al. (2016) reported that the lipid content was significantly boosted up to 56% (dcw) in nitrogen-starved acetate-supplemented heterotrophic cultures compared with 15% (dcw) control for the chlorophycean microalga *C. vulgaris* Beijerinck var. *vulgaris* NIES-227. In a recent study, Kwak et al. (2016) found that under nitrogen starvation with 200 mM salt stress, the microalga *Chlamydomonas reinhardtii* CC125 showed a high lipid content of 54% (dcw) compared with only 25% (dcw) recorded under the control condition. Wang et al. (2016) also reported that the green microalga *C. pyrenoidosa* showed an increased lipid accumulation up to 52% (dcw) under nitrate starvation compared with 34% control.

In addition to nitrogen limitation and deficiency, various other stress conditions may also improve the cellular lipid accumulation in microalgae (Table 5.1). Ruangsomboon (2012) found that under phosphorous limitation, the microalga *Botryococcus braunii* showed a high lipid content of 54% (dcw) compared with 17% (dcw) recorded under the control condition. Gorain et al. (2013) reported that under calcium starvation, the lipid content was boosted up to 40% and 37%, respectively, compared with 12% and 11% (dcw) control for the microalgae *C. vulgaris* and *S. obliquus*. In another report (Ren et al., 2014), the microalga *Scenedesmus* sp. Showed a lipid content of 47% (dcw) compared with 10% control under a low dosage of calcium. Han et al. (2015) showed an elevated lipid content of 32% (dcw) under the high light intensity of 10,000 lux for the green microalga *Chlorella* sp. Bajwa and Bishnoi (2015) found that NaCl stress could improve lipid accumulation in *C. pyrenoidosa* (Table 5.1). A very recent study has also revealed that proper optimization of various metal stresses with ethylenediaminetetraacetic acid (EDTA) could be effective for high lipid accumulation in the chlorophycean microalga *Acutodesmus obliquus* (Singh et al., 2016). It was also demonstrated that the lipid content was increased up to 51% (dcw) under sulfur deficiency compared with only 15% (dcw) found in the control condition for the microalga *Parachlorella kessleri* (Ota et al., 2016).

5.2.2 CO_2 biofixation and lipid accumulation potential of microalgae

The U.S. Environmental Protection Agency (USEPA) reported that CO_2 accounted for 84% of all U.S. greenhouse gas emissions by human activities, and currently one-fifth of the global CO_2 emissions are only from the transport sector (Anastas, 2011). According to a contemporary report of the World Bank, the huge growth of cement industries is also responsible for CO_2 emissions to the atmosphere (World Bank Report). One possible solution for the reduction of atmospheric CO_2 concentration could be the use of green photosynthetic microalgae, which biologically fix CO_2 via photosynthesis, and subsequent utilization of the resulting biomass for production of lipids, and consecutively renewable biodiesel (Balat and Balat, 2010). It is noteworthy here that the European renewable energy policy-making agencies have set a target to reduce 50% of greenhouse gas emissions by 2017 with the help of biofuels (Hennecke et al., 2013). Various reports have also shown that the CO_2 level selected for the optimum growth of microalgae is below 20%, and thermal power stations' exhaust gasses contain 10%–20% CO_2, which is ~500 times higher than the atmospheric CO_2 concentration (Kodama et al., 1993; Zhao and Su, 2014). In a study, it was reported that the maximum biomass yield and CO_2 fixation rate were 2.34 g L^{-1} and 0.373 g L^{-1} day^{-1}, respectively, for the mixed microalgal cultures, grown under 15% CO_2 supplementation (Singh et al., 2015). Nascimento et al. (2015) observed that the two green microalgae, *C. vulgaris* and *B. terribilis*, grown under the elevated level of CO_2 showed a maximum CO_2 biofixation rate of 0.61 under 5%–10% CO_2 concentration. In continuation, Zhao et al. (2015) also reported that the biomass yield and CO_2 biofixation rate were found to be 2.07 g L^{-1} and 0.50 g L^{-1} day^{-1} for the microalgae *Chlorella* sp. grown under 0.1 volume of air per volume of liquid per minute (vvm) air input with 10% CO_2 supplementation (Table 5.2). A very recent report revealed that the biomass yield and the corresponding CO_2 biofixation rate were recorded as 0.50, 0.34, 0.30 g L^{-1} and 0.13, 0.12, 0.09 g L^{-1} day^{-1} for the microalgae *N. limnetica*, *B. braunii*, and *Stichococcus bacillaris*, respectively, grown in a 30-L photobioreactor (PBR) under 0.1 vvm air input with 3% CO_2 sparging (Parupudi et al., 2016).

It is striking that the elevated level of CO_2 sparging can enhance the cellular lipid content in microalgae. Table 5.2 provides an overview of these studies. Chiu et al. (2009) found an increase in the lipid pool up to 50% (dcw) compared with 31% control for the microalga *N. oculata* NCTU-3 grown under 2% CO_2 concentration. Tang et al. (2011) reported that the two green microalgae, *S. obliquus* and *C. pyrenoidosa*, grew well under 10% CO_2 concentration in 1-L Erlenmeyer flasks, and the maximum biomass yield was recorded as 1.84 and 1.55 g L^{-1}, respectively, whereas the maximum lipid yield was found to be 350 and 372 mg L^{-1}, respectively. The rise in lipid content from 15% to 18.5% (dcw) for *S. obliquus*, and 20.9% to 24% (dcw) for *C. pyrenoidosa* were recorded. Following this, Jiang et al. (2011) also found that the maximum lipid yield was 1.34 g L^{-1} (lipid content was 60% dcw) for *Nannochloropsis* sp. grown under 15% CO_2 with N-starvation and high light intensity (compared with 23.7% control). Another report showed that under biphasic nitrogen starvation with 2%–3% CO_2 sparging, *C. vulgaris* showed an enhanced lipid content of 53% (dcw) compared with 24.6% control (Mujtaba et al., 2012).

Moreover, many researchers have found that the increased lipid yield is generally associated with the high biomass production in microalgae under an elevated level of CO_2 sparging. Anjos et al. (2013) demonstrated

Table 5.2 Effects of CO_2 sparging on microalgal CO_2 biofixation and biomass and lipid yield

Test organism	CO_2 concentration (%)	CO_2 biofixation rate (g L^{-1} day^{-1})	Biomass yield (g L^{-1})[a]	Lipid yield (mg L^{-1})[a]	Reference
Chlorella sp.	2	NR	1.45 (0.68)	NR	Chiu et al. (2009)
S. obliquus	10	0.29	1.84 (1.05)	342.1 (159.1)	Tang et al. (2011)
C. pyrenoidosa		0.26	1.55 (0.87)	372.0 (181.8)	
Nannochloropsis sp.	15% CO_2 under N starvation + high light intensity + wastewater	NR	2.23 (0.16)	1338.2 (37.9)	Jiang et al. (2011)
C. vulgaris	2% CO_2 + reduced dose of nitrate	0.43	2.03 (NR)	925.7 (NR)	Yeh and Chang (2011)
Scenedesmus sp.	4	NR	1.40 (NR)	NR	Pegallapati et al. (2012)
C. vulgaris	2%–3% CO_2 sparging under N starvation	NR	1.60 (1.28)	848.1 (315.0)	Mujtaba et al. (2012)
C. vulgaris (strain P12)	6.5	2.22	10.0 (NR)	1100.0 (NR)	Anjos et al. (2013)
Nannochloris eucaryotum	100	NR	0.35 (NR)	56.7 (NR)	Concas et al. (2013)
Mixed microalgae	15	0.37	2.34 (0.75)	NR	Singh et al. (2015)
C. vulgaris	5	0.61	1.04 (NR)	198.6 (NR)	Nascimento et al. (2015)
B. terribilis	10	0.61	1.51 (NR)	455.1 (NR)	
B. braunii		0.56	1.56 (NR)	496.1 (NR)	
Chlorella sp.	10	0.50	2.07 (NR)	NR	Zhao et al. (2015)
N. limnetica	3	0.13	0.50 (0.22)	0.17 (0.07)	Parupudi et al. (2016)
B. braunii		0.12	0.34 (0.16)	0.10 (0.04)	
S. bacillaris		0.09	0.30 (0.15)	0.08 (0.04)	
Desmodesmus sp. 3Dp86E-1	20	NR	4.50 (3.10)	NR	Solovchenko et al. (2016)
S. obliquus (Turpin) Kütz.	15	0.77	7.01 (1.35)	2.0 (0.16)	Bagchi and Mallick (2016)

[a] Values in the parentheses represent biomass and lipid yield for control cultures without CO_2 sparging; NR = not reported.

that the lipid yield was found to be 1.1 g L^{-1} for *C. vulgaris* strain P12 grown under 6% CO_2 sparging in a bubble column photobioreactor. The maximum biomass yield was recorded as 1.04, 1.51, and 1.56 g L^{-1} for the three microalgal species, *C. vulgaris, B. terribilis,* and *B. braunii,* whereas the maximum lipid yield was found to be 198.6, 455.1, and 496.1 mg L^{-1}, respectively, under 20% CO_2 sparging (Nascimento et al., 2015). However, with 100% CO_2 sparging, the biomass and lipid yield were recorded as 0.35 g L^{-1} and 56.7 mg L^{-1}, respectively, for the microalga *Nannochloris eucaryotum* (Concas et al., 2013). A detailed review is presented in Table 5.2. In a recent report, Mohsenpour and Willoughby (2016) found a high biomass and lipid yield of 2.25 g L^{-1} and 827.1 mg L^{-1} under 15% CO_2 supplementation for the green microalga *G. membranacea CCAP 1430/3*. In this report, the maximum CO_2 biofixation rate was recorded as 0.21 g L^{-1} day^{-1}.

Parupudi et al. (2016) also reported that, under 3% CO_2 supplementation, the lipid yield was found to be enhanced up to 170, 100, and 80 mg L^{-1} compared with 70, 40, and 39.6 mg L^{-1} found in control cultures for the three microalgae *N. limnetica, B. braunii,* and *S. bacillaris,* respectively. In a recent study, the locally isolated chlorophycean microalga *Sc. obliquus* (Turpin) Kützing depicted the maximum biomass and lipid yields of 7.01 and 2 g L^{-1} grown under mixotrophic growth condition with 15% CO_2 sparging at 0.6 vvm air flow and 1% glucose supplementation followed by biphasic nitrogen starvation in a photobioreactor system. The maximum CO_2 biofixation rate was also recorded as 0.77 g L^{-1} day^{-1} (Bagchi and Mallick, 2016).

5.3 Microalgae cultivation: Strategies for biomass and lipid production

5.3.1 Photobioreactors

Microalgae can be cultivated in PBRs with high process control. Depending on the configuration, PBRs are classified into four major types, namely, vertical or columnar, tubular, annular, and flat plate reactors (Brennan and Owende, 2010). The basic principle behind the construction of PBRs is to reduce the light path length to increase the light energy accumulation by every single cell of alga (Borowitzka, 1999). For this, mixing is achieved by mechanical stirring and airlift pumping, which provide simultaneous benefits of mixing, aeration, and high gaseous exchange (Amaro et al., 2011).

PBRs are also suitable for outdoor cultivation, especially tubular and flat plate PBRs, which have a wide surface area that is largely exposed to light energy. Researchers also found that the tubular/bubble-column reactors are the most suitable to produce high-value bioproducts from algae such as astaxanthin (Norsker et al., 2011). Due to the closed environment, PBRs have certain advantages like cultivation, proliferation and maintenance of monoculture, low space requirements, control over temperature, pH, CO_2 sparging, and light intensity as per the requirement (Harun et al., 2010; Mutanda et al., 2011).

It was reported that high biomass yield up to 6.0 g L^{-1} could be achieved for various microalgae cultivated in tubular PBRs (Davis et al., 2011). It was also projected that high surface-to-volume ratios for well-designed PBRs may able to produce biomass yield up to 10.0 g L^{-1} (Stephens et al., 2010). In a recent report, Soman and Shastri (2015) suggested that PBRs should be operated with optimized dark-light cycles with an optimum surface-to-volume ratio. Moreover, hydrodynamics based on certain simulations would be helpful to improve the overall biomass productivity.

5.3.2 *Raceway ponds*

5.3.2.1 *General outline*

Microalgae mass cultivation in raceway ponds has now been considered as the most promising means for large-scale biodiesel production in terms of less capital investment and least operational cost compared to the engineered closed PBR systems. Microalgae mass cultivation carries immense importance because an enormous amount of biomass is required for commercial production of biodiesel from microalgae.

Researchers observed that the practice of microalgae mass cultivation in raceways as a source of biomass for biodiesel production was successfully initiated in the early 1950s with the chlorophycean microalga *Chlorella*, and then moved toward large-scale biomass generation in high-rate algal ponds (HRAPs) by using Oswald's large-raceway-pond designs (Oswald and Golueke, 1960). Since that time, it has been noticed that the open raceways are generally more preferred than the PBRs for microalgae mass cultivation. However, achieving high productivity and maintaining monoalgal species are real shortcomings of raceway pond systems. Open ponds are generally highly vulnerable to contamination by other microorganisms, such as other algal species or bacteria. Thus, researchers usually choose closed systems for monocultures. Open systems also do not offer control over temperature and lighting. The growing season is largely reliant on climate variance and, aside from tropical areas, is limited to the warmer months only. But in tropical areas, most of the algal strains are unable to grow well under the scorching sunlight in the summers when the temperature reaches beyond 45°C. Therefore, it is very necessary to maintain the favorable atmosphere for algal growth year-round by enclosing the raceways with the transparent or translucent poly barrier that is commonly called polyhouse. Polyhouse farming requires the construction by a metal structure covered by thick polythene sheets. Various environmental parameters such as temperature, light, air flow, humidity, pH, CO_2, and dissolved oxygen (DO) can be regularly monitored and

Figure 5.1 View of *S. obliquus* cultures grown in closed raceway ponds at IIT Kharagpur, India.

many of them can be controlled easily. Figure 5.1 presents the pictorial view of microalgal mass cultivation in closed raceway ponds at the Indian Institute of Technology Kharagpur, India.

5.3.2.2 Biomass and lipid production under raceway pond cultivation

Chinnasamy et al. (2010) used four raceway ponds of 950-L capacity to cultivate a consortium of 15 native algal isolates with carpet industry effluents and municipal sewage. With 6% CO_2 sparging, the annual biomass productivity was projected as 9.2–17.8 tons ha^{-1} $year^{-1}$. The corresponding lipid content was found to be 6.8% (dcw). Ashokkumar and Rengasamy (2012) cultivated the microalga, *B. bruanii* Kütz. AP103 in a 2000-L-capacity raceway pond with 15-cm depth and reported a biomass yield and annual biomass productivity of 1.8 g L^{-1} and 68.4 tons ha^{-1} $year^{-1}$, respectively, with a lipid content of 19% (dcw). These values were profoundly higher than many previous studies. Chiaramonti et al. (2013) cultured the green microalga, *Nannochloropsis* sp. F&M-M24 in 10-m-long and 3000-L-capacity raceway ponds, and reported an areal biomass productivity of 14.1 g m^{-2} day^{-1}. In this study, the maximum lipid content was also reported around 30% (dcw). Further, the chlorophycean microalga *C. vulgaris* was grown in a 14.62 billion liters capacity raceway pond at 30-cm depth. In this study, it was observed that the biomass yield was 0.5 g L^{-1} with a significantly high areal productivity of 15 g m^{-2} day^{-1}. The lipid content was found to be 25% (dcw) (Rogers et al., 2014). In continuation, Raes et al. (2014)

reported that *Tetraselmis* sp. MUR-233 was cultivated well in a mass cultivation process of 20-cm culture depth with a mixing speed of 22 cm s^{-1} under CO_2 sparging, and the maximum areal biomass productivity was recorded as 7.2 g m^{-2} day^{-1} (Table 5.3). In this case, the maximum lipid content was also found to be 42% and 46.5% (dcw), respectively, without and with CO_2 supply.

In another study, the microalga *Nannochloropsis* sp. 75B1 was grown in 8000-L-volume indoor raceway ponds with cultures supplemented with flue gas from coal-fired power plants and a maximum biomass yield and lipid content of 0.34 g L^{-1} and 25.7% (dcw), respectively, were found (Zhu et al., 2014). Following this, Bhowmick et al. (2014) demonstrated maximum areal biomass productivity and lipid content of 8.1 g m^{-2} day^{-1} and 12% (dcw), respectively, for the marine microalga *C. variabilis* PTA-ATCC 12198 grown in a 400-L raceway pond with 150-L working volume. A recent report showed an areal biomass productivity of 6.2–14.0 g m^{-2} day^{-1} for the microalga *Nannochloropsis* sp. cultured in a raceway pond with a wide paddle wheel mixing and CO_2 supply (de Vree et al., 2015). For the microalga *Desmodesmus* sp. MCC34, grown in a 20-m-long raceway pond for 18 days, the annual biomass productivity was reported as 13.95 tons ha^{-1} $year^{-1}$ (Nagappan and Verma, 2016). In a very recent report, Wen et al. (2016) cultured the green microalga *Graesiella* sp. WBG-1 in a 20-m-long, 12-m-wide, and 20-cm-deep raceway pond with an area of 200 m^2 with a culture volume of 40,000 L. The paddle wheel was adjusted to maintain a uniform liquid flow velocity of 45 cm s^{-1} throughout the raceway pond. Pure-grade CO_2 was also supplemented into the culture. The maximum areal biomass and lipid productivities were recorded as 8.7 and 2.9 g m^{-2} day^{-1}, respectively in the month of June. Table 5.3 provides a detailed review on this aspect.

From the above reports discussed here, it has been depicted that a wide range of values for the annual biomass and lipid productivities were obtained by various researchers. Biomass productivity ranging from 9.2–45.0 t ha^{-1} $year^{-1}$ has been reported by many researchers (Chinnasamy et al., 2010; Chiaramonti et al., 2013; Raes et al., 2014; Huang et al., 2015; Nagappan and Verma, 2016; Wen et al., 2016). The recent work published by Nagappan and Verma (2016) showed that the annual biomass productivity could be estimated to be 13.95 t ha^{-1} $year^{-1}$. Nevertheless, a much higher value (68.4 t ha^{-1} $year^{-1}$) has been reported by Ashokkumar and Rengasamy (2012) with the microalga *B. bruanii* Kütz. AP103 cultivated in 2,000-L raceway ponds. Based on the findings of the researchers and analysis of the facts, it can be noted that the biomass productivity of the test microalga would be enhanced further by supplementation of CO_2 from flue gas in well-designed raceways. It has also been observed from the consultation of various reports published to date (Table 5.3) that lipid productivity ranging from 0.63–13.54 t ha^{-1} $year^{-1}$ could be achieved by different microalgae under raceway pond cultivation (calculation was

Table 5.3 Biomass and lipid productivity of some microalgal species under raceway pond cultivation

Test organism	Operational description	Biomass productivity	Lipid productivity	References
A consortium of 15 native microalgae	Four 950-L-capacity raceway ponds, wastewater containing 85%–90% carpet industry effluents with 10%–15% municipal sewage and 6% CO_2 sparging	9.2–17.8 t ha^{-1} $year^{-1}$	0.63–1.21 t ha^{-1} $year^{-1}$	Chinnasamy et al. (2010)
B. bruanii Kütz. AP103	6.1 × 1.52 × 0.3 m raceway ponds, culture volume 2000 L	68.4 t ha^{-1} $year^{-1}$	12.99 t ha^{-1} $year^{-1}$	Ashokkumar and Rengasamy (2012)
Nannochloropsis sp. F&M-M24	10 × 2 × 0.15 m raceway, culture volume 3000 L	14.1 g m^{-2} day^{-1}	4.23 g m^{-2} day^{-1}	Chiaramonti et al. (2013)
Tetraselmis suecica F&M-M33		8.4 g m^{-2} day^{-1}	2.52 g m^{-2} day^{-1}	
C. vulgaris	Volume 14.62 billion liters, 30-cm depth	15 g m^{-2} day^{-1}	3.75 g m^{-2} day^{-1}	Rogers et al. (2014)
Tetraselmis sp. MUR-233	1.0 × 1.0 × 0.20 m raceway ponds, culture volume 200 L	7.2 g m^{-2} day^{-1}	3.35 g m^{-2} day^{-1}	Raes et al. (2014)
C. variabilis (PTA-ATCC 12198)	2.5 × 0.32 × 0.1 m raceway, culture volume 150 L	8.1 g m^{-2} day^{-1}	0.97 g m^{-2} day^{-1}	Bhowmick et al. (2014)
Nannochloropsis sp.	Raceway pond: 25.4 m^2, culture volume 4,730 L with 2% CO_2 sparging	6.2–14.0 g m^{-2} day^{-1}	NR	de Vree et al. (2015)
C. pyrenoidosa	10 × 0.93 × 0.15 m raceway ponds, 5% CO_2 sparging	8.5–14.5 g m^{-2} day^{-1}	NR	Huang et al. (2015)
Desmodesmus sp. MCC34	20-m-long raceway ponds, culture volume 1000 L	13.95 t ha^{-1} $year^{-1}$	NR	Nagappan and Verma (2016)
Graesiella sp. WBG-1	20 × 12 × 0.2 m raceway, 200-m^2 raceway pond area with culture volume of 40,000 L, 1% CO_2 sparging	8.7 g m^{-2} day^{-1}	2.9 g m^{-2} day^{-1}	Wen et al. (2016)

Note: NR = not reported.

made on the basis of leaving 45 days of a year for maintenance of the system). In consultation of the strategies discussed in Section 5.3, lipid productivity for microalgae could be enhanced further by reducing the growth cycle with the input of CO_2, which in turn would result in increased number of cultivation cycles *vis-a-vis* rise in lipid productivity. The overall analysis clearly revealed that the annual lipid productivity of different microalgae was found to be profoundly higher than that of *J. curcas*, one of the most acclaimed energy crops, where the oil productivity was projected to a maximum value of only ~1.67 t ha^{-1} $year^{-1}$ (Khan et al., 2009).

5.4 *Qualitative assessment of microalgal biodiesel*

5.4.1 *Lipid extraction from algal biomass*

Lipids are a diverse group of naturally occurring biomolecules made up primarily of fats, waxes, fat-soluble vitamins, and some other compounds like monoglycerides, diglycerides, triglycerides, and sterols. The main function of lipids is to store energy and to act as a structural module in cell membranes of the organisms. Like microalgal harvesting and drying, several methods are now in use by different researchers for microalgal lipid extraction, such as cell disruption by a bead beater or sonication followed by solvent extraction mainly with chloroform-methanol, pretreatment of biomass with propanol followed by sonication, oil extraction using *n*-hexane in a Soxhlet apparatus, and lyophilized biomass followed by two-stage solvent extraction, among others. Various cell-disruption techniques (autoclaving, bead beating, microwave treatment, sonication, and NaCl treatment) were compared by Lee et al. (2010); the microwave pretreatment emerged as the most efficient for lipid extraction from microalgae. By contrast, a report showing that pretreatment of algal biomass did not register any significant rise in lipid recovery is also available (Mandal et al., 2013). It is also to be noted that the best and widely accepted method for total lipid extraction from microalgae is the Bligh and Dyer's (1959) lipid extraction method in which the proteins are generally precipitated in the middle of the two distinct liquid phases. Researchers strongly suggested that this method is still extensively used and followed for the large-scale lipid extraction from microalgae (Kumar et al., 2015).

5.4.2 *Biodiesel production and analysis of FAMEs*

Microalgal biodiesel contains various kinds of fatty acid methyl esters, that is, saturated, monounsaturated and polyunsaturated fatty acid methyl esters. The fatty acid composition of algae can vary both quantitatively and qualitatively with their physiological states and culture conditions, and the properties of biodiesel are mainly determined by its fatty acid esters (Knothe, 2005). Converti et al. (2009) analyzed the fatty acid methyl esters in biodiesel produced from *N. oculata* and *C. vulgaris*. The most abundant one was methyl palmitate, which was 62% and 66%, respectively, in *N. oculata* and *C. vulgaris* biodiesel. However, the high concentration of methyl linolenate (18%) in *N. oculata* was unable to meet the requirement of European legislation for biodiesel. Damiani et al. (2010) produced biodiesel from the microalga *Haematococcus pluvialis* using potassium hydroxide (KOH) as the catalyst.

The major constituent of *H. pluvialis* biodiesel was methyl palmitate followed by linoleic, oleic, and linolenic acid methyl esters. Chinnasamy et al. (2010) also produced biodiesel by a two-step transesterification process (acid-catalyzed followed by base-catalyzed) from a consortium of 15 native microalgae. The biodiesel was predominated with palmitic, oleic, linoleic, and linolenic acid methyl esters. Moazami et al. (2012) reported a high percentage of methyl oleate for *Nannochloropsis* sp. biodiesel. Mallick et al. (2012) showed that the biodiesel consists of ~82% saturated fatty acids, with palmitate and stearate as major components for *C. vulgaris*. Liu et al. (2013) found that the main FAME components were methyl palmitate and oleate in the case of *Coelastrum* sp. HA-1 with the acid-catalyzed transesterification. Another report depicted that the biodiesel produced from *S. obliquus* biomass was rich with palmitic, oleic, and linolenic acid methyl esters.

Methyl linolenate was found to be more than 20%, which was not recommended for a good-quality biodiesel (Abomohra et al., 2014). Hena et al. (2015) also reported more than 60% of saturated and monounsaturated fatty acid methyl esters obtained from a consortium of microalgal biodiesel. In this study, they used acid catalysis followed by the base catalysis for transesterification. The study conducted by Bagchi et al. (2015) also showed elevated saturated and monounsaturated fatty acid contents of >85% of the total FAME yield in the biodiesel obtained from a locally isolated microalga *Scenedesmus* sp. A recent study by Bhattacharya et al. (2016) reported that *C. variabilis* biodiesel was also rich in saturated and monounsaturated fatty acid methyl esters with 4.8% of methyl lenolenate.

The fatty acid composition has an imperative role in determining the quality of the biodiesel with respect to fuel properties. For a high-quality biodiesel, a higher level of saturation is required for better oxidation stability, and thus the industries can store the biodiesel for a long period of time. It must also be considered that for a good-quality biodiesel, the concentration of linolenic acid content should not exceed the upper limit of 12% (European Standards, 2013). The high linolenic acid content in oil can lead to oxidative instability, which may increase the level of rancidity in the biodiesel (Guldhe et al., 2014). Frankel (1998) reported that the oxidative instability rate for the fatty acid methyl esters was only 1 for oleate, whereas it was 41 for linoleate and 98 for linolinate. Thus, it can be seen that the presence of a small quantity of highly unsaturated fatty acid component may have disproportionately tough effects on the quality of biodiesel.

5.4.3 Determination of fuel properties of microalgal biodiesel

Determination of the fuel properties of microalgal biodiesel is another area that needs to be addressed. Therefore, it is essentially needed to check and standardize various fuel properties, namely, density, viscosity, iodine value, acid value, saponification value, calorific value, cetane index, ash, and water content to ensure the best possible qualitative biodiesel production without any difficulties. Density and viscosity are two most important fuel parameters of a lubricant regarding fuel distribution and atomization. Understanding these two parameters promotes the ability to reduce wear, improve fuel economy, and have a higher horsepower in an efficient engine. The viscosity of biodiesel depends on the fatty acid composition of the oil from which it is made. Therefore, it increases with the increasing level of saturation and long chain length of the fatty acids (Knothe, 2005). Another very essential fuel-quality-determining parameter is the high cetane index that is generally observed for saturated fatty acid methyl esters such as methyl palmitate or stearate (Knothe et al., 2003).

Saponification value is generally associated with the average fatty acids' chain lengths. The increasing chain length of the fatty acids resulted in a relatively low saponification value because in long-chain fatty acids, fewer functional carboxylic groups are present compared with the short-chain fatty acid esters. Researchers predicted saponification and iodine values ranging from 188.6–207.8 mg KOH/ g oil and 60.5–136.7 g iodine/100 g oil, respectively, for a good-quality biodiesel (Gopinath et al., 2009). There are many documented reports in which the researcher found that biodiesel made from microalgae also met international or national prescribed standards of biodiesel and diesel fuel. The microalgae biodiesel made up from *C. protothecoides* showed a viscosity and acid value of 5.2 cSt and 0.37 mg KOH/g oil, respectively, which were found to be comparable with the international biodiesel standards (Xu et al., 2006). In a continuation, Chen et al. (2012) determined the properties of biodiesel for the green microalga *C. protothecoides* and found the kinematic viscosity, acid value, and iodine value as 4.43 mm^2/s, 0.29 mg KOH/g oil, and 112.2 g $I_2/100$ g oil, respectively, which were within the specified international diesel parameter limits. One recent report showed that the biodiesel obtained from *C. variabilis* BTA4109 and *C. sorokiniana* were found to be suitable strains satisfying the criteria prescribed by the EN 14214 international biodiesel standards for the good-quality biodiesel production (Sinha et al., 2016).

5.5 Concluding remarks

From the above discussion, it can be concluded that the world's energy security scenario would remain susceptible until nonconventional, indigenously produced renewable biofuels to replace or supplement petroleum-based conventional fossil fuels. Indian Railways has recently introduced 5% biodiesel blending with normal diesel for some passenger trains and rail freight transport. In its annual report, the officials of Indian Railways have also stated that the organization will give due consideration to harness energies from indigenous renewable biomass sources (Indian Railways, 2015). In the case of developed countries like the United States, biodiesel production was 285 million gallons during the months of October and November 2016. There are a total of 97 biodiesel plants with a total production capacity of 2.3 billion gallons per year (http://www.eia.gov/biofuels/biodiesel/production/).

Algal biodiesel seems to be the best possible substitute for liquid fossil fuels that use algae as its source of energy-rich oils. In the near future, algal fuel may be very crucial to reduce transportation sector emissions and could help find alternative options for aviation, railways, and commercial vehicle fuels. Several private companies and government agencies are putting effort into reducing capital and operating costs to make algae fuel production commercially viable. However, the major bottleneck for commercial biodiesel production from microalgae is its high price. Louw et al. (2016) assessed the techno-economic details of algal biodiesel production in the United States. This study estimated the maximum commercial production costs of algal biodiesel range from US\$7.5 L^{-1} in case of open raceway ponds to US\$72 L^{-1} for closed photobioreactors. The cost of media utilization, harvesting, drying, and solvent utilizations in lipid extraction and transesterification are the major challenges for economic feasibility of algal biodiesel production.

Improvement in various downstream processing techniques gives them the potential to reduce overall energy utilization, but algal cultivation remains a decisive factor for the commercialization of algal biodiesel. Nevertheless, the utilization of municipal wastewater as the growth medium would definitely reduce the cost of microalgal biomass production. Furthermore, the cost can also be reduced by using an algal refinery approach like a petroleum refinery in which each and every component is converted to produce useful products such as biodiesel, bioethanol, glycerol, beta-carotene, protein-rich algal feed for fishes, and omega-3 fatty acids.

References

Abomohra, A. E.-F., El-Sheekh, M., and D. Hanelt. 2014. Pilot cultivation of the chlorophyte microalga *Scenedesmus obliquus* as a promising feedstock for biofuel. *Biomass and Bioenergy* 64: 237–244.

Amaro, H. M., Guedes, A. C., and F. X. Malcata. 2011. Advances and perspectives in using microalgae to produce biodiesel. *Applied Energy* 88: 3402–3410.

Anastas, T. P. 2011. Draft plan to study the potential impacts of hydraulic fracturing on drinking water resources. U.S. Environmental Protection Agency, Washington, D.C., 1–120, https://yosemite.epa.gov/sab/sabproduct.nsf/0/D3483AB445AE61418525775900603E79/$File/Draft+Plan+to+Study+the+Potential+Impacts+of+Hydraulic+Fracturing+on+Drinking+Water+Resources-February+2011.pdf. Accessed on September 15, 2016

Anjos, M., Fernandes, B. D., Vicente, A. A., Teixeira, J. A., and G. Dragone. 2013. Optimization of CO_2 bio-mitigation by *Chlorella vulgaris*. *Bioresource Technology* 139: 149–154.

Ashokkumar, V. and R. Rengasamy. 2012. Mass culture of *Botryococcus braunii* Kutz. under open raceway pond for biofuel production. *Bioresource Technology* 104: 394–399.

Bagchi, S. K. and N. Mallick. 2016. CO_2 biofixation and lipid accumulation potential of an indigenous microalga *Scenedesmus obliquus* (Turpin) Kutzing (GA 45) for biodiesel production. *RSC Advances* 6: 29889–29898.

Bagchi, S. K., Rao, P. S., and N. Mallick. 2015. Development of an oven drying protocol to improve biodiesel production for an indigenous chlorophycean microalga *Scenedesmus* sp. *Bioresource Technology* 180: 207–213.

Bajwa, K. and N. R. Bishnoi. 2015. Osmotic stress induced by salinity for lipid overproduction in batch culture of *Chlorella pyrenoidosa* and effect on others physiological as well as physicochemical attributes. *Journal of Algal Biomass Utilization* 6: 26–34.

Balat, M. and H. Balat. 2010. Progress in biodiesel processing. *Applied Energy* 87: 1815–1835.

Behrens, P. W. and D. J. Kyle. 1996. Microalgae as a source of fatty acids. *Journal of Food Lipids* 3: 259–272.

Bhattacharya, S., Maurya, R., S. K. Mishra et al. 2016. Solar driven mass cultivation and the extraction of lipids from *Chlorella variabilis*: A case study. *Algal Research* 14: 137–142.

Bhowmick, G. D., Subramanian, G., Mishra, S., and R. Sen. 2014. Raceway pond cultivation of a marine microalga of Indian origin for biomass and lipid production: A case study. *Algal Research* 6: 201–209.

Bligh, E. G. and W. J. Dyer. 1959. A rapid method of total lipid extraction and purification. *Canadian Journal of Biochemistry and Physiology* 37: 911–917.

Borowitzka, M. A. 1999. Commercial production of microalgae: Ponds, tanks, tubes and fermenters. *Journal of Biotechnology* 70: 313–321.

Brennan, L. and P. Owende. 2010. Biofuels from microalgae—A review of technologies for production, processing, and extractions of biofuels and co-products. *Renewable and Sustainable Energy Reviews* 14: 557–577.

Central Intelligence Agency (CIA). 2015. The world factbook. https://www.cia.gov/library/publications/the-world-factbook/geos/in.html. Accessed on June 06, 2016.

Chen, Y. H., Huang, B. Y., Chiang, T. H., and T. C. Tang. 2012. Fuel properties of microalgae (*Chlorella protothecoides*) oil biodiesel and its blends with petroleum diesel. *Fuel* 94: 270–273.

Chiaramonti, D., Prussi, M., and D. Casini. 2013. Review of energy balance in raceway ponds for microalgae cultivation: Re-thinking a traditional system is possible. *Applied Energy* 102, 101–111.

Chinnasamy, S., Bhatnagar, A., Hunt, R. W., and K. C. Das. 2010. Microalgae cultivation in a wastewater dominated by carpet mill effluents for biofuel applications. *Bioresource Technology* 101: 3097–3105.

Chisti, Y. 2007. Biodiesel from microalgae. *Biotechnology Advances* 25: 294–306.

Chisti, Y. 2008. Biodiesel from microalgae beats bioethanol. *Trends in Biotechnology* 26: 126–131.

Chiu, S. Y., Kao, C. Y., M. T. Tsai et al. 2009. Lipid accumulation and CO_2 utilization of *Nannochloropsis oculata* in response to CO_2 aeration. *Bioresource Technology* 100: 833–838.

Concas, A., Lutzu, G. A., Locci, A., and G. Cao. 2013. *Nannochloris eucaryotum* growth in batch photobioreactors: Kinetic analysis and use of 100% (v/v) CO_2. *Advances in Environmental Research* 2: 19–33.

Converti, A., Casazza, A. A., Ortiz, E. Y., Perego, P., and M. Del Borghi. 2009. Effect of temperature and nitrogen concentration on the growth and lipid content of *Nannochloropsis oculata* and *Chlorella vulgaris* for biodiesel production. *Chemical Engineering and Processing: Process Intensification* 48: 1146–1151.

Damiani, M. C., Popovich, C. A., Constenla, D., and P. I. Leonardi. 2010. Lipid analysis in *Haematococcus pluvialis* to assess its potential use as a biodiesel feedstock. *Bioresource Technology* 101: 3801–3807.

Davis, R., Aden, A., and P. T. Pienkos. 2011. Techno-economic analysis of autotrophic microalgae for fuel production. *Applied Energy* 88: 3524–3531.

de Godos, I., Blanco, S., Garcia-Encina, P. A., Becares, E., and R. Munoz. 2009. Long-term operation of high rate algal ponds for the bioremediation of piggery wastewaters at high loading rates. *Bioresource Technology* 100: 4332–4339.

de Vree, J. H., Bosma, R., Janssen, M., Barbosa, M. J., and R. H. Wijffels. 2015. Comparison of four outdoor pilot-scale photobioreactors. *Biotechnology for Biofuels* 8: 215–226.

Dorado, M. P., Ballesteros, E., J. A. de Almeida et al. 2002. An alkali-catalyzed transesterification process for high free fatty acid waste oils. *Transactions of the American Society of Agricultural Engineers* 45: 525–529.

European Standards. 2013. CSN EN 14214+A1: Liquid petroleum products - Fatty acid methyl esters (FAME) for use in diesel engines and heating applications - Requirements and test methods. Released in 2013.

Fakhry, E. M. and D. M. E. Maghraby. 2015. Lipid accumulation in response to nitrogen limitation and variation of temperature in *Nannochloropsis salina*. *Botanical Studies* 56: 1–8.

Fan, J., Cui, Y., Wan, M., Wang, W., and Y. Li. 2014. Lipid accumulation and biosynthesis genes response of the oleaginous *Chlorella pyrenoidosa* under three nutrition stressors. *Biotechnology for Biofuels* 7: 2–14.

Frankel, E. N. 1998. *Lipid Oxidation*, Vol. 10, The Oily Press, Dundee, Scotland.

Ghadge, S. V. and H. Raheman. 2006. Process optimization for biodiesel production from mahua (*Madhuca indica*) oil using response surface methodology. *Bioresource Technology* 97: 379–384.

Global Agricultural Information Network (GAIN). 2016. *India Biofuels Annual*, New Delhi, India, pp. 1–29.

Gopinath, A., Puhan, S., and G. Nagarajan. 2009. Theoretical modeling of iodine value and saponification value of biodiesel fuels from their fatty acid composition. *Renewable Energy* 34: 1806–1811.

Gorain, P. C., Bagchi, S. K., and N. Mallick. 2013. Effects of calcium, magnesium and sodium chloride in enhancing lipid accumulation in two green microalgae. *Environmental Technology* 34: 1887–1894.
Greenwell, H. C., Laurens, L. M. L., Shields, R. J., Lovitt, R. W., and K. J. Flynn. 2010. Placing microalgae on the biofuels priority list: A review of the technological challenges. *Journal of Royal Society Interface* 7: 703–726.
Guldhe, A., Singh, B., Rawat, I., Ramluckan, K., and F. Bux. 2014. Efficacy of drying and cell disruption techniques on lipid recovery from microalgae for biodiesel production. *Fuel* 128: 46–52.
Han, F., Pei, H., W. Hu et al. 2015. Optimization and lipid production enhancement of microalgae culture by efficiently changing the conditions along with the growth-state. *Energy Conversion and Management* 90: 315–322.
Harun, R., Singh, M., Forde, G. M., and M. K. Danquah. 2010. Bioprocess engineering of microalgae to produce a variety of consumer products. *Renewable and Sustainable Energy Reviews* 14: 1037–1047.
Hena, S., Fatimah, S., and S. Tabassum. 2015. Cultivation of algae consortium in a dairy farm wastewater for biodiesel production. *Water Resources and Industry* 10: 1–14.
Hennecke, A. M., Faist, M., J. Reinhardt et al. 2013. Biofuel greenhouse gas calculations under the European Renewable Energy Directive—A comparison of the BioGrace tool vs. the tool of the Roundtable on Sustainable Biofuels. *Applied Energy* 102: 55–62.
Huang, J., Qu, X., M. Wan et al. 2015. Investigation on the performance of raceway ponds with internal structures by the means of CFD simulations and experiments. *Algal Research* 10: 64–71.
Indian Railways. 2015. Indian Railways Report. *Rail Bandhu*.
Jiang, L., Luo, S., Fan X., Yang, Z., and R. Guo. 2011. Biomass and lipid production of marine microalgae using municipal waste water and high concentration of CO_2. *Applied Energy* 88: 3336–3341.
Khan, S. A., Rashmi, Hussain, M. Z., Prasad, S., and U. C. Banerjee. 2009. Prospects of biodiesel production from microalgae in India. *Renewable and Sustainable Energy Reviews* 13: 2361–2372.
Knothe, G. 2005. Dependence of biodiesel fuel properties on the structure of fatty acid alkyl esters. *Fuel Processing Technology* 86: 1059–1070.
Knothe, G., Matheaus, A. C., and T. W. Ryan. 2003. Cetane numbers of branched and straight chain fatty esters determined in an ignition quality tester. *Fuel* 82: 971–975.
Kodama, M., Ikemoto, H., and S. Miyachi. 1993. A new species of highly CO_2-tolerant fast-growing marine microalga suitable for high-density culture. *Journal of Marine Biotechnology* 1: 21–25.
Kosinkova, J., Doshi, A., J. Maire et al. 2015. Measuring the regional availability of biomass for biofuels and the potential for microalgae. *Renewable and Sustainable Energy Reviews* 49: 1271–1285.
Krawczyk, T. 1996. Biodiesel alternative fuel makes inroads but hurdles remain. *Inform* 7: 801–829.
Krienitz, L. 2009. Algae. In: Likens, G. (Ed.) *Encyclopedia of Inland Waters*, Vol. I. Elsevier, Oxford, pp. 103–113.
Kumar, R. R., Rao, P. H., and M. Arumugam. 2015. Lipid extraction methods from microalgae: A comprehensive review. *Frontiers in Energy Research* 2: 1–9.

Kwak, H. S., Kim, J. Y. H., H. M. Woo et al. 2016. Synergistic effect of multiple stress conditions for improving microalgal lipid production. *Algal Research* 19: 215–224.
Lee, J. Y., Yoo, C., and S. Y. Jun. 2010. Comparison of several methods for effective lipid extraction from microalgae. *Bioresource Technology* 101: 575–577.
Li, Y., Horsman, M., Wu, N., Lan, C. Q., and N. Dubois-Calero. 2008. Biofuels from microalgae. *Biotechnology Progress* 24: 815–820.
Li, Y. X., Zhao F. J., D. D. Yu. et al. 2015. Effect of nitrogen limitation on cell growth, lipid accumulation and gene expression in *Chlorella sorokiniana*. *Brazilian Archives of Biology and Technology* 58: 462–467.
Liu, Z., Liu, C., Y. Hou et al. 2013. Isolation and characterization of a marine microalga for biofuel production with astaxanthin as a co-product. *Energies* 6: 2759–2772.
Louw, T. M., Griffiths M. J., Jones S. M. J., and S. T. L. Harrison. 2016. Technoeconomics of algal biodiesel. In: Bux, F. and Y. Chisti (Eds.), *Algae Biotechnology*, Springer International Publishing, Switzerland, pp. 111–141.
Mallick, N., Mandal, S., Singh, A. K., Bishai, M., and A. Dash. 2012. Green microalga *Chlorella vulgaris* as a potential feedstock for biodiesel. *Journal of Chemical Technology and Biotechnology* 87: 137–145.
Mandal, S. and N. Mallick. 2009. Microalga *Scenedesmus obliquus* as a potential source for biodiesel production. *Applied Microbiology and Biotechnology* 84: 281–291.
Mandal, S. and N. Mallick. 2011. Waste utilization and biodiesel production by the green microalga *Scenedesmus obliquus*. *Applied and Environmental Microbiology* 77: 374–377.
Mandal, S. Patnaik, R., Singh, A. K., and N. Mallick. 2013. Comparative assessment of various lipid extraction protocols and optimization of transesterification process for microalgal biodiesel production. *Environmental Technology* 34: 2009–2018.
McMillen, S., Shaw, P., Jolly, N., Goulding, B., and V. Finkle. 2005. Biodiesel: Fuel for thought, fuel for Connecticut's future. Connecticut Center for Economic Analysis Report, University of Connecticut, Connecticut, United States, 56 pp.
Meher, L. C., Vidya Sagar, D., and S. N. Naik. 2006. Technical aspects of biodiesel production by transesterification—A review. *Renewable and Sustainable Energy Reviews* 10: 248–268.
Moazami, N., Ashori, A., R. Ranjbar et al. 2012. Large-scale biodiesel production using microalgae biomass of *Nannochloropsis*. *Biomass and Bioenergy* 39: 449–453.
Moazami, N., Ranjbar, R., Ashori, A., Tangestani, M., and A. S. Nejad. 2011. Biomass and lipid productivities of marine microalgae isolated from the Persian Gulf and the Qeshm Island. *Biomass and Bioenergy* 35: 1935–1939.
Mohsenpour, S. F. and N. Willoughby. 2016. Effect of CO_2 aeration on cultivation of microalgae in luminescent photobioreactors. *Biomass and Bioenergy* 85: 168–177.
Mujtaba, G., Choi, W., Lee, C. G., and K. Lee. 2012. Lipid production by *Chlorella vulgaris* after a shift from nutrient-rich to nitrogen starvation conditions. *Bioresource Technology* 123: 279–283.
Mutanda, T., Karthikeyan, S., and F. Bux. 2011. The utilization of post-chlorinated municipal domestic wastewater for biomass and lipid production by *Chlorella* spp. under batch conditions. *Applied Biochemistry and Biotechnology* 164: 1126–1138.

Nagappan, S. and S. K. Verma. 2016. Growth model for raceway pond cultivation of *Desmodesmus* sp. MCC34 isolated from a local water body. *Engineering in Life Sciences* 16: 45–52.

Nascimento, I. A., Cabanelas, I. T. D., J. N. Santos et al. 2015. Biodiesel yields and fuel quality as criteria for algal-feedstock selection: effects of CO_2-supplementation and nutrient levels in cultures. *Algal Research* 8: 53–60.

Nigam, P. S. and A. Singh. 2011. Production of liquid biofuels from renewable resources. *Progress in Energy and Combustion Science* 37: 52–68.

Norsker, N. H., Barbosa, M. J., Vermue, M. H., and R. H. Wijffels. 2011. Microalgal production—A close look at the economics. *Biotechnology Advances* 29: 24–7.

Oswald, W. J. and C. G. Golueke. 1960. Biological transformation of solar energy. *Advances in Applied Microbiology* 2: 223–262.

Ota, S., Oshima, K., T. Yamazaki et al. 2016. Highly efficient lipid production in the green alga *Parachlorella kessleri*: Draft genome and transcriptome endorsed by whole-cell 3D ultrastructure. *Biotechnology for Biofuels* 9: 13–22.

Parupudi, P., Kethineni, C., P. B. Dhamole et al. 2016. CO_2 fixation and lipid production by microalgal species. *Korean Journal of Chemical Engineering* 33: 587–593.

Pegallapati, A. K., Arudchelvam, Y., and N. Nirmalakhandan. 2012. Energy-efficient photobioreactor configuration for algal biomass production. *Bioresource Technology* 126: 266–273.

Petroleum Planning and Analysis Cell (PPAC). 2013. All India study on sectoral demand of diesel & petrol. *PPAC Report*, New Delhi, India, pp. 17–18.

Raes, E. J., Isdepsky, A., Muylaert, K., Borowitzka, M. A., and N. R. Moheimani. 2014. Comparison of growth of *Tetraselmis* in a tubular photobioreactor (Biocoil) and a raceway pond. *Journal of Applied Phycology* 26: 247–255.

Rao, P. P. and K. V. Gopalkrishnan. 1991. Vegetable oils and their methyl esters as fuels diesel engines. *Indian Journal of Technology* 29: 292–297.

Ren, H. Y., Liu, B. F., F. Kong et al. 2014. Enhanced lipid accumulation of green microalga *Scenedesmus* sp. by metal ions and EDTA addition. *Bioresource Technology* 169: 763–767.

Ren, H., Liu, B., Ma, C., Zhao, L., and N. Ren. 2013. A new lipid-rich microalga *Scenedesmus* sp. strain R-16 isolated using Nile red staining: Effects of carbon and nitrogen sources and initial pH on the biomass and lipid production. *Biotechnology for Biofuels* 6: 143–152.

Renewable Energy Policy Network for the 21st Century (REN21). 2015. Renewables: Global status report. REN21 Secretariat, Paris, p. 250.

Rogers, J. N., Rosenberg, J. N., and B. J. Guzman. 2014. A critical analysis of paddlewheel-driven raceway ponds for algal biofuel production at commercial scales. *Algal Research* 4: 76–88.

Ruangsomboon, S. 2012. Effect of light, nutrient, cultivation time and salinity on lipid production of newly isolated strain of the green microalga, *Botryococcus braunii* KMITL 2. *Bioresource Technology* 109: 261–265.

Scott, S. A., Davey, M. P., J. S. Dennis et al. 2010. Biodiesel from algae: Challenges and prospects. *Current Opinion in Biotechnology* 21: 277–286.

Sheehan, J., Dunahay, T., Benemann, J., and P. Roessler. 1998. A look back at the U.S. Department of Energy's Aquatic Species Program—Biodiesel from algae. National Renewable Energy Laboratory Report, U.S. Department of Energy, Colorado, USA, 323 pp.

Shen, X. F., Liu, J. J., and A. S. Chauhan. 2016. Combining nitrogen starvation with sufficient phosphorus supply for enhanced biodiesel productivity of *Chlorella vulgaris* fed on acetate. *Algal Research* 17: 261–267.

Shukla, M. and D. W. Dhar. 2014. Biotechnological potentials of microalgae: Past and present scenario. *Vegetos: An International Journal of Plant Research* 26: 229–237.

Singh, J. and S. Gu. 2010. Commercialization potential of microalgae for biofuels production. *Renewable and Sustainable Energy Reviews* 14: 2596–2610.

Singh, P., Guldhe, A., Kumari, S., Rawat, I., and F. Bux. 2016. Combined metals and EDTA control: An integrated and scalable lipid enhancement strategy to alleviate biomass constraints in microalgae under nitrogen limited conditions. *Energy Conversion and Management* 114: 100–109.

Singh, D., Yadav, K., Deepshikha, and R. S. Singh. 2015. Biofixation of carbon dioxide using mixed culture of microalgae. *Indian Journal of Biotechnology* 14: 228–232.

Sinha, S. K., Gupta A., and R. Bharalee. 2016. Production of biodiesel from freshwater microalgae and evaluation of fuel properties based on fatty acid methyl ester profile. *Biofuels* 7: 105–121.

Solovchenko, A., Gorelova, O., I. Selyakh et al. 2016. Nitrogen availability modulates CO_2 tolerance in a symbiotic chlorophyte. *Algal Research* 16:177–188.

Soman, A. and Y. Shastri. 2015. Optimization of novel photobioreactor design using computational fluid dynamics. *Applied Energy* 140: 246–255.

Stephens, E., Ross, I. L., J. H. Mussgnug et al. 2010. Future prospects of microalgal biofuel production systems. *Trends in Plant Science* 15: 554–564.

Tanaka, N. 2011. Technology roadmap: Biofuels for transport. International Energy Agency (IEA), Paris, France. pp. 1–52.

Tang, D., Han, W., Li, P., Miao, X., and J. Zhong. 2011. CO_2 biofixation and fatty acid composition of *Scenedesmus obliquus* and *Chlorella pyrenoidosa* in response to different CO_2 levels. *Bioresource Technology* 102: 3071–3076.

Thrush, S. F., Hewitt, J. E., Gibbs, M., Lundquist, C., and A. Norkko. 2006. Functional role of large organisms in intertidal communities: Community effects and ecosystem function. *Ecosystems* 9: 1029–1040.

Thurman, H. V. 1997. *Introductory Oceanography*, Prentice-Hall, Upper Saddle River, New Jersey, USA.

U.S. Energy Information Administration (EIA). 2014. India is increasingly dependent on imported fossil fuels as demand continues to rise. http://www.eia.gov/todayinenergy/detail.php?id=17551. Accessed on February 15, 2017.

Wang, X., Shen, Z., and X. Miao. 2016. Nitrogen and hydrophosphate affects glycolipids composition in microalgae. *Scientific Reports* 6: 30145–30153.

Watanabe, Y. and H. Saiki. 1997. Development of a photobioreactor incorporating *Chlorella* sp. for removal of CO_2 in stack gas. *Energy Conversion and Management* 38: S499–S503.

Wen, X., Du, K., and Z. Wang. 2016. Effective cultivation of microalgae for biofuel production: A pilot-scale evaluation of a novel oleaginous microalga *Graesiella* sp. *WBG-1. Biotechnology for Biofuels* 9: 123–134.

Woertz, I. C., Futon, L., and T. J. Lundquist. 2009. Nutrient removal and greenhouse gas abatement with CO_2-supplemented algal high rate ponds. *Proceedings of the Water Environment Federation*, 2009(7), 7924–7936.

World Bank Report. World Development Indicators: Trends in greenhouse gas emissions (http://wdi.worldbank.org/table/3.9).
Worldwatch Institute. 2011. //www.worldwatch.org/biofuels-make-comeback-despite-tough-economy. Accessed on July 15, 2016.
Xiao, Y., Zhang, J., J. Cui et al. 2015. Simultaneous accumulation of neutral lipids and biomass in *Nannochloropsis oceanica* IMET1 under high light intensity and nitrogen replete conditions. *Algal Research* 11: 55–62.
Xu, H., Miao, X., and Q. Wu. 2006. High quality biodiesel production from a microalgae *Chlorella protothecoides* by heterotrophic growth in fermenters. *Journal of Biotechnology* 126: 499–507.
Yeh, K. L. and J. S. Chang. 2011. Nitrogen starvation strategies and photobioreactor design for enhancing lipid content and lipid production of a newly isolated microalga *Chlorella vulgaris* ESP-31: Implications for biofuels. *Biotechnology Journal* 6: 1358–1366.
Zhao, B. and Y. Su. 2014. Process effect of microalgal-carbon dioxide fixation and biomass production: A review. *Renewable and Sustainable Energy Reviews* 31: 121–132.
Zhao, B., Su, Y., Zhang, Y., and G. Cui. 2015. Carbon dioxide fixation and biomass production from combustion flue gas using energy microalgae. *Energy* 89: 347–357.
Zhu, B., Sun, F., Yang, M., L., Lu, Yang, G. and Pan, K. 2014. Large-scale biodiesel production using flue gas from coal-fired power plants with *Nannochloropsis* microalgal biomass in open raceway ponds. *Bioresource Technology* 174: 53–59.

chapter six

Bio-oil production

Jon Alvarez, Maider Amutio, Gartzen Lopez, Martin Olazar, and Javier Bilbao

Contents

6.1 Introduction: Background and driving forces

The technological development of the different biomass waste to energy routes play an essential role in the future energy production, driven by growing energy demand and environmental policies in the developed countries facing the challenges of progressive oil depletion and the regulation of CO_2 emissions (Demirbas et al., 2009; Foley et al., 2017).

Considering the availability of wastes (agricultural, forestry, timber industry, and urban) and crops specifically grown for waste-to-energy purposes, biomass has promising prospects for its exploitation as a raw material for the production of biofuels (Huber et al., 2007; Carpenter et al., 2014), high value-added biomaterials (food additives, pharmaceuticals), and chemicals (surfactants, monomers, solvents, fertilizers) (FitzPatrick et al., 2010). However, the technological development for biomass waste-to-energy is still incipient (Powell, 2017), and the associated industry (biorefinery) has little capacity for developing its own economy, while at the same time competing with the valorization of fossil fuels (coal and natural gas) with their huge reserves (Partridge et al., 2017). In the current scenario of transition toward a sustainable energy market, the biomass waste-to-energy economy is considered a powerful and complementary alternative to be exploited in a coordinated manner with fossil sources, using common infrastructures, and seeking the maximum synergy in the production of chemicals (Ma et al., 2015).

The key stages of the main routes for large-scale biomass waste-to-energy involve the following processes: (1) fermentation (Nguyen et al., 2017), (2) gasification (Baruah and Baruah, 2014), and (3) fast pyrolysis, whose products are bioethanol, syngas, and bio-oil, respectively. These raw materials produce a wide range of fuels, synthetic raw materials, and derivatives, traditionally produced from petroleum. Bio-oil's interest as an intermediate raw material is based on the advantages of fast pyrolysis, which are as follows: (1) its application to different kinds of biomasses (lignin, sewage sludge), to their blends, and to biomass and other material blends (plastic residues and waste tires) (Kim and Parker, 2008; Ben and Ragauskas, 2011); (2) the capacity for operating in a delocalized manner to obtain a bio-oil that can be fed into a larger-scale catalytic process or even into a refinery unit (Anex et al., 2010; Balat, 2010; Bridgwater, 2012); and (3) the high level of technological development, with simple equipment, low investment in infrastructures, versatility, and no emissions (Butler et al., 2011; Meier et al., 2013).

6.2 Fundamentals of fast biomass pyrolysis

Pyrolysis is thermal degradation in the absence of oxygen and is one of the most energy-efficient and cheapest processes for the production of liquid fuels from biomass (Chiaramonti et al., 2007). The objective of pyrolysis is the attainment of a liquid fraction (bio-oil), being the char collection (priority in conventional pyrolysis) of complementary interest, while the gaseous stream may be used to satisfy the energy demand of the unit and may also be recirculated to reduce the fluidizing agent requirements. Likewise, the char may be valued for the production of activated carbons to be used as adsorbent to remove gaseous effluents and contaminated liquids (Cotoruelo et al., 2012; Alvarez et al., 2015c).

Fast pyrolysis is performed at intermediate temperatures (450–550°C), with very high heating rates (10^3–10^4 °C s^{-1}) and low residence time of volatiles (<1 s), which are rapidly cooled after leaving the reactor, so the bio-oil yield is in the 60–75 wt% range.

Fast biomass pyrolysis elapses in three successive stages, the first being the evaporation of free moisture, followed by the primary decomposition and secondary reactions. Primary decomposition generally occurs between 200 and 400°C, in which the individual components (mainly cellulose, hemicellulose, and lignin, as well as extractives) are decomposed, resulting in the release of volatiles. It should be pointed out that at high heating rates (>100°C s^{-1}), rearrangement reactions are usually hindered. According to the literature (Collard and Blin, 2014), in the primary conversion stage three phenomena take place: (1) char formation where the biomass is converted into solid with aromatic polycyclic structure; (2) depolymerization, which consists of the release of monomer structures from the polymers; and (3) fragmentation, in which the linkage of covalent bonds in the polymer, as well as in the monomer units, results in the formation of gaseous compounds and a diversity of organic compounds. Secondary reactions involve those volatiles that are unstable at operating temperature, as well as the carbonaceous matrix above 400°C. Throughout the second and third phenomena mentioned above, a large number of reactions occur, including dehydration, isomerization, aromatization, cracking, decarboxylation, and decarbonylation, among others.

Moreover, the degradation pathways of the cellulose, hemicellulose, and lignin are different. Hemicellulose (an amorphous polymer whose main monomer is xylose) has the lowest thermal stability and decomposes between 250 and 350°C, whereas cellulose (a crystalline polymer made of repeat units of the monomer glucose) has a higher thermal stability (due to its crystalline structure) and pyrolyzes in the 325–400°C range. Lignin (constituted by oxygenated polyaromatic compounds) is the most stable component and decomposes in a wider temperature range, 150–700°C (Amutio et al., 2013a). Furthermore, when other types of biomasses with a more heterogeneous structure than lignocellulosic ones are pyrolyzed, for example, sewage sludge, it is even more difficult to clarify its complex degradation behavior. In this case, pyrolysis starts with moisture evaporation and extractive release in the 100–200°C range, followed by the carbohydrate and lipid degradation around 255 and 300°C, respectively. Besides, the mass loss observed between 360 and 525°C corresponds to protein decomposition with the contribution of lignin pyrolysis (Alvarez et al., 2015b).

Accordingly, the relative mass ratio of each component, as well as the inorganic matter (whose composition and content vary widely depending on the biomass type and growth environment), will affect the bio-oil formation and chemical composition. As reported in the literature, hemicellulose and cellulose contribute to higher bio-oil production, favoring the former for the formation of furans, ketones, and aldehydes, and the latter for saccharides and aldehydes (Wang et al., 2011; Amutio et al., 2013a). Lignin leads to higher char formation and its decomposition gives way to various phenols as well as carbon monoxide and methane (Collard and Blin, 2014). However, the strong interactions among the corresponding pyrolysis mechanisms will result in more complex product distribution, enhancing the production of some compounds and inhibiting others (Peters, 2011). In addition, the ashes of lignocellulosic biomass are composed of alkaline and alkaline earth metals (AAEM), as well as Si, Fe, and Al, whose content vary depending on the biomass type (lower in woody biomasses and higher in agricultural residues such as corn stove or rice husk) (Oasmaa et al., 2016). In the case of sewage sludge, the ash content is much higher due to the presence of heavy metals, such as Zn, Cu, Ni, and Cd, among others, existing in the wastewater (Fytili and Zabaniotou, 2008). The ash present in the char catalyze the cracking reactions involving volatiles, thus resulting in an increase in gas yield and a decrease in bio-oil yield, whose high water content could lead in some cases to phase separation (Oasmaa et al., 2015).

6.3 Fast biomass pyrolysis technologies

6.3.1 Reactors

At a laboratory scale, pyrolysis microreactors or pyrolysis gas chromatography (PyGC) systems are an attractive microscale technique for studying biomass pyrolysis. Small amounts of biomass and catalyst are required, and ease of operation makes this apparatus a widely applied laboratory tool for rapid screening of various biomass types and catalysts (Gamliel et al., 2015).

Despite the fact that the initial investment for reactor configuration represents approximately 10%–15% of the total capital cost in the pyrolysis process, most research and development efforts have been focused on developing and testing various reactor technologies for maximizing bio-oil yield with a wide variety of feedstock (Meier et al., 2013; Garcia-Perez et al., 2015). The reactors developed to the present (Figure 6.1) have different fluid dynamics: (1) bubbling fluidized bed (BFB) (DeSisto et al., 2010; Li et al., 2015a; Black et al., 2016; Kim, 2016); (2) circulating fluidized bed (CFB) (dilute or dense phase) (Ding et al., 2012; Sun et al., 2015; Cai and Liu, 2016); (3) ablative (hot wall, rotatory, and cyclonic) (Lédé et al., 2007; Sandstrom et al., 2016); (4) screw kiln (Ingram et al., 2008; Kelkar et al., 2015; Solar et al., 2016); and (5) conical spouted bed (Amutio et al., 2012c; Alvarez et al., 2014, 2015b).

The BFB is the most developed technology. This reactor can work at a large scale in an isothermal regime due to the good heat transfer rate between phases without segregation and low residence time of the

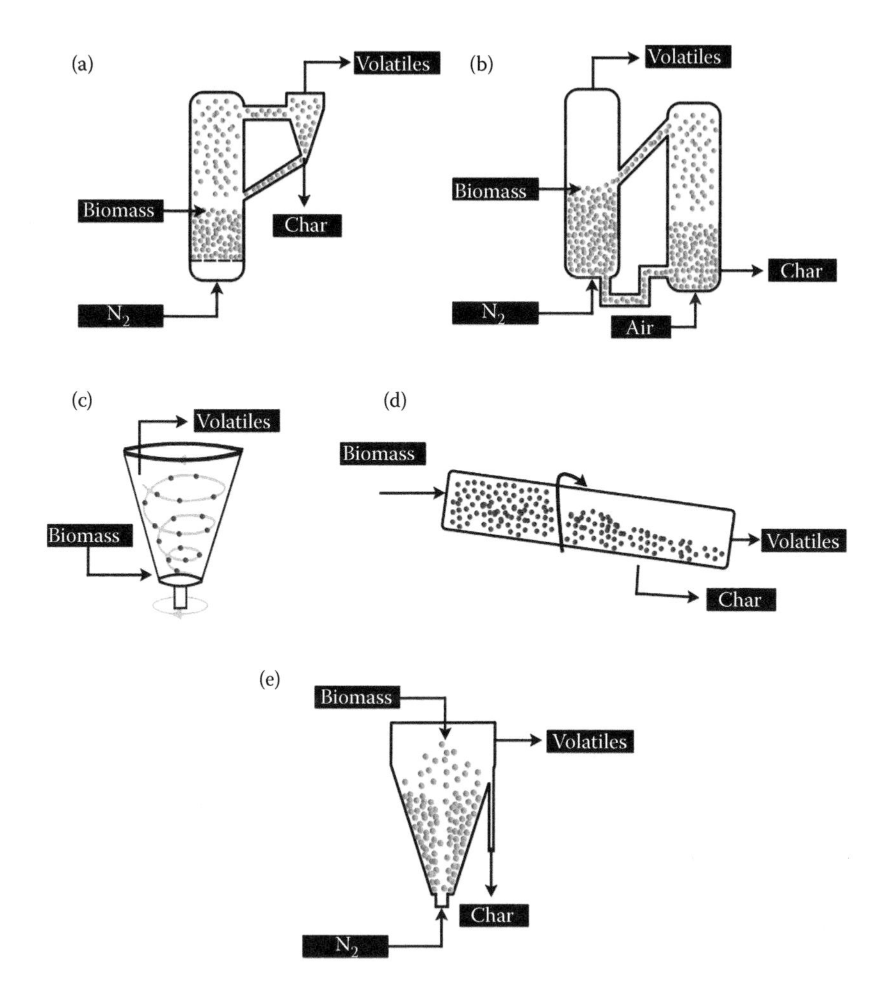

Figure 6.1 Fast pyrolysis reactors: (a) bubbling fluidized bed; (b) circulating fluidized bed; (c) ablative reactor; (d) screw kiln reactor; and (e) conical spouted bed reactor.

volatiles (which minimizes gas formation). Nevertheless, it requires high amounts of energy to heat the sand used to aid fluidization. In the CFBs, the heat transfer is not so efficient given that it is performed by convection and with no relative velocity between phases.

In ablative or cyclonic reactors, the biomass is brought in contact with the reactor hot wall, thus conduction is the main heat transfer mechanism (Oasmaa et al., 2016). Apart from having lower energy requirements, the pyrolysis products are not contaminated by sand. However, the pyrolysis rate is limited by the heat transmission within the particle, which gives way to lower bio-oil yields.

The screw or auger reactor does not use sand to transfer heat and operates in a simpler way, which eases product separation. Additionally, the vapor residence time (usually between 5 and 30 s) can be modified by changing the heated zone through which vapors pass prior to entering the condenser train (Bridgwater, 2012). However, this technology has some drawbacks associated with the energy consumption from the mechanical drive.

The conical spouted bed reactor (CSBR) is an alternative to conventional fluidized beds that has been used for the fast pyrolysis of several types of biomasses, including pinewood (Amutio et al., 2012c), rice husk (Alvarez et al., 2014), forest shrub wastes (Amutio et al., 2103b, 2015), and sewage sludge (Alvarez et al., 2015b), as well as other waste materials, such as tires (López et al., 2010) or plastics (Artetxe et al., 2015). In this reactor, three different regions must be considered (Figure 6.2): the spout zone is the central region where particles move upward, as in a transported bed; the annulus is the surrounding area where particles move downward, as in a moving bed; and, finally, there is a fountain region where particles fall from the spout onto the annulus zone. The vigorous cyclic particle movement allows handling larger particles than those allowed in the fluidized beds, of irregular texture, fine materials, and sticky solids with no agglomeration and segregation problems (which facilitates the use of catalysts *in situ*) (Olazar et al., 1993; Aguado et al., 2005). Moreover, the countercurrent displacement of the solids in

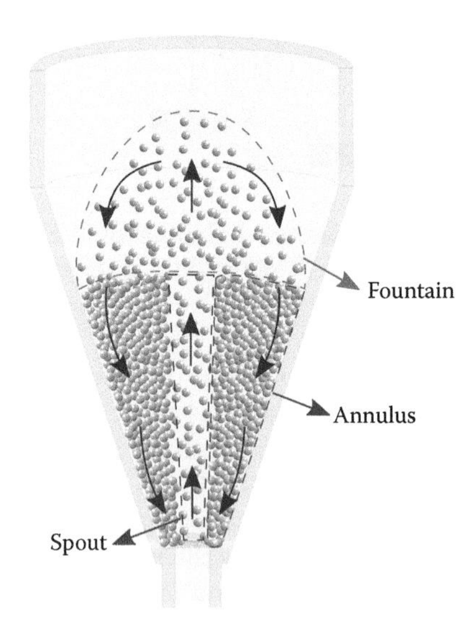

Figure 6.2 Spouted bed reactor regions. (From Artetxe, M., G. Lopez, M. Amutio et al. 2015. *Waste Manage.* 45: 126–133.)

the annulus and the gas in the spout zone leads to high heat and mass transfer rates between phases (Makibar et al., 2011). In addition, very low gas residence times are attained (as low as milliseconds in the dilute spouted bed regime) (Fernandez-Akarregi et al., 2013), thereby minimizing gas formation. It is noteworthy that this reactor has a simple design (distributor plate is not required) and requires lower volumes than fluidized beds for the same capacity, simplifying the scaling up of the pyrolysis process. Currently, a 25 kg h^{-1} pilot plant is operative (Fernandez-Akarregi et al., 2013).

Apart from the previously mentioned reactor configurations for conventional pyrolysis process, increased attention is being paid to microwave and solar-type reactors in which the heat transfer mechanism is more efficient and would increase bio-oil yield and quality. In microwave-assisted pyrolysis, the heating of biomass occurs in a microwave field, which acts as a radiation energy source and accelerates, thus favoring biomass devolatilization and bio-oil formation (Kabir and Hameed, 2017). Nevertheless, apart from the large amount of electric power required, the electromagnetic fields in microwave ovens are not uniform, which leads to nonhomogeneous heating, thereby lowering the bio-oil yield and hindering its scaling up (Huang et al. 2016).

The main advantage of solar reactors is that high heating rates and temperatures may be attained using a renewable energy form. These reactors are made of quartz because high temperatures (>700°C) are generated by a parabolic solar concentrator placed around the reactor to concentrate the solar radiation (Jahirul et al., 2012; Morales et al., 2014). However, this technology requires high investment and technological development is still needed for their scaling up (Nzihou et al., 2012).

At the industrial scale, three types of reactors, namely, rotating cone, circulating fluidized bed, and bubbling fluidized bed, have been implemented by companies such as Ensyn, Dynamotive, BTG, and Fortum, among others (Table 6.1). However, some of them have closed down their plants after operating some years.

6.3.2 *Operating conditions*

Apart from reactor configuration and biomass feedstock, there are other parameters (temperature, residence time, particle size, or condensation system) influencing bio-oil yield and composition. With regard to temperature, the highest bio-oil yields in fluidized and spouted bed reactors have been attained in the 450–550°C range (Figure 6.3). Moreover, short residence time favors bio-oil formation given that rapid removal of organic vapors hinders secondary reactions. In addition, very high heat and mass transfer rates are required and big particle sizes will lead to temperature gradients across the particle, thus reducing bio-oil production (Westerhof

Table 6.1 The current state (2016) of fast pyrolysis industrial plants for capacity higher than 1000 kg/h

Technology	Organization	Feedstock	Location	Capacity (kg h^{-1})	Bio-oil production (kg h^{-1})	Current status	Start-up date
Bubbling fluidized bed	Fortum and Valmet	Harvest residues	Joensuu, Finland	10,500	6000	Operational	2015
	Dynamotive	Demolition construction wood	Guelph, Canada	7500	4800	Shut down	2007
	Dynamotive	Erie Flooring and Wood Products	West Lorne, Canada	4150	2700	Dormant	2008
Circulating fluidized bed	Ensyn and Red Arrow	Wood residues	Wisconsin, United States	1250–1600	Chemicals and heating fuels	Operational	1995–2002
	Ensyn and Fibria	Eucalyptus residue	Aracruz (Brazil)	17,000	11,000	Design phase	2018
	Ensyn	Wood and agricultural residues	Ontario, Canada	2700	1500	Operational	2006
	Ensyn and Arbec Forest	Slash and other forest residues	Quebec, Canada	7500	5000	Under construction	2017
	Green Fuel Nordic	Wood	Iisalmi, Finland	40 dry m^3/h	10,000	Project (20 biorefineries)	–
Rotatory cone	BTG-EMPYRO	Clean wood residue	Hengelo, Nerthelands	5000	3200	Operational	2014
	Genting	Empty fruit brunches	Kuala Lumpur, Malaysia	2000	Not provided	Dormant	2005
Auger	ABRI Tech	Agricultural residues	Ontario, Canada	2000	1300	Dormant	2009

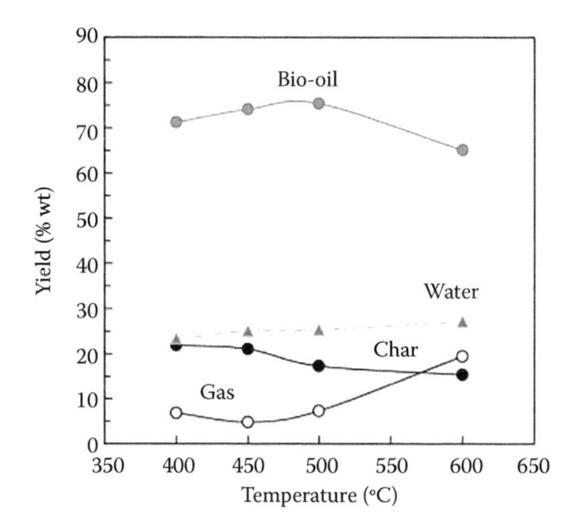

Figure 6.3 Effect of temperature on product yields in fast pinewood pyrolysis (on a wet basis) performed in a CSBR. (From Amutio, M., G. Lopez, M. Artetxe et al. 2012c. *Resour. Conserv. Recycl.* 59: 23–31.)

et al., 2012). Normally, particles below 2 mm are under a kinetic-controlled regime. However, particle size reduction involves energy consumption and increases the cost of the fast biomass pyrolysis operation (Kan et al., 2016).

In order to improve the overall economy of the process for its successful industrial implementation, several techniques have been proposed in the literature, with the most significant ones being those of copyrolysis and vacuum and/or oxidative pyrolysis. The copyrolysis of biomass with other feedstocks (coal or plastic wastes), as well as copyrolysis with different kinds of biomasses, is an interesting strategy, not only for the economy of scale, but also for contributing to the joint valorization of these wastes. Additionally, copyrolysis may improve the characteristics of the bio-oil, for example, increase its yield, reduce the water content, and increase the heating value due to the synergetic interactions between the individual mechanisms of pyrolysis (Abnisa and Daud, 2014; Alvarez et al., 2015a; Hassan et al., 2016). The synergetic effect between coal and biomass results from H and OH radicals generated from the biomass pyrolysis transferring to the coal structure in order to facilitate the cleavage of the double bond of the coal, thus enhancing the volatile yields while lowering the char yields (Yuan et al., 2012; Hassan et al., 2016). Additionally, biomass with high ash content would enhance the cracking of aromatic compounds in the coal (Song et al., 2014). The synergies between plastics and biomass are attributed to radical interaction during copyrolysis. Plastics and synthetic polymers act as hydrogen donors in the thermal coprocessing with biomass because of their high hydrogen content compared to biomass (Oyedun et al., 2014).

It is noteworthy that industrial implementation of biomass pyrolysis requires suitable reactors that may treat different types of heterogeneous and irregular materials that are geographically delocalized and whose composition is also influenced by seasonal variations jointly. Thus, the CSBR has been proven to be suitable for jointly coprocessing lignocellulosic biomass with sewage sludge (Alvarez et al., 2015a) or plastics (Arregi et al., 2017).

Biomass is pyrolyzed under vacuum in order to favor devolatilization, decrease the pyrolysis temperature, and minimize cracking of volatiles. An additional advantage is that it eases the condensation of pyrolysis products (Tripathi et al., 2016). Vacuum pyrolysis is usually carried out at 400–550°C and a total pressure of 2–25 kPa. Amutio et al. (2011) studied pinewood sawdust vacuum pyrolysis in a CSBR in continuous mode. These authors concluded that fluidizing agent requirements were lower under vacuum, and therefore energy costs were considerably reduced. Additionally, they observed that bio-oil composition was scarcely affected by vacuum down to 0.25 atm.

Moreover, fast biomass pyrolysis is endothermic and heat is typically transferred by either fluidizing inert sand or, alternatively, by circulating sand particles (Li et al., 2015b). In this scenario, one of the greatest challenges for scaling up the pyrolysis processes lies in minimizing the energy required to heat the raw material and the fluidizing gas to the process temperature (Amutio et al., 2012a). Accordingly, oxidative pyrolysis, involving the addition of a low concentration of oxygen, allows attaining an autothermal regime, which will reduce operating costs and therefore improve process feasibility (Amutio et al., 2012b). In addition, carefully controlled partial oxidative pyrolysis with low oxygen concentrations will improve the bio-oil quality by increasing the concentration of monomers and reducing the concentration of oligomers or pyrolytic lignin (Li et al., 2014, Li et al., 2015b). Figure 6.4 shows the mechanism proposed by Amutio et al. (2012b) for the oxidative pyrolysis, which consisted of six simultaneous reactions originating from two types of reactants, three corresponding to the heterogeneous oxidation of the pyrolyzable and combustible fraction of each biomass component, and the remaining reactions corresponding to the combustion of the nonpyrolyzable and combustible fraction of each component.

The aim of use of *in situ* catalysts in fast biomass pyrolysis is to lower the reaction temperature and at the same time reduce the oxygen functionalities of bio-oil structures through dehydration, decarbonylation, and decarboxylation reactions (Atutxa et al., 2005; Kabir and Hameed, 2017). Additionally, cracking reactions of rearranging hydrocarbon molecules are also promoted in the presence of a catalyst, thus significantly

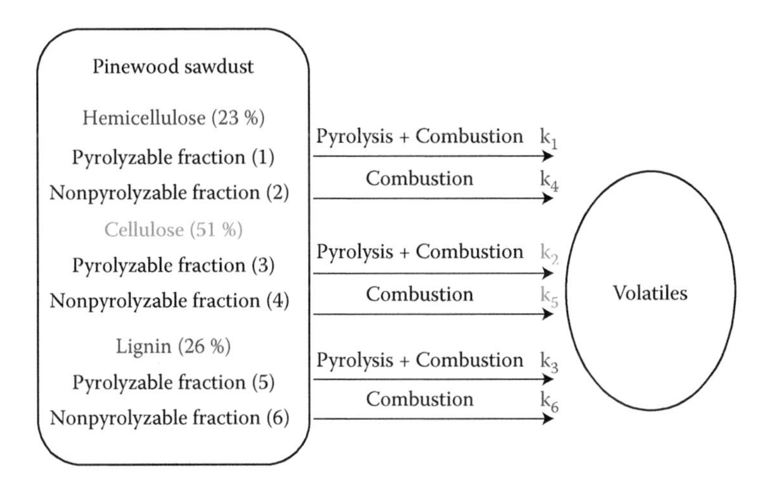

Figure 6.4 Kinetic scheme for pinewood oxidative pyrolysis. (From Amutio, M., G. Lopez, R. Aguado et al. 2012b. *Fuel* 95: 305–311.)

altering the bio-oil composition. Consequently, the aromatic content and the heating value of the bio-oil are increased in the presence of catalysts, such as HZSM-5 zeolites (Thangalazhy-Gopakumar et al., 2012), mesoporous aluminosilicates (Al-MCM-41) (Lappas et al., 2015), alkaline catalysts (NaOH, KOH, KCO_3, $CaOH_3$, MgO) (Auta et al., 2014), and ZSM-5 (Zhang et al., 2014), and low-cost base catalysts (Kabir and Hameed, 2017). However, the presence of these catalysts would also increase the water content and reduce the yield of the liquid fraction due to the promotion of deoxygenation reactions, thereby increasing the gas content. It should be pointed that catalyst activity is closely related to the biomass constituents and to the bio-oil composition produced in the thermal pyrolysis.

Apart from using the catalyst directly in the pyrolysis reactor (*in situ* operation), it may also be placed in another reactor connected in line with the pyrolysis process in order to separately transform the pyrolysis primary vapors (*ex situ* operation, Figure 6.5) (Section 6.3.3) or it may be used

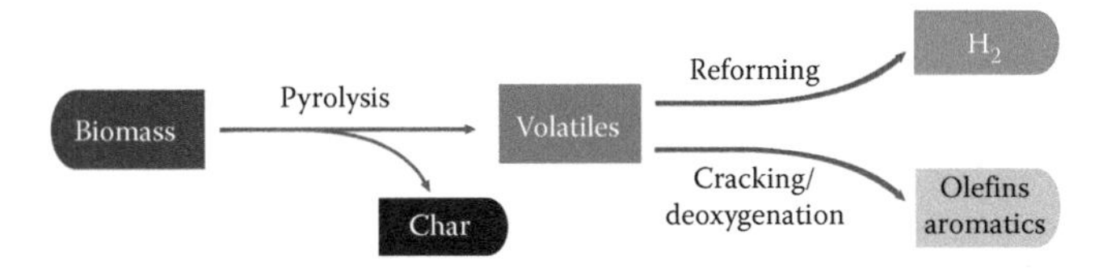

Figure 6.5 Biomass valorization by two in-line steps of pyrolysis and catalytic conversion.

in a separate process to valorize the bio-oil after being condensed in an additional stage (Section 6.4).

6.3.3 *In-line catalytic conversion of pyrolysis volatiles*

The advantages of valorizing the pyrolysis volatiles in line in a continuous catalytic reactor over those using a catalyst *in situ* are as follows: (1) catalyst efficiency is higher, and therefore a lower amount is required to attain the same conversion level; (2) optimum conditions, and particularly different temperatures, can be established in the pyrolysis and catalytic reactors; and (3) the catalyst deactivation is lower because the oxygenated stream is lighter and with lower capacity for coke formation. Moreover, regarding the bio-oil valorization strategies (Section 6.4), the bio-oil is neither condensed nor stored in the inline catalytic converter. Furthermore, the bio-oil vaporization stage, which is required in the catalytic conversion, is also avoided, thereby reducing the energy consumption and the problems associated with the repolymerization of bio-oil phenolic compounds.

Among the different processes for the catalytic upgrading of pyrolysis vapors, steam reforming and cracking have been the most studied in the literature (Bimbela et al., 2013; Cheng et al., 2016; Artetxe et al., 2017). Regarding steam reforming, Arregi et al. (2016) proved that fast biomass pyrolysis in a CSBR and in-line reforming in a fluidized bed was a feasible alternative for the production of H_2 (117 g of H_2 kg^{-1} of biomass), attaining a full conversion of volatiles with a steam–biomass ratio of 4 on a Ni commercial catalyst on a Ni commercial catalyst at 600°C. Through on-line catalytic cracking of pyrolysis vapors in fixed bed reactors on a HZSM-5 catalyst, Huang et al. (2015) obtained the highest hydrocarbon content (58.63%) and the highest C_8–C_{12} content (48.03%) in the oily phase of the product at 500°C in both pyrolysis and catalytic reactors.

6.4 *Bio-oil*

6.4.1 *Composition and properties*

Bio-oil is a dark brown, polar, and hydrophilic liquid, consisting of a mixture of water and oxygenates (from which more than 300 have been identified) that can be grouped into several families (Garcia-Perez et al., 2007; Jacobson et al., 2013): (1) hydroxyacetaldehydes, (2) hydroxyketones, (3) sugars, (4) carboxylic acids, and (5) phenolic compounds. A more detailed classification distinguishes acids, alcohols, aldehydes, esters, ketones, phenols, guaiacols, syringols, sugars, furans, alkenes, aromatics, nitrogenated compounds, and various oxygenated compounds. Table 6.2 shows the typical composition of the bio-oil derived from lignocellulosic biomass and sewage sludge. The latter has a very diverse composition due to

Table 6.2 Typical characteristics of the bio-oil produced from lignocellulosic biomass and sewage sludge

Characteristics	Lignocellulosic bio-oil	Sewage sludge bio-oil
Chemical composition (wt%)		
Oxygenated compounds		
Phenols	5–25	10–28
Acids	3–7	2–28
Alcohols	1–9	2–8
Ketones	2–9	2–8
Furans	3–7	1–3
Sugars	10–30	1–4
Others	10–20	2–15
Nitrogenated compounds		
Amines	–	1–3
Amides	–	1–4
Nitriles	–	1–4
Others	0–1.5	1–15
Hydrocarbons/Extractives	4–6	0.3–20
Sulfur compounds	–	0.1–5
Lignin-derived fraction	15–25	–
Physical properties		
Water content (wt%)	20–40	20–33
Viscosity (cSt at 40°C)	15–1000	20–1500
Ash (wt%)	0–0.25	0–0.8
Density (kg m^{-3})	1–1.3	1.05–1.2
pH	2–4	6–10
Lower heating value (LHV) (MJ kg^{-1})	14–19	18–25
Ultimate analysis (wt%)		
Carbon	32–48	30–70
Hydrogen	7–8.5	7–9
Nitrogen	0–0.9	4–10
Sulfur	<0.05	0.4–2.5
Oxygen	40–55	12–39

Note: Information for lignocellulosic bio-oil is from Chiaramonti et al. (2007), Amutio et al. (2012c), Oasmaa et al. (2010, 2016), and Cheng et al. (2016); information for sewage sludge bio-oil is from Fullana et al. (2003), Cao et al. (2010), Fonts et al. (2012), Trinh et al. (2013), Zuo et al. (2013), Leng et al. (2015), Alvarez et al. (2016), and Arazo et al. (2017).

the heterogeneity of the initial sewage sludge, whose composition depends on the treatment processes to which it has been subjected. Furthermore, the average molecular weight of bio-oil is in the 370–1000 g mol^{-1} range and the oligomeric species derived from lignin and cellulose lead to aerosols with molecular weights of up to 5000 g mol^{-1}. Consequently, a combination of several techniques is required for a detailed bio-oil characterization. The main one is gas chromatography–mass spectrometry (GC–MS), but it needs to be complemented with Fourier-transform infrared spectroscopy (FTIR), ^{1}H and ^{13}C nuclear magnetic resonance (NMR), liquid chromatography–mass spectrometry (LC–MS), gel permeation chromatography (GPC), high-performance liquid chromatography (HPLC), or high-resolution mass spectrometry (HRMS) (Garcia-Perez et al., 2007; Kanaujia et al., 2013; Stas et al., 2014; Yang et al., 2015; Hao et al., 2016; Joseph et al., 2016; Oasmaa et al., 2016). According to Michailof et al. (2016), an approximate classification of the bio-oil components considering the analytical method or technique used for their detection allows their grouping into four fractions: (1) medium-polar monomers detectable by GC, approximately 40 wt%; (2) polar monomers detectable directly by HPLC (or GC after derivation), approximately 10–15 wt%; (3) water, approximately 25–30 wt%, determined by the Karl–Fischer method; and (4) oligomeric material, approximately 20 wt%, determined by GPC or HRMS.

The bio-oil water content is in the 21–27 wt% range for wood, but it is higher (39–51 wt%) for agricultural residues constituted by herbaceous materials, which also contribute to a higher oxygen content in the bio-oil (approximately 45–50 wt%) (Oasmaa et al., 2010). This water is the result of the original content in the feedstock and dehydration reactions that occur during pyrolysis (Mohan et al., 2006). Water content increases with the content of AAEMs (Na, K, Mg, and Ca) in the biomass, and has both positive and negative effects on bio-oil properties. On the one hand, it causes difficulties in ignition and contributes to lower energy density and flame temperature, but on the other, water enhances the atomization properties by reducing viscosity, and therfore thermal NO_x and unburned particulate emissions (Lehto et al., 2014; Yang et al., 2015)

The solids present in the bio-oil are the char and heavy components of the organic fraction. The bio-oil produced in a fluidized bed reactor may contain inorganic materials used for promoting fluidization and they should be separated by cyclones or other devices due to their negative effect. Thus, during storage they tend to agglomerate in larger particles and act as catalysts in the aging reactions, while in combustion they obstruct injection systems, hinder bio-oil pumping and atomization, and form carbonaceous particulate emissions (Liao et al., 2013; Molinder et al., 2016). The char particles leaving the reactor are mostly retained in cyclones; however, these systems are not effective at retaining fine particles (<10 μm). Accordingly, hot vapor filtration (Chen et al., 2011) and microfiltration (Javaid et al., 2010) have been useful techniques for trapping the finest particles and improving bio-oil quality.

The presence of organic acids, mainly acetic acid and formic acid, give the bio-oil an acid character (pH = 2–4), making it corrosive for building materials like carbon steel (Czernik and Bridgwater, 2004). Nevertheless, the bio-oil derived from sewage sludge has a basic character (pH = 6–10) due to the presence of NH_3 and organic compounds containing N (Fonts et al., 2012).

The viscosity of the bio-oil can vary in a wide range (10–1000 cSt at 40°C) depending on the feedstock, the water content of the bio-oil, the operating conditions, and the extent to which the oil has aged (Meier et al., 2013; Yang et al., 2015). Moderate preheating (below 80°C) is required to reduce this viscosity in order to facilitate pumping and atomization (Lu et al., 2009). The density of the bio-oil is approximately 1.2 kg m^{-3}, higher than that of conventional fuels (0.8–1 kg m^{-3}), and therefore the energy density corresponds to 50%–60% of oil-derived fuels (Lu et al., 2009; Yang et al., 2015). The parameters affecting density are mainly pyrolysis temperature and water content (No, 2014).

Moreover, due to the high oxygen content, the lower heating value of the bio-oil derived from lignocellulosic biomass is in the 14–18 MJ kg^{-1} range, and of that from sewage sludge is between 18 and 25 MJ kg^{-1}, much lower than that from conventional fuels (41–43 MJ kg^{-1}) (Fonts et al., 2012; Alvarez et al., 2016).

6.4.2 *Nature, phase separation, and stability*

Bio-oil is considered a microemulsion composed of a continuous phase constituted by an aqueous solution of holocellulose decomposed products and small products from lignin decomposition, and a discontinuous phase composed of macromolecules derived from lignin pyrolysis (pyrolytic lignin), with this being stabilized by the continuous phase. The microemulsion stabilization is produced by hydrogen bonds and the formation of nanomicelles and micromicelles, which tend to disappear at temperatures higher than 60°C (Fratini et al., 2006). Alternatively, an extra phase (called the upper layer or extractive rich layer) may be formed from biomass extractives, depending on the biomass type (Oasmaa et al., 2016).

Pyrolysis oils differ in their ability to dissolve water, which is higher when the oxygen content increases (Oasmaa et al., 2016). The water content dissolved in the bio-oil is in the 22–31 wt% range, and for values higher than these sedimentation of the hydrophobic components derived from lignin occurs, leading to a phase separation (Oasmaa and Czernik, 1999).

Bio-oil is an unstable product because some of the species are still highly active (Yang et al., 2015). Therefore, during storage and transportation, bio-oil undergoes an aging process, leading to an increase in viscosity and water content, loss of volatiles, and phase separation due to the microemulsion rupture and the chemical reactions leading to changes in physical and chemical properties that increase the average molecular weight (Bridgwater, 2012). According to Hilten and Das (2010), the changes that occur during bio-oil aging include a decrease in carbonyl groups (aldehydes and ketones) and an increase in bigger molecules of different polarity composed of lignin fragments, extractives, and solids residue. Temperature has the strongest influence on these reactions, thus favoring bio-oil instability by increasing the average molecular weight and total acid number (Oasmaa et al., 2015; Haverly et al., 2016).

6.4.3 *Improving bio-oil stability*

In order to increase stability during storage and at the same time improve its properties for the subsequent valorization, different treatments have been proposed: (1) addition of polar solvents (methanol), (2) emulsion with other fuels, (3) hydrodeoxygenation (HDO), (4) solids removal, and (5) torrefaction before the pyrolysis or thermal treatment of the bio-oil (accelerated aging).

The addition of methanol is an effective treatment to reduce the initial viscosity, thus attenuating bio-oil aging (Diebold and Czernik, 1997). Furthermore, it also promotes the solubility of the hydrophobic compounds (high molecular weight lignin derivatives) and hinders its polymerization because acetalization and esterification reactions are promoted with the formation of stable components (Fei et al., 2014; Yang et al., 2015). Nevertheless, the use of organic solvents to improve bio-oil stability is limited by the decrease in the flash point and the cost of solvents (Oasmaa et al., 2016).

Another approach to decrease bio-oil viscosity and improve stability is to form emulsions with mineral oils, such as diesel, using surfactant agents (Martin et al., 2014; Yang et al., 2015). However, the high consumption of energy in stirring and the cost of surfactants are the main challenges for using this stabilization method.

The improvement of bio-oil stability by reducing the oxygen content can be attained through HDO to use it directly as fuel or for its catalytic valorization. HDO takes place at a moderate temperature range (300–600°C) under high hydrogen atmosphere in the presence of heterogeneous catalysts. Ni—Co and Co—Mo supported on SiO_2 and SiO_2—Al_2O_3 are the most conventional transition metal sulfide catalysts used in HDO, although Pt-, Rd-, and Ru-supported ones have higher activity and stability in the reaction medium (with high water contents) (Saidi et al., 2014; Patel and Kumar, 2016; Cordero-Lanzac et al., 2017). The possible reactions during the HDO process are as follows (Gollakota et al., 2016): (1) dehydration, (2) decarboxylation where oxygen is removed, (3) hydrogenation of unsaturated compounds, (4) hydrogenolysis related to the breakup of C—O bonds and water release, and (5) hydrocracking of C—C bonds of high molecular weight derived compounds. The bio-oil obtained has adequate viscosity, is less corrosive, and is more stable (Zacher et al., 2014).

Moreover, in order to reduce the oxygen content and obtain a stable bio-oil, a thermal treatment between 200 and 350°C can be performed, either to the bio-oil (accelerated aging) (Bertero et al., 2011) or to the raw biomass before pyrolysis run (torrefaction). When the biomass is previously torrified, the energy density is higher, hygroscopicity is lowered, and bio-oils contain lower oxygen content and acidity (Meng et al., 2015; Kan et al., 2016). In addition, de Miguel Mercader et al. (2010) showed that a high-pressure thermal postreatment (200 bar and 200–300°C) to the bio-oil led to an increase in water content, thus reducing the oxygen content of the organic phase and increasing the energy content (from 14.1 to 28.4 MJ kg^{-1}).

6.4.4 *Bio-oil catalytic valorization*

Among the different bio-oil valorization routes, the most relevant ones are (Jacobson et al., 2013) (1) component extraction, (2) use as fuel, and (3) a catalytic process for the production of H_2, chemicals, and fuels.

Bio-oil contain a wide range of organic components, including a range of high-value-added chemicals, including (Effendi et al., 2008) (1) phenol and derivatives (formulation of resins), (2) volatile organic acids (acetic acid for antifreeze synthesis), and (3) levoglucosan, hydroxyacetaldehydes, and other additives (pharmaceutical industry, fertilizer synthesis). Although the separation methods are still costly and the content of some compounds is low, there is a potential market for some industries using bio-oil as an environmentally friendly and renewable feedstock (Kim, 2015).

Additionally, bio-oil has features of environmental interest (null content of S and low content of N) that make it potentially attractive as fuel, with low net CO_2 emissions (following an adequate biomass exploitation policy), zero SO_x emission, and lower NO_x emission than fossil fuels. It has been tested that bio-oil could be technically suitable for replacing heavy fuel in district heating, although the boilers originally designed for heavy fossil fuels must be adapted to improve the combustion and reduce particulate emissions (Lehto et al., 2014).

However, the direct combustion of bio-oil, even after being conditioned, presents serious drawbacks (high water content, low calorific value, low cetane number, high viscosity, acidity, and instability), which requires its emulsion with another petroleum-derived fuel, such as diesel, or biodiesel (No, 2014). Likewise, when large-scale production of bioethanol from lignocellulosic biomass is feasible, the bio-oil/bio-ethanol blends will also be of great interest given that bioethanol has proven to be suitable for keeping the bio-oil stable during its storage. This mixture (with 20% bio-oil) can be injected at high pressure (25 bar) without problems of stability in the combustion (Nguyen and Honnery, 2008).

Catalytic valorization pathways on continuous regime are more interesting for large-scale bio-oil valorization. These routes can be directed (Figure 6.6) toward H_2 production by means of reforming and/or to fuels and chemicals (olefins and aromatics) through cracking and deoxygenation. Reactors and processes can be designed specifically for each target and bio-oil can also be cofed together with the common feed in refinery units, such as those of fluid catalytic cracking (FCC) and hydrocracking or to the reactor of methanol into olefins (MTO).

This strategy allows separation of the process of pyrolysis and that of catalytic valorization. The bio-oil can be produced in a geographically delocalized way, using small mobile units of fast pyrolysis, which can be energy self-sufficient by combusting the gas and char. The bio-oil would be stored and transported for its valorization in a catalytic process of the scale required for its viability.

In order to avoid the problems associated with bio-oil vaporization, such as phenolic compound repolymerization, which lead to blockage of feeding pipes and reactor flow, as well as a rapid catalyst deactivation (Cheng et al., 2016), Gayubo et al. (2010b) proposed a two-stage system (Figure 6.7). In the former, the solid product derived from the polymerization of the phenolic components in the bio-oil (pyrolytic lignin, with similar properties and composition as that of commercial lignin) is separated, and the volatile stream is upgraded in an on-line catalytic reactor, either to produce H_2 (with a reforming catalyst) or fuels and/or chemicals (with a cracking-deoxygenation catalyst).

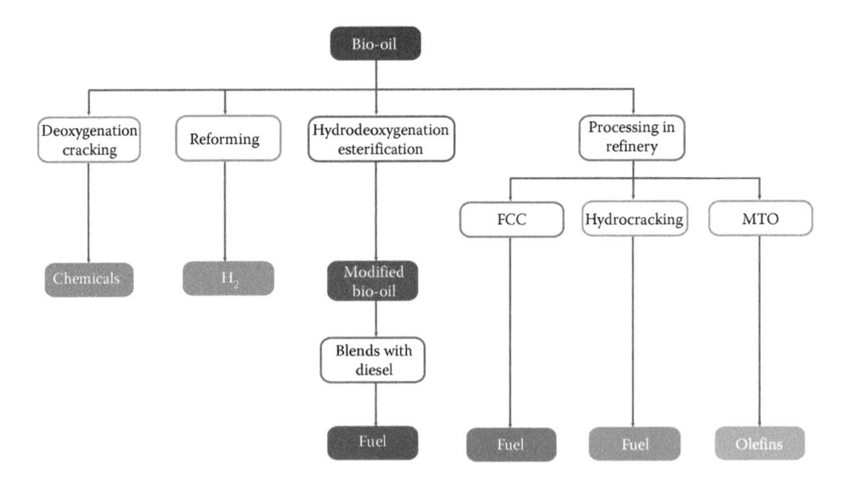

Figure 6.6 Catalytic valorization routes of bio-oil.

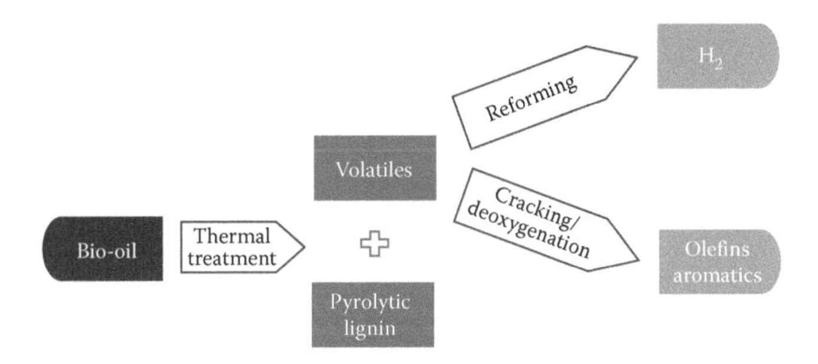

Figure 6.7 Two-stage system for the production of H_2 or chemicals from raw bio-oil.

Moreover, the bio-oil can be transported to refinery units for its catalytic conversion, in which separation of chemical products or their integral valorization may be carried out.

6.4.5.1 *Reforming*

The higher technological capacity and economic viability for producing bio-oil than obtaining other oxygenates derived from biomass [methanol, dimethyl ether (DME), ethanol] makes bio-oil an interesting H_2 vector for large-scale production. In addition, there is no need for separating water from the bio-oil, given that it hinders the rapid deactivation caused by the coke produced by polymerization of the phenolic compounds in the bio-oil. Remiro et al. (2013a,b) and Valle et al. (2014) have worked on the steam reforming of both the aqueous fraction of the bio-oil and the raw bio-oil using the two-stage strategy shown in Figure 6.7.

The widest used catalysts in the bio-oil steam reforming are Ni/Al_2O_3 modified with Ca or Mg, La_2O_3, or MgO. Novel metal catalysts, such as Ru–Mg–Al_2O_3, supported on monoliths or on γ-Al_2O_3 have also been used (Gollakota et al., 2016). Moreover, the use of a material capable of capturing CO_2 together with the catalyst, such as dolomite, allows CO_2 sequestration, which, apart from being interesting from an environmental perspective, generates a relevant synergy by displacing the thermodynamic equilibrium of reforming and water-gas shift (WGS) reactions, increasing H_2 yield (Remiro et al., 2013c).

6.4.5.2 *Cracking and deoxygenation*

Catalytic cracking of bio-oil is a versatile strategy to obtain hydrocarbons with a suitable composition for being used as fuel and raw materials (light olefins or BTX aromatics), either in units designed for this purpose or, as is more interesting, for large-scale bio-oil valorization in refinery FCC units by cofeeding the bio-oil with the standard feed.

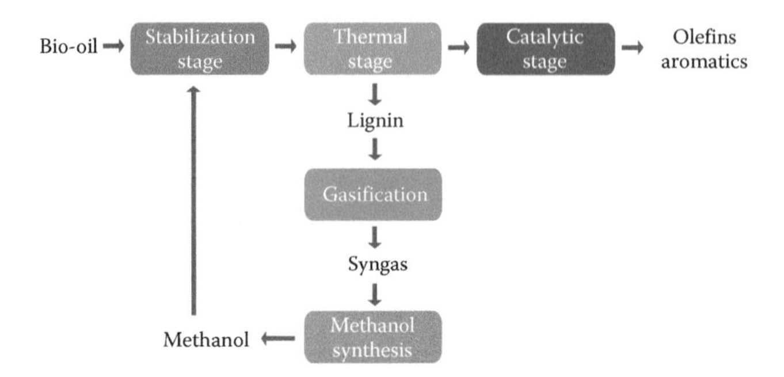

Figure 6.8 Two-stage process for the catalytic valorization of whole raw bio-oil with integration of pyrolytic lignin valorization.

HZSM-5 zeolite catalysts have been shown to be effective in the deoxygenation of bio-oil in a two-stage process (Figure 6.7). By selecting a suitable SiO_2/Al_2O3 ratio in the zeolite or passivating the strong acid sites (by silication or metals incorporation), stable and selective catalysts (avoiding dealumination) can be obtained for the production of olefins (Gayubo et al., 2010a) and aromatics (Valle et al., 2010a,b). These authors propose (Figure 6.8) integrating the two-stage process for the production of hydrocarbons, together with the valorization of the pyrolytic separated from the lignin to methanol. Methanol has proven to be an excellent bio-oil stabilizer during its storage, and cofeeding bio-oil and methanol mixtures involves remarkable synergies by attenuating catalyst deactivation (Gayubo et al., 2009). The synergy of this cofeeding occurs due to the relationship between the hydrocarbon formation mechanisms, that is, the intermediate furans formed from the bio-oil oxygenated aromatics' cracking and deoxygenation are part of the reactive hydrocarbon pool in the formation of olefins (primary products), together with intermediate polyaromatic benzene derivatives of methanol transformation (Gayubo et al., 2010a; Valle et al., 2010b; Zhang et al., 2012). These advantages are promising for the future of bio-oil and methanol cofeeding in the MTO process.

Coking is the main cause of catalyst deactivation in catalytic cracking. The main precursors for coke formation (especially on zeolite catalysts) are lignin-derived phenolic compounds, acetic acid, and acetaldehydes (Gayubo et al., 2005; Ibáñez et al., 2012; Valle et al., 2012). However, the acidity attenuation (acid strength and number of acid sites) of zeolites and the increase of H/C ratio in the feedstock (cofeeding methanol with bio-oil) contributes to lowering coke formation.

6.4.5.3 Cracking under FCC conditions

The cofeeding of raw bio-oil into the FCC unit with the standard feedstock [vacuum gas oil (VGO)] is considered an interesting strategy for fuel production from biomass, given that it reduces oil consumption and global emission of greenhouse gases (Talmadge et al., 2014; Corma and Sauvanaud, 2013). The cofeeding strategy is encouraged by the considerable technological development of fast biomass pyrolysis and the large capacity of FCC units, which are being modified to improve versatility for the processing of heavier streams.

Although there are limitations in the experimental representation of the real conditions in an FCC industrial unit, as in the characteristics of the commercial FCC catalyst and the process conditions (temperature, contact time), laboratory-scale studies have demonstrated that high bio-oil conversions are attained, with the synergy between the blends of hydrocarbons and oxygenates being significant in the cracking stages (Bertero and Sedran, 2013; Naik et al., 2014). Ibarra et al. (2016a) performed the catalytic cracking of a blend of raw bio-oil (20 wt%) and gas oil (VGO, 80 wt%) in a riser simulator reactor under similar conditions (temperature, reaction time, and catalyst/feed mass ratio) as in an FCC unit reactor and using a commercial FCC equilibrium catalyst. The results showed that cofeeding had a favorable synergy because it promoted the formation of C_3–C_4 and gasoline (C_5–C_{12}) hydrocarbon lumps, attenuating the formation of CO_2, CO, and coke. The deposited coke has an intermediate composition between that corresponding to the VGO and bio-oil, with a lower development of polyaromatics due to the presence of steam (Ibarra et al., 2016b)].

6.5 Conclusions

Fast pyrolysis is a key process for the sustainable production of energy and raw materials. A proper understanding of its fundamental aspects and mechanisms has driven the technological development of different reaction systems for producing bio-oil, fuels, and raw materials through a two-stage system (pyrolysis and in-line catalytic transformation).

Among the reactors proposed for fast pyrolysis, fluidized beds and conical spouted beds have proven to be suitable for attaining high bio-oil yields, and they can be implemented at medium and large scale with simple equipment. Vacuum pyrolysis requires further research in order to decrease the inert gas flow rate. Other initiatives that require greater attention are microwave heating and oxidative pyrolysis because they provide solutions for energy requirements in pyrolysis.

The integration of pyrolysis with the use of catalysts, either *in situ* or *ex situ*, leads to different possibilities for improving product quality and distribution. Thus, the *in situ* use of acid cracking and deoxygenation catalysts in the pyrolysis reactor allows fine-tuning the bio-oil composition to facilitate its use as fuel or for subsequent valorization. However, the main drawback of this alternative is the reduction in bio-oil yield.

The *ex situ* use of the catalyst by means of a two-stage system (the volatiles released in the pyrolysis stage are transformed in line in a catalytic reactor) allows selection of the catalysts and the operating conditions in the catalytic stage, selectively producing syngas with a high H_2 concentration (with a reforming catalyst), or a hydrocarbon stream with a high BTX or light olefin content, depending on the cracking and deoxygenation catalyst and reaction conditions.

There are several possibilities for bio-oil valorization: (1) separating the oxygenated compounds of commercial interest, (2) use as fuel, and (3) catalytic transformation. This last strategy is the one with the best prospective for large-scale implementation, and is also interesting for combining the delocalized production of bio-oil in small and transportable pyrolysis units with its large-scale catalytic valorization. This catalytic valorization can be performed using bespoke processes (biorefinery) to produce H_2, hydrocarbons (to be used as fuel), or raw materials (BTX or light olefins). Especially interesting are the encouraging results obtained in the joint valorization of bio-oil with the usual feed in FCC refinery units, whose greater capacity and versatility make this initiative possible.

References

Abnisa, F. and W. M. A. W. Daud. 2014. A review on co-pyrolysis of biomass: An optional technique to obtain a high-grade pyrolysis oil. *Energy Convers. Manage.* 87: 71–85.

Aguado, R., R. Prieto, M. J. S. Josex et al. 2005. Defluidization modelling of pyrolysis of plastics in a conical spouted bed reactor. *Chem. Eng. Process* 44: 231–235.

Alvarez, J., M. Amutio, G. Lopez et al. 2015a. Fast co-pyrolysis of sewage sludge and lignocellulosic biomass in a conical spouted bed reactor. *Fuel* 159: 810–818.

Alvarez, J., M. Amutio, G. Lopez et al. 2015b. Sewage sludge valorization by flash pyrolysis in a conical spouted bed reactor. *Chem. Eng. J.* 273: 173–183.

Alvarez, J., G. Lopez, M. Amutio et al. 2014. Bio-oil production from rice husk fast pyrolysis in a conical spouted bed reactor. *Fuel* 128: 162–169.

Alvarez, J., G. Lopez, M. Amutio et al. 2015c. Physical activation of rice husk pyrolysis char for the production of high surface area activated carbons. *Ind. Eng. Chem. Res.* 54: 7241–7250.

Alvarez, J., G. Lopez, M. Amutio et al. 2016. Characterization of the bio-oil obtained by fast pyrolysis of sewage sludge in a conical spouted bed reactor. *Fuel Process. Technol.* 149: 169–175.

Amutio, M., G. Lopez, R. Aguado et al. 2011. Effect of vacuum on lignocellulosic biomass flash pyrolysis in a conical spouted bed reactor. *Energy Fuels* 25: 3950–3960.

Amutio, M., G. Lopez, R. Aguado et al. 2012a. Biomass oxidative flash pyrolysis: Autothermal operation, yields and product properties. *Energy Fuels* 26: 1353–1362.

Amutio, M., G. Lopez, R. Aguado et al. 2012b. Kinetic study of lignocellulosic biomass oxidative pyrolysis. *Fuel* 95: 305–311.
Amutio, M., G. Lopez, J. Alvarez et al. 2013a. Flash pyrolysis of forestry residues from the Portuguese central inland region within the framework of the BioREFINA-Ter project. *Bioresour. Technol.* 129: 512–518.
Amutio, M., G. Lopez, J. Alvarez et al. 2013b. Pyrolysis kinetics of forestry residues from the Portuguese Central Inland Region. *Chem. Eng. Res. Des.* 91: 2682–2690.
Amutio, M., G. Lopez, J. Alvarez et al. 2015. Fast pyrolysis of eucalyptus waste in a conical spouted bed reactor. *Bioresour. Technol.* 194: 225–232.
Amutio, M., G. Lopez, M. Artetxe et al. 2012c. Influence of temperature on biomass pyrolysis in a conical spouted bed reactor. *Resour. Conserv. Recycl.* 59: 23–31.
Anex, R. P., A. Aden, F. K. Kazi et al. 2010. Techno-economic comparison of biomass-to-transportation fuels via pyrolysis, gasification, and biochemical pathways. *Fuel* 89: S29–S35.
Arazo, R. O., D. A. D. Genuino, M. D. G. de Luna et al. 2017. Bio-oil production from dry sewage sludge by fast pyrolysis in an electrically-heated fluidized bed reactor. *Sustainable Environ. Res.* 27: 7–14.
Arregi, A., M. Amutio, G. Lopez et al. 2017. Hydrogen-rich gas production by continuous pyrolysis and in-line catalytic reforming of pine wood waste and HDPE mixtures. *Energy Convers. Manage.* 136: 192–201.
Arregi, A., G. Lopez, M. Amutio et al. 2016. Hydrogen production from biomass by continuous fast pyrolysis and in-line steam reforming. *RSC Adv.* 25975–25985.
Artetxe, M., J. Alvarez, M. A. Nahil et al. 2017. Steam reforming of different biomass tar model compounds over Ni/Al_2O_3 Catalysts. *Energy Convers. Manage.* 136: 119–126.
Artetxe, M., G. Lopez, M. Amutio et al. 2015. Styrene recovery from polystyrene by flash pyrolysis in a conical spouted bed reactor. *Waste Manage.* 45: 126–133.
Atutxa, A., R. Aguado, A. G. Gayubo et al. 2005. Kinetic description of the catalytic pyrolysis of biomass in a conical spouted bed reactor. *Energy Fuels* 19: 765–774.
Auta, M., L. M. Ern, B. H. Hameed. 2014. Fixed-bed catalytic and non-catalytic empty fruit bunch biomass pyrolysis. *J. Anal. Appl. Pyrolysis* 107: 67–72.
Bai, X., K. H. Kim, R. C. Brown et al. 2014. Formation of phenolic oligomers during fast pyrolysis of lignin. *Fuel* 128: 170–179.
Balat, M. 2010. Thermochemical routes for biomass-based hydrogen production. *Energy Sources, Part A* 32: 1388–1398.
Baruah, D. and D. C. Baruah. 2014. Modeling of biomass gasification: A review. *Renewable Sustainable Energy Rev.* 39: 806–815.
Ben, H. and A. J. Ragauskas. 2011. Pyrolysis of kraft lignin with additives. *Energy Fuels* 25: 4662–4668.
Bertero, M., G. De La Puente, and U. Sedran. 2011. Effect of pyrolysis temperature and thermal conditioning on the coke-forming potential of bio-oils. *Energy Fuels* 25: 1267–1275.
Bertero, M. and U. Sedran. 2013. Upgrading of bio-oils over equilibrium FCC catalysts. Contribution from alcohols, phenols and aromatic ethers. *Catal. Today* 212: 10–15.

Bimbela, F., M. Oliva, J. Ruiz et al. 2013. Hydrogen production via catalytic steam reforming of the aqueous fraction of bio-oil using nickel-based coprecipitated catalysts. *Int. J. Hydrogen Energy* 38: 14476–14487.

Black, B. A., W. E. Michener, K. J. Ramirez et al. 2016. Aqueous stream characterization from biomass fast pyrolysis and catalytic fast pyrolysis. *ACS Sustainable Chem. Eng.* 4: 6815–6827.

Bridgwater, A. V. 2012. Review of fast pyrolysis of biomass and product upgrading. *Biomass Bioenergy* 38: 68–94.

Butler, E., G. Devlin, D. Meier et al. 2011. A review of recent laboratory research and commercial developments in fast pyrolysis and upgrading. *Renewable Sustainable Energy Rev.* 15: 4171–4186.

Cai, W. and R. Liu. 2016. Performance of a commercial-scale biomass fast pyrolysis plant for bio-oil production. *Fuel* 182: 677–686.

Cao, J. P., X. Y. Zhao, K. Morishita et al. 2010. Fractionation and identification of organic nitrogen species from bio-oil produced by fast pyrolysis of sewage sludge. *Bioresour. Technol.* 101: 7648–7652.

Carpenter, D., T. L. Westover, S. Czernik et al. 2014. Biomass feedstocks for renewable fuel production: A review of the impacts of feedstock and pretreatment on the yield and product distribution of fast pyrolysis bio-oils and vapors. *Green Chem.* 16: 384–406.

Chen, T., C. Wu, R. Liu et al. 2011. Effect of hot vapor filtration on the characterization of bio-oil from rice husks with fast pyrolysis in a fluidized-bed reactor. *Bioresour. Technol.* 102: 6178–6185.

Cheng, S., L. Wei, X. Zhao et al. 2016. Application, deactivation, and regeneration of heterogeneous catalysts in bio-oil upgrading. *Catalysts* 6: 195–219.

Chiaramonti, D., A. Oasmaa, Y. Solantausta. 2007. Power generation using fast pyrolysis liquids from biomass. *Renewable Sustainable Energy Rev.* 11: 1056–1086.

Collard, F. X. and J. Blin. 2014. A review on pyrolysis of biomass constituents: mechanisms and composition of the products obtained from the conversion of cellulose, hemicelluloses and lignin. *Renewable Sustainable Energy Rev.* 38: 594–608.

Cordero-Lanzac, T., R. Palos, J. M. Arandes et al. 2017. Stability of an acid activated carbon based bifunctional catalyst for the raw bio-oil hydrodeoxygenation. *Appl. Catal. B* 203: 389–399.

Corma, A. and L. Sauvanaud. 2013. FCC testing at bench scale: New units, new processes, new feeds. *Catal. Today* 218–219: 107–114.

Cotoruelo, L. M., M. D. Marques, F. J. Diaz et al. 2012. Lignin-based activated carbons as adsorbents for crystal violet removal from aqueous solutions. *Environ. Prog. Sustainable Energy* 31: 386–396.

Czernik, S. and A. V. Bridgwater. 2004. Overview of applications of biomass fast pyrolysis oil. *Energy Fuels* 18: 590–598.

de Miguel Mercader, F.,. M. J. Groeneveld, S. R. A. Kersten et al. 2010. Pyrolysis oil upgrading by high pressure thermal treatment. *Fuel* 89: 2829–2837.

Demirbas, M. F., M. Balat, H. Balat. 2009. Potential contribution of biomass to the sustainable energy development. *Energy Convers. Manage.* 50: 1746–1760.

DeSisto, W. J., N. Hill, S. H. Beis et al. 2010. Fast pyrolysis of pine sawdust in a fluidized-bed reactor. *Energy Fuels* 24: 2642–2651.

Diebold, J. P. and S. Czernik. 1997. Additives to lower and stabilize the viscosity of pyrolysis oils during storage. *Energy Fuels* 11: 1081–1091.

Ding, T., S. Li, J. Xie et al. 2012. Rapid pyrolysis of wheat straw in a bench-scale circulating fluidized-bed downer reactor. *Chem. Eng. Technol.* 35: 2170–2176.

Effendi, A., H. Gerhauser, and A. V. Bridgwater. 2008. Production of renewable phenolic resins by thermochemical conversion of biomass: A review. *Renewable Sustainable Energy Rev.* 12: 2092–2116.

Fei, W. T., R. H. Liu, W. Q. Zhou et al. 2014. Influence of methanol additive on bio-oil stability. *Int. J. Agric. Biol. Eng.* 7: 83–92.

Fernandez-Akarregi, A. R., J. Makibar, G. Lopez et al. 2013. Design and operation of a conical spouted bed reactor pilot plant (25 kg/H) for biomass fast pyrolysis. *Fuel Process. Technol.* 112: 48–56.

FitzPatrick, M., P. Champagne, M. F. Cunningham et al. 2010. A biorefinery processing perspective: Treatment of lignocellulosic materials for the production of value-added products. *Bioresour. Technol.* 101: 8915–8922.

Foley, A., B. M. Smyth, T. Puksec et al. 2017. A review of developments in technologies and research that have had a direct measurable impact on sustainability considering the Paris agreement on climate change. *Renewable Sustainable Energy Rev.* 68: 835–839.

Fonts, I., G. Gea, M. Azuara et al. 2012. Sewage sludge pyrolysis for liquid production: A review. *Renewable Sustainable Energy Rev.* 16: 2781–2805.

Fratini, E., M. Bonini, A. Oasmaa et al. 2006. SANS analysis of the microstructural evolution during the aging of pyrolysis oils from biomass. *Langmuir* 22: 306–312.

Fullana, A., J. A. Conesa, R. Font et al. 2003. Pyrolysis of sewage sludge: Nitrogenated compounds and pretreatment effects. *J. Anal. Appl Pyrolysis* 68–69: 561–575.

Fytili, D. and A. Zabaniotou. 2008. Utilization of sewage sludge in EU application of old and new methods: A review. *Renewable Sustainable Energy Rev.* 12: 116–140.

Gamliel, D. P., S. Du, G. M. Bollas et al. 2015. Investigation of *in situ* and *ex situ* catalytic pyrolysis of miscanthus × giganteus using a PyGC–MS microsystem and comparison with a bench-scale spouted-bed reactor. *Bioresour. Technol.* 191: 187–196.

Garcia-Perez, M., A. Chaala, H. Pakdel et al. 2007. Characterization of bio-oils in chemical families. *Biomass Bioenergy* 31: 222–242.

Garcia-Perez, M., J. A. Garcia-Nunez, M. R. Pelaez-Samaniego et al. 2015. Sustainability, business models, and techno-economic analysis of biomass pyrolysis technologies. In: *Innovative Solutions in Fluid-Particle Systems and Renewable Energy Management*, Tannous, K. (Ed.), 298–342, IGI Global, Hershey, PA.

Gayubo, A. G., A. T. Aguayo, A. Atutxa et al. 2005. Undesired components in the transformation of biomass pyrolysis oil into hydrocarbons on an HZSM-5 zeolite catalyst. *J. Chem. Technol. Biotechnol.* 80: 1244–1251.

Gayubo, A. G., B. Valle, A. T. Aguayo et al. 2009. Attenuation of catalyst deactivation by cofeeding methanol for enhancing the valorisation of crude bio-oil. *Energy Fuels* 23: 4129–4136.

Gayubo, A. G., B. Valle, A. T. Aguayo et al. 2010a. Olefin production by catalytic transformation of crude bio-oil in a two-step process. *Ind. Eng. Chem. Res.* 49: 123–131.

Gayubo, A. G., B. Valle, A. T. Aguayo et al. 2010b. Pyrolytic lignin removal for the valorization of biomass pyrolysis crude bio-oil by catalytic transformation. *J. Chem. Technol. Biotechnol.* 85: 132–144.

Gollakota, A. R. K., M. Reddy, M. D. Subramanyam et al. 2016. A review on the upgradation techniques of pyrolysis oil. *Renewable Sustainable Energy Rev.* 58: 1543–1568.

Hao, N., H. Ben, C. G. Yoo et al. 2016. Review of NMR characterization of pyrolysis oils. *Energy Fuels* 30: 6863–6880.

Hassan, H., J. K. Lim, B. H. Hameed. 2016. Recent progress on biomass co-pyrolysis conversion into high-quality bio-oil. *Bioresour. Technol.* 221: 645–655.

Haverly, M. R., K. V. Okoren, R. C. Brown. 2016. Thermal stability of fractionated bio-oil from fast pyrolysis. *Energy Fuels* 30: 9419–9426.

Hilten, R. N. and K. C. Das. 2010. Comparison of three accelerated aging procedures to assess bio-oil stability. *Fuel* 89: 2741–2749.

Huang, Y. F., P. T. Chiueh, S. L. Lo. 2016. A review on microwave pyrolysis of lignocellulosic biomass. *Sustainable Environ. Res.* 26: 103–109.

Huang, Y., L. Wei, J. Julson et al. 2015. Converting pine sawdust to advanced biofuel over HZSM-5 using a two-stage catalytic pyrolysis reactor. *J. Anal. Appl. Pyrolysis* 111: 148–155.

Huber, G. W., P. O'Connor, and A. Corma. 2007. Processing biomass in conventional oil refineries: Production of high quality diesel by hydrotreating vegetable oils in heavy vacuum oil mixtures. *Applied Catalysis A: General* 329: 120–129.

Ibáñez, M., B. Valle, J. Bilbao et al. 2012. Effect of operating conditions on the coke nature and HZSM-5 catalysts deactivation in the transformation of crude bio-oil into hydrocarbons. *Catal. Today* 195: 106–113.

Ibarra, A., E. Rodríguez, U. Sedran et al. 2016a. Synergy in the cracking of a blend of bio-oil and vacuum gasoil under fluid catalytic cracking conditions. *Ind. Eng. Chem. Res.* 55: 1872–1880.

Ibarra, A., A. Veloso, J. Bilbao et al. 2016b. Dual coke deactivation pathways during the catalytic cracking of raw bio-oil and vacuum gasoil in FCC conditions. *Appl. Catal. B* 182: 336–346.

Ingram, L., D. Mohan, M. Bricka et al. 2008. Pyrolysis of wood and bark in an auger reactor: Physical properties and chemical analysis of the produced bio-oils. *Energy Fuels* 22: 614–625.

Jacobson, K., K. C. Maheria, and A. K. Dalai. 2013. Bio-oil valorization: A review. *Renewable Sustainable Energy Rev.* 23: 91–106.

Jahirul, M. I., M. G. Rasul, A. A. Chowdhury et al. 2012. Biofuels production through biomass pyrolysis—A technological review. *Energies* 5: 4952–5001.

Javaid, A., T. Ryan, G. Berg et al. 2010. Removal of char particles from fast pyrolysis bio-oil by microfiltration. *J. Memb. Sci.* 363: 120–127.

Joseph, J., M. J. Rasmussen, J. P. Fecteau et al. 2016. Compositional changes to low water content bio-oils during aging: An NMR, GC/MS, and LC/MS study. *Energy Fuels* 30: 4825–4840.

Kabir, G. and B. H. Hameed. 2017. Recent progress on catalytic pyrolysis of lignocellulosic biomass to high-grade bio-oil and bio-chemicals. *Renewable Sustainable Energy Rev.* 70: 945–967.

Kan, T., V. Strezov, and T. J. Evans. 2016. Lignocellulosic biomass pyrolysis: A review of product properties and effects of pyrolysis parameters. *Renewable Sustainable Energy Rev.* 57: 1126–1140.

Kanaujia, P. K., Y. K. Sharma, U. C. Agrawal et al. 2013. Analytical approaches to characterizing pyrolysis oil from biomass. *TrAC: Trends Anal. Chem.* 42: 125–136.

Kelkar, S., C. M. Saffron, L. Chai et al. 2015. Pyrolysis of spent coffee grounds using a screw-conveyor reactor. *Fuel Process. Technol.* 137: 170–178.
Kim, J. S. 2015. Production, separation and applications of phenolic-rich bio-oil—A review. *Bioresour. Technol.* 178: 90–98.
Kim, S. W. 2016. Pyrolysis conditions of biomass in fluidized beds for production of bio-oil compatible with petroleum refinery. *J. Anal. Appl. Pyrolysis* 117: 220–227.
Kim, Y. and W. Parker. 2008. A technical and economic evaluation of the pyrolysis of sewage sludge for the production of bio-oil. *Bioresour. Technol.* 99: 1409–1416.
Lappas, A. A., K. G. Kalogiannis, E. F. Iliopoulou et al. 2015. Catalytic pyrolysis of biomass for transportation fuels. In: *Advances in Bioenergy: The Sustainability Challenge*, 45–56, John Wiley & Sons, UK.
Lédé, J., F. Broust, F. T. Ndiaye et al. 2007. Properties of bio-oils produced by biomass fast pyrolysis in a cyclone reactor. *Fuel* 86: 1800–1810.
Lehto, J., A. Oasmaa, Y. Solantausta et al. 2014. Review of fuel oil quality and combustion of fast pyrolysis bio-oils from lignocellulosic biomass. *Appl. Energy* 116: 178–190.
Leng, L., X. Yuan, X. Chen et al. 2015. Characterization of liquefaction bio-oil from sewage sludge and its solubilization in diesel microemulsion. *Energy* 82: 218–228.
Li, D., F. Berruti, and C. Briens. 2014. Autothermal fast pyrolysis of birch bark with partial oxidation in a fluidized bed reactor. *Fuel* 121: 27–38.
Li, D., C. Briens, and F. Berruti. 2015a. Improved lignin pyrolysis for phenolics production in a bubbling bed reactor—Effect of bed materials. *Bioresour. Technol.* 189: 7–14.
Li, D., C. Briens, and F. Berruti. 2015b. Oxidative pyrolysis of kraft lignin in a bubbling fluidized bed reactor with air. *Biomass Bioenergy* 76: 96–107.
Liao, H., Q. Lu, Z. Zhang et al. 2013. Overview of methods to remove solid particles from biomass fast pyrolysis oils. *Adv. Mat. Res.* 608–609: 265–268.
López, G., M. Olazar, R. Aguado et al. 2010. Continuous pyrolysis of waste tyres in a conical spouted bed reactor. *Fuel* 89: 1946–1952.
Lu, Q., W. Z. Li., and X.-F. Zhu. 2009. Overview of fuel properties of biomass fast pyrolysis oils. *Energy Convers. Manage.* 50: 1376–1383.
Ma, R., Y. Xu, and X. Zhang. 2015. Catalytic oxidation of biorefinery lignin to value-added chemicals to support sustainable biofuel production. *ChemSusChem* 8: 24–51.
Makibar, J., A. R. Fernandez-Akarregi, I. Alava et al. 2011. Investigations on heat transfer and hydrodynamics under pyrolysis conditions of a pilot-plant draft tube conical spouted bed reactor. *Chem. Eng. Process* 50: 790–798.
Martin, J. A., C. A. Mullen, and A. A. Boateng. 2014. Maximizing the stability of pyrolysis oil/diesel fuel emulsions. *Energy Fuels* 28: 5918–5929.
Meier, D., B. van de Beld, A. V. Bridgwater et al. 2013. State-of-the-art of fast pyrolysis in IEA bioenergy member countries. *Renewable Sustainable Energy Rev.* 20: 619–641.
Meng, J., A. Moore, D. C. Tilotta et al. 2015. Thermal and storage stability of bio-oil from pyrolysis of torrefied wood. *Energy Fuels* 29: 5117–5126.
Michailof, C. M., K. G. Kalogiannis, T. Sfetsas et al. 2016. Advanced analytical techniques for bio-oil characterization. *Wiley Interdiscip. Rev: Energy Environ* 5: 614–639.

Mohan, D., C. U. Pittman Jr., and P. H. Steele. 2006. Pyrolysis of wood/biomass for bio-oil: A critical review. *Energy Fuels* 20: 848–889.

Molinder, R., L. Sandström, and H. Wiinikka. 2016. Characteristics of particles in pyrolysis oil. *Energy Fuels* 30: 9456–9462.

Morales, S., R. Miranda, D. Bustos et al. 2014. Solar biomass pyrolysis for the production of bio-fuels and chemical commodities. *J. Anal. Appl. Pyrolysis* 109: 65–78.

Naik, D. V., V. Kumar, B. Prasad et al. 2014. Catalytic cracking of pyrolysis oil oxygenates (aliphatic and aromatic) with vacuum gas oil and their characterization. *Chem. Eng. Res. Des.* 92: 1579–1590.

Nguyen, D. and D. Honnery. 2008. Combustion of bio-oil ethanol blends at elevated pressure. *Fuel* 87: 232–243.

Nguyen, Q. A., J. Yang, and H. J. Bae. 2017. Bioethanol production from individual and mixed agricultural biomass residues. *Ind. Crops Prod.* 95: 718–725.

No, S. Y. 2014. Application of bio-oils from lignocellulosic biomass to transportation, heat and power generation—A review. *Renewable Sustainable Energy Rev.* 40: 1108–1125.

Nzihou, A., G. Flamant, and B. Stanmore. 2012. Synthetic fuels from biomass using concentrated solar energy—A review. *Energy* 42: 121–131.

Oasmaa, A. and S. Czernik. 1999. Fuel oil quality of biomass pyrolysis oils—state of the art for the end users. *Energy Fuels* 13: 914–921.

Oasmaa, A., I. Fonts, M. R. Pelaez-Samaniego et al. 2016. Pyrolysis oil multiphase behavior and phase stability: A review. *Energy Fuels* 30: 6179–6200.

Oasmaa, A., Y. Solantausta, V. Arpiainen et al. 2010. Fast pyrolysis bio-oils from wood and agricultural residues. *Energy Fuels* 24: 1380–1388.

Oasmaa, A., T. Sundqvist, E. Kuoppala et al. 2015. Controlling the phase stability of biomass fast pyrolysis bio-oils. *Energy Fuels* 29: 4373–4381.

Olazar, M., M. J. San Jose, F. J. Penas et al. 1993. Stability and hydrodynamics of conical spouted beds with binary mixtures. *Ind. Eng. Chem. Res* 32: 2826–2834.

Oyedun, A. O., C. Z. Tee, S. Hanson et al. 2014. Thermogravimetric analysis of the pyrolysis characteristics and kinetics of plastics and biomass blends. *Fuel Process. Technol.* 128: 471–481.

Partridge, T., M. Thomas, B. H. Harthorn et al. 2017. Seeing futures now: Emergent US and UK views on shale development, climate change and energy systems. *Global Environ. Change* 42: 1–12.

Patel, M. and A. Kumar. 2016. Production of renewable diesel through the hydroprocessing of lignocellulosic biomass-derived bio-oil: A review. *Renewable Sustainable Energy Rev.* 58: 1293–1307.

Peters, B. 2011. Prediction of pyrolysis of pistachio shells based on its components hemicellulose, cellulose and lignin. *Fuel Process. Technol.* 92: 1993–1998.

Powell, J. B. 2017. Application of multiphase reaction engineering and process intensification to the challenges of sustainable future energy and chemicals. *Chem. Eng. Sci.* 157: 15–25.

Remiro, A., B. Valle, A. T. Aguayo et al. 2013a. Operating conditions for attenuating $Ni/La_2O_3–\alpha Al_2O_3$ catalyst deactivation in the steam reforming of bio-oil aqueous fraction. *Fuel Process. Technol.* 115: 222–232.

Remiro, A., B. Valle, A. T. Aguayo et al. 2013b. Steam reforming of raw bio-oil in a fluidized bed reactor with prior separation of pyrolytic lignin. *Energy Fuels* 27: 7549–7559.

Remiro, A., B. Valle, B. Aramburu et al. 2013c. Steam reforming of the bio-oil aqueous fraction in a fluidized bed reactor with in situ CO_2 capture. *Ind. Eng. Chem. Res* 52: 17087–17098.

Saidi, M., F. Samimi, D. Karimipourfard et al. 2014. Upgrading of lignin-derived bio-oils by catalytic hydrodeoxygenation. *Energy Environ. Sci.* 7: 103–129.

Sandstrom, L., A. C. Johansson, H. Wiinikka et al. 2016. Pyrolysis of Nordic biomass types in a cyclone pilot plant—Mass balances and yields. *Fuel Process. Technol.* 152: 274–284.

Solar, J., I. de Marco, B. M. Caballero et al. 2016. Influence of temperature and residence time in the pyrolysis of woody biomass waste in a continuous screw reactor. *Biomass Bioenergy* 95: 416–423.

Song, Y., A. Tahmasebi, and J. Yu. 2014. Co-pyrolysis of pine sawdust and lignite in a thermogravimetric analyzer and a fixed-bed reactor. *Bioresour. Technol.* 174: 204–211.

Stas, M., D. Kubicka, J. Chudoba et al. 2014. Overview of analytical methods used for chemical characterization of pyrolysis bio-oil. *Energy Fuels* 28: 385–402.

Sun, Y., B. Jin, W. Wu et al. 2015. Effects of temperature and composite alumina on pyrolysis of sewage sludge. *J. Environ. Sci.* 30: 1–8.

Talmadge, M. S., R. M. Baldwin, M. J. Biddy et al. 2014. A perspective on oxygenated species in the refinery integration of pyrolysis oil. *Green Chem.* 16: 407–453.

Thangalazhy-Gopakumar, S., S. Adhikari, S. A. Chattanathan et al. 2012. Catalytic pyrolysis of green algae for hydrocarbon production using H-ZSM-5 catalyst. *Bioresour. Technol.* 118: 150–157.

Trinh, T. N., P. A. Jensen, D. J Kim et al. 2013. Influence of the pyrolysis temperature on sewage sludge product distribution, bio-oil, and char properties. *Energy Fuels* 27: 1419–1427.

Tripathi, Manoj, J. N. Sahu, and P. Ganesan. 2016. Effect of process parameters on production of biochar from biomass waste through pyrolysis: A review. *Renewable Sustainable Energy Rev.* 55: 467–481.

Valle, B., B. Aramburu, A. Remiro et al. 2014. Effect of calcination/reduction conditions of Ni/La_2O_3-αAl_2O_3 catalyst on its activity and stability for hydrogen production by steam reforming of raw bio-oil/ethanol. *Appl. Catal. B* 147: 402–410.

Valle, B., P. Castano, M. Olazar et al. 2012. Deactivating species in the transformation of crude bio-oil with methanol into hydrocarbons on a HZSM-5 catalyst. *J. Catal.* 285: 304–314.

Valle, B., A. G. Gayubo, A. T. Aguayo et al. 2010a. Selective production of aromatics by crude bio-oil valorization with a nickel-modified HZSM-5 zeolite catalyst. *Energy Fuels* 24: 2060–2070.

Valle, B., A. G. Gayubo, A. Alonso et al. 2010b. Hydrothermally stable HZSM-5 zeolite catalysts for the transformation of crude bio-oil into hydrocarbons. *Appl. Catal. B* 100: 318–327.

Wang, S., X. Guo, K. Wang et al. 2011. Influence of the interaction of components on the pyrolysis behavior of biomass. *J. Anal. Appl. Pyrolysis* 91: 183–189.

Westerhof, R. J. M., H. S. Nygard, W. P. M. Van Swaaij et al. 2012. Effect of particle geometry and microstructure on fast pyrolysis of beech wood. *Energy Fuels* 26: 2274–2280.

Yang, Z, A. Kumar, and R. L. Huhnke. 2015. Review of recent developments to improve storage and transportation stability of bio-oil. *Renewable Sustainable Energy Rev.* 50: 859–870.

Yuan, S., Z. Dai, Z. Zhou et al. 2012. Rapid co-pyrolysis of rice straw and a bituminous coal in a high-frequency furnace and gasification of the residual char. *Bioresour. Technol.* 109: 188–197.

Zacher, A. H., M. V. Olarte, D. M. Santosa et al. 2014. A review and perspective of recent bio-oil hydrotreating research. *Green Chem.* 16: 491–515.

Zhang, H., T. R. Carlson, R. Xiao et al. 2012. Catalytic fast pyrolysis of wood and alcohol mixtures in a fluidized bed reactor. *Green Chem.* 14: 98–110.

Zhang, H., M. Luo, R. Xiao et al. 2014. Catalytic conversion of biomass pyrolysis-derived compounds with chemical liquid deposition (CLD) modified ZSM-5. *Bioresour. Technol.* 155: 57–62.

Zuo, W., B. Jin, Y. Huang et al. 2013. Pyrolysis of high-ash sewage sludge in a circulating fluidized bed reactor for production of liquids rich in heterocyclic nitrogenated compounds. *Bioresour. Technol.* 127: 44–48.

chapter seven

Bioethanol production

Advances in technologies and raw materials

Carlos Ariel Cardona Alzate, Carlos Andrés García, and Sebastián Serna Loaiza

Contents

7.1 Introduction

Biomass is used today for multiple purposes, but mostly for the generation of energy through the production of energy vectors, mainly biofuels at different scales and purposes. For 2022 alone, an overall production of 1.6×10^5 million liters is expected (Cremonez et al., 2015). Biofuels can be really competitive for countries with difficulties in the energy supply matrix or with low oil reserves.

In recent years, different energy policies in many countries together with industry and research have focused on available and efficient raw materials for bioethanol production such as sugars (Dias et al., 2012; Wirawan et al., 2012), glycerol (Chaudhary, 2011; Jang et al., 2012), starch (Bai et al., 2008; Jamai et al., 2007), lignocellulosics (Moncada et al., 2013; Triana et al., 2011; Zabed et al., 2016), syngas (Devarapalli et al., 2016; Liu et al., 2014), and even materials obtained in microalgae as well as other raw materials. Among these routes, thermochemical, biological, biochemical, and hybrid routes are found (Naik et al., 2010; Yuan and Eden, 2015; Yue et al., 2014). Bioethanol, being a petrol additive or substitute, has interested the scientific world as well as governments because of different strategic policies to alleviate the oil-dependent energy matrix in many countries. In the industry, bioethanol is derived from alcoholic fermentation of sucrose or simple sugars, which are produced mainly from energy crops but in most cases represent an alternative business at large scale for farmers.

7.2 Biomass as raw material for bioethanol production

7.2.1 Definition and types of biomass

Biomass as defined by the Organisation for Economic Co-operation and Development (OECD, 2002) as the quantity of living material of plant or animal origin present at a given time within a given area. It is a very inclusive definition and does not imply that biomass is only lignocellulosics, as was established in most of the lignocellulosic publications last decade. Among the different sources of biomass to produce bioethanol, four types can be distinguished today at different levels of technological maturity: natural biomass, agroindustrial residuary, residues, and energy crops (FAO, 2010; Goyal et al., 2008). Natural biomass comes from noncrop areas, which means forests or forestry zones where there is no human intervention for plant cultivation. It is not a positive environmental practice, but from this some residual parts such as logs and coniferous and hardwood branches derived from the cleaning and maintenance of forests and plantations can be used in the very long-term future for producing ethanol. "Residue" can be a very wide classification but includes carbon dioxide, glycerol, and other residues not coming directly from the forest or agribusiness. The agroindustrial residuary biomass mainly comes from crops and forestry as well as from the industry processing these materials, usually not causing food competition. Finally, biomass from energy crops is that grown specifically for the production of bioethanol such as sugarcane, cereals, oilseeds, and some residues of these crops (Goyal et al., 2008).

Natural biomass from residues and energy crops are composed mainly of cellulose, hemicellulose, and lignin. Due to this composition, they have been called lignocellulosic biomass and they have been the most studied residues today for producing bioethanol based on its high availability in the world.

7.2.2 Sugars

Sugars (from sugarcane, sugar beets, molasses, and fruits) can be converted into ethanol directly. Real advances in using dedicated crops for sugars as raw material for bioethanol during the last 10 years are practically not noted. Given that fact, sugars and starch are not deeply analyzed in this chapter. The main sugary feedstock for ethanol production is sugarcane in the form of either cane juice or molasses (by-product of sugar mills). Approximately 79% of ethanol in Brazil is produced from fresh sugarcane juice and the remaining percentage is from cane molasses (Sanchez and Cardona, 2008). Quintero et al. (2008) performed the economic and environmental assessment of the bioethanol production from sugarcane and corn. When the economic and environmental criteria are analyzed together, the sugarcane-based process presented a better performance than the corn-based process. In this sense, the use of sugarcane juice for bioethanol production evidenced a higher economic profitability and low environmental impact than the corn-based process (Lin and Tanaka, 2006). One interesting topic in sugary feedstocks is the use of multicomponent residual mixtures without sugars concentration or fractionation. Siqueira et al. (2008), for example, evaluated the production of bioethanol from soybean molasses at the laboratory, pilot, and industrial scale. The production of bioethanol from soybean molasses evidenced satisfactory yields at different scales with small decreases from 169.8 L at the laboratory scale to 163.6 and 162.7 L of absolute ethanol per ton of dry molasses obtained at the pilot and industrial scales, respectively.

7.2.3 Starch

Fermentation of starch is a complex process because starch must be converted into sugars prior to microbial conversion to bioethanol. The general overview of the starch pretreatment is described as follows: starch is first hydrolyzed by adding α-amylase, and then cooked at high temperature (90–110°C) for the breakdown of starch kernels. The product of the first stage of pretreatment is submitted to saccharification at lower temperatures (60–70°C) in order to produce glucose through glucoamylases (Sanchez and Cardona, 2008). Finally, the C6 sugar fraction is fermented to ethanol by different microorganisms, producing CO_2 as a coproduct.

Different starchy materials have been reported to produce bioethanol such as corn, wheat, starch and potatoes, and cassava root. In the United States, ethanol is produced almost exclusively from corn, whereas wheat is used to produce ethanol in Europe with a similar process as that of corn. Nevertheless, cassava is the tuber that has gained the most interest, mainly in Asia, not only in bioethanol production, but also for production of syrups, due to is availability and low cost.

Several authors have studied the bioconversion of starchy materials (e.g., corn, cassava, and wheat) to bioethanol. Patzek (2006) studied the mean stoichiometric yield of ethanol from wet milling corn. From this procedure, an ethanol yield of 0.364 kg of ethanol per 1 kg of dry corn was calculated. Kosugi et al. (2009) evaluated the production of ethanol from pretreated cassava pulp using *Saccharomyces cerevisiae* K7. The ethanol yield varies from 0.30 to 0.50 kg of ethanol per 1 kg of glucose depending on the cassava pulp initial concentration (5–30 w/v%). An increase in the cassava pulp initial concentration reduces the ethanol yield due to the presence of inhibitory compounds from the hydrothermal pretreatment. Murphy and Power (2008) performed the economic and environmental assessment of the potential bioethanol production using wheat in Ireland. When wheat is used as starchy raw material, an ethanol yield of 0.295 kg of ethanol per 1 kg of wheat was obtained.

7.2.4 Lignocellulosic biomass

Lignocellulosic biomass is the raw material that has been studied and demonstrated in the research and industry community for producing bioethanol the most often in recent years. It is a more complex substrate than starch.

Figure 7.1 shows the internal structure of lignocellulosic biomass. This is composed mainly of a complex matrix of cellulose, hemicellulose, and lignin (Alvira et al., 2010; Khoo, 2015; Maurya et al., 2013; Moreno et al., 2014; Mosier et al., 2005). Cellulose represents between 40 and 50 wt% of these materials; it is a homopolysaccharide of 1,4-D-glucopyranose units

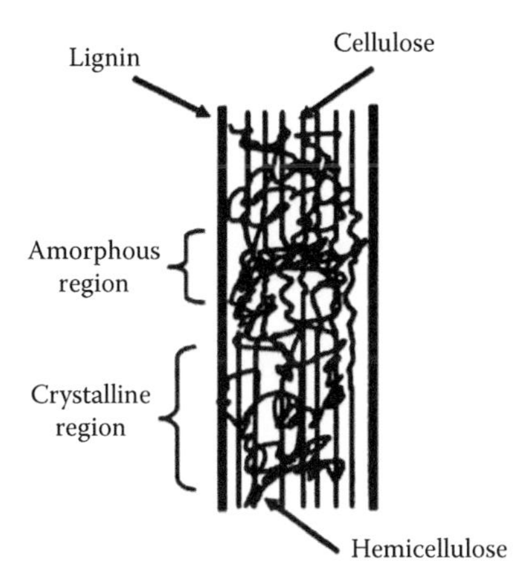

Figure 7.1 Internal structure of lignocellulosic biomass. (From Mosier, N., C. Wyman, B. Dale et al., *Bioresour. Technol.*, 96, 6, 673–686, 2005.)

linked together by β-1,4 glycosidic bonds, which are insoluble in water and in different organic solvents (Goyal et al., 2008; Mood et al., 2013; Quintero and Cardona, 2009). These units associate forming cellulose fibers with crystalline and amorphous structure. The distribution of the amorphicity varies depending on the nature of the material (Mosier et al., 2005).

On the other hand, hemicellulose represents approximately 20–40 wt% of the material; it is a heteropolysaccharide of short side chains consisting of different sugar polymers such as hexoses (D-glucose, D-mannose, D-galactose), pentoses (D-xylose, L-arabinose), and uronic acids and is easy to hydrolyze (Goyal et al., 2008; Mood et al., 2013; Quintero and Cardona, 2009). Hemicellulose is located within the wall of vegetable cells and is responsible for the rigidity because it serves as the connection between lignin and cellulose. After cellulose and hemicellulose, lignin is one of the most abundant polymers in nature, because it is located in the cell wall. This is an aromatic polymer (phenylpropane units) with multiple ramifications and substitutions of organic compounds, which make it insoluble in water. Its function is to give the plant impermeability, structural resistance, and resistance to the attack of microbes and oxidant agents due to the recalcitrant structure formed. Lignin is adjacent to the cellulose fibers, forming the lignocellulosic complex (Goyal et al., 2008; Karimi and Taherzadeh, 2016; Mood et al., 2013; Quintero and Cardona, 2009).

The presence of each one of these polymers in biomass varies according to the nature of the biomass, as shown in Table 7.1. These materials have in common a cellulose content between 40 and 55 wt%, but for agricultural and agroindustrial residues the lignin content normally is below 20 wt%. This low content of lignin contributes to the fact that

Table 7.1 Composition of cellulose, hemicellulose, and lignin of different lignocellulosic materials

Lignocellulosic biomass	Lignin (%)	Hemicellulose (%)	Cellulose (%)	Reference
	Agricultural and agroindustrial waste			
Sweet sorghum	21	27	45	Anwar et al. (2014)
Corncob	15	35	45	
Corn straws	11–19	21–31	35–39	
Wheat straw	16–21	26–32	29–35	
Sugarcane bagasse	20	25	42	
Plantain residues	14	14.8	13.2	
Bagasse	23.33	16.52	54.87	
Walnut shells	30–40	25–30	25–30	
Corn fodder	19	26	38	
Rice straw	18	24	32.1	
Palm oil peel	51.5	22.3	20.5	Saka et al. (2008)
Empty palm oil fruit bunches	23.8	23	33	Daza Serna (2015)
Rice husk	19.40	18.47	42.20	Abdullah et al. (2010)
Grapeseed	47.13	19.03	13.37	Ruales Salcedo (2015)
Softwood	25–35	25–35	45–50	Ruales Salcedo (2015)
Spruce	29	26	43	Olsson and Hahn-Hagerdal (1996)
Pine	29	26	44	
Pinus banksiana	28.6	25.6	41.6	Conde-Mejia et al. (2012)
Pinus pinaster	30.2	17.6	42.9	
Fir	28.4	26.5	43.9	
Hardwood	18–25	24–40	40–55	Anwar et al. (2014)
Birch	21	39	40	Olsson and Hahn-Hagerdal (1996)
Willow	21	23	37	
Aspen	16	29	51	
Eucalyptus viminalis	31	14.1	41.7	Conde-Mejia et al. (2012)

these materials are considered good substrates for obtaining value-added products. Just to exemplify one of multiple examples, a banana peel has been used to produce ethanol (Oberoi et al., 2011). However, in other raw materials with high lignin content, such as hardwoods, the composition represents a physical and chemical barrier that hinders their applicability as substrate, especially in the process of enzymatic hydrolysis (Alvira et al., 2010).

Therefore, structural and compositional features of lignocellulosic biomass represent a very important role in enzymatic digestibility, given that the hydrolysis yields of cellulose in biomass are lower than 20% (Mosier et al., 2005). In this sense, a stage before the enzymatic hydrolysis is necessary in order to increase the accessibility of the enzyme and enhance the digestibility of the cellulose (Sanchez and Cardona, 2008; Alvira et al., 2010; Mood et al., 2013).

The effective utilization of the lignocellulosic feedstock is not always practical because of its seasonal availability, scattered stations, and the high costs of transportation and storage of such large amounts of organic material (Lin and Tanaka, 2006). Besides these limitations, other disadvantages such as technological issues, lack of biorefinery approach, and high crystallinity of the raw material, among others, hinder the use of lignocellulosic biomass at an industrial level. In general, prospective lignocellulosic materials for fuel ethanol production can be divided into six main groups: crop residues (cane bagasse, corn stover, wheat straw, rice straw, rice hulls, barley straw, sweet sorghum bagasse, olive stones, and pulp), hardwood (aspen, poplar), softwood (pine, spruce), cellulose wastes (newsprint, waste office paper, recycled paper sludge), herbaceous biomass (alfalfa hay, switchgrass, reed canary grass, coastal Bermuda grass, timothy grass), and municipal solid wastes (MSWs) (Sanchez and Cardona, 2008).

In tropical countries, one of the major lignocellulosic materials found in great quantities is sugarcane bagasse. Cardona et al. (2010) studied the effect of the sugarcane bagasse pretreatment and detoxification method and the strain for fermentation in bioethanol production. The ethanol yield varies from 0.18 to 0.48 kg of ethanol per 1 kg reducing sugars depending on the pretreatment method (e.g., acid hydrolysis, alkaline treatment, saccharification, and steam treatment), detoxification method (e.g., electrodialysis, ion exchange resin, activated charcoal, enzymes, neutralization), and the microorganisms for fermentation (e.g. *S. cerevisiae, Candida shehatae, Pachysolen tannophilus*).

Forest plantations are one of the major sources of lignocellulosic biomass that can be used to produce sugar-rich fractions, and then ethanol. Moncada et al. (2016) evaluated the production of bioethanol using *Pinus patula* as raw material considering different pretreatment methods (e.g., acid hydrolysis, alkaline treatment, and hot water). The highest ethanol yield was obtained when the alkaline hydrolysis was used as a pretreatment method with a value of 0.42 kg of ethanol per 1 kg of reducing sugars after 72 h.

7.2.5 Pretreatment of lignocellulosic biomass

The biomass pretreatment is one of the main steps to produce bioethanol. The purpose of this stage is to adapt the materials for the subsequent hydrolysis and fermentation operations. In the case of sugars mechanically obtained from sugarcane, for example, additional acid or enzymatic hydrolysis is widely used as pretreatment to increase the glucose content in the mixture with sucrose (Cheng et al., 2008). In the case of starch, gelatinization is also considered as pretreatment to provide accessibility to enzymes (Bai et al., 2008). However, the main problem appears in lignocellulosics in which the recalcitrance is the main barrier for obtaining sugars given the low accessibility of enzymes to the holocellulose (hemicellulose and cellulose).

Using lignocellulosic biomass implies the need of a stage previous to the transformation in a production process. Its objective is to enhance the subsequent stages of the process, which directly affect the productivity and yield of the product of interest. This pretreatment stage must consider not only physical treatments, but technologies that are able to divide the recalcitrant structure (Karimi and Taherzadeh, 2016). This division will reduce the resistance of the lignocellulosic material to enzymatic or microbial attack (Roberto, 2009). Kumar et al. (2009) and Alvira et al. (2010) considered a series of key factors for an effective pretreatment of lignocellulosic biomass, some of which are mentioned as follows:

- Enhancing the formation of sugars or the capacity of posterior formation of sugars via hydrolysis.
- Avoiding the degradation or loss of the carbohydrates.
- Avoiding the formation of compounds that inhibit posterior stages as hydrolysis and fermentation.
- Profitability.

7.2.5.1 Pretreatment technologies

Different pretreatment technologies have been identified to dissociate the matrix of lignocellulosic biomass, which are classified into physical, chemical, physicochemical, and biological pretreatments (Alvira et al., 2010; Conde-Mejia et al., 2012; Gupta, 2014; Kumar et al., 2009; Mood et al., 2013; Mosier et al., 2005). These technologies are shown in Figure 7.2 and some will be explained as follows.

The physical pretreatments include mechanic-type technologies such as mills and extruders to reduce the size of the particles as well as the

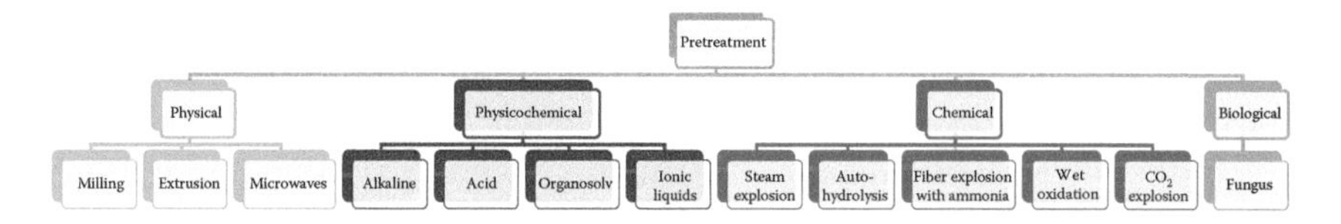

Figure 7.2 Technologies for the pretreatment of lignocellulosic biomass.

crystallinity. The reduction of particle size leads to an increase of the available surface and a reduction of the degree of polymerization. The increase of the specific surface, the reduction of the polymerization degree, and the cutting increase the performance of the hydrolysis of the lignocellulosic biomass. These have been used to condition the bagasse, alfalfa, wood, and forestry residues. The reduction of size in sieves lower than No. 40 has not generated a favorable result. Although it is a traditional technique, it represents a huge demand of energy given the small particle size and the operation costs associated with the high volumes of production of these materials (Quintero Suarez, 2011). On the other hand, extrusion is a relatively new thermomechanical pretreatment that may be applied to lignocellulosic biomass. Its action is based on the rotation of one or two endless screws within a compartment, producing strong friction between the biomass, the screw, and the walls of the module (Duque Garcia, 1998; Mood et al., 2013). This has been used to treat the bagasse of sugarcane and corn, enhancing the production of sugars in the enzymatic hydrolysis, increasing the surface area and the cellulose exposure, and decreasing the crystallinity (Coimbra et al., 2016; da Silva et al., 2013).

In chemical pretreatments, the use of acid or alkaline chemical agents has been emphasized. In acid pretreatments, the hemicellulose fraction of the biomass is solubilized, mainly making the cellulose more accessible to the enzyme. Acids such as concentrated or dilute H_2SO_4, HCl, HNO_3, and H_3PO_4 can be used. However, the use of concentrated acids is less attractive due to the formation of inhibitory compounds and high operating and maintenance costs. Therefore, dilute H_2SO_4 is generally used in concentrations of 0.5–5 w/v%, temperatures lower than 160°C, and solid loads between 10% and 40% (Jung and Kim, 2015). Currently, dilute acid is the most used pretreatment for lignocellulosic biomass (Mood et al., 2013).

During alkaline pretreatment, the first reactions that take place are solvation and saponification. This causes the biomass to expand and become more accessible to enzymes and bacteria. Alkaline solutions such as NaOH, $Ca(OH)_2$ (lime), or ammonia are used to remove part of the lignin and hemicellulose. In this process, different concentrations can be used, pretreatment times can be longer, and temperatures can be lower compared to the pretreatment with dilute acid. The use of this pretreatment prevents the formation of fermentation inhibitors and increases the accessibility of the enzyme to cellulose. It has been used in hardwoods because they are lignocellulosic materials with high lignin content (Balan et al., 2009; Goshadrou et al., 2011; Shafiei et al., 2015; Taherzadeh and Karimi, 2008; Zhang et al., 2012; Zhang and Wu, 2014).

In the physical-chemical pretreatments, steam explosion and hot liquid water or autohydrolysis have been extensively investigated. In steam explosion, pressures between 0.69 and 4.85 MPa and temperatures of 160–260°C are handled for a few seconds or minutes. It is then necessary to cool rapidly and wash the solids to remove possible inhibitory products. It has been used to treat bagasse, softwoods, eucalyptus, and rice straw, among others. This process hydrolyzes approximately 80% of the hemicellulose and its energy costs are lower compared to the use of mills (Rocha et al., 2012a,b; Sanchez and Cardona, 2005). The objective of autohydrolysis is to mainly solubilize hemicellulose to have better access to cellulose and to prevent the formation of inhibitors. To avoid the formation of inhibitors, the pH should be maintained between 4 and 7 during the pretreatment. The temperatures being handled are between 160 and 190°C for approximately 46 min (Jeong and Lee, 2015; Kim et al., 2009; Sanchez and Cardona, 2005; Sun et al., 2016; Yu et al., 2013).

7.2.6 Glycerol

Glycerol is produced abundantly as a by-product of both soap manufacturing and biodiesel production. In this sense, glycerol has become a potential feedstock in the production of various chemicals via fermentation processes (Chaudhary, 2011). One of the advantages of using glycerol as a raw material is the fact that microorganisms can degrade this carbon source directly to produce bioethanol. Several authors have studied the production of bioethanol from glycerol using different strains for fermentation. Ito et al. (2005) studied the production of bioethanol from glycerol-containing wastes after the biodiesel process using the strain *Enterobacter aerogenes* HU-101 at different glycerol concentrations. Bioethanol yield varies from 0.56 to 0.96 mol of ethanol per 1 mol of glycerol-containing wastes (close to technical quality) and from 0.8 to 1.0 mol of ethanol per 1 mol crude glycerol. According to the authors, it shows that the ethanol yield decreases when the glycerol concentration increases. Additionally, the production of bioethanol from glycerol-containing wastes is hindered by the presence of salts in the medium. It is an important raw material to be considered in the industry, especially for feedstocks balancing in biodiesel plants based on ethanol and oils, where 1 ton of glycerol is obtained per 10 ton of biodiesel.

7.2.7 Sugars and starch from algae and microalgae

Algae are simple organisms containing chlorophyll and they use light for photosynthesis. They can grow phototrophically or heterotrophically.

Algae are classified as microalgae and macroalgae. Microalgae, as the name implies, are prokaryotic or eukaryotic photosynthetic microorganisms. They can survive in hard conditions with their unicellular or simple colony structures (Ozcimen and Inan, 2015). Because they are photosynthetic organisms, they can produce high amounts of lipids, proteins, and carbohydrates in a short time. Lipids are the main components of the internal structure of the microalgae. However, there is a potential for carbohydrates in the structure of algae that can be used for ethanol production after various hydrolysis processes. The bioethanol production from algae has several advantages, such as (1) no competition for food in either land or water; (2) relatively high content of carbohydrates (e.g., starch and sugars) in the algae cell; (3) algae do not have lignin and the hemicellulose content is relatively low, which could increase the hydrolysis and fermentation yields; (4) algae have the ability to capture CO_2 from the atmosphere; (5) algae can grow rapidly in different environments (e.g., freshwater, saline water, municipal waste water); and (6) microalgae cells have very fast productivity and harvesting cycles (Li et al., 2014).

Due to the several advantages of the bioethanol production from algae, different studies have been carried out in order to establish suitable conditions for the fermentation of microalgae. Ho et al. (2013) studied the effect of different pretreatment methods in the bioethanol production from microalgae *Chlorella vulgaris* FSP-E using the recombinant *Zymomonas mobilis*. The highest bioethanol yield (0.233 g of ethanol per 1 g of algae) and ethanol concentration (11.66 g per L) were obtained when acid hydrolysis was used as the pretreatment method. However, an open question in microalgae bioethanol is the internal competition between the added value of the products that can be obtained from its processing. Microalgae can be used to obtain omega-3 and omega-6 as well as other very high added-value metabolites (10 to 20 times more expensive than ethanol). The only solution as for other cases like this is the use of the biorefinery concept to compensate the failings of bioethanol prices in the market.

7.2.8 Syngas from biomass

Most lignocellulosic biomass contains a large proportion of matter that cannot be converted economically to ethanol by microorganisms or just to sugars through pretreatments and hydrolysis. An alternative might be to gasify this biomass in order to produce synthesis gas that can be used as an intermediate platform in the production of ethanol and other valuable compounds. Syngas is composed mainly of nitrogen, hydrogen, carbon monoxide, carbon dioxide, and methane. From this gaseous mixture, CO and H_2 are used as substrates for microbial conversion into added-value products. As a consequence, it is expected that syngas fermentation plays an important role in the conversion of lignocellulosic biomass, wastes,

Table 7.2 Overall representation of the described raw materials to produce bioethanol

Raw material (RM)	Pretreatment	Substrate conversion (%)	Fermentative strain	Ethanol concentration (g/L)	Ethanol yield (g EtOH/g dry RM)	Reference
Sugarcane	Milling, dilute acid hydrolysis	90	*S. cerevisiae*	Non-determined	0.203	Quintero et al. (2008)
Soy molasses	–	Non-determined	*S. cerevisiae*	40.01	43.0	Siqueira et al. (2008)
Cassava pulp	Hydrothermal, enzymatic saccharification	79	*S. cerevisiae* K7	18.6	0.50	Kosugi et al. (2009)
Wheat	Liquefaction, hydrothermal	76	*S. cerevisiae*	Non-determined	0.295	Power et al. (2008)
Sugarcane bagasse	Alkaline treatment, enzymatic saccharification	55	Recombinant *S. cerevisiae*	12.9	0.258	Hernández-Salas et al. (2009)
	Acid hydrolysis, electrodialysis	52	*P. tannophilus* DW06	19	0.34	Cheng et al. (2008)
P. patula	Alkaline hydrolysis, enzymatic saccharification	60	*S. cerevisiae*	15.50	0.42	Moncada et al. (2016)
Crude glycerol	–	100	*E. aerogenes* HU-101	4.3	0.86	Ito et al. (2005)
Glycerol from biodiesel process	Diluted with synthetic medium	100	*E. aerogenes* HU-101	3.35	0.67	Ito et al. (2005)
Microalgae (*Ch. vulgaris*)	Enzymatic	Non-determined	*Z. mobilis*	3.55	0.178	Ho et al. (2013)
Microalgae (*Chlamydomonas reinhardtii*)	Acid hydrolysis	58	*S. cerevisiae*	14.6	0.292	Nguyen et al. (2009)
Synthesis gas	No pretreatment	Non-determined	*C. ljungdahlii*	48.0	Non-determined	Klasson et al. (1992)
Synthesis gas	No pretreatment	Non-determined	*C. ljungdahlii*	12.0	0.141	Younesi et al. (2005)

and further residues that cannot be used for direct fermentation (Henstra et al., 2007).

Syngas fermentation into ethanol and other bioproducts offers several advantages such as (1) utilization of the whole biomass (including lignin), (2) elimination of complex pretreatment steps and costly enzymes, (3) higher specificity of the biocatalyst, (4) independence of the H_2:CO ratio (Munasinghe and Khanal, 2011), (5) greater resistance to catalyst poisoning, and (6) lower energy costs (Henstra et al., 2007).

Different microorganisms (e.g., *Clostridium ljungdahlii, C. autoethanogenum, Acetobacterium woodii, C. carboxidivorans,* and *Peptostreptococcus productus*) have been used to produce liquid fuels from synthesis gas (Munasinghe and Khanal, 2011). The main limitations of syngas fermentation is the low productivity and poor solubility of gaseous substrates in the liquid phase.

Different authors have studied the production of ethanol and other by-products from syngas fermentation using the microorganism *C. ljungdahlii* that was one of the first bacteria to be able to convert CO and H_2 into ethanol and acetate, Table 7.2 (Klasson et al., 1992). The ethanol concentration varies from 12 to 48 g/L depending on the reactor volume and the operation mode (batch or continuous) (Klasson et al., 1992; Younesi et al., 2005).

7.3 Technologies

7.3.1 Fermentation

In a broad sense, fermentation is a process of growing microorganisms in a given growth medium, generally with the aim of producing a specific chemical product. This process occurs in yeast, bacteria, and sometimes in muscle cells. Regarding ethanol production, this is a biological process through which bioethanol is obtained by means of a microorganism that produces extracellular ethanol.

The production of ethanol via fermentation necessarily starts with sugar. For example, when glucose is used as the main sugar, it is converted to pyruvate by glycolysis enzymes and it also produces adenosine triphosphate (ATP) and nicotinamide adenine dinucleotides (NADHs). Then, pyruvate is decarboxylated by pyruvate decarboxylase to acetaldehyde, which is later reduced to ethanol by alcohol dehydrogenase using the NADH previously generated (Jarboe et al., 2009). This is the metabolic route applied by *S. cerevisiae,* and it is very common in yeasts and fungi. *S. cerevisiae* is the most used yeast at the industrial scale and it is able to ferment six-carbon sugars, but hardly ferments five-carbon sugars (pentoses) because it cannot express the xylanase enzyme to convert xylose to xylulose (Tian et al., 2008).

Z. mobilis, *Acetobacter pasteurianus*, and *Zymobacter palmae* are some of the microorganisms that are able to ferment five-carbon sugars, and *Z. mobilis* and *Zymob. palmae* are the only bacteria that during normal anaerobic growth produce ethanol as the main fermentation product (Jarboe et al., 2009). Even when most bacteria have a broad substrate range compared to yeast, ethanol is not the single product of their metabolism, which complicates the downstream processing (Girio et al., 2010). Some of the key factors for ethanol production during fermentation are a high growth of cells and the number of cells. However, these factors may be limited by many attributes such as the exhaustion of substrate in the broth, the presence of inhibitory furan compounds (furfural and hydroxymethylfurfural), or even high concentrations of ethanol (Olsson and Hahn-Hagerdal, 1996).

There are mainly three operation modes for fermentation: batch, fed-batch, and continuous fermentation. Some of the factors that determine the application of a given operation mode are the type and availability of the feedstock, process economics, and the kinetic behavior of the microorganism. These operation modes will be described in the following.

7.3.1.1 *Batch fermentation operation*

A batch operation can be considered as the simplest operation mode for fermentation. It consists of a closed system in which the microorganism is inoculated to a fermentation media with a defined amount of substrate and nutrients. The fermentation proceeds until the microorganism consumes all the substrate. No nutrients are added after the start, except for pH control. The main characteristic of this mode is that the microorganism works in an environment that goes from high substrate concentration to high product concentration (Zabed et al., 2016). However, the batch regime alone using only one fermenter is not enough to support the large-scale bioethanol production (from 200 to 300 thousand liters a day). Blocks of a number of serial and parallel batch fermenters are preferred today to simulate the continuous (high productivity) regime through a system in which well-organized actions of charge, operation, discharge, and maintenance, among others, are programmed in Gantt diagrams.

7.3.1.2 *Continuous fermentation operation*

The main characteristic of this operation mode is that during the entire fermentation time there is an inlet stream of nutrients and substrate and an outlet stream of products, biomass, and nonconsumed nutrients. This type of fermentation is normally performed in continuous stirred tank bioreactors or plug flow reactors. Compared to batch operation mode, it is possible to obtain higher productivities, but they generally occur at low dilution rates including additional challenges in microorganism stability (metabolic and genetic) as well as immobilization (Balat, 2011; Balat and Balat, 2009).

7.3.1.3 Fed-batch fermentation operation

This operation mode consists of a mixture of a batch and continuous fermentation, supplying new substrate and nutrients in each period and then allowing the system to operate as a batch. Under this mode, the microorganism operates continuously with low concentrations of the substrate, while the concentration of the product increases. This system is widely used at the industrial scale because it offers the advantages of both batch and continuous processes (Jamai et al., 2007). With this operation mode, it is possible to achieve the maximum concentration of viable cells, extending the lifetime of the culture and obtaining a higher concentration of the product (Zabed et al., 2016).

7.3.2 Process integration

S. cerevisiae is the most used microorganism at the industrial scale for the production of ethanol, taking advantage of its ability to assimilate sugars as glucose, mannose, fructose, and disaccharides such as sucrose and maltose. This microorganism assimilates the hexoses by the Embden–Meyerhof glycolytic route through anaerobic fermentations, and the resulting pyruvate is converted into ethanol by alcoholic fermentation (Boles and Hollenberg, 1997). However, it does not have the ability to assimilate xylose because it does not possess the pentose-phosphate pathway in its metabolism and requires the enzymatic battery to produce xylose isomerase (Gunsalus et al., 1955). Additionally, this microorganism also does not have the capacity to produce the enzymatic cellulase battery (cellobiohydrolases, endoglucanases, and β-glucosidases) for the assimilation of cellulose. In recent years, modifications have been made in certain microorganisms through metabolic engineering to add metabolic pathways (Jarboe et al., 2009). These microorganisms include yeasts such as *S. cerevisiae* and bacteria such as *Z. mobilis* and *Escherichia coli*, among others. In the case of *S. cerevisiae*, modified strains have been reported to produce cellulases (cellobiohydrolases, endoglucanases, and β-glucosidases) in percentages around 10% of the cellular protein content (Penttila et al., 1987; Schmidt, 2004).

Considering the advances that have been made on the genetic modification of microorganisms to add new metabolic routes and therefore the ability to assimilate new substrates, a series of modifications have been proposed that seek to reduce the required stages in the fermentation process and the costs associated with them, as well as better yields in terms of a lower production of inhibitory compounds. The main process configurations for the production of bioethanol from lignocellulosic biomass are separate hydrolysis and fermentation, simultaneous saccharification and fermentation (SSF), simultaneous saccharification and cofermentation (SSCF), and consolidated bioprocesses (CBP) (Ask et al., 2012).

Other processes in which reactions and separations are integrated, such as extractive fermentation or membrane fermentation, are not considered here because of the space limit but are well discussed in Anwar et al. (2014).

7.3.2.1 Separated hydrolysis and fermentation

Separated hydrolysis and fermentation is the most used processing scheme for ethanol production. It consists of two separate stages:

1. Pretreatment of the raw material (acid and enzymatic hydrolysis) to convert the cellulose and hemicellulose into glucose and xylose, respectively. Some processes also consider a detoxification stage the decreases the concentration of fermentation inhibitors.
2. Fermentation of the sugar-rich stream obtained in the previous stage and subsequent distillation of the broth to recover the ethanol.

This scheme applies for processes that consider the fermentation of glucose only, or both glucose and xylose [separate hydrolysis and cofermentation (SHCF)].

This scheme offers the possibility of operating both stages, pretreatment and fermentation, at their respective optimal temperature and pH conditions. However, it implies more capital costs associated with the requirement of separate vessels to perform each stage. Another major disadvantage may be the high concentrations of sugars from the hydrolysis that may result in inhibitions to the fermentation (Kumar et al., 2015; Zabed et al., 2016).

7.3.2.2 Simultaneous saccharification and fermentation

SSF is a process that combines enzymatic hydrolysis with fermentation to obtain ethanol in a single step (Ballesteros et al., 2004). When the sugars produced are used by the microorganisms to produce ethanol, a symbiotic effect appears (Watanabe et al., 2012).

SSF has great advantages with respect to other fermentative processes. Some of the advantages compared to separate enzymatic hydrolysis and fermentation (SHF) is the use of a single equipment for the fermentation and saccharification, reducing both the residence times and the capital costs of the process (Oliva et al., 2008). Another prominent advantage is the reduction of inhibitory compounds from enzymatic hydrolysis (mainly the same sugars inhibiting the saccharification and consumed *in situ* by the microorganisms), which improves the overall performance of the process (Ballesteros et al., 2004). Due to its great advantages, SSF has been widely investigated for the production of ethanol and butanol from lignocellulosic and starchy raw materials (Ballesteros et al., 2004). Only for starchy materials is SSF is becoming a normal industrial practice, especially in Asia.

SSF can have disadvantages that limit its use at industrial levels in comparison with SHF (Olofsson et al., 2008). The pH and temperature of the process generally do not agree and a compromise should be found (Olofsson et al., 2008).

7.3.2.3 Simultaneous saccharification and cofermentation

SSCF is based on the use of an enzymatic complex to decompose cellulose into hexoses at conditions that also solubilize the hemicellulose into pentoses (usually with acids) (Zhang et al., 2009). These sugars produced *in situ* together with pentoses after pretreatment are consumed simultaneously by a specialized microorganism to produce ethanol (Olofsson et al., 2010).

Usually, after the enzymatic hydrolysis the microorganism first consumes the glucose, and when this compound is at low concentrations the microorganism consumes the xylose (Olofsson et al., 2010). Some options for solving this problem are the gradual addition of cellulases to the process, or a prehydrolysis, so that the process starts with a low concentration of glucose, forcing the microorganism to consume the two substrates simultaneously (Jin et al., 2013).

The production of bioethanol by means of SSCF presents great benefits such as the use of less equipment, less reaction time, less pollution, integral use of raw materials, and high efficiencies for ethanol production (Jin et al., 2010). However, the last advances in SSCF are not enough to use this process at the industry level today. Only demonstration units use it, and big projects such as Crescentino in the north of Italy are beginning to consider this option.

7.3.2.4 Consolidated bioprocessing

In the previous process (SSCF), it was possible to observe how two different stages of process are still maintained. For this reason, the objective must be a single process of a stage in which the ethanol is directly obtained from the cellulose and the production process of the respective enzymes and the fermentation of the substrate to produce the ethanol is made by a single microorganism (van Zyl et al., 2007). Therefore, no additional hydrolytic enzyme needs to be added (Khramtsov et al., 2011). This process is called consolidated bioprocess.

With this process scheme, the production cost of ethanol from cellulose can be reduced, making CBP a potential process for ethanol production from lignocellulose (Lynd et al., 2006). During recent years, some microorganisms have been evaluated under a CBP scheme. *S. cerevisiae* is the most advanced option to date (Olson et al., 2012). Some of the difficulties with the use of *S. cerevisiae* in a CBP are (1) adverse effects of the coexpression of multiple recombinant proteins, (2) modulation of simultaneous coexpression of multiple genes at the transcription level, and (3) misfolding of some of the secretory proteins (Xu et al., 2009).

7.4 Bioethanol in a biorefinery context

Traditionally, bioethanol is produced in a standalone pathway that involves the conversion of fermentable sugars into ethanol. However, different waste streams are generated and some are disposed directly into the environment (e.g., stillage). The environmental concerns related to the waste streams have recently increased, and therefore several authors have started to evaluate different alternatives for the valorization of these streams in order to reduce the emissions related to its final disposal. Additionally, bioethanol is not a high-added-value product and some feasibility problems can appear at reduced scales. In this sense, the biorefinery concept seems to be a promising alternative for the proper valorization of these streams into bioenergy and/or biochemicals. Moncada et al. (2014) performed the techno-economic and environmental assessment of the oil palm in Colombia. As a result, the environmental impact was inverse to the economic performance. Therefore, this demonstrates that both economic and environmental objectives can be improved following an integrated processing sequence to obtain multiple products, not only biofuels but also bio-based materials. The inclusion of other products enhances the environmental and economic metrics of biorefinery systems by using streams in a more reasonable way.

7.4.1 Example: Palm oil biorefinery

Palm is one of the most important oleochemical feedstocks in Malaysia, Indonesia, Nigeria, and Colombia. Currently, in Colombia palm oil is used for biodiesel production. From this process, glycerol is obtained as the main by-product of the process and its valorization can be an excellent opportunity not only to raise the profitability of current biodiesel processes but also to produce other chemicals from a bio-based raw material.

The processing of palm for oil extraction leads to the formation of several by-products and residues that have an economical potential. The empty fruit bunch (EFB) is the solid residue with the highest production. Due to its high moisture content (approximately 68%), this material is not appropriate as a solid fuel, and therefore it is mostly used as manure. However, the relative high lignocellulosic content of the raw material can be potentially converted into different biofuels and chemicals (e.g., ethanol and sugar-derived products).

The following biorefinery configuration considers the production of biodiesel, polyhydroxybutyrate (PHB), and ethanol. Crude oil from oil palm is directly used to produce biodiesel and glycerol as a main by-product through the transesterification reaction. Subsequently, glycerol is purified and used as raw material in the production of PHB using *Cupriavidus necator* as the fermenting microorganism. On the other hand, the lignocellulosic residue from the oil palm processing is submitted into a pretreatment stage in order to obtain fermentable sugars that are converted into bioethanol through recombinant *Z. mobilis*. Figure 7.3 presents the flowsheet of the oil palm biorefinery configuration to produce biodiesel, bioethanol, and PHB.

From the simulation procedure, the economic and environmental performance of the biorefinery was improved in comparison to the standalone ethanol production. The economic margin of the biorefinery was 64.5%, which was 1.33-fold higher than the standalone production of ethanol. In contrast, the environmental impact was 156.42 potential environmental impact (PEI)/t product, which was 1.27-fold lower than the standalone production of biodiesel.

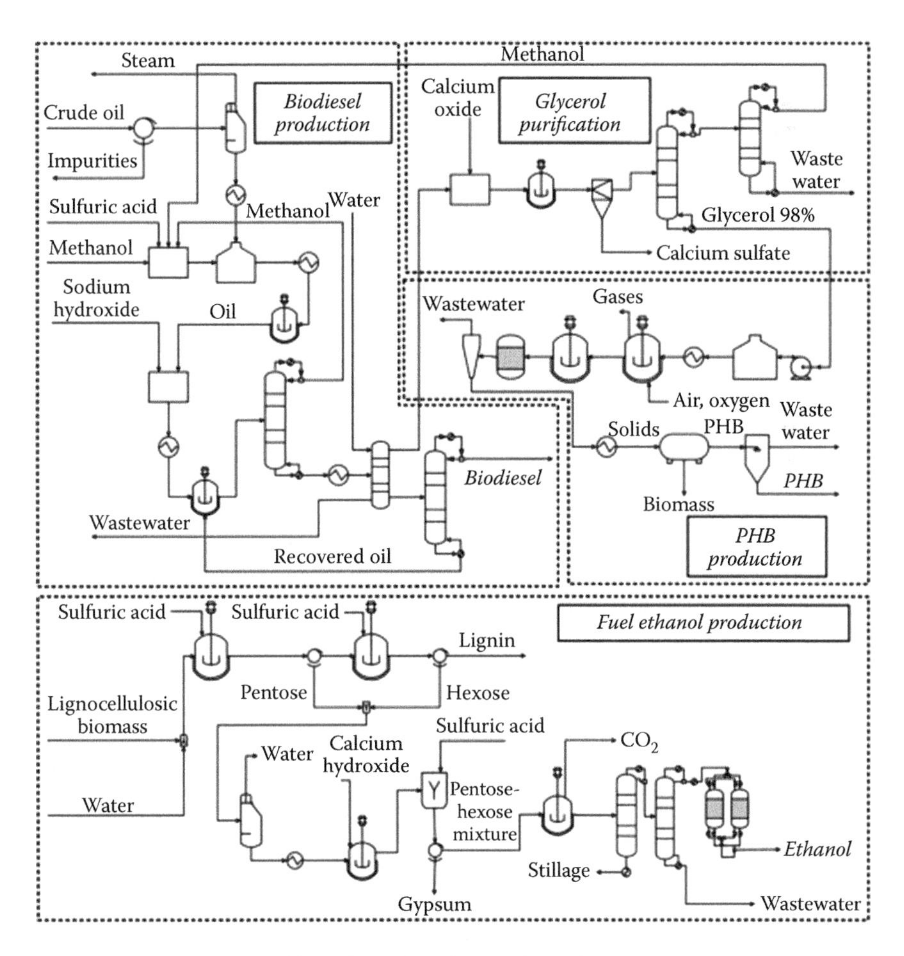

Figure 7.3 Production of bioethanol under the biorefinery concept.

7.5 Bioethanol industry

Ethanol can be produced from different sources such as sugars, starch, grains, lignocellulosic biomass, and algae. Different companies exist worldwide that produce bioethanol from any of the previously mentioned renewable energy sources, aiming to improve the sustainability of the biofuels production. Abengoa Bioenergy produces bioethanol in distributed plants in Europe, the United States, and Brazil. Grains (corn, sorghum, and wheat) and sugarcane are used to produce ethanol through a series of chemical treatments. The ethanol is used as transportation fuel to produce ethyl tert-butyl ether (ETBE) or as a mixture with gasoline, the most common being E85, E15, E10, and E5 (gasoline with a bioethanol percentage of 85%, 15%, 10%, and 5%, respectively). As the main coproduct from the bioethanol production process, dry distillers grains (DDGs) are obtained with a high protein content. DDG is used for animal feed production. Abengoa Bioenergy has three ethanol plants in Brazil with a total capacity of 235 million liters of ethanol per year. Additionally, sugarcane bagasse is destined for cogeneration with a capacity of 913.500 MWh.

Companies like Pacific Ethanol, POET, and Archer Daniels Midland (ADM) produce ethanol from corn in the United States. Ethanol, DDG, corn gluten feed, corn gluten meal, corn germ, corn oil, distiller yeast, and carbon dioxide are the main products of the processing plant. The facilities of Pacific Ethanol have an ethanol production of 515 million gallons per year, and 1 million tons per year of coproducts are obtained.

Due to food security concerns, other companies are dealing with the production of ethanol from lignocellulosic residues in order to exploit different residues from the agriculture and forestry sectors. SEKAB and Alpena Biorefinery are two companies that produce cellulosic ethanol. SEKAB uses a wide range of raw materials (e.g., woodchips, switchgrass, wheat, cottonwood, corn stover, paper, corn, sugarcane) to obtain three products: ethanol, lignin, and biogas. Currently, the company has a demonstration-scale biorefinery that consists of a pretreatment stage (acid hydrolysis and neutralization), SSF, and distillation. Lignin is obtained from the solid residue after the enzymatic saccharification and biogas is produced from the bottom residue of the distillation. On the other hand, Alpena Biorefinery uses wood residues from the manufacture of decorative panels as raw material for the production of ethanol and potassium acetate. The production process is similar to SEKAB's process, but a reaction stage and osmosis reverse equipment are required in order to separate the potassium acetate. Currently, the company produces 945,000 gal. of standard denatured ethanol per year and 700,000 gal. per year of potassium acetate.

The production of ethanol from algae has emerged as a promising alternative due to high growth productivity of the algae. Algenol patented a technology for the production of biofuels primarily using algae, sunlight, carbon dioxide, and saltwater, all on nonarable land. The company has a unique two-step, sustainable process that first produces ethanol directly from the algae and then converts the spent algae biomass into green crude (renewable crude oil that can be refined to fuels).

7.6 Simultaneous fermentation and separation

As observed in industry, biorefineries are not having as much of an effect on the ethanol price as the improvements in some standalone processes can have. One example is the simultaneous saccarification and fermentation that can conceptually be considered the last technological way to complement the consolidated bioprocesses based on multipurpose microorganisms that cannot selectively separate the products. Many fermentations are inhibited due to high concentrations of product. This phenomenon is known as end-product inhibition and it is the case of ethanol fermentation.

The tolerance of the microorganisms to ethanol depends on the strain and certain operation conditions. However, ethanol fermenting microorganisms present strong inhibition around 12% v/v of ethanol (Luong, 1985). Some of the problems associated with alcohols are the increase in membrane fluidity and the decrease of enzymatic activity associated with sugar transport, which therefore inhibits cell growth and decrease production rates of ethanol (Park and Geng, 1992).

This situation has led to looking for solutions associated with the *in situ* removal of the product in order to obtain higher amounts of ethanol. There are two main types of simultaneous fermentation-separation schemes: membrane and solvent based. The membrane-based technique consists of locating a membrane inside a fermenter that selectively separates ethanol and water from the rest of the broth.

Pervaporation is one of the most studied examples of this technique (Fu et al., 2016, Fan et al., 2014a,b; van der Bruggen and Luis, 2014), consisting in coupling a membrane normally composed of polydimethylsiloxane (PDMS) to the fermentation broth and the membrane selectively separates the ethanol and water into a vacuum-created vapor phase. However, the recovered ethanol normally reaches a concentration of 30% and additional downstream processing is still necessary to obtain fuel-grade ethanol (Fu et al., 2016). In the case of solvent-based schemes, a water-immiscible solvent is added to the fermentation broth in order to create two phases and to remove the ethanol from the culture medium in the organic phase. This scheme implies the separation of the organic phase, recovering the solvent and further purifying the ethanol (Palacios-Bereche et al., 2014). In some cases, coupling of the system to a vacuum flask is also considered to ease the separation of the volatile compounds. Multiple solvents have been studied (*n*-dodecanol, olefin alcohol, polypropylene glycol, and dibutyl phthalate, among others) (Kollerup and Daugulis, 1986; Minier and Goma, 1982), but the most important feature is that they are 100% biocompatible (nontoxic and no inhibition).

The importance of this type of simultaneous fermentation-separation schemes is that obtaining higher concentrations of ethanol is critical to reducing the energy consumption in downstream processes (distillation) and to decrease the environmental impact associated with stillage (Vane, 2008). It is extremely essential to enhance the performance of the integrated fermentation-pervaporation process to generate higher ethanol concentration on the permeate side of the membrane. Table 7.3 presents a detailed description of some of the most studied simultaneous fermentation-separation schemes for ethanol production.

7.7 Bioethanol benefits and considerations

Ethanol is a renewable, domestically produced transportation fuel. Even when used in low-level blends, such as E10 (10% ethanol, 90% gasoline), ethanol helps reduce petroleum use in transportation and greenhouse gas (GHG) emissions. Additionally, ethanol production creates jobs in rural areas where employment opportunities are needed. Like any alternative fuel, there are some considerations to take into account when contemplating the use of ethanol.

One of the main concerns about bioethanol production is the energy balance, which is a comparison between the total amount of energy put into the process and the release energy by burning the ethanol fuel. This balance considers the full life cycle of the biofuel including the crop cultivation, transportation, and production of the required energy, and the use of oils and agrochemicals. According to LBP (2015), the production of bioethanol from sugarcane in Brazil has a favorable energy balance, varying from 8.3 to 10.2, which means that one unit of fossil-fuel energy is required to produce 8.3 energy units of bioethanol.

The carbon dioxide captured when the feedstock crops are grown to produce ethanol offsets the carbon dioxide released by a vehicle when ethanol is burned. This differs from petroleum-based production of gasoline and diesel. No emissions are offset when these petroleum products are burned. On a life-cycle analysis basis, GHG emissions are reduced on average by 40% with corn-based ethanol produced from dry mills, and up to 108% if cellulosic feedstocks are used (U.S. DOE, 2017). However, some authors have questioned the benefits of bioethanol production in the reduction of GHG emissions because the effect of the land-use changes in most of the studies is not considered. The authors considered that a "biofuel carbon debt is created when ethanol-producing countries convert undisturbed ecosystems to biofuels production. Nevertheless, it is

Table 7.3 Simultaneous fermentation and separation schemes for bioethanol production

Process	Description	Results	Ref.
Continuous and closed-circulating fermentation (CCCF) system with a pervaporation membrane	CCCF system with a PDMS membrane in which broth is continuously pumped to the membrane module. Permeate is condensed and collected, and the retentate is recirculated to the broth.	Glucose consumption rate: 6.06 g L^{-1} h^{-1}. Ethanol productivity: 2.31 g L^{-1} h^{-1}. Ethanol yield: 0.38. Total ethanol produced: 609.8 g L^{-1}. Ethanol *in situ* removal promoted the cell second growth. Accumulation of secondary metabolites in the broth became inhibitors.	Fan et al. (2014b)
Batch, fed-batch, and continuous fermentation coupled to a pervaporation membrane in a closed-circulating fermentation (CCF) system	CCF system with a PDMS membrane. Broth is continuously pumped to the membrane module. Permeate is condensed and collected and the retentate is recirculated to the broth. Fed-batch was run for five cycles and continuous operation, new broth is fed to the system.	Batch fermentation: Ethanol concentration increased. With pervaporation, at 40 h, the concentration was 63.8 g L^{-1}. Without pervaporation, this value was achieved after 50 h. Fed-batch: There is still accumulation of ethanol in broth that inhibits fermentation, but *in situ* removal has better results of ethanol productivity. Continuous: Ethanol concentration remained around 100 g L^{-1}. Permeate ethanol purity accounted to 40–50 wt%.	Fu et al. (2016)
CCCF with a pervaporation membrane	PDMS membrane. Broth continuously pumped to the membrane module. Permeate is condensed and collected, and the retentate is recirculated to the broth. Pervaporation operated in cycles of 11 h per day.	Three sequential 500-h cycles of CCCF experiments. Glucose volumetric consumption of 3.8 g L^{-1} h^{-1} and ethanol volumetric productivity of 1.39 g L^{-1} h^{-1}. Specific glucose utilization rate of 0.32 h^{-1} and ethanol yield rate of 0.13 h^{-1}.	Chen et al. (2012)

(Continued)

Table 7.3 (Continued) Simultaneous fermentation and separation schemes for bioethanol production

Process	Description	Results	Ref.
CCCF with an immobilized cell coupled to a pervaporation membrane	Closed-circulating system for ethanol fermentation coupled to a cell-immobilized bed fermenter (cell density in immobilized bed of 1.76×10^{10} cells g^{-1} gel) with pervaporation using a composite PDMS membrane. Continuous fermentation experiment carried out for 250 h.	Ethanol concentration in the broth was maintained at 43 g L^{-1}. Glucose utilization and ethanol productivity were 23.26 g L^{-1} h^{-1} and 9.6 g L^{-1} h^{-1}, respectively. Average ethanol concentration in permeate was 23.1 wt%. Ethanol productivity decreases with increasing bed length but closed circulating system prevents inhibition.	Ding et al. (2011)
Hybrid-simple batch membrane fermenter (H-SBMF) using PDMS pervaporation membrane	Batch fermentation with outlet stream, separation, and recirculation of biomass. Liquid stream passes through pervaporation membrane and retentate is recirculated.	H-SBMF achieves a higher ethanol production in the range of 10%–13% compared to conventional batch fermenter. Increase in productivity of 150%. Decrease in steam consumption due to the adaptation of the membrane to the fermenter, achieving almost 17% steam reduction.	Leon et al. (2016)
Batch fermentation with *in situ* gas stripping	Batch fermentation of sugarcane molasses with gas stripping (4 and 6 L min^{-1}) and sugar concentration (170 and 250 g L^{-1}). Gas is condensed and then collected. Carbon dioxide used as carrier gas.	Gas stripping was effective and ethanol concentration was maintained below threshold of toxicity. Maximum concentration of 53.11 g L^{-1} of ethanol was obtained. Ethanol removal allows higher sugar concentrations than those used industrially.	Ponce et al. (2016)

(Continued)

Table 7.3 (Continued) Simultaneous fermentation and separation schemes for bioethanol production

Process	Description	Results	Ref.
Batch fermentation with *in situ* gas stripping	Batch fermentation of sugarcane molasses with gas stripping (2 volumes of air per volume of medium per minute [vvm]) and sugar concentration (170 and 250 g L^{-1}). Gas is condensed and then collected. Carbon dioxide used as carrier gas.	Removal of ethanol clearly reduced its inhibitory effect. Removed ethanol in the CO_2 stream could be recovered by absorption in a wash column. Maximum concentration of ethanol obtained was 72 g L^{-1}.	Sonego et al. (2014)
Extractive fermentation with solvents	Ethanol fermentation with anaerobic bacterium *Z. anaerobia*. For this microorganism, ethanol presents inhibition at 20 g L^{-1}.	Oleyl alcohol, isooctadecanol and 2-octyl-1-dodecanol were the three solvents that were found to be nontoxic to *Z. anaerobia*.	Abdul Aziz et al. (2010)
Extractive fermentation with solvents	*In situ* batch extractive fermentations were carried out in a 2-L stirred bioreactor (1:1 volume ratio of solvent and fermentation broth). Glucose concentration of 150 g L^{-1}.	Ethanol productivity was enhanced by continuous *in situ* extraction with water-immiscible organic solvents. Using isooctadecanol, the maximum ethanol concentration was 75 g L^{-1}.	Ajit et al. (2017)
Retentostatic extractive vacuum (REV) system	Batch, fed-batch, and continuous fermentation coupled to a flash tank operated at vacuum conditions to evaporate ethanol. Ethanol extraction was performed intermittently every 12 h.	Cell concentrations reached values around 22 g L^{-1}, with xylose concentration of 100 g L^{-1}. Maximum ethanol productivity of 1 g L^{-1} h^{-1}. Titer inside the reactor was low (20–35 g L^{-1}). Flash system operated properly, leading to a condensed alcoholic solution ranging from 167 to 240 g L^{-1}.	Farias et al. (2017)
REV system	A vacuum flash tank is linked to the fermentation process to remove the ethanol from the fermentative medium.	Vacuum extractive fermentation, coupled with a partial condenser, showed the highest production of ethanol, but power requirement increases significantly due to the refrigeration system.	Palacios-Bereche et al. (2014)

very difficult to determine the indirect effect of the land use for sugarcane production since not all soil carbon losses can be attributed to sugarcane crop" (Ros et al., 2010).

Even though all automotive fossil fuels emit aldehydes, one of the drawbacks of the use of hydrated ethanol, in ethanol-only engines, is the increase in aldehyde emissions compared with gasoline or gasohol (e.g., E5, E10, and E85). The emissions of formaldehyde and acetaldehyde are significantly high, and although both aldehydes are found in the environment, additional emissions may be considered due to their role in smog formation (LBP, 2015).

7.8 Problems and threats of ethanol: Comparison with butanol

One of the most studied and used liquid biofuels is bioethanol. It can be used as a fuel for vehicles in its pure form, but it is usually used as a gasoline additive to increase octane and improve vehicle emissions. Bioethanol is widely used in the United States and Brazil. On the other hand, butanol may be used as a fuel in an internal combustion engine (Kumar and Gayen, 2011). Due to its longer hydrocarbon chain, which causes it to be fairly nonpolar, it is more similar to gasoline than ethanol. Butanol has been demonstrated to work in vehicles designed for use with gasoline without modification (Demirbas, 2009).

The lower heating value (LHV) of a fuel is defined as the amount of heat released by combusting a specified quantity (initially at 25°C) and returning the temperature of the combustion products to 150°C (Oak Ridge National Laboratory, 2012). The LHVs for ethanol and butanol are 29.85 and 37.33 MJ/kg, respectively. In terms of energy content, butanol has a higher capacity to provide energy. However, the current production process to obtain ethanol requires less energy than butanol and the net balance still favors ethanol (Pfromm et al., 2010). Another index used to compare biofuels is the octane rating, which refers to the capacity of a biofuel to withstand compression before detonating (igniting). In broad terms, high-performance gasoline engines require higher compression ratios, and hence higher octane rating, and the use of low-octane gasoline may lead to the problem of engine knocking (Dabelstein et al., 2007). In this regard, octane ratings (measured as Research Octane Number) for ethanol and butanol are 108.6 and 92, respectively (Eyidogan, 2010). The octane rating of *n*-butanol is lower than the ethanol; hence, in terms of octane rating, ethanol has better performance.

However, the production of butanol does not seem advantageous at this time compared to ethanol because of the low yield of lower heating value per mass of processed biomass (Pfromm et al., 2010). Ethanol still presents technical problems associated with bound humidity, which decreases the performance and corrosiveness to engines (Renewable Energy UK, 2014).

7.9 Conclusions

Advances in new raw materials and technologies have not been great in recent years. Most of the research and industry efforts have been related to lignocellulosic biomass. A lot of demonstration and commercial plants were implemented based on pretreatment technologies and enzymes developments. However, most of the demonstration plants last year (2016), such as the University of Florida plant in the north of Greensville and the commercial Abengoa plant in the center of the United States, are closing. These events demonstrate the difficulties of new bioethanol plants using new raw materials and that research developments are not enough. Issues related to logistics that cannot be considered just a bioethanol problem are the main barrier affecting the above-mentioned projects. One real strategy to prevent these problems and reduce risks is the biorefinery approach in which a multiproduct portfolio including bioethanol reduces uncertainties regarding the supply chain (because the existing portfolio is totally integrated into the biorefinery) and the market prices or competitors are fully considered (for example, biobutanol as direct ethanol competitor is produced in the same biorefinery).

References

Abdul Aziz, A., A. Zaid Sulaiman, and Y. Chisti, Extractive fermentation for enhanced productivity of bioethanol by *Zymomonas anaerobia*, *J. Biotechnol.*, 150S, S135–S136, 2010.

Abdullah, S., S. Yusup, M. Ahmad, A. Ramli, and L. Ismail, Thermogravimetry study on pyrolysis of various lignocellulosic biomass for potential hydrogen production, *Int. J. Chem. Biol. Eng.*, 4, 12, 750–754, 2010.

Ajit, A., A. Z. Sulaiman, and Y. Chisti, Production of bioethanol by *Zymomonas mobilis* in high-gravity extractive fermentations, *Food Bioprod. Process.*, 102, 123–135, 2017.

Alvira, P., E. Tomás-Pejó, M. Ballesteros, and M. J. Negro, Pretreatment technologies for an efficient bioethanol production process based on enzymatic hydrolysis: A review, *Bioresour. Technol.*, 101, 13, 4851–4861, 2010.

Anwar, Z., M. Gulfraz, and M. Irshad, Agro-industrial lignocellulosic biomass a key to unlock the future bio-energy: A brief review, *J. Radiat. Res. Appl. Sci.*, 7, 2, 163–173, 2014.

Ask, M., K. Olofsson, K. Di Felice et al., Challenges in enzymatic hydrolysis and fermentation of pretreated *Arundo donax* revealed by a comparison between SHF and SSF, *Process Biochem.*, 47, 10, 1452–1459, 2012.

Bai, F. W., W. A. Anderson, and M. Moo-Young, Ethanol fermentation technologies from sugar and starch feedstocks, *Biotechnol. Adv.*, 26, 1, 89–105, 2008.

Balan, V., B. Bals, S. P. S. Chundawat, D. Marshall, and B. E. Dale, Lignocellulosic biomass pretreatment using AFEX, *Methods Mol. Biol.*, 581, 61–77, 2009.
Balat, M., Production of bioethanol from lignocellulosic materials via the biochemical pathway: A review, *Energy Convers. Manag.*, 52, 2, 858–875, 2011.
Balat, M. and H. Balat, Recent trends in global production and utilization of bioethanol fuel, *Appl. Energy*, 86, 11, 2273–2282, 2009.
Ballesteros, M., J. M. Oliva, M. J. Negro, P. Manzanares, and I. Ballesteros, Ethanol from lignocellulosic materials by a simultaneous saccharification and fermentation process (SFS) with *Kluyveromyces marxianus* CECT 10875, *Process Biochem.*, 39, 12, 1843–1848, 2004.
Boles, E. and C. P. Hollenberg, The molecular genetics of hexose transport in yeasts, *FEMS Microbiol. Rev.*, 21, 1, 85–111, 1997.
Cardona, C.A., J. A. Quintero, and I. C. Paz, Production of bioethanol from sugarcane bagasse: Status and perspectives, *Bioresour. Technol.*, 101, 13, 4754–4766, 2010.
Chaudhary, N., Biosynthesis of ethanol and hydrogen by glycerol fermentation using *Escherichia coli*, *Adv. Chem. Eng. Sci.*, 1, 83–89, 2011.
Chen, C., X. Tang, Z. Xiao et al., Ethanol fermentation kinetics in a continuous and closed-circulating fermentation system with a pervaporation membrane bioreactor, *Bioresour. Technol.*, 114, 707–710, 2012.
Cheng, K. K., B. Y. Cai, J. A. Zhang et al., Sugarcane bagasse hemicellulose hydrolyzate for ethanol production by acid recovery process, *Biochem. Eng. J.*, 38, 1, 105–109, 2008.
Clark, J. and F. Deswarte, Introduction to chemicals from biomass, in *Wiley Series in Renewable Resources*, C. V. Stevens, Ed., West Sussex, United Kingdom: John Wiley & Sons, Inc., 2008.
Coimbra, M. C., A. Duque, F. Saez et al., Sugar production from wheat straw biomass by alkaline extrusion and enzymatic hydrolysis, *Renew. Energy*, 86, 1060–1068, 2016.
Conde-Mejia, C., A. Jimenez-Gutierrez, and M. El-Halwagi, A comparison of pretreatment methods for bioethanol production from lignocellulosic materials, *Process Saf. Environ. Prot.*, 90, 3, 189–202, 2012.
Cremonez, P. A., M. Feroldi, A. Feiden et al., Current scenario and prospects of use of liquid biofuels in South America, *Renew. Sustain. Energy Rev.*, 43, 352–362, 2015.
Dabelstein, W., A. Reglitzky, A. Schutze, and K. Reders, Automotive fuels, in *Ullmann's Encyclopedia of Industrial Chemistry*, Weinheim, Germany: Wiley-VCH, 2007.
da Silva, A. S., R. S. S. Teixeira, T. Endo, E. P. S. Bon, and S.-H. Lee, Continuous pretreatment of sugarcane bagasse at high loading in an ionic liquid using a twin-screw extruder, *Green Chem.*, 15, 1991–2001, 2013.
Daza Serna, L.V., Assessment of non-conventional pretreatments for agriculture wastes utilization, Master's thesis, Universidad Nacional de Colombia sede Manizales, Manizales, Colombia, 2015.
Demirbas, A., Biofuels securing the planet's future energy needs, *Energy Convers. Manag.*, 50, 9, 2239–2249, 2009.
Devarapalli, M., H. K. Atiyeh, J. R. Phillips, R. S. Lewis, and R. L. Huhnke, Ethanol production during semi-continuous syngas fermentation in a trickle bed reactor using *Clostridium ragsdalei*, *Bioresour. Technol.*, 209, 56–65, 2016.
Dias, M. O. S., T. L. Junquera, O. Cavallett et al., Integrated versus stand-alone second generation ethanol production from sugarcane bagasse and trash, *Bioresour. Technol.*, 103, 1, 152–161, 2012.

Ding, W. W., Y. T. Wu, X. Y. Tang, L. Yuan, and Z. Y. Xiao, Continuous ethanol fermentation in a closed-circulating system using an immobilized cell coupled with PDMS membrane pervaporation, *J. Chem. Technol. Biotechnol.*, 86, 1, 82–87, 2011.

Duque Garcia, A., Pretratamiento de extrusion reactiva para la produccion de bioetanol a partir de paja de cebada, Ph.D. thesis, Universidad de Valladolid, Valladolid, Spain, 1998.

Eyidogan, M., Impact of alcohol–gasoline fuel blends on the performance and combustion characteristics of an SI engine, *Fuel*, 89, 10, 2713–2720, 2010.

Fan, S., Z. Xiao, X. Tang et al., Inhibition effect of secondary metabolites accumulated in a pervaporation membrane bioreactor on ethanol fermentation of *Saccharomyces cerevisiae*, *Bioresour. Technol.*, 162, 8–13, 2014a.

Fan, S., Z. Xiao, Y. Zhang et al., Enhanced ethanol fermentation in a pervaporation membrane bioreactor with the convenient permeate vapor recovery, *Bioresour. Technol.*, 155, 229–234, 2014b.

Farias, D., D. I. P. Atala, and F. M. Maugeri Filho, Improving bioethanol production by *Scheffersomyces stipitis* using retentostat extractive fermentation at high xylose concentration, *Biochem. Eng. J.*, 121, 171–180, 2017.

Food and Agriculture Organization of the United Nations (FAO), Bioenergy and global food security. The BEFS analytical framework, Rome, 2010.

Fu, C., D. Cai., S. Hu et al., Ethanol fermentation integrated with PDMS composite membrane: An effective process, *Bioresour. Technol.*, 200, 648–657, 2016.

Girio, F. M., C. Fonseca, F. Carvalheiro et al., Hemicelluloses for fuel ethanol: A review, *Bioresour. Technol.*, 101, 13, 4775–4800, 2010.

Goshadrou, A., K. Karimi, and M. J. Taherzadeh, Bioethanol production from sweet sorghum bagasse by *Mucor hiemalis*, *Ind. Crops Prod.*, 34, 1, 1219–1225, 2011.

Goyal, H. B., D. Seal, and R. C. Saxena, Bio-fuels from thermochemical conversion of renewable resources: A review, *Renew. Sustain. Energy Rev.*, 12, 2, 504–517, 2008.

Gunsalus, I. C., B. L. Horecker, and W. A. Wood, Pathways of carbohydrate metabolism in microorganisms, *Bacteriol. Rev.*, 19, 2, 79, 1955.

Gupta, V. K., *Bioenergy Research: Advances and Applications*. London: Elsevier B.V., 2014.

Henstra, A. M., J. Sipma, A. Rinzema, and A. J. Stams, Microbiology of synthesis gas fermentation for biofuel production, *Curr. Opin. Biotechnol.*, 18, 3, 200–206, 2007.

Hernández-Salas, J. M. et al., Comparative hydrolysis and fermentation of sugarcane and agave bagasse, *Bioresour. Technol.*, 100(3), 1238–1245, 2009.

Ho, S. H., S. W. Huang, C. Y. Chen et al., Bioethanol production using carbohydrate-rich microalgae biomass as feedstock, *Bioresour. Technol.*, 135, 191–198, 2013.

Ito, T., Y. Nakashimada, K. Senba, T. Matsui, and N. Nishio, Hydrogen and ethanol production from glycerol-containing wastes discharged after biodiesel manufacturing process, *J. Biosci. Bioeng.*, 100, 3, 260–265, 2005.

Jamai, L., K. Ettayebi, J. El Yamani, and M. Ettayebi, Production of ethanol from starch by free and immobilized *Candida tropicalis* in the presence of α-amylase, *Bioresour. Technol.*, 98, 14, 2765–2770, 2007.

Jang, Y. S., A. Malaviya, C. Cho, J. Lee, and S. Y. Lee, Butanol production from renewable biomass by clostridia, *Bioresour. Technol.*, 123, 653–663, 2012.

Jarboe, L. R., K. T. Shanmugam, and L. O. Ingram, Ethanol, in *Encyclopedia of Microbiology,* 3rd ed., M. Schlaechter, Ed., Kidlington, Oxford, United Kingdom: Academic Press, 2009, pp. 295–304.

Jeong, S.-Y. and J.-W. Lee, Hydrothermal treatment, in *Pretreatment of Biomass. Processes and Technologies,* A. Pandey, S. Negi, P. Binod, and C. Larroche, Eds., Kidlington, Oxford, United Kingdom: Academic Press, 2015, pp. 61–74.

Jin, M., M. W. Lau, V. Balan, and B. E. Dale, Two-step SSCF to convert AFEX-treated switchgrass to ethanol using commercial enzymes and *Saccharomyces cerevisiae* 424A (LNH-ST), *Bioresour. Technol.*, 101, 21, 8171–8178, 2010.

Jin, M., C. Sarks, C. Gunawan et al., Phenotypic selection of a wild *Saccharomyces cerevisiae* strain for simultaneous saccharification and co-fermentation of AFEX TM pretreated corn stover, *Biotechnol. Biofuels,* 6, 108, 1–14, 2013.

Jung, Y. H. and K. H. Kim, Acidic pretreatment, in *Pretreatment of Biomass. Processes and Technologies,* A. Pandey, S. Negi, P. Binod, and C. Larroche, Eds., 2015, pp. 27–50.

Karimi, K. and M. J. Taherzadeh, A critical review of analytical methods in pretreatment of lignocelluloses: Composition, imaging, and crystallinity, *Bioresour. Technol.*, 200, 1008–1018, 2016.

Khoo, H. H., Review of bio-conversion pathways of lignocellulose-to-ethanol: Sustainability assessment based on land footprint projections, *Renew. Sustain. Energy Rev.*, 46, 100–119, 2015.

Khramtsov, N., McDade, L., Amerik, A. et al., Industrial yeast strain engineered to ferment ethanol from lignocellulosic biomass, *Bioresour. Technol.*, 102, 17, 8310–8313, 2011.

Kim, Y., R. Hendrickson, N. S. Mosier, and M. R. Ladisch, Liquid hot water pretreatment of cellulosic biomass, *Methods Mol. Biol.*, 581, 93–102, 2009.

Klasson, K. T., C. M. D. Ackerson, E. C. Clausen, and J. L. Gaddy, Biological conversion of synthesis gas into fuels, *Int. J. Hydrogen Energy,* 17, 4, 281–288, 1992.

Kollerup, F. and A. Daugulis, Ethanol production by extractive fermentation–solvent identification and prototype development, *Can. J. Chem.*, 64, 4, 598–606, 1986.

Kosugi, A., A. Kondo, M. Ueda et al., Production of ethanol from cassava pulp via fermentation with a surface-engineered yeast strain displaying glucoamylase, *Renew. Energy,* 34, 5, 1354–1358, 2009.

Kumar, G., P. Bakonyi, S. Periyasamy et al., Lignocellulose biohydrogen: Practical challenges and recent progress, *Renew. Sustain. Energy Rev.*, 44, 1, 728–737, 2015.

Kumar, P., D. M. Barrett, M. J. Delwiche et al., Methods for pretreatment of lignocellulosic biomass for efficient hydrolysis and biofuel production, *Ind. Eng. Chem. Res.*, 48, 8, 3713–3729, 2009.

Kumar, M. and K. Gayen, Developments in biobutanol production: New insights, *Appl. Energy,* 88, 6, 1999–2012, 2011.

LBP, *Brazil Energy Policy, Laws and Regulations Handbook,* vol. 1, Brazil 2015.

Leon, J.A., R. Palacios-Bereche, and S. A. Nebra, Batch pervaporative fermentation with coupled membrane and its influence on energy consumption in permeate recovery and distillation stage, *Energy,* 109, 77–91, 2016.

Li, K., S. Liu, and X. Liu, An overview of algae bioethanol production, *Int. J. Energy Res.*, 38, 8, 1–13, 2014.

Lin, Y. and S. Tanaka, Ethanol fermentation from biomass resources: Current state and prospects, *Appl. Microbiol. Biotechnol.*, 69, 6, 627–642, 2006.

Liu, K., H. K. Atiyeh, B. S. Stevenson et al., Continuous syngas fermentation for the production of ethanol, *n*-propanol and *n*-butanol, *Bioresour. Technol.*, 151, 69–77, 2014.

Luong, J. H. T., Kinetics of ethanol inhibition in alcohol fermentation, *Biotechnol. Bioeng.*, 27, 3, 280–285, 1985.

Lynd, L. R., P. J. Weimer, G. Wolfaardt, and Y.-H. P. Zhang, Cellulose hydrolysis by *Clostridium thermocellum*: A microbial perspective, in *Cellulosome: Molecular Anatomy and Physiology of Proteinaceous Machines*, I. A. Kataeva, Ed., Hauppage, NY: Nova Science Publishers, 2006, pp. 95–117.

Maurya, D. P., A. Singla, and S. Negi, An overview of key pretreatment processes for biological conversion of lignocellulosic biomass to bioethanol, *3 Biotech*, 3, 5, 597-609, 2013.

Minier, M. and G. Goma, Ethanol production by extractive fermentation, *Biotechnol. Bioeng.*, 24, 7, 1565–1579, 1982.

Moncada, J., C. A. Cardona, J. C. Higuita, J. J. Velez, and F. E. Lopez-Suarez, Wood residue (*Pinus patula* bark) as an alternative feedstock for producing ethanol and furfural in Colombia: Experimental, techno-economic and environmental assessments, *Chem. Eng. Sci.*, 140, 309–318, 2016.

Moncada, J., L. G. Matallana, and C. A. Cardona, Selection of process pathways for biorefinery design using optimization tools: A Colombian case for conversion of sugarcane bagasse to ethanol, poly-3-hydroxybutyrate (PHB), and energy, *Ind. Eng. Chem. Res.*, 52, 11, 4132–4145, 2013.

Moncada, J., J. Tamayo, and C. A. Cardona, Evolution from biofuels to integrated biorefineries: Techno-economic and environmental assessment of oil palm in Colombia, *J. Clean. Prod.*, 81, 51–59, 2014.

Mood, S. H., A. H. Golfeshan, M. Tabatabei et al., Lignocellulosic biomass to bioethanol, a comprehensive review with a focus on pretreatment, *Renew. Sustain. Energy Rev.*, 27, 77–93, 2013.

Moreno, J., R. Moral, J. L. Garcias-Morales, J. A. Pascual, and M. P. Bernal, Eds., *De Residuo a Recurso: El camino hacia la sostenibilidad I Residuos organicos 1 Residuos agrícolas*. Madrid: Red Española De Compostaje, 2014.

Mosier, N., C. Wyman, B. Dale et al., Features of promising technologies for pretreatment of lignocellulosic biomass, *Bioresour. Technol.*, 96, 6, 673–686, 2005.

Munasinghe, P. C. and S. K. Khanal, Biomass-derived syngas fermentation into biofuels, *Biofuels*, 101, 13, 79–98, 2011.

Murphy, J. D. and N. M. Power, How can we improve the energy balance of ethanol production from wheat? *Fuel*, 87, 10–11, 1799–1806, 2008.

Naik, S. N., V. V. Goud, P. K. Rout, and A. K. Dalai, Production of first and second generation biofuels: A comprehensive review, *Renew. Sustain. Energy Rev.*, 14, 578–597, 2010.

Nguyen, M. T., P. C. Seung, L. Jinwon, H. L. Jae, and J. S. Sang, Hydrothermal acid pretreatment of *Chlamydomonas reinhardtii* biomass for ethanol production, *J. Microbiol. Biotechnol.*, 19, 2, 161–166, 2009.

Oak Ridge National Laboratory—Center for Transportation Analysis. 2012. "Lower and Higher Heating Values of Gas, Liquid and Solid Fuels." *Biomass Energy Data Book*. http://cta.ornl.gov/bedb/appendix_a/Lower_and_Higher_Heating_Values_of_Gas_Liquid_and_Solid_Fuels.pdf. Accessed March 16, 2017.

Oberoi, H. S., P. V. Vadlani, L. Saida, S. Bansal, and J. D. Hughes, Ethanol production from banana peels using statistically optimized simultaneous saccharification and fermentation process, *Waste Manag.*, 31, 7, 1576–1584, 2011.

Oliva, J. M., M. Ballesteros, and L. Olsson, Comparison of SHF and SSF processes from steam-exploded wheat straw for ethanol production by xylose-fermenting and robust glucose-fermenting *Saccharomyces cerevisiae* strains, *Biotechnol. Bioeng.*, 100, 6, 1122–1131, 2008.

Olofsson, K., M. Bertilsson, and G. Liden, A short review on SSF—An interesting process option for ethanol production from lignocellulosic feedstocks, *Biotechnol. Biofuels*, 1, 7, 1–14, 2008.

Olofsson, K., M. Wiman, and G. Liden, Controlled feeding of cellulases improves conversion of xylose in simultaneous saccharification and co-fermentation for bioethanol production, *J. Biotechnol.*, 145, 168–175, 2010.

Olson, D. G., J. E. McBride, A. J. Shaw, and L. R. Lynd, Recent progress in consolidated bioprocessing, *Curr. Opin. Biotechnol.*, 23, 3, 396–405, 2012.

Olsson, L. and B. Hahn-Hagerdal, Fermentation of lignocellulosic hydrolysates for ethanol production, *Enzyme Microb. Technol.*, 18, 5, 312–331, 1996.

Organisation for Economic Co-operation and Development (OECD), Glossary of statistical terms–Biomass, 2002. https://stats.oecd.org/glossary/detail.asp?ID=216. Accessed May 5, 2017.

Ozcimen, D. and B. Inan, An overview of bioethanol production from algae, in *Biofuels—Status and Perspective*, K. Biernat, Ed., Rijeka, Croatia: InTech, 2015.

Palacios-Bereche, R., A. Ensinas, M. Modesto, and S. A. Nebra, New alternatives for the fermentation process in the ethanol production from sugarcane: Extractive and low temperature fermentation, *Energy*, 70, 595–604, 2014.

Park, C.-H. and Q. Geng, Simultaneous fermentation and separation in the ethanol and ABE fermentation, *Sep. Purif. Rev.*, 21, 2, 127–174, 1992.

Patzek, T. W., A statistical analysis of the theoretical yield of ethanol from corn starch, *Nat. Resour. Res.*, 15, 3, 205–212, 2006.

Penttila, M. E., L. Andre, M. Saloheimo, P. Lehtovaara, and J. K. Knowles, Expression of two *Trichoderma reesei* endoglucanases in the yeast *Saccharomyces cerevisiae*, *Yeast*, 3, 3, 175–185, 1987.

Pfromm, P. H., V. Amanor-Boadu, R. Nelson, P. Vadlani, and R. Madl, Bio-butanol vs. bio-ethanol: A technical and economic assessment for corn and switchgrass fermented by yeast or *Clostridium acetobutylicum*, *Biomass and Bioenergy*, 34, 4, 515–524, 2010.

Ponce, G. H. S. F., J. M. Neto, S. M. de Jesus et al., Sugarcane molasses fermentation with in situ gas stripping using low and moderate sugar concentrations for ethanol production: Experimental data and modeling, *Biochem. Eng. J.*, 110, 152–161, 2016.

Power, N., J. D. Murphy, and E. McKeogh, What crop rotation will provide optimal first-generation ethanol production in Ireland, from technical and economic perspectives? *Renew. Energy*, 33, 7, 1444–1454, 2008.

Quintero, J. A. and C. A. Cardona, Nuevas tecnologaas en la produccion de bioetanol a partir de biomasa lignocelulosica, in *Avances Investigativos en la Produccioon de Biocombustibles*, 1st ed., C. A. Cardona and C. E. Orrego Alzate, Eds., Manizales, Colombia: Universidad Nacional de Colombia—Sede Manizales, 2009, pp. 171–198.

Quintero, J.A., M. I. Montoya, O. J. Sanchez, O. H. Giraldo, and C. A. Cardona, Fuel ethanol production from sugarcane and corn: Comparative analysis for a Colombian case, *Energy*, 33, 3, 385–399, 2008.
Quintero Suarez, J. A., Design and evaluation of fuel alcohol production from lignocellulosic raw materials, Ph.D. thesis, Universidad Nacional de Colombia sede Manizales, Manizales, Colombia, 2011.
Renewable Energy UK, A brief introduction to butanol/biobutanol as an alternative to ethanol. Butanol vs Ethanol Fuel of the Future, 2014, http://www.REUK.co.uk. Accessed May 5, 2017.
Roberto A. A., Evaluacion experimental del proceso safes de etanol a partir de material lignocelulosico, Ph.D. thesis, Universidad Nacional de Colombia sede Manizales, Manizales, Colombia, 2009.
Rocha, G. J. M., A. R. Goncalves, B. R. Oliveira, E. G. Olivares, and C. E. V. Rossell, Steam explosion pretreatment reproduction and alkaline delignification reactions performed on a pilot scale with sugarcane bagasse for bioethanol production, *Ind. Crops Prod.*, 35, 1, 274–279, 2012a.
Rocha, G. J. M., C. Martin, V. F. N. da Silva, E. O. Gomez, and A. R. Goncalves, Mass balance of pilot-scale pretreatment of sugarcane bagasse by steam explosion followed by alkaline delignification, *Bioresour. Technol.*, 111, 447–452, 2012b.
Ros, J. P. M., K. P. Overmars, E. Stehfest et al., Identifying the indirect effects of bio-energy production, Bilthoven, Netherlands: PBL Bilthoven, 2010.
Ruales Salcedo, A. V., Evaluacion del potencial energetico y bioactivo de los residuos generados por la produccion y transformacion de la uva, Master's thesis, Universidad Nacional de Colombia sede Manizales, Manizales, Colombia, 2015.
Saka, S., M. V. Munusamy, M. Shibata, Y. Tono, and H. Miyafuji, Chemical constituents of the different anatomical parts of the oil palm (*Elaeis guineensis*) for their sustainable utilization, In: *Seminar Proceedings – Natural Resources & Energy Environment JSPS-VCC Program on Environmental Science, Engineering and Ethics (Group IX)*, 24–25 November 2008, Kyoto, Japan.
Sanchez, O. J. and C. A. Cardona, Produccion biotecnologica de alcohol carburante I: Obtencion a partir de diferentes materias primas, *Interciencia*, 30, 11, 671–678, 2005.
Sanchez, O. J. and C. A. Cardona, Trends in biotechnological production of fuel ethanol from different feedstocks, *Bioresour. Technol.*, 99, 13, 5270–5295, 2008.
Schmidt, F. R., Recombinant expression systems in the pharmaceutical industry, *Appl. Microbiol. Biotechnol.*, 65, 4, 363–372, 2004.
Shafiei, M., K. Rajeev, and K. Karimi, Pretreatment of lignocellulosic biomass, in *Lignocellulose-Based Bioproducts, 1st edition*, K. Karimi, Ed., Cham, Germany: Springer International Publishing, 2015, pp. 85–154.
Siqueira, P. F., S. G. Karp, J. C. Carvalho et al., Production of bio-ethanol from soybean molasses by *Saccharomyces cerevisiae* at laboratory, pilot and industrial scales, *Bioresour. Technol.*, 99, 17, 8156–8163, 2008.
Sonego, J. L. S., D. A. Lemos, G. Y. Rodriguez, A. J. G. Cruz, and A. C. Badino, Extractive batch fermentation with CO_2 stripping for ethanol production in a bubble column bioreactor: Experimental and modeling, *Energy & Fuels*, 28, 12, 7552–7559, 2014.
Sun, S., S. Sun, X. Cao, and R. Sun, The role of pretreatment in improving the enzymatic hydrolysis of lignocellulosic materials, *Bioresour. Technol.*, 199, 49–58, 2016.

Taherzadeh, M. J. and K. Karimi, Pretreatment of lignocellulosic wastes to improve ethanol and biogas production: A review, *Int. J. Mol. Sci.*, 9, 9, 1621–51, 2008.

Tian, S., J. Zang, Y. Pan et al., Construction of a recombinant yeast strain converting xylose and glucose to ethanol, *Front. Biol. China*, 3, 2, 165–169, 2008.

Triana, C. F., J. A. Quintero, R. A. Agudelo, C. A. Cardona, and J. C. Higuita, Analysis of coffee cut-stems (CCS) as raw material for fuel ethanol production, *Energy*, 36, 7, 4182–4190, 2011.

U.S. Department of Energy, Alternative Fuels Data Center, 2017. https://www.afdc.energy.gov/. Accessed May 05, 2017.

van der Bruggen, B. and P. Luis, Pervaporation as a tool in chemical engineering: A new era? *Curr. Opin. Chem. Eng.*, 4, 47–53, 2014.

Vane, L. M., Separation technologies for the recovery and dehydration of alcohols from fermentation broths, *Biofuels, Bioprod. Biorefining*, 2, 6, 553–588, 2008.

van Zyl, W. H., L. Lynd, R. den Haan, and J. McBride, Consolidated bioprocessing for bioethanol production using *Saccharomyces cerevisiae, Adv. Biochem. Eng./Biotechnol.*, 108, 205–235, 2007.

Watanabe, I., N. Miyata, A. Ando et al., Ethanol production by repeated-batch simultaneous saccharification and fermentation (SSF) of alkali-treated rice straw using immobilized *Saccharomyces cerevisiae* cells, *Bioresour. Technol.*, 123, 695–698, 2012.

Wirawan, F., C.-L. Cheng, W.-C. Kao, D.-J. Lee, and J.-S. Chang, Cellulosic ethanol production performance with SSF and SHF processes using immobilized *Zymomonas mobilis, Appl. Energy*, 100, 19–26, 2012.

Xu, Q., A. Singh, and M. E. Himmel, Perspectives and new directions for the production of bioethanol using consolidated bioprocessing of lignocellulose, *Curr. Opin. Biotechnol.*, 20, 3, 364–371, 2009.

Younesi, H., G. Najafpour, and A. R. Mohamed, Ethanol and acetate production from synthesis gas via fermentation processes using anaerobic bacterium, *Clostridium ljungdahlii, Biochem. Eng. J.*, 27, 2, 110–119, 2005.

Yu, Q., X. Zhuang, S. Lv et al., Liquid hot water pretreatment of sugarcane bagasse and its comparison with chemical pretreatment methods for the sugar recovery and structural changes, *Bioresour. Technol.*, 129, 592–598, 2013.

Yuan, Z. and M. R. Eden, Recent advances in optimal design of thermochemical conversion of biomass to chemicals and liquid fuels, *Curr. Opin. Chem. Eng.*, 10, 70–76, 2015.

Yue, D., F. You, and S. W. Snyder, Biomass-to-bioenergy and biofuel supply chain optimization: Overview, key issues and challenges, *Comput. Chem. Eng.*, 66, 36–56, 2014.

Zabed, H., J. N. Sahu, A. N. Boyce, and G. Faruq, Fuel ethanol production from lignocellulosic biomass: An overview on feedstocks and technological approaches, *Renew. Sustain. Energy Rev.*, 66, 751–774, 2016.

Zhang, Y., Y. Y. Liu, J. L. Xu et al., High solid and low enzyme loading based saccharification of agricultural biomass, *BioResources*, 7, 1, 345–353, 2012.

Zhang, J., X. Shao, O. V Townsend, and L. R. Lynd, Simultaneous saccharification and co-fermentation of paper sludge to ethanol by *Saccharomyces cerevisiae* RWB222—Part I: Kinetic modeling and parameters, *Biotechnol. Bioeng.*, 104, 5, 920–931, 2009.

Zhang, H. and S. Wu, Dilute ammonia pretreatment of sugarcane bagasse followed by enzymatic hydrolysis to sugars, *Cellulose*, 21, 3, 1341–1349, 2014.

chapter eight

Current status and challenges in biobutanol production

Manish Kumar, Tridib Kumar Bhowmick, Supreet Saini, and Kalyan Gayen

Contents

8.1 Introduction

The superiority of biobutanol as a fuel for vehicles over other biofuels has been reported numerous times in recent years (Kumar and Gayen, 2011). Therefore, fermentative production of butanol has attracted the attention of various new research groups in academia and the industry. In a review, we have already summarized the development and challenges in fermentative production of biobutanol until 2011 (Kumar and Gayen, 2011). In this chapter, we will highlight the further developments in this field along with the outstanding challenges. In spite of several advantages of butanol as fuel for vehicles, there are still some challenges in acetone–butanol–ethanol (ABE) fermentation in terms of low butanol titer, low productivity, and low yield, high cost of pretreatment of cellulosic feedstocks, the presence of toxic compounds in feedstock hydrolysates, product inhibition, and costly and low efficiency recovery processes, which make biobutanol less competitive than other biofuels like bioethanol and biodiesel (Zheng et al., 2015). These challenges are being addressed using several strategies. Details of these strategies are discussed in different sections of this chapter.

ABE fermentation has been in development for the past 100 years. Over time, factors, such as demand of products and cost of feedstocks, have made this fermentation beneficial in different forms (Moon et al., 2015). For instance, in the early twentieth century, ABE fermentation was used for acetone production, which was the one of the precursors to generating explosives during World War I. During that time, food-based feedstocks were used to operate the ABE fermentation plants (Kumar and Gayen, 2011). More recently, however, we are already facing food-related crises in most parts of the world (Tenenbaum, 2008). Therefore, novel renewable materials, such as lignocellulosic material and algae biomass (John et al., 2011) are being investigated as feedstock of biobutanol fermentation.

Lignocellulosic biomass is available in abundance, although it has a complex structure of cellulose, hemicellulose, and lignin (Baral and Shah, 2014). In natural form, butanol producers cannot consume cellulosic feedstocks with their metabolism. Therefore, before being used in the fermentation process, lignocellulosic feedstocks are hydrolyzed to simple fermentable sugars, which are also called the pretreatment process. Inauspiciously, current pretreatment methods are costly and inefficient in order to make biobutanol production economical (Qureshi et al., 2016).

Another promising feedstock, algae biomass, does not contain lignin, which makes algae biomass an advantageous feedstock for biobutanol production. However, one of the major issues with algae biomass is high cultivation cost (John et al., 2011). Along with feedstock and pretreatment, product recovery is also a critical factor in terms of the production cost of biobutanol. In the following sections of this chapter, we discuss several challenges with biobutanol production and how several strategies are being used to address these challenges.

8.2 Amelioration in the selection of feedstocks for ABE fermentation

Selection of feedstock for ABE fermentation is always considered a critical factor to improve the economy of the process. Various renewable raw materials have been examined for enhancing the efficiency and feasibility of the ABE fermentation process in this highly energy-dependent era. In this section, we have summarized the research endeavors that have taken place in the last few years for testing different feedstocks for butanol production (Table 8.1). The focus of research was mostly on lignocellulosic (Baral and Shah, 2014), microalgae, and macroalgae biomasses (Wei et al., 2013). Properties, such as low cost, dependency on photosynthesis for growth, and nonfood competitiveness, make these feedstocks more suitable and feasible for the production of fermentation-based biofuels (Baral and Shah, 2014). However, the use of algae and lignocellulosic biomasses as raw materials for the fermentation process also illustrate several challenges. Lignocellulosic materials contain a very complex structure including three main constituents, namely, cellulose, hemicellulose, and lignin. Therefore, the primary challenge is to reduce these complex constituents to simple or reducing sugars because most of the butanol producers consume monosaccharaides as the primary source of carbon for their metabolism. There are several pretreatment methods that have been examined for converting complex sugars in fermentable sugars (Kujawska et al., 2015). A detailed summary of these pretreatment methods is discussed in Section 8.3 of this chapter. Another challenge with lignocellulosic materials is low butanol productivity because of the presence of several toxic compounds, such as furfural, hydroxylmethylfurfural, and other lignin-degradation products in lignocellulosic hydrolysates for microorganism growth (Weber et al., 2010).

Another important feedstock for butanol production, algae biomass, exhibits several advantages, such as (1) containing a high amount of lipids and carbohydrates, (2) being able to grow on wastewater and salty water, (3) not needing fertile land to grow, (4) not containing the complex constituents like lignin (John et al., 2011), and (5) being able to fix the carbon from

Table 8.1 Renewable and cost-effective feedstocks for biobutanol production

Feedstock	Main constituents	Other uses	Year of study	References
Spoilage date palm fruit	Sugars, proteins, and lipids	Waste	2012	Abd-Alla and Elsadek El-Enany (2012)
Rice bran, sesame oil cake, and deoiled rice bran	Sugars, starch, proteins, and lipids	Edible	2012, 2016	Al-Shorgani et al. (2012) and Rajagopalan et al. (2016)
Rice straw	Lignocellulosic material	Waste	2012, 2013–2015	Cheng et al. (2012), Gottumukkala et al. (2013), Ranjan et al. (2013), Amiri et al. (2014) and Kiyoshi et al. (2015)
Sugarcane bagasse	Lignocellulosic material	Waste	2012, 2016	Cheng et al. (2012) and Pang et al. (2016)
Microalgae and macroalgae biomass, and microalgae-based diesel residue	Lipids, proteins, and carbohydrates	Photosynthetic microorganisms	2012, 2015–2016	Efremenko et al. (2012), Huesemann et al. (2012), Cheng et al. (2014), Castro et al. (2015) and Wang et al. (2016)
Cassava bagasse	Lignocellulosic material	Waste	2012, 2016	Lu et al. (2012), Wang et al. (2016), and Zhang et al. (2016)
Sugarcane molasses	Sugars	Raw material of several fermentation-based industries	2012	Ni et al. (2012) and Wei et al. (2013)
Sugar maple and aspen	Lignocellulosic material	Woody material	2012, 2015	Sun and Liu (2012) and Mechmech et al. (2015)
Corn fiber	Lignocellulosic material	Waste	2013	Guo et al. (2013)

(*Continued*)

Table 8.1 (Continued) Renewable and cost-effective feedstocks for biobutanol production

Feedstock	Main constituents	Other uses	Year of study	References
Willow biomass	Lignocellulosic material	Woody material	2013	Han et al. (2013b)
Corn stover	Lignocellulosic material	Waste	2013, 2014, 2016	He and Chen (2013), Qureshi et al. (2014), Ding et al. (2016), Dong et al. (2016), and Xu et al. (2016)
Felled oil palm trunk and oil palm empty fruit bunch	Sugars and lignocellulosic material	Raw material of logging industries	2013, 2015	Ibrahim et al. (2015)
Distiller's dried grains with solubles (DDGS)	Lignocellulosic material	Coproduct of ethanol production process from corn	2013	Wang et al. (2013)
Wheat straw	Lignocellulosic material	Waste	2014	Bellido et al. (2014)
Switchgrass and phragmites	Lignocellulosic material	Waste	2014	Gao et al. (2014)
Sugarcane juice	Sugars	Raw material of sugar industries	2014	Jiang et al. (2014)
Glycerol	–	Raw material of several industries	2014–2016	Lin et al. (2015) and Yadav et al. (2014)
Barley straw and grain	Starch and lignocellulosic material	Barley straw (waste) and grain (edible)	2014	Yang et al. (2015a,b)
Food waste	Sugars and starch	Waste	2015	Huang et al. (2015)

(*Continued*)

Table 8.1 (Continued) Renewable and cost-effective feedstocks for biobutanol production

Feedstock	Main constituents	Other uses	Year of study	References
Paper mill sludge	Lignocellulosic material	Waste	2016	Guan et al. (2016)
Sweet sorghum bagasse	Lignocellulosic material	Waste	2016	Jafari et al. (2016) and Qureshi et al. (2016)
Sweet sorghum juice	Sugars	Raw material of sugar industries	2016	Sirisantimethakom et al. (2016)
Agroindustrial wastes [suspended brewery liquid waste (BLW), starch industry wastewater (SIW), and apple pomace ultrafiltration sludge (APUS)]	Sugars and starch	Waste	2016	Maiti et al. (2016)
Carbon monoxide	Gas	Waste		Fernandez-Naveira et al. (2016)

carbon dioxide in the presence of sunlight and inorganic nutrients faster than the lignocellulosic biomass (Subhadra and Edwards, 2010). Recently, several studies have employed lignocellulosic and algae biomasses to examine their feasibility for biobutanol production (Table 8.1).

8.2.1 *Lignocellulosic materials*

8.2.1.1 *Rice straw*

Cheng et al. (2012) fermented two lignocellulosic materials, namely, rice straw and sugarcane bagasse, to butanol using mixed culture of four *Clostridium* species, namely, *Clostridium saccharoperbutylacetonicum, C. butylicum, C. beijerinckii,* and *C. acetobutylicum.* Comparative results from this study suggested that rice straw (maximum butanol titer = 2.93 g/L, productivity = 0.86 g/L day, and yield = 0.49 mol butanol/mol reducing sugar) as a feedstock for butanol production is better than sugarcane bagasse (maximum butanol titer = 1.95 g/L, productivity = 0.61 g/L day, and yield = 0.37 mol butanol/mol reducing sugar). In another study, it was found that supplementing yeast extract and calcium carbonate in rice straw hydrolysate elevated the butanol production up to 5.52 g/L using *C. sporogenes* (Gottumukkala et al., 2013). Moreover, a higher maximum titer of 13.5 g/L butanol was reported on the consumption of rice straw in a 12-day-long fermentation process (Ranjan et al., 2013).

8.2.1.2 *Sugarcane bagasse*

Sugarcane bagasse is the lignocellulosic part of the sugarcane plant and is mostly used in the paper industry as a raw material. Recently, some studies have examined sugarcane bagasse as feedstock for ABE fermentation (Cheng et al., 2012; Pang et al., 2016). Pang et al. (2016) employed a series of pretreatment methods in separate experiments and improved the butanol concentration up to 21.11 g/L in a fed-batch system. This was significantly higher than the previous effort by Cheng et al. (2012). Still, more efforts on this feedstock will be beneficial for countries such as Brazil, India, Thailand, China, the United States, and other tropical and subtropical countries, which produce a significant amount of sugarcane bagasse annually (Das et al., 2004).

8.2.1.3 *Cassava bagasse*

Cassava is a very important crop and one of the main sources of starch in Southeast Asia. Industrial processing of cassava produces a waste by-product, called cassava bagasse, which is generally used as animal feed. Cassava bagasse was first reported as feedstock for butanol production in 2012 (Lu et al., 2012). This study reported a maximum butanol titer of 76.4 g/L in a fed-batch fermentation equipped with a gas stripping recovery system. In another endeavor, cassava bagasse was fermented in batch immobilized coculture fermentation and achieved a maximum butanol titer of 13.26 g/L (Zhang et al., 2016).

8.2.1.4 Sugar maple and aspen

Apart from the lignocellulosic wastes from different industries, direct wood logs of trees like sugar maple and aspen were tested in ABE fermentation (Sun and Liu, 2012; Mechmech et al., 2015). Both of these raw materials are very significant in terms of their abundant availability; however, more investigations are required to improve the productivity of butanol production from these feedstocks. Low titers of 7 and 3.95 g/L were reported on the consumption of only sugar maple and both sugar maple and aspen in batch fermentation experiments, respectively (Sun and Liu, 2012; Mechmech et al., 2015).

8.2.1.5 Corn stover and corn fiber

Corn crops contain lignocellulosic biomass in different forms, such as corn stover and corn fiber, which remain wastes after harvesting cereal grains. Recently, a significant number of studies have been conducted to produce biobutanol from corn stover and fiber. In an absorbed fermentation system, a maximum titer of 10.11 g/L was achieved using corn stover as feedstock (He and Chen, 2013). Qureshi et al. (2014) employed a simultaneous saccharification, fermentation, and recovery (SSFR) process to elevate the yield of acetone–butanol–ethanol up to 0.39 g/g, which is very close to theoretical yield on the consumption of corn stover. This study claimed a substantial improvement at the process level; further natural progression in this field can be toward perturbation in the genome of the microorganism though metabolic engineering. This effort might allow raising the theoretical yield of butanol. Some other studies have achieved a maximum titer of butanol between 7.9 and 12.3 g/L using different pretreatment and recovery methods in fermentation of the corn stover batch system (Xu et al., 2016; Dong et al., 2016; Ding et al., 2016). Similarly, 9.3 g/L of butanol was produced using *C. beijerinckii* mutant from corn fiber (Guo et al., 2013).

In the last few years, various studies have investigated more lignocellulose-based raw materials apart from the previously described feedstocks. Some examples of these lignocellulosic materials are willow biomass, distiller's dried grains with solubles (DDGS) (Wang et al., 2013), wheat straw (Bellido et al., 2014), switchgrass and phragmites (Gao et al., 2014), barley straw (Yang et al., 2015a,b), paper mill sludge (Guan et al., 2016), and sweet sorghum bagasse (Qureshi et al., 2016; Jafari et al., 2016). Because of the high abundance and reproducibility of natural resources, lignocellulosic materials are still considered to be the feedstocks with most potential for ABE fermentation. However, barriers, like the difficult hydrolysis to monosugars, low productivity, and the presence of toxic substances, need to be resolved in order to use lignocellulose-based materials in ABE fermentation at the industrial level.

8.2.2 Microalgae and macroalgae biomasses

The main constituents in microalgae and macroalgae biomasses are carbohydrates, lipids, and proteins. Auspiciously, algae biomass is rich in fermentable carbohydrates (Wei et al., 2013). Vassilev and Vassileva (2016) give a detailed elemental composition of the variety of algae. Largely, algae biomass is produced through two methods, namely, harvesting wild stocks and produced by aquaculture in industry. However, the current focus of algae-producing industries is to fulfill the requirement of human food consumption (Wei et al., 2013). Recently, algae biomass has attracted the attention of the scientific community as a feedstock for ABE fermentation due to the several advantages (as described above in this section). Some initial research efforts have already been conducted to check the feasibility of converting algae biomass to liquid biofuels (Efremenko et al., 2012; Huesemann et al., 2012; Cheng et al., 2014; Narayanasamy and Murugesan, 2014; Castro et al., 2015; Wang et al., 2016).

Several thermal pretreated microalgae strains of *Arthrospira platensis, Nannochloropsis* sp., *Dunaliella tertiolecta, D. salina, Galdieria partita, Chlorella vulgaris,* and *Cosmarium* sp. were fermented using *C. acetobutylicum* in an immobilized cells fermentation system. This demonstration indicated the suitability of *C. acetobutylicum* for fermenting microalgae as well as the potentiality of algae biomass as feedstock for biobutanol production (Efremenko et al., 2012). In a similar attempt using *C. acetobutylicum,* specific substrates (free glucose, laminarin-derived glucose, and mannitol) from an aqueous extract of macroalgae, *Saccharina* sp., have been identified as principal substrates for ABE fermentation (Huesemann et al., 2012).

In the view of betterment in utilization of more substrates of microalgae biomass in biofuel production, carbohydrates and lipids present in microalgae were suggested to be converted into fermentation-based biofuels and biodiesel, respectively. After extracting oil from the microalgae biomass, the remaining carbohydrate-rich residue was fermented mainly in butanol using *C. acetobutylicum.* Results suggested that it is possible to ferment microalgae-based biodiesel residues to biobutanol; however, more efforts are required to improve the production yield (Cheng et al., 2014). Furthermore, despite the various advantages of algae biomass over lignocellulosic materials as feedstock for biobutanol production, the yield and productivity of butanol from algae biomass lag far behind the outcomes from utilization of lignocellulosic materials. Primarily, more research endeavors should be conducted to track down and improve the limiting factors involved in low sugars utilization in algae biomass hydrolysates in ABE fermentation.

8.2.3 Glycerol

In addition to lignocellulosic and algae biomasses, glycerol is also a feedstock of interest for biobutanol production and recently various research groups tested the feasibility of glycerol for this process. Advantageously, glycerol is an inexpensive compound and is produced as a by-product in bioethanol, biodiesel, and several chemical production processes. Moreover, reduction in the price of crude glycerol over the years makes this feedstock more attractive for ABE fermentation (Clomburg and Gonzalez, 2013). To date, only a few microorganisms have been reported to have the natural ability to ferment glycerol to butanol. *C. pasteurianum* can grow and produce butanol on glycerol as a sole carbon source, but some strains of *C. acetobutylicum* and *C. beijerinckii* can convert glycerol to butanol in the presence of glucose or yeast extract in media (Taconi et al., 2009; Yadav et al., 2014). Yadav et al. (2014) reported a maximum butanol titer of 13.57 g/L in 96 h of fermentation time using *C. acetobutylicum* KF15879.

In another study, a maximum butanol titer of 29.8 g/L was achieved on the consumption of glycerol supplemented with butyric acid by *C. pasteurianum*. These results illustrate the potentiality of glycerol for biobutanol production. However, it has been noticed that there is very limited literature available related to fermentation of glycerol to butanol in recent years. This focuses our attention toward attempting more research efforts to optimize the utilization of glycerol in producing biobutanol.

8.2.4 Other feedstocks

Aside from the preceding three categories of feedstocks, there are some more feedstocks, which were investigated in recent years for butanol production. These feedstocks are either primarily used as human food or food wastes. One of the major concerns with human-food-based feedstocks is that these can be feasible for biobutanol production depending on geographical location and their productivity. Some examples of such feedstocks are sweet sorghum juice (Sirisantimethakom et al., 2016), sugarcane juice (Jiang et al., 2014), sugarcane molasses (Ni et al., 2012), agroindustrial wastes [suspended brewery liquid waste (BLW), starch industry wastewater (SIW), and apple pomace ultrafiltration sludge (APUS)] (Maiti et al., 2016), food waste (Huang et al., 2015), spoilage date palm fruit (Abd-Alla and Elsadek El-Enany, 2012), rice bran, sesame oil cake, and deoiled rice bran (Al-Shorgani et al., 2012; Rajagopalan et al., 2016).

8.3 Amelioration in pretreatment of complex feedstocks for biobutanol fermentation

Pretreatment is one of the vital factors in fermentation of complex feedstocks to butanol because it is a mandatory step in the process and affects the economy of the production. In this section, we discuss improvements in different pretreatment methods, which have recently been investigated in various studies (Table 8.2). Some examples of such methods are thermal decomposition (Efremenko et al., 2012), acid hydrolysis (Han et al., 2013a; Castro et al., 2015), alkaline pretreatment (Kiyoshi et al., 2015; Dong et al., 2016), enzymatic hydrolysis (Li et al., 2014), and their different combinations (He and Chen, 2013; Wang et al,. 2013, 2016; Amiri et al., 2014; Gao et al., 2014; Ibrahim et al., 2015; Mechmech et al., 2015; Yang et al., 2015b; Ding et al., 2016; Dong et al., 2016; Guan et al., 2016; Jafari et al., 2016; Pang et al., 2016; Qureshi et al., 2016; Xu et al., 2016; Zhang et al., 2016). It has been noticed that these pretreatment methods were used mainly to hydrolyze lignocellulosic and algae biomasses in the last 5–6 years.

8.3.1 Thermal pretreatment

Efremenko et al. (2012) investigated thermal decomposition as a pretreatment method using several combinations of temperature, pressure, and the presence of acid to pretreat (or extract the simple fermentable sugars from algae biomass) biomasses of various species of microalgae. This study's results suggested that the best combination was a temperature of 108°C and pressure of 0.5 atm in the presence of 0.1-mM sulfuric acid in terms of maximum production of butanol and ethanol (0.45 and 0.30 g/L, respectively).

8.3.2 Acid hydrolysis

Different concentrations of sulfuric acid at different temperature ranges and retention times were tested to liberate the fermentable sugars from microalgae and willow biomass (Han et al., 2013a; Castro et al., 2015). A diluted microalgae biomass was treated with different acid concentrations (0.0, 0.35, 0.50, 0.70, 1.00, and 1.50 M) at various temperature ranges (25–30°C, 45–55°C, and 80–90°C) for a number of retention times (40, 80, and 120 min). The temperature range of 80–90°C was found to be more effective than the other two ranges and had the highest sugar yield using 1.5-M sulfuric acid for 40–120 min as well as using 1.0-M sulfuric acid for 120 min. These results indicate interconnectivity among these factors (acid concentration, temperature, and retention time) toward controlling the efficiency of hydrolysis of an algae biomass (Castro et al., 2015). In another study, acid hydrolysis was used to pretreat the willow biomass

Table 8.2 Different pretreatment methods for feedstocks in biobutanol fermentation

Pretreatment method	Feedstock	Organism used for fermentation	Year of study	References
Thermal pretreatment	Microalgae biomass	*C. acetobutylicum*	2012	Efremenko et al. (2012)
Acid hydrolysis	Willow biomass	*C. beijerinckii*	2013	Han et al. (2013b)
Acid hydrolysis	Microalgae biomass	*C. acetobutylicum*	2015	Castro et al. (2015)
Alkaline pretreatment	Rice straw	*C. thermocellum* and *C. saccharoperbutylacetonicum*	2015	Kiyoshi et al. (2015)
Alkaline pretreatment	Corn stover	*C. saccharobutylicum*	2016	Dong et al. (2016)
Enzymatic hydrolysis	Cassava	*C. acetobutylicum*	2014	Li et al. (2014)
Acid pretreatment and enzymatic hydrolysis	Barley straw	*C. acetobutylicum*	2015	Yang et al. (2015a))
Alkaline pretreatment and enzymatic hydrolysis	Oil palm empty fruit bunch	*C. acetobutylicum*	2015	Ibrahim et al. (2015)
Alkaline pretreatment and enzymatic hydrolysis	Paper mill sludge	*C. acetobutylicum*	2016	Guan et al. (2016)
Alkaline pretreatment and enzymatic hydrolysis	Switchgrass and phragmites	*C. saccharobutylicum*	2014	Gao et al. (2014)
Steam and acid pretreatment	Sugar maple and aspen	*C. acetobutylicum*	2015	Mechmech et al. (2015)
Steam and alkaline pretreatment	Corn stover	*C. acetobutylicum*	2013	He and Chen (2013)
Hot water pretreatment and enzymatic hydrolysis	Sweet sorghum bagasse	*C. beijerinckii*	2016	Qureshi et al. (2016)

(*Continued*)

Table 8.2 (Continued) Different pretreatment methods for feedstocks in biobutanol fermentation

Pretreatment method	Feedstock	Organism used for fermentation	Year of study	References
Alkaline twin-screw extrusion pretreatment	Corn stover	*C. acetobutylicum*	2014	Zhang et al. (2014)
Alkaline and ionic liquid pretreatment and enzymatic hydrolysis	Corn stover	*C. saccharobutylicum*	2016	Ding et al. (2016)
Electrolyzed water pretreatment and enzymatic hydrolysis	DDGS	*C. beijerinckii*	2013	Wang et al. (2013)
Alkaline pretreatment, and acids and enzymatic hydrolysis	Micro-algae biomass	*C. acetobutylicum*	2016	Wang et al. (2016)
Alkaline pretreatment, and acids and enzymatic hydrolysis	Sugarcane bagasse	*C. acetobutylicum*	2016	Pang et al. (2016)
Deep eutectic solvent pretreatment and enzymatic hydrolysis	Corn stover	*C. saccharobutylicum*	2016	Xu et al. (2016)
Ethanol pretreatment and acid and enzymatic hydrolysis	Rice straw	*C. acetobutylicum*	2014	Amiri et al. (2014)
Acetone pretreatment and enzymatic hydrolysis	Sweet sorghum bagasse	*C. acetobutylicum*	2016	Jafari et al. (2016)

using a two-stage procedure. In the first stage, lignocellulosic material was treated with diluted acid at 30°C for 1 h, and in second stage the temperature was raised to 105°C for 1 h. A maximum butanol concentration of 4.5 g/L was achieved in the fermentation of willow hydrolysate. Han et al. suggested the reason of low butanol concentration may be the presence of inhibitors, which can interfere with the switch from acidogenesis to solventogenesis in the metabolic pathway of the butanol producer (Han et al. 2013a).

8.3.3 Alkaline pretreatment

In two different studies, sodium hydroxide (1% w/v) at 100 and 120°C for 1 h was used to treat the lignocellulosic materials (rice straw and corn stover) (Kiyoshi et al., 2015; Dong et al., 2016). Pretreated rice straw was fermented using a coculture of *C. thermocellum* and *C. saccharoperbutylacetonicum* at 60°C for the first 24 h, then the temperature was decreased to 30°C for the remaining fermentation time. *C. thermocellum* and *C. saccharoperbutylacetonicum* were used to perform cellulolytic and butanol-producing events during the fermentation, respectively. As a result in this study, a maximum butanol titer of 6.9 g/L was reported (Kiyoshi et al., 2015). Similarly, alkali-pretreated lignocellulosic material (corn stover) was employed in a simultaneous saccharification and fermentation process using *C. saccharobutylicum* DSM 13864 and an external cellulase enzyme. Results demonstrated that a significantly higher butanol titer of 12.3 g/L was achieved. Despite producing a high butanol titer, the major drawback of this study was the use of a costly commercial enzyme, which can be resolved using a cellulase-secreting organism separately or simultaneously with the fermentation process (Dong et al., 2016).

8.3.4 Enzymatic hydrolysis

In order to hydrolyze the feedstock, starch materials, which contain a less complex structure than lignocellulosic materials, are used. In a recent effort, using only enzymatic hydrolysis, 68.3% of the polysaccharides were converted in monosaccharides (Li et al., 2014). During hydrolysis, two enzymes, namely, alpha-amylase (activity = 3,500,000 U/mL) and glucoamylase (activity = 2,800,000 U/mL), were used to hydrolyze the cassava flour to generate fermentable simple sugars.

8.3.5 Combinations of two pretreatment methods

8.3.5.1 Acid and enzymatic hydrolysis

Before enzymatic hydrolysis, barley straw was treated with sulfuric acid at two different temperatures, 30 and 121°C. Two commercial enzymes (cellulase and xylanase) were used with different loading concentrations to hydrolyze the acid pretreated barley straw. Using 8% dry matter loading (w/v) of enzymes, Yang et al. achieved a high reducing sugar concentration of 42.5 g/L, which was fermented to butanol of 7.9 g/L, acetone of 2.4 g/L, and ethanolof 0.5 g/L (Yang et al., 2015b).

8.3.5.2 Alkaline pretreatment and enzymatic hydrolysis

Recently, a combination of alkaline pretreatment and enzymatic hydrolysis was employed in several studies for hydrolyzing the lignocellulosic materials. During these studies, the first step was to treat lignocellulosic materials with 1%–2% sodium hydroxide at different temperatures (121 and 60°C) and retention times (from 30 min to 24 h) depending on the type of feedstock. In the second step, commercial cellulase enzymes were used for the saccharification process (Gao et al., 2014; Ibrahim et al., 2015; Guan et al., 2016).

8.3.5.3 Steam and hot water pretreatment with acid and alkaline pretreatment

Instead of using any cellulase to hydrolyze woody material and corn stover, a combination of steam and hot water pretreatment and acid and alkaline hydrolysis was used to generate simple fermentable sugars. Finally, the hydrolysates were used for butanol production in a culture of *C. acetobutylicum* (He and Chen, 2013; Mechmech et al., 2015; Qureshi et al., 2016).

8.3.5.4 Ionic water/liquid and enzymatic pretreatment

Ionic liquids (ILs) are suggested as an efficient pretreatment agent for reducing the crystallinity, hemicellulose, and lignin, and increasing the porosity and surface area of lignocellulosic materials. It reduces the barrier for the next step of pretreatment, that is, enzymatic hydrolysis (Behera et al., 2014). Ding et al. (2016) examined ionic liquid of 1-butyl-3-methylimidazolium chloride ([Bmim][Cl]) as a pretreatment agent to treat the corn stover. Ding et al. claimed that alkali-IL pretreatment was found to be superior among all tested approaches (only IL and only alkali pretreatment). Likewise, in the fermentation of DDGS to butanol, electrolyzed waster was used as a pretreatment catalyst. In this study, Ding et al also examined the potentiality of electrolyzed water as pretreatment agent compared with the outcomes from other pretreatment methods, for instance, acid hydrolysis, alkaline pretreatment, and hot water. In terms of releasing total reduced sugars, electrolyzed water performed better than other methods (Wang et al., 2013). These current efforts have proposed a new alternative method for pretreatment combined with enzyme hydrolysis for lignocellulosic materials. Moreover, the electrolyzed water pretreatment method is not only superior in terms of performance, but is also very cost-effective compared to other methods.

8.3.6 Combinations of acid, alkaline, and enzymatic hydrolysis

Recently, a few studies have used acid, alkaline, and enzymatic hydrolysis to enhance the efficacy of the pretreatment process. Microalgae biomass and sugarcane bagasse were treated with the combination of the previous three pretreatment methods in two separate studies. These studies claimed high hydrolysis rate; however, Pang et al. and Wang et al did not perform any controlled experiments in order to compare with other combinations of pretreatment methods (Pang et al., 2016; Wang et al., 2016).

8.3.7 Combination of solvent pretreatment and enzymatic hydrolysis

Similar to the ionic liquid pretreatment method, a novel combination of solvent pretreatment and enzymatic hydrolysis were used for the hydrolysis of lignocellulosic materials. For instance, different concentrations of ethanol and acetone were investigated in this direction. Nevertheless, more studies are required to examine the solvent pretreatment and the possibility of its usability instead of other pretreatment methods (Amiri et al., 2014; Jafari et al., 2016; Xu et al., 2016).

8.4 Amelioration in downstream and recovery processes

Recovery processes in ABE fermentation is another critical factor, not only to purify the products, but also because it greatly affects the economy of the production process. Moreover, recovery processes integrated with the fermentation process stimulate the performance of the butanol producer by reducing the toxicity due to the presence of solvents in the media. Out of many recovery methods, some have been recently tested in various efforts to increase the productivity of ABE fermentation (Table 8.3). Some examples are gas stripping (Xue et al., 2012, 2013, 2016b; Ezeji et al., 2013; Chen et al., 2014), liquid–liquid extraction (Bankar et al., 2011; Dhamole et al., 2012; Yen and Wang, 2013), pervaporation (Wu et al., 2012; Chen et al., 2013; Xue et al., 2014), adsorption (Liu et al., 2014; Xue et al., 2016a), and membrane distillation (Lin et al., 2015). In this section, we will discuss some currently investigated recovery processes and their advantages and disadvantages in ABE fermentation. In a techno-economic analysis, distillation has been found to be a less economic method for ABE fermentation than the membrane recovery method (Ezeji et al., 2010). Other cost-effective methods (gas stripping, liquid–liquid extraction, pervaporation, and adsorption) also have limitations to be resolved, for instance, low recovery rate in gas stripping, substrate loss in adsorption, toxicity for cells in liquid–liquid

Table 8.3 Recovery methods integrated with biobutanol fermentation process

Solvent extraction process	Type of fermentation process	Year of study	References
Gas stripping	Fed-batch fermentation	2012	Xue et al. (2012)
Gas stripping	Continuous fermentation	2013	Ezeji et al. (2013)
Gas stripping	Batch fermentation	2013	Xue et al. (2013)
Gas stripping	Batch and fed-batch fermentation	2014	Xue et al. (2014)
Gas stripping	Batch and fed-batch fermentation	2014	Chen et al. (2014)
Gas stripping and pervaporation	Batch and fed-batch fermentation	2016	Xue et al. (2016b)
Liquid–liquid extraction	Continuous fermentation	2012	Bankar et al. (2011)
Liquid–liquid extraction using biodiesel	Batch and fed-batch fermentation	2013	Yen and Wang (2013)
Liquid–liquid extraction using nonionic surfactants	Batch fermentation	2012	Dhamole et al. (2012)
Pervaporation using polydimethylsiloxane (PDMS)/ ceramic composite membrane	Batch fermentation	2012	Wu et al. (2012)
Pervaporation using PDMS membrane	Continuous fermentation	2013	Chen et al. (2013)
Pervaporation using zeolite-mixed PDMS membranes	Batch and fed-batch fermentation	2015	Xue et al. (2016a)
Adsorption	Batch and fed-batch fermentation	2016	Xue et al. (2016b)
Fixed-bed adsorption	Batch and fed-batch fermentation	2014	Liu et al. (2014)
Recovery by vacuum	Batch fermentation	2012	Mariano et al. (2012)
Polytetrafluoroethene (PTFE) membrane distillation	Batch fermentation	2015	Lin et al. (2015)
Vapor stripping–vapor permeation (VSVP) process using PDMS membranes	Batch fermentation	2016	Xue et al. (2016b)

extraction, and loss of intermediate fermentation in pervaporation (Kujawska et al., 2015).

8.5 Investigations based on amendment in biobutanol fermentation driven by chemical supplementation in media

Supplementation of several chemicals in the ABE fermentation media was found to be beneficial to enhancing the production of butanol. For instance, additional supplementation of calcium carbonate, ammonium sulfate, and butyric acids in sweet sorghum juice media supported the butanol production reaching 15.46 g/L and productivity reaching 0.43 g/L h, while butanol production and productivity in fermentation on controlled sorghum juice media was recorded as only 0.64 g/L and 0.02 g/L h, respectively (Sirisantimethakom et al., 2016). In order to understand the mechanism behind the effect of calcium on ABE production, Han et al. (2013b) suggested that calcium directly and indirectly enhances the activities of enzymes involved in acidogenesis and solventogenesis of the butanol producer. In this way, the organism grows faster and produces more acids and solvents.

Similar to calcium carbonate, the addition of 0.001 g/L zinc sulfate in media helped solventogenesis to begin more quickly and achieved the approximately similar concentration of acetone–butanol–ethanol 24 h earlier than media without zinc sulfate (Wu et al., 2013). A faster phase shift as well as increased production and productivity were also recorded by the addition of yeast extract (2.5 g/L) in cassava meal medium (Li et al., 2012). These studies indicate that during the use of any renewable raw material for ABE fermentation, media optimization is an effective step to perform in terms of enhancing the butanol production and productivity. This is because chemical supplementation in media was found to be effective at enhancing the titer of butanol as well as reducing the fermentation time or metabolic transition time from acidogenesis to solventogenesis.

8.6 Amelioration in metabolic engineering of ABE fermentation

Various efforts at the genetic level have been made in order to improve the butanol yield and ratio in ABE fermentation through overexpression, knock out, and insertion of genes encoding key enzymes in acid- and solvent-producing pathways in the metabolism of a butanol-producing organism (Higashide et al., 2011; Cho et al., 2012; Lan and Liao, 2012; Lee et al., 2012; Krivoruchko et al., 2013; Bhandiwad et al., 2014; Generoso et al., 2015; Si et al., 2014; Saini et al., 2015; Swidah et al., 2015; Yu et al.,

2015; Gong et al., 2016; Yan et al. 2016). Genetic manipulations have been performed in *Clostridia* (*C. acetobutylicum, C. beijerinckii, C. saccharoperbutylacetonicum, C. saccharoperbutylicum, C. sporogenes, C. perfringens, C. pasteurianum, C. carboxidivorans, C. tetanomorphum, C. aurantibutyricum,* and *C. cadaveris*) as well as in *Clostridia* hosts (*Bacillus subtilis, C. tyrobutyricum, Escherichia coli, Lactobacillus brevis, L. buchneri, Lactococcus lactis, Pseudomonas putida,* and *Saccharomyces cerevisiae*) (Zheng et al., 2015).

In contrast to the improvement at the process level (pretreatment, fermentation, recovery processes), genetic-level efforts are still not very effective at enhancing the butanol yield in ABE fermentation. Therefore, more endeavors are required to manipulate the metabolic network of *Clostridia*. Moreover, discovery of novel hosts for adapting the butanol-producing pathway can provide a milestone in this field.

8.7 Future prospective and summary

Currently, various efforts are taking place to utilize lignocellulosic and algae biomasses by a *Clostridia* butanol producer. Furthermore, cheap and abundant substrate, such as carbon dioxide and carbon monoxide, should be investigated for butanol production. Some natural and genetically modified organisms (*C. carboxidivorans* and *C. ljungdahlii*) have been examined to ferment the waste gaseous substrates in butanol (Durre, 2016; Fernandez-Naveira et al., 2016). However, more research in this direction could allow us to produce butanol along with eliminate toxic waste gases from the environment. In order to reduce the cost of the pretreatment process for cellulosic and algae biomasses, identification of a cheaper physical or chemical hydrolysis process rather than using costly enzymes (for instance, cellulase) can be more effective. Another very important concern is optimization of media for ABE fermentation. Recently, some reports suggested supplementation of chemicals, such as calcium carbonate, zinc sulfate, and yeast extract in media based on cellulosic material, have enhanced the productivity of butanol. These results direct us toward new investigations involving supplementation of more chemical substrates in renewable-material-based media for enhancing the production of butanol.

Still, our understanding of mechanisms behind physiological events (acidogenesis, solventogenesis, phase transition, and sporulation) in the metabolism of butanol producers is very poor. This is one of the reasons for insubstantial progress in the area of metabolic engineering of butanol producers. Therefore, the research efforts in order to enrich the understanding of physiological events and regulatory circuits involved in metabolism can allow us to design an efficient mutant for higher butanol production. Likewise, recently introduced nonacetone-producing strains and mixed cultures can be considered for further investigations (Cheng et al., 2012; Gottumukkala et al., 2013; Sabra et al., 2014).

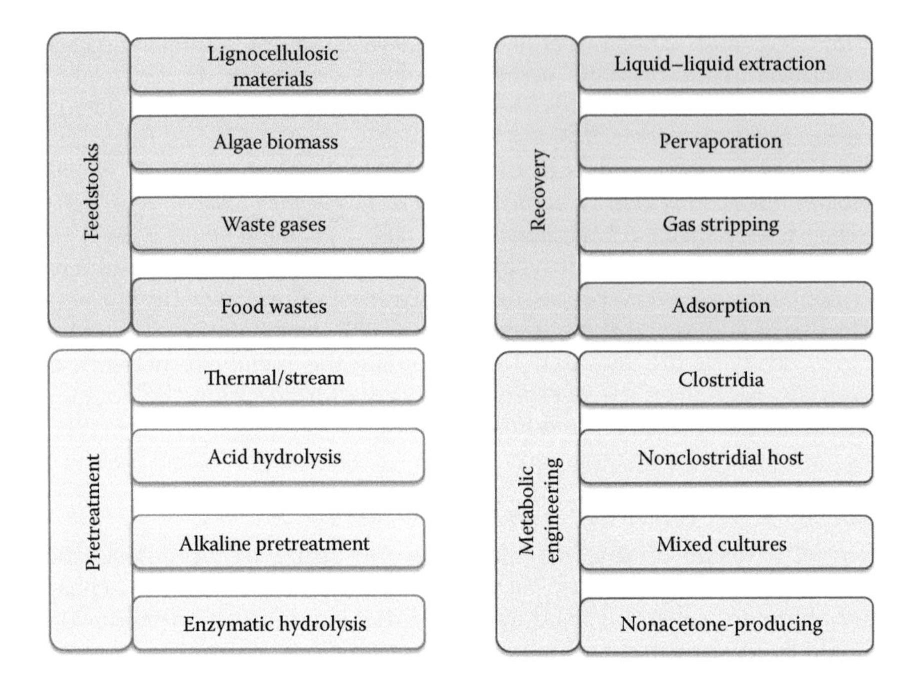

Figure 8.1 A schematic summary of this chapter.

In summary, this chapter discussed the recent developments in ABE fermentation during the last 5 years. We highlighted the improvements in selection of feedstocks, pretreatment methods, recovery methods, strategies of metabolic engineering, and the effect of supplementation of chemical substance in ABE fermentation media. A graphical summary can be seen in Figure 8.1.

References

Abd-Alla, M. H. and A. W. Elsadek El-Enany. 2012. Production of acetone-butanol-ethanol from spoilage date palm (*Phoenix dactylifera l.*) fruits by mixed culture of *Clostridium acetobutylicum* and bacillus subtilis. *Biomass and Bioenergy* 42: 172–78.

Al-Shorgani, N. K. N., M. S. Kalil, and W. M. W. Yusoff. 2012. Biobutanol production from rice bran and de-oiled rice bran by *Clostridium saccharoperbutylacetonicum* n1-4. *Bioprocess and Biosystems Engineering* 35: 817–26.

Amiri, H., K. Karimi, and H. Zilouei. 2014. Organosolv pretreatment of rice straw for efficient acetone, butanol, and ethanol production. *Bioresource Technology* 152: 450–56.

Bankar, S. B, S. A. Survase, R. S. Singhal, and T. Granstrom. 2011. Continuous two stage acetone–butanol–ethanol fermentation with integrated solvent removal using *Clostridium acetobutylicum* B 5313. *Bioresource Technology* 106: 110–16.

Baral, N. R. and A Shah. 2014. Microbial inhibitors: formation and effects on acetone–butanol–ethanol fermentation of lignocellulosic biomass. *Applied Microbiology and Biotechnology* 98: 9151–72.
Behera, S., R. Arora, N. Nandhagopal, and S. Kumar. 2014. Importance of chemical pretreatment for bioconversion of lignocellulosic biomass. *Renewable and Sustainable Energy Reviews* 36: 91–106.
Bellido, C., M. L. Pinto, M. Coca, G. Gonzalez-Benito, and M. T. Garcia-Cubero. 2014. Acetone–butanol–ethanol (ABE) production by *Clostridium beijerinckii* from wheat straw hydrolysates: Efficient use of penta and hexa carbohydrates. *Bioresource Technology* 167: 198–205.
Bhandiwad, A., A. J. Shaw, A. Guss et al. 2014. Metabolic engineering of *Thermoanaerobacterium saccharolyticum* for *n*-butanol production. *Metabolic Engineering* 21: 17–25.
Castro, Y. A., J. T. Ellis, C. D. Miller, and R. C. Sims. 2015. Optimization of wastewater microalgae saccharification using dilute acid hydrolysis for acetone, butanol, and ethanol fermentation. *Applied Energy* 140: 14–19.
Chen, Y., H. Ren, D. Liu et al. 2014. Enhancement of *n*-butanol production by *in situ* butanol removal using permeating–heating–gas stripping in acetone–butanol–ethanol fermentation. *Bioresource Technology* 164: 276–84.
Chen, C., Z. Xiao, X. Tang et al. 2013. Acetone–butanol–ethanol fermentation in a continuous and closed-circulating fermentation system with PDMS membrane bioreactor. *Bioresource Technology* 128: 246–51.
Cheng, C. L., P. Y. Che, B. Y. Chen et al. 2012. Biobutanol production from agricultural waste by an acclimated mixed bacterial microflora. *Applied Energy* 100: 3–9.
Cheng, H. S., L. M. Whang, K. C. Chan et al. 2014. Biological butanol production from microalgae-based biodiesel residues by *Clostridium acetobutylicum*. *Bioresource Technology* 184: 379–85.
Cho, J. H., M. H. Eom, J. A. Im et al. 2012. Enhanced butanol production obtained by reinforcing the direct butanol-forming route in *Clostridium acetobutylicum*. *Mbio* 3(5): 1–9.
Clomburg, J. M. and R. Gonzalez. 2013. Anaerobic fermentation of glycerol: A platform for renewable fuels and chemicals. *Trends in Biotechnology* 31: 20–28.
Das, P., A. Ganesh, and P. Wangikar. 2004. Influence of pretreatment for deashing of sugarcane bagasse on pyrolysis products. *Biomass and Bioenergy* 27: 445–57.
Dhamole, P. B., Z. Wang, Y. Liu, B. Wang, and H. Feng. 2012. Extractive fermentation with non-ionic surfactants to enhance butanol production. *Biomass and Bioenergy* 40: 112–19.
Ding, J. C., G. C. Xu, R. Z. Han, and Y. Ni. 2016. Biobutanol production from corn stover hydrolysate pretreated with recycled ionic liquid by *Clostridium saccharobutylicum* DSM 13864. *Bioresource Technology* 199: 228–34.
Dong, J. J., J. C. Ding, Y. Zhang et al. 2016. Simultaneous saccharification and fermentation of dilute alkaline-pretreated corn stover for enhanced butanol production by *Clostridium saccharobutylicum* DSM 13864. *FEMS Microbiology Letters* 363: 1–6.
Durre, P. 2016. Butanol formation from gaseous substrates. *FEMS Microbiology Letters* 363: 1–7.
Efremenko, E. N., A. B. Nikolskaya, I. V. Lyagin et al. 2012. Production of biofuels from pretreated microalgae biomass by anaerobic fermentation with immobilized *Clostridium acetobutylicum* cells. *Bioresource Technology* 114: 342–48.

Ezeji, T., C. Milne, N. D. Price, and H. P. Blaschek. 2010. Achievements and perspectives to overcome the poor solvent resistance in acetone and butanol-producing microorganisms. *Applied Microbiology and Biotechnology* 85: 1697–1712.

Ezeji, T. C., N. Qureshi, and H. P. Blaschek. 2013. Microbial production of a biofuel (acetone–butanol–ethanol) in a continuous bioreactor: Impact of bleed and simultaneous product removal. *Bioprocess and Biosystems Engineering* 36: 109–16.

Fernandez-Naveira, A., H. N. Abubackar, M. C. Veiga, and C. Kennes. 2016. Efficient butanol–ethanol (B-E) production from carbon monoxide fermentation by *Clostridium carboxidivorans*. *Applied Microbiology and Biotechnology* 100: 3361–70.

Gao, K., S. Boiano, A. Marzocchella, and L. Rehmann. 2014. Cellulosic butanol production from alkali-pretreated switchgrass (*Panicum virgatum*) and phragmites (*Phragmites australis*). *Bioresource Technology* 174: 176–81.

Generoso, W. C., V. Schadeweg, M. Oreb, and E. Boles. 2015. Metabolic engineering of *Saccharomyces cerevisiae* for production of butanol isomers. *Current Opinion in Biotechnology* 33: 1–7.

Gong, F., G. Bao, C. Zhao et al. 2016. Fermentation and genomic analysis of acetone-uncoupled butanol production by *Clostridium tetanomorphum*. *Applied Microbiology and Biotechnology* 100: 1523–29.

Gottumukkala, L. D., B. Parameswaran, S. K. Valappil et al. 2013. Biobutanol production from rice straw by a non acetone producing *Clostridium sporogenes* BE01. *Bioresource Technology* 145: 182–87.

Guan, W., S. Shi, M. Tu, and Y. Y. Lee. 2016. Acetone–butanol–ethanol production from kraft paper mill sludge by simultaneous saccharification and fermentation. *Bioresource Technology* 200: 713–21.

Guo, T., A. Y. He, T. F. Du et al. 2013. Butanol production from hemicellulosic hydrolysate of corn fiber by a *Clostridium beijerinckii* mutant with high inhibitor-tolerance. *Bioresource Technology* 135: 379–85.

Han, S. H., D. H. Cho, Y. H. Kim, and S. J. Shin. 2013a. Biobutanol production from 2-year-old willow biomass by acid hydrolysis and acetone–butanol–ethanol fermentation. *Energy* 61: 13–17.

Han, B., V. Ujor, L. B. Lai, V. Gopalan, and T. C. Ezeji. 2013b. Use of proteomic analysis to elucidate the role of calcium in acetone–butanol–ethanol fermentation by *Clostridium beijerinckii* NCIMB 8052. *Applied and Environmental Microbiology* 79:282–93.

He, Q., and H. Chen. 2013. Improved efficiency of butanol production by absorbed lignocellulose fermentation. *Journal of Bioscience and Bioengineering* 115: 298–302.

Higashide, W., Y. Li, Y. Yang, and J. C. Liao. 2011. Metabolic engineering of *Clostridium cellulolyticum* for production of isobutanol from cellulose. *Applied and Environmental Microbiology* 77: 2727–33.

Huang, H., V. Singh, and N. Qureshi. 2015. Butanol production from food waste: A novel process for producing sustainable energy and reducing environmental pollution. *Biotechnology for Biofuels* 8: 147.

Huesemann, M. H., L. J. Kuo, L. Urquhart, G. A. Gill, and G. Roesijadi. 2012. Acetone–butanol fermentation of marine macroalgae. *Bioresource Technology* 108: 305–9.

Ibrahim, M. F., S. Abd-Aziz, M. E. M. Yusoff, L. Y. Phang, and M.A. Hassan. 2015. Simultaneous enzymatic saccharification and ABE fermentation using pretreated oil palm empty fruit bunch as substrate to produce butanol and hydrogen as biofuel. *Renewable Energy* 77: 447–55.

Jafari, Y., H. Amiri, and K. Karimi. 2016. Acetone pretreatment for improvement of acetone, butanol, and ethanol production from sweet sorghum bagasse. *Applied Energy* 168: 216–25.

Jiang, W., J. Zhao, Z. Wang, and S. T. Yang. 2014. Stable high-titer *n*-butanol production from sucrose and sugarcane juice by *Clostridium acetobutylicum* JB200 in repeated batch fermentations. *Bioresource Technology* 163: 172–79.

John, R. P., G. S. Anisha, K. M. Nampoothiri, and A. Pandey. 2011. Micro and macroalgal biomass: A renewable source for bioethanol. *Bioresource Technology* 102: 186–93.

Kiyoshi, K., M. Furukawa, T. Seyama et al. 2015. Butanol production from alkali-pretreated rice straw by co-culture of *Clostridium thermocellum* and *Clostridium saccharoperbutylacetonicum*. *Bioresource Technology* 186: 325–28.

Krivoruchko, A., C. Serrano-Amatriain, Y. Chen, V. Siewers, and J. Nielsen. 2013. Improving biobutanol production in engineered *Saccharomyces cerevisiae* by manipulation of acetyl-CoA metabolism. *Journal of Industrial Microbiology and Biotechnology* 40: 1051–56.

Kujawska, A., J. Kujawski, M. Bryjak, and W. Kujawski. 2015. ABE fermentation products recovery methods—A review. *Renewable and Sustainable Energy Reviews* 48: 648–61.

Kumar, M., and K. Gayen. 2011. Developments in biobutanol production: New insights. *Applied Energy* 88: 1999–2012.

Lan, E. I., and J. C. Liao. 2012. ATP drives direct photosynthetic production of 1-butanol in cyanobacteria. *Proceedings of the National Academy of Sciences of the United States of America* 109: 6018–23.

Lee, J., Y. S. Jang, S. C. Choi et al. 2012. Metabolic engineering of *Clostridium acetobutylicum* ATCC 824 for isopropanol-butanol-ethanol fermentation. *Applied and Environmental Microbiology* 78: 1416–23.

Li, J., X. Chen, B. Qi et al. 2014. Efficient production of acetone–butanol–ethanol (ABE) from cassava by a fermentation-pervaporation coupled process. *Bioresource Technology* 169: 251–57.

Li, X., Z. Li, J. Zheng, Z. Shi, and L. Li. 2012. Yeast extract promotes phase shift of bio-butanol fermentation by *Clostridium acetobutylicum* ATCC824 using cassava as substrate. *Bioresource Technology* 125: 43–51.

Lin, D.-S., H.-W. Yen, W. C. Kao et al. 2015. Bio-butanol production from glycerol with *Clostridium pasteurianum* CH4: The effects of butyrate addition and in situ butanol removal via membrane distillation. *Biotechnology for Biofuels* 8: 168.

Liu, D., Y. Chen, F.-Y. Ding et al. 2014. Biobutanol production in a *Clostridium acetobutylicum* biofilm reactor integrated with simultaneous product recovery by adsorption. *Biotechnology for Biofuels* 7: 1–13.

Lu, C., J. Zhao, S.-T. Yang, and D. Wei. 2012. Fed-batch fermentation for n-butanol production from cassava bagasse hydrolysate in a fibrous bed bioreactor with continuous gas stripping. *Bioresource Technology* 104: 380–87.

Maiti, S., S. J. Sarma, S. K. Brar et al. 2016. Agro-industrial wastes as feedstock for sustainable bio-production of butanol by *Clostridium beijerinckii*. *Food and Bioproducts Processing* 98: 217–26.

Mariano, A. P., N. Qureshi, R. M. Filho, and T. C. Ezeji. 2012. Assessment of in situ butanol recovery by vacuum during acetone butanol ethanol (ABE) fermentation. *Journal of Chemical Technology and Biotechnology* 87: 334–40.

Mechmech, F., H. Chadjaa, M. Rahni et al. 2015. Improvement of butanol production from a hardwood hemicelluloses hydrolysate by combined sugar concentration and phenols removal. *Bioresource Technology* 192: 287–95.

Moon, H. G., Y. S. Jang, C. Cho et al. 2015. One hundred years of *clostridial* butanol fermentation. *FEMS Microbiology Letters* 363(3): 1–15.

Narayanasamy, L. and T. Murugesan. 2014. Degradation of alizarin yellow R using UV/H_2O_2 advanced oxidation process. *Environmental Progress & Sustainable Energy*, 33(2): 482–489.

Ni, Y., Y. Wang, and Z. Sun. 2012. Butanol production from cane molasses by *Clostridium saccharobutylicum* DSM 13864: Batch and semicontinuous fermentation. *Applied Biochemistry and Biotechnology* 166: 1896–1907.

Pang, Z. W., W. Lu, H. Zhang et al. 2016. Butanol production employing fed-batch fermentation by *Clostridium acetobutylicum* GX01 using alkali-pretreated sugarcane bagasse hydrolysed by enzymes from *Thermoascus aurantiacus* QS 7-2-4. *Bioresource Technology* 212: 82–91.

Qureshi, N., S. Liu, S. Hughes et al. 2016. Cellulosic butanol (ABE) biofuel production from sweet sorghum bagasse (SSB): Impact of hot water pretreatment and solid loadings on fermentation employing *Clostridium beijerinckii* P260. *Bioenergy Research* 9: 1167–79.

Qureshi, N., V. Singh, S. Liu et al. 2014. Process integration for simultaneous saccharification, fermentation, and recovery (SSFR): Production of butanol from corn stover using *Clostridium beijerinckii* P260. *Bioresource Technology* 154: 222–28.

Rajagopalan, G., J. He, and K. L. Yang. 2016. One-pot fermentation of agricultural residues to produce butanol and hydrogen by *Clostridium* strain BOH3. *Renewable Energy* 85: 1127–34.

Ranjan, A., S. Khanna, and V. S. Moholkar. 2013. Feasibility of rice straw as alternate substrate for biobutanol production. *Applied Energy* 103: 32–38.

Sabra, W., C. Groeger, P. N. Sharma, and A. P. Zeng. 2014. Improved *n*-butanol production by a non-acetone producing *Clostridium pasteurianum* DSMZ 525 in mixed substrate fermentation. *Applied Microbiology and Biotechnology* 98: 4267–76.

Saini, M., M. H. Chen, C. C. Chiang, and Y. P. Chao. 2015. Potential production platform of *n*-butanol in *Escherichia coli*. *Metabolic Engineering* 27: 76–82.

Si, T., Y. Luo, H. Xiao, and H. Zhao. 2014. Utilizing an endogenous pathway for 1-butanol production in *Saccharomyces cerevisiae*. *Metabolic Engineering* 22: 60–68.

Sirisantimethakom, L., L. Laopaiboon, P. Sanchanda, J. Chatleudmongkol, and P. Laopaiboon. 2016. Improvement of butanol production from sweet sorghum juice by *Clostridium beijerinckii* using an orthogonal array design. *Industrial Crops and Products* 79: 287–94.

Subhadra, B. and M. Edwards. 2010. An integrated renewable energy park approach for algal biofuel production in United States. *Energy Policy* 38: 4897–4902.

Sun, Z., and S. Liu. 2012. Production of *n*-butanol from concentrated sugar maple hemicellulosic hydrolysate by *Clostridia acetobutylicum* ATCC824. *Biomass and Bioenergy* 39: 39–47.

Swidah, R., H. Wang, P. J. Reid et al. 2015. Butanol production in *S. cerevisiae* via a synthetic ABE pathway is enhanced by specific metabolic engineering and butanol resistance. *Biotechnology for Biofuels* 8: 97.

Taconi, K. A., K. P. Venkataramanan, and D. T. Johnson. 2009. Growth and solvent production by *Clostridium pasteurianum* ATCC 6013 utilizing biodiesel-derived crude glycerol as the sole carbon source. *Environmental Progress & Sustainable Energy* 28: 100–110.

Tenenbaum, D. J. 2008. Food vs. fuel diversion of crops could cause more hunger. *Environmental Health Perspectives* 116: 254–57.

Vassilev, S. V. and C. G. Vassileva. 2016. Composition, properties and challenges of algae biomass for biofuel application: An overview. *Fuel* 181: 1–33.

Wang, Y., W. Guo, C. L. Cheng et al. 2016. Enhancing bio-butanol production from biomass of *Chlorella vulgaris* JSC-6 with sequential alkali pretreatment and acid hydrolysis. *Bioresource Technology* 200: 557–64.

Wang, X., Y. Wang, B. Wang et al. 2013. Biobutanol production from fiber-enhanced DDGS pretreated with electrolyzed water. *Renewable Energy* 52: 16–22.

Weber, C., A. Farwick, F. Benisch et al. 2010. Trends and challenges in the microbial production of lignocellulosic bioalcohol fuels. *Applied Microbiology and Biotechnology* 87: 1303–15.

Wei, N., J. Quarterman, and Y. S. Jin. 2013. Marine macroalgae: An untapped resource for producing fuels and chemicals. *Trends in Biotechnology* 31: 70–77.

Wu, H., X. P. Chen, G. P. Liu et al. 2012. Acetone–butanol–ethanol (ABE) fermentation using *Clostridium acetobutylicum* XY16 and *in situ* recovery by PDMS/ceramic composite membrane. *Bioprocess and Biosystems Engineering* 35: 1057–65.

Wu, Y. D., C. Xue, L. J. Chen, and F. W. Bai. 2013. Effect of zinc supplementation on acetone–butanol–ethanol fermentation by *Clostridium acetobutylicum*. *Journal of Biotechnology* 165: 18–21.

Xu, G. C., J. C. Ding, R. Z. Han, J. J. Dong, and Y. Ni. 2016. Enhancing cellulose accessibility of corn stover by deep eutectic solvent pretreatment for butanol fermentation. *Bioresource Technology* 203: 364–69.

Xue, C., G. C. Du, J. X. Sun et al. 2014. Characterization of gas stripping and its integration with acetone–butanol–ethanol fermentation for high-efficient butanol production and recovery. *Biochemical Engineering Journal* 83: 55–61.

Xue, C., F. Liu, M. Xu et al. 2016a. Butanol production in acetone–butanol–ethanol fermentation with *in situ* product recovery by adsorption. *Bioresource Technology* 219: 158–68.

Xue, C., F. Liu, M. Xu et al. 2016b. A novel *in situ* gas stripping-pervaporation process integrated with acetone–butanol–ethanol fermentation for hyper *n*-butanol production. *Biotechnology and Bioengineering* 113: 120–29.

Xue, C., J. Zhao, C. Lu et al. 2012. High-titer *n*-butanol production by *Clostridium acetobutylicum* JB200 in fed-batch fermentation with intermittent gas stripping. *Biotechnology and Bioengineering* 109: 2746–56.

Xue, C., J. Zhao, F. Liu et al. 2013. Two-stage in situ gas stripping for enhanced butanol fermentation and energy-saving product recovery. *Bioresource Technology* 135: 396–402.

Yadav, S., G. Rawat, P. Tripathi, and R. K. Saxena. 2014. A novel approach for biobutanol production by *Clostridium acetobutylicum* using glycerol: A low cost substrate. *Renewable Energy* 71: 37–42.

Yan, Y., A. Basu, T. Li, and J. He. 2016. Direct conversion of xylan to butanol by a wild-type *Clostridium* species strain G117. *Biotechnology and Bioengineering* 113: 1702–10.

Yang, M., S. Kuittinen, J. Zhang et al. 2015a. Co-fermentation of hemicellulose and starch from barley straw and grain for efficient pentoses utilization in acetone–butanol–ethanol production. *Bioresource Technology* 179: 128–35.

Yang, M., J. Zhang, S. Kuittinen et al. 2015b. Enhanced sugar production from pretreated barley straw by additive xylanase and surfactants in enzymatic hydrolysis for acetone–butanol–ethanol fermentation. *Bioresource Technology* 189: 131–37.

Yen, H. W. and Y. C. Wang. 2013. The enhancement of butanol production by *in situ* butanol removal using biodiesel extraction in the fermentation of ABE (acetone–butanol–ethanol). *Bioresource Technology* 145: 224–28.

Yu, L., M. Xu, I. C. Tang, and S. T. Yang. 2015. Metabolic engineering of *Clostridium tyrobutyricum* for *n*-butanol production through co-utilization of glucose and xylose. *Biotechnology and Bioengineering* 112: 2134–41.

Zhang, Y, T. Hou, B. Li et al. 2014. Acetone–butanol–ethanol production from corn stover pretreated by alkaline twin-screw extrusion pretreatment. *Bioprocess and Biosystems Engineering* 37: 913–21.

Zhang, S., C. Qu, X. Huang et al. 2016. Enhanced isopropanol and *n*-butanol production by supplying exogenous acetic acid via co-culturing two *Clostridium* strains from cassava bagasse hydrolysate. *Journal of Industrial Microbiology and Biotechnology* 43: 915–25.

Zheng, J., Y. Tashiro, Q. Wang, and K. Sonomoto. 2015. Recent advances to improve fermentative butanol production: Genetic engineering and fermentation technology. *Journal of Bioscience and Bioengineering* 119: 1–9.

section three

Gaseous biofuels

chapter nine

Biohydrogen

A zero-carbon fuel for the future

Sinu Kumari and Debabrata Das

Contents

9.1 Introduction

Hydrogen has emerged as a potential fuel for the future. It is a colorless, odorless, tasteless, and flammable gas. Among the known fuels, it has the highest energy density (142 KJ/g) (Das and Veziroglu, 2001). During combustion, hydrogen produces water as the only by-product. Thus, it is an efficient energy carrier and a clean and environmentally benign fuel. A variety of microbial processes can be used to produce hydrogen. Four major types of hydrogen production processes are water-splitting photosynthesis, photofermentation by photosynthetic bacteria, dark fermentation, and microbial electrolysis cells (MECs). The hydrogen production through the biological process is called biohydrogen. It can be produced from different organic wastes by the fermentation process. Biological hydrogen production processes are not only environmentally friendly but also inexhaustible (Benemann, 1997).

Hans Gaffron, a scientist at the University of Chicago, observed that green algae (*Chlamydomonas reinhardtii*) have the capability of producing hydrogen instead of oxygen, but he was unable to find the reason for this (Kirakosyan and Kaufman, 2009). Further, in the 1990s Anastasios Melis from The University of California, Berkeley also observed the production of hydrogen from algae in sulfur-deprived media. He found that hydrogenase enzyme is responsible for hydrogen production, which becomes inactive in the presence of oxygen (Tanawade et al., 2011). Initially, the algae were used for the production of hydrogen. Hydrogen gas, which is produced by biological processes, becomes very interesting and promising because the biological processes can be operated at ambient temperature and atmospheric pressure with minimal energy consumption, and therefore become more environmentally friendly (Azwar et al., 2014).

Depending on the requirement of light energy, the biological hydrogen production processes can be categorized into light-dependent and light-independent hydrogen production process. Further, the light-dependent process can be divided into direct and indirect biophotolysis mediated by cyanobacteria and green algae, respectively, and photofermentation mediated by photofermentative bacteria. Light-independent processes can be divided into dark fermentation mediated mostly by fermentative bacteria and microbial electrolysis cell mediated by electrogenic bacteria.

9.2 Different processes of biohydrogen production

9.2.1 Biophotolysis

9.2.1.1 Direct biophotolysis

In the direct biophotolysis process, the hydrogen production occurs via photosynthetic reaction. In this process, solar energy converts water to oxygen and hydrogen (Equation 9.1). Moreover, hydrogen production is possible only under special condition because the Fe-hydrogenase activity is extremely sensitive to oxygen.

$$2H_2O + \text{light energy} \rightarrow 2H_2 + O_2 \tag{9.1}$$

In the direct biophotolysis process, two photosystems are present: Photosystem I (PSI) and Photosystem II (PSII). PSI is responsible for either CO_2 reduction using photons from water or hydrogen production in the presence of hydrogenase enzyme. PSII is responsible for water splitting and oxygen evolution. Due to the absence of hydrogenase enzyme in green plants, only CO_2 reduction takes place, while green and blue-green algae contain hydrogenase, which leads to hydrogen production. In this process, PSII absorbs light energy and generates electrons, and then these electrons transferred to ferredoxin (Fd) via PSI and generate hydrogen. The process is shown in Figure 9.1a.

The biophotolysis process is very sensitive to the presence of oxygen. A very small amount of oxygen can inhibit the hydrogen production. The required maximum partial pressure is equivalent to 1 atm of O_2. This is a thousandfold higher than the maximum tolerance concentration. Hence, the sensitivity of O_2 for hydrogenase enzyme is the key problem (Hallenbeck, 2002). A maximum hydrogen production rate of 0.07 mmol/L h is reported (Melis et al., 2000; Kosourov et al., 2002). It was observed that green algae were able to produce hydrogen in the anaerobic condition in the dark. When these algae were shifted to light, after a certain period of time, hydrogen production decreased (due to photosynthetic evolution of CO_2 and O_2) (Das and Veziroglu, 2001).

9.2.1.2 Indirect biophotolysis

In the case of indirect biophotolysis, the sensitivity of hydrogen productions is avoided by use of temporal and spatial separation of oxygen evolution and hydrogen production. That means that the hydrogen

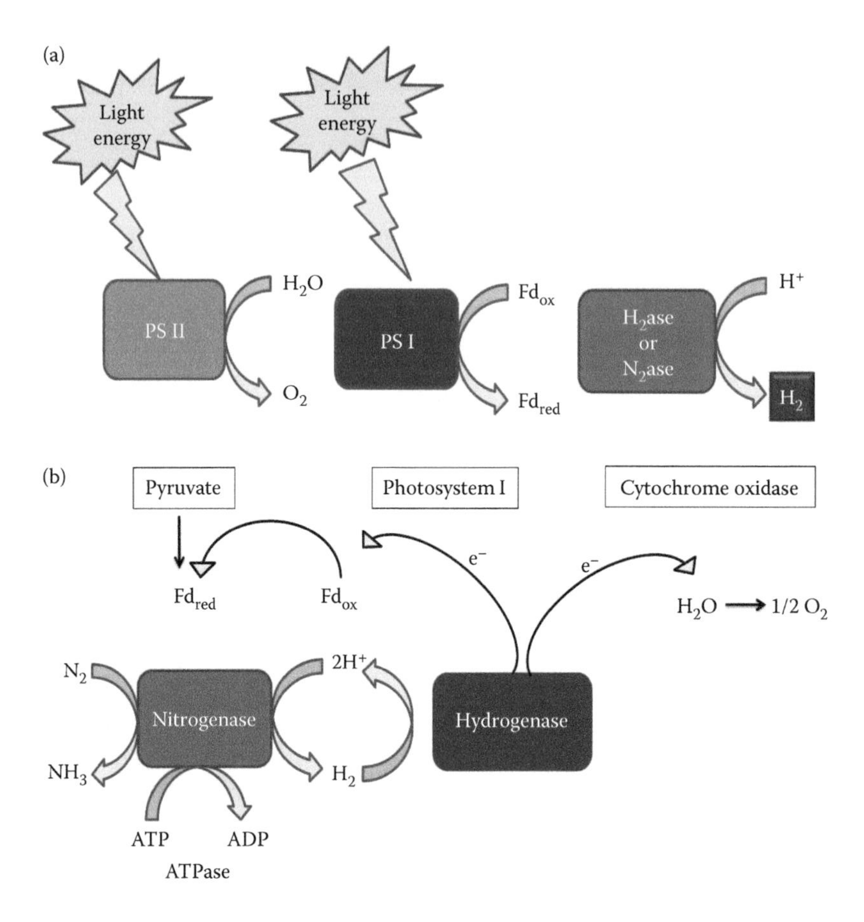

Figure 9.1 (a) One-stage direct biophotolysis for hydrogen production; (b) Relationship between nitrogenase-catalyzed hydrogen production and hydrogenase-catalyzed hydrogen uptake by *Cyanobacteria* sp.

production occurs in two stages: first the accumulation of carbohydrates by photosynthesis, and then utilization of stored carbohydrates for hydrogen production through dark fermentation. Cyanobacteria and green algae are gram-positive bacteria that are mainly present in the marine environment. Cyanobacteria and green algae have the capability to fix nitrogen and generate hydrogen by splitting water. The relationship between nitrogenase-catalyzed hydrogen production and hydrogenase-catalyzed hydrogen uptake by Cyanobacteria is presented in Figure 9.1b.

The hydrogen production through indirect biophotolysis in Cyanobacteria occurs through the following reactions (Equations 9.2 and 9.3) (Levin, 2004):

$$12H_2O + 6CO_2 \rightarrow C_6H_{12}O_6 + 6O_2 \quad (9.2)$$

$$C_6H_{12}O_6 + 12H_2O \rightarrow 12H_2 + 6CO_2 \quad (9.3)$$

9.2.2 *Photofermentation*

Solar energy can be converted into biochemical energy with the help of a photosynthetic apparatus of hydrogenesis to molecular H_2 (Jordan et al., 2001). This is one of the promising approaches for sustainable energy generation. A photosynthetic mechanism of H_2 production has been known since the early 1940s (Gaffron, 1942). It is carried out by two principal classes of photosynthetic bacteria: the purple nonsulfur bacteria (PNS) and the green sulfur bacteria (GS).

In the case of PNS, the PSII-type reaction center does not reduce ferredoxin, but can generate adenosine triphosphate (ATP) through cyclic electron flow. The electrons required for hydrogen production through nitrogenase come from carbohydrate-rich substances. No oxygen evolution was observed in this process (Dasgupta et al., 2009). The uptake hydrogenase activity influences the total amount of hydrogen produced. Photobiological hydrogen can be produced from various types of organic wastes as well as the spent medium of the dark fermentative hydrogen production process. The spent medium mainly contains acetate, butyrate, propionate, and ethanol. These reduced end metabolites can be further utilized by the photosynthetic bacteria using sunlight as an energy source. This integration process improves the overall hydrogen yield (Das and Veziroglu, 2001). The photosynthetic organisms are mostly photoautotrophs and obligate anaerobes. The overall reactions of dark fermentation and photofermentation are presented in Equations 9.4 and 9.5

$$\text{Dark fermentation: } C_6H_{12}O_6 + 2H_2O \rightarrow 2CH_3COOH + 2CO_2 + 4H_2 \quad (9.4)$$

$$\text{Photofermentation: } 2CH_3COOH + 4H_2O \rightarrow 4CO_2 + 8H_2 \quad (9.5)$$

In the case of GS bacteria, a PSI-type reaction center present. The carbohydrate-rich substances are oxidized and then transfer the electrons to reduce ferredoxin via FeS proteins. The reduced ferredoxin donates electrons for carbon fixation (dark reaction) as well as for H_2 production. The general schematic representation of photofermentative hydrogen production is presented in Figure 9.2.

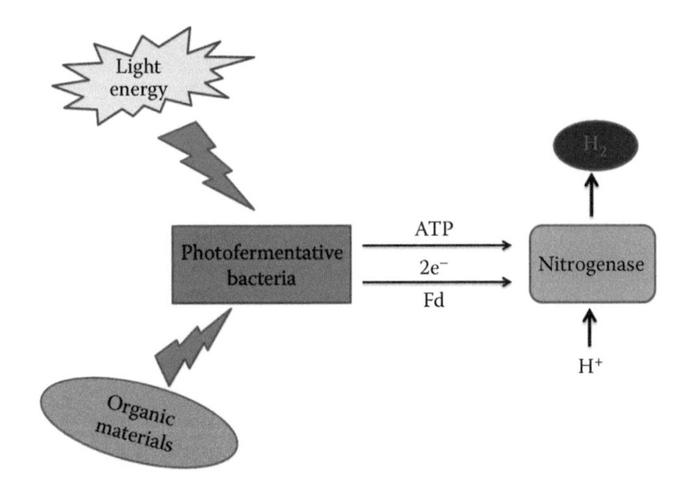

Figure 9.2 General representation of photofermentative hydrogen production.

Photosynthetic bacteria can also produce hydrogen using carbon mono oxide (CO) by a microbial shift reaction (Das and Veziroglu, 2001). The equation is given as follows:

$$CO + H_2O \rightarrow CO_2 + H_2 \tag{9.6}$$

9.2.3 *Dark fermentation*

Dark fermentation is the reduction of carbohydrate rich substrate into molecular hydrogen by fermentation process. It is an anaerobic process. Diverse group of bacteria has capability to produce hydrogen through a series of biochemical reaction processes. In the anaerobic condition, TCA cycle is blocked. Therefore, the excess cellular reductants disposed off by the formation of reduced metabolites i.e. acetate, butyrate, ethanol etc. Moreover, hydrogen is also a reduced metabolic end product produced to maintain the cellular redox potential.

In the dark fermentation process, organic rich substances converted to glucose by hydrolysis, further this glucose converted to pyruvate by glycolysis process. This pyruvate can either go for hydrogen production along with volatile fatty acids (VFAs) production or can be enter into the TCA cycle for energy generation.

The *Clostridium* sp. and thermophilic bacteria contain pyruvate-ferredoxin oxidoreductase (PFOR) enzyme which oxidizes pyruvate to acetyl coenzyme A (acetyl-CoA) (Furdui and Ragsdale, 2000). Further, acetyl-CoA is converted to acetyl phosphate. This leads to generation of ATP and acetate. In the process of oxidation of pyruvate to acetyl-CoA, the ferredoxin (Fe) gets reduced. Further, reduced ferredoxin is oxidized

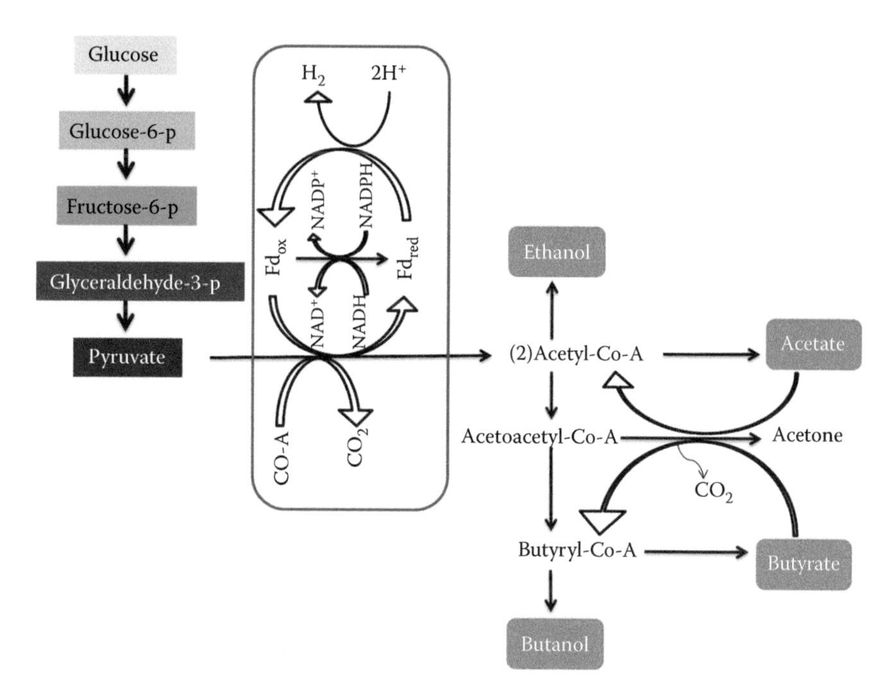

Figure 9.3 General representation of dark fermentative hydrogen production.

by [FeFe] hydrogenase. This leads to formation of molecular hydrogen (Figure 9.3). The overall reaction is shown in Equations 9.7 and 9.8.

$$\text{Pyruvate} + \text{CoA} + 2\text{Fd}_{(\text{ox})} \rightarrow \text{Acetyl-CoA} + 2\text{Fd}_{(\text{red})} + \text{CO}_2 \qquad (9.7)$$

$$2\text{H}^+ + \text{Fd}_{(\text{red})} \rightarrow \text{H}_2 + \text{Fd}_{(\text{ox})} \qquad (9.8)$$

Whereas, in case of facultative anaerobic bacteria (*E. coli*, *Enterobacter* sp. etc.), pyruvate format lyase (PFL) enzyme present which catalyzes pyruvate to acetyl-CoA and formate (Equation 9.9).

$$\text{Pyruvate} + \text{CoA} \rightarrow \text{Acetyl-CoA} + \text{HCOOH} \qquad (9.9)$$

Further, formate is converted to carbon dioxide and hydrogen. This reaction is catalyzed by formate hydrogen lyase (FHL) enzyme (Equation 9.10)

$$\text{HCOOH} \rightarrow \text{CO}_2 + \text{H}_2 \qquad (9.10)$$

The maximum four moles of hydrogen per mol of glucose can be obtained when pyruvate is oxidized to acetate as the sole end metabolite whereas two moles of hydrogen per mol of glucose can be obtained when pyruvate is oxidized to butyrate as the sole end metabolite (Hawkes et al., 2002). The overall reaction of hydrogen production from glucose is shown in Equations 9.11 and 9.12. For mixed acids fermentation pathway, concentration of acetate to butyrate ratio plays an important role in hydrogen production process (Khanna et al., 2011).

$$C_6H_{12}O_6 + 2H_2O \rightarrow 2CH_3COOH + 2CO_2 + 4H_2 \quad (9.11)$$

$$C_6H_{12}O_6 \rightarrow CH_3CH_2CH_2COOH + 2CO_2 + 2H_2 \quad (9.12)$$

Thus, yield of dark fermentative hydrogen production depends on oxidation of pyruvate.

9.2.4 *Microbial electrolysis cell*

Microbial electrolysis cell is a recent technique of producing hydrogen from a variety of substrates (Croese et al., 2011; Kiely et al., 2011). It is basically a modified microbial fuel cell (MFC), which have been studied for decades, with much recent research and patent activity on these devices (Maza-Marquez et al., 2016). The anodic reactions are the same in both devices. Electrogenic microbes have the capability to degrade various types of organic substrates to produce electrons and protons at anodes or cathodes by a series of biochemical reactions. At an anode, the oxidation of acetate takes place under standard conditions (25°C, 1 atm pressure, and pH 7). Further, hydrogen production takes place at a cathode by applying small external voltage (Logan et al., 2008). Many electrogenic bacteria, that is, *Pseudomonas, Geobacter,* and *Shewanella,* have the capability to transfer electrons at an anode. The overall reaction of hydrogen production can be represented by the following Equations 9.13, 9.14, and 9.15:

$$C_6H_{12}O_6 + 2H_2O \rightarrow 2CH_3COOH + 2CO_2 + 4H_2 \quad (9.13)$$

$$\text{Anode: } CH_3COOH + 2H_2O \rightarrow 2CO_2 + 8e^- + 8H^+ \quad (9.14)$$

$$\text{Cathode: } 8e^- + 8H^+ \rightarrow 4H_2 \quad (9.15)$$

The minimum theoretical voltage required to produce hydrogen is −410 mV, but the voltage produced by oxidation of organic matters is approximately −300 mV. Hence, external voltage higher than −110 mV is required to produce hydrogen at a cathode (Liu et al., 2005).

The comparison of different biohydrogen production processes and energy obtained by different biohydrogen production processes are presented in Tables 9.1 and 9.2.

Table 9.1 Advantages and disadvantages of different biohydrogen production processes

Biohydrogen production process	Advantages	Disadvantages
Biophotolysis by Cyanobacteria	Has ability to produce hydrogen from water. Nitrogenase enzyme is mainly involved in hydrogen production. Can fixed nitrogen from the atmosphere.	Presence of uptake hydrogenase enzyme. Presence of O_2 inhibits the nitrogenase activity.
Biophotolysis by green algae	Can produce hydrogen from water. Solar conversion efficiency is tenfold higher than green plants.	Requires light as a source of energy for hydrogen production. Presence of little O_2 can inhibit the hydrogen production.
Photofermentation	Can produce hydrogen using different organic wastes by utilizing broad spectrum of light energy.	Requires light energy for hydrogen production. Presence of CO_2 in the gas. Fermentation effluent discharge is the major concern. Needs some treatment step before discharge, otherwise can create water pollution.
Dark fermentation	Can produce hydrogen using different organic wastes Does not require any light energy for hydrogen production. Effluent can be further used for gaseous energy production.	Fermentative effluent discharge can cause water pollution.

9.3 Microbiology

9.3.1 Microorganisms for biophotolysis

9.3.1.1 Cyanobacteria

Cyanobacteria are mainly unicellular, photoautotrophic, filamentous prokaryotes. Cyanobacteria are also known as blue-green algae. The photosynthetic mechanism of cyanobacteria is similar to green plants. It produces ATP and oxygen through photosynthesis by using CO_2 as the carbon source. It has the ability to produce hydrogen. The

Table 9.2 Energy obtained by different hydrogen production processes in terms of unit cost

Hydrogen production process	Conversion efficiency (%)	Unit cost of energy content of fuel (US$/MBTU)	References
Gasoline	–	6	Bockris (1981)
Photochemical	1	21	Bockris (1981)
Thermal decomposition	–	13	Bockris (1981)
Advanced electrolysis	–	10	Bockris (1981)
Hydrogen from coal, biomass	–	4	Bockris (1981)
Photobiological hydrogen production	~10	~10	Benemann (1997)
Fermentative hydrogen production	~10	~40	Tashino and Ishiwata (1994)

hydrogen production and oxygen evolution are separated by compartmentalization (heterocyst) or by temporal separation. Heterocyst is mostly present in *Anabaena* and *Nostoc* sp. and is mainly responsible for maintaining anaerobic conditions. Nitrogenase enzyme is present inside the heterocyst, which is the key enzyme for hydrogen production. In the case of nonheterocyst strains, hydrogen production occurs mostly by bidirectional hydrogenase. The major drawback of cyanobacteria-mediated biophotolysis is the presence of the uptake hydrogenase, which oxidizes the molecular hydrogen. Cyanobacteria have the capability to produce hydrogen both by direct biophotolysis (electrons comes from splitting of water) and indirect biophotolysis (electrons comes from reserve carbohydrate). A few studies have been performed using various cyanobacteria for hydrogen production (Sveshnikov et al., 1997; Tsygankov et al., 1998; Happe et al., 2000; Serebryakova et al., 2000).

9.3.1.2 Green algae

Green algae are mostly aerobic eukaryotes. They carry out photosynthesis in the presence of oxygen like higher plants. And in the anoxia condition, they can evaluate hydrogen. Gaffron and Rubin (1942) first observed that green alga, that is, *Scenedesmus obliquus,* has the ability to produce hydrogen in the presence of light under anaerobic conditions. It was observed that the hydrogenase enzyme was responsible for hydrogen production in green algae. In other green alga, that is, *C. reinhardtii,* sustainable hydrogen production was observed under low-sulfur conditions. In sulfur-deprived media, the net oxygen evolution decreases due to inactivation of Photosystem II.

9.3.2 *Microorganisms for photofermentation*

Photosynthetic bacteria can be broadly classified into two distinct groups: anoxygenic phototrophic bacteria and oxygenic cyanobacteria. The anoxygenic phototrophic bacteria are well known under their previous name purple bacteria, which was from their red, brown, and pinkish colors. These are further subdivided into the following categories: (1) PNS bacteria (Thiorhodaceae or Rhodospirillaceae), (2) purple sulfur bacteria (Thiorhodaceae or Chromatiaceae), (3) green sulfur bacteria (Chlorobiaceae), and (4) multicellular filamentous green bacteria.

9.3.2.1 *Purple nonsulfur bacteria*

PNS bacteria are photoautotrophic bacteria. The hydrogen production takes place in the absence of oxygen because in the presence of oxygen, it accepts electrons instead of protons. The nitrogenase becomes inactive in the presence of oxygen. The carbon and electrons come from external carbon sources. The hydrogen evolution takes place during the nitrogen starvation condition, which is catalyzed by nitrogenase. The hydrogen yield depends on the uptake nitrogenase activity. Das and Veziroglu (2001) reported that PNS bacteria are the most suitable for photofermentative hydrogen production. They can use organic or industrial waste as a carbon source for hydrogen production, which is economically feasible (Das and Veziroglu, 2001). The most commonly used PNS bacteria is *Rhodobacter sphaeroides*. It is a gram-negative, rod-shaped photoheterotrophic bacterium belonging to Proteobacteria, and is mostly present in sludge, mud, soil, and organic-rich water habitats.

9.3.2.2 *Purple sulfur bacteria*

Purple sulfur bacteria are obligate anaerobic photoautotrophic bacteria mainly present in stagnant water. Most of the purple sulfur bacteria are diazotrophs. They convert nitrogen to ammonia. During nitrogen starvation, hydrogen evolution takes place. Examples of common purple sulfur bacteria are *Allochromatium vinosum* and *Thiocapsa roseopersicina*.

9.3.2.3 *Green bacteria*

Green bacteria are photoautotrophic, obligate anaerobic bacteria. In the case of green bacteria, FeS reduces ferredoxin and transfers electrons for hydrogen production. Hydrogen production takes place in nitrogen-limited conditions catalyzed by nitrogenase similar to purple bacteria. In green bacteria, ribulose-1,5-biphosphate carboxylase/oxygenase (RuBisCo) is not present. Therefore, the CO_2 fixation takes place by reductive tricarboxylic acid (TCA) cycle. Hence, there is no competitive inhibition for hydrogen production by CO_2 fixation. Common examples of green bacteria are *Chlorobium phaeovibrioides* 1930 and *Desulfuromonas acetoxidans* 5071.

9.3.3 *Microorganisms for dark fermentation*

Hydrogen can be produced from a variety of microorganisms. These microorganisms are mostly present in the environment and can produce hydrogen in anaerobic conditions. The microorganisms can be classified on the basis of oxygen sensitivity and growth temperature. The microorganisms that require a strict anaerobic environment are called obligate anaerobic microorganisms. Those organisms that can grow in either the presence or absence of oxygen are called facultative anaerobes. Further microorganism can be classified on the basis of growth temperature. The organisms that require ambient temperature for growth and production are called mesophiles, while the organisms that require higher temperature for their growth are called thermophiles. The selection of microorganism depends on environmental conditions and substrate used for the fermentation process.

9.3.3.1 *Mesophiles*

The microorganisms that can grow at ambient temperatures are mesophiles. Mesophilic bacteria can be facultative or obligate. Facultative bacteria can produce ATP in the presence of oxygen by means of respiration. On the other hand, in the absence of oxygen, it generates ATP by the anaerobic fermentation process. *Enterobacteriaceae* sp. is the common example of facultative anaerobes. Facultative organisms have several advantages over obligates anaerobes. First, they are easy to handle during the anaerobic fermentation process. Second, they can sustain the high concentration of hydrogen. The enzyme involved in hydrogen production is mostly Fe-Fe hydrogenase or format hydrogen lyase (FHL). The very well-known mesophilic bacteria are *Enterobacter cloacae* IIT-BT 08 (presently known as *Klebsiella pneumoniae* IIT-BT 08) isolated by Kumar and Das (2000) from leaf extract, *En. aerogenes* E.82005 isolated from a leaf of *Mirabilis jalapa* (Tanisho and Ishiwata, 1994), *Escherichia coli*, *Citrobacter* sp., and so on. Some mesophilic organisms, such as *Bifidobacterium* sp., have the ability to degrade complex molecules like starch to small molecules. These small molecules can further utilized by *Enterobacter* or *Clostridium* sp. to produce hydrogen. The maximum hydrogen yield of 2.2 mol H_2/mol glucose can be obtained using *En. cloacae* IIT-BT 08 (Kumar and Das, 2000). Mishra et al. (2015) studied the effect of pure culture and coculture of *K. pneumoniae* IIT-BT 08 and *Citrobacter freundii* IIT-BT L139 on hydrogen production using distillery effluent. Improved hydrogen production was observed using coculture (*K. pneumoniae* IIT-BT 08: *Ci. freundii* IIT-BT L139, 1:1) as compared to pure culture (Mishra et al., 2015).

9.3.3.2 *Thermophiles*

Thermophiles are mostly anaerobes. They are mostly present in the heated regions of Earth. Because they are obligate anaerobes, they require reducing agents such as L-cysteine hydrochloride to remove any trace amount of oxygen present in the culture media (van de Werken et al., 2008). Thermophiles have the ability to utilize a broad range of substrate such as sugar-based substrate, lignocellulosic-based substrate, and pectin-containing biomass (O-Thong et al., 2008). Typical examples of these groups of organisms are *Thermoanaerobacter, Thermotoga,* and *Caldicellulosiruptor,* of which *Caldicellulosiruptor* contains cellulolytic enzyme that can directly degrade cellulose at high temperatures. They can produce hydrogen near stoichiometric yield that is 4 mol hydrogen per 1 mol of glucose. The major end metabolites formed after fermentation are acetate, butyrate, lactate, and ethanol (van Niel et al., 2003).

9.3.3.3 Mixed culture

Complex substrates are mainly composed of various types of carbohydrates. Therefore, a cocktail of microorganisms is required to degrade the complex biomass for fermentative hydrogen production. A mixed consortium can enhance the degradation of the complex substrate by trading metabolites or by exchanging dedicated molecular signals. Thus, use of a consortium can enhance substrate utilization efficiency (Guwy et al., 1997). Various studies have been performed using mixed microbial consortia with complex substrate, which does not require sterile conditions. On the other hand, pure cultures require sterile conditions for fermentative hydrogen production, which is an energy-intensive process.

Mixed microbial consortia can be developed from sewage sludge, soybean meal, anaerobic digester plant, municipal sewage, organic waste, or kitchen wastewater. The major drawback of developing hydrogen-producing mixed consortia is the presence of methanogens. Therefore, some pretreatment steps (acid, alkali, heat shock, freeze-drying, or chemicals pretreatment) are required to suppress the methanogens and to enrich the hydrogen-producing bacteria (Giordano et al., 2014; Mohammadi et al., 2011; Shaw et al., 2008). Mostly, *Clostridium* sp. is present in the mixed culture. Hydrogen production at higher temperatures is favorable for reaction kinetics because it prevents contamination from hydrogen-consuming bacteria (Zhang et al., 2003).

The potentiality of various microorganisms toward biological hydrogen production is shown in Table 9.3.

9.4 Factors influencing dark fermentative biohydrogen production

9.4.1 Medium pH

The pH is responsible for the stability of acidogenic bacteria inside the fermenter. It regulates the metabolic pathway as well as the dominance of

Table 9.3 Potentiality of different microorganisms for dark fermentative biohydrogen production using various carbon sources

Substrate	Inoculum	Mode of operation	Maximum hydrogen yield	References
Nitrogen-starved medium	*Anabaena cylindrical*	Continuous	31.3 mmol/g dry cell	Weissmen and Bonemann (1977)
Allen and Arnon medium	*A. variabilis*	Continuous	14.4 mmol/g dry cell	Markov et al. (1997)
Fermented cow dung	*Rhodopseudomonas capsulata*	Continuous	7.2 mmol/g dry cell	Fascetti et al. (1998)
Lactate with other nitrogen source	*R. capsulata*	Continuous	0.13 mol/g dry cell	Hillmer and Gest (1977)
Vegetable starch	*Rhodopseudomonas* sp.	Continuous	31.2 mmol/g dry cell	Vrati and Verma (1983)
Sugarcane juice	*Rhodopseudomonas* sp.	Continuous	48 mmol/g dry cell	Singh et al. (1994)
Lactic acid fermentable waste	*Rb. sphaeroides*	Continuous	0.14 mol/g dry cell	Sasikala et al. (1991)
Distillery wastewaster	*Rb. sphaeroides*	Continuous	11 mmol/g dry cell	Sasikala et al. (1992)
Lactate liquor municipal solid waste	*Rb. sphaeroides*	Continuous	0.11 mol/g dry cell	Fascetti et al. (1998)
L-arabinose	*En. cloacae* IIT-BT 08	Batch	1.50 mol/mol arabinose	Kumar and Das (2000)
Cellobiose	*En. cloacae* IIT-BT 08	Batch	5.40 mol/mol cellobiose	Kumar and Das (2000)
Glucose	*Thermoanaerobacterium thermosaccharolyticum*	Batch	2.40 mol/mol glucose	Ueno et al. (2001)
Glycerol	*En. aerogenes* HU-101	Batch	0.60 mol/mol glucose	Nakashimada et al. (2002)
Glucose	*Clostridium acetobutylicum*	Batch	2.00 mol/mol glucose	Chin et al. (2003)
Lactose	*Cl. thermolaticum*	Continuous	3.00 mol/mol lactose	Collect (2004)

(*Continued*)

Table 9.3 (Continued) Potentiality of different microorganisms for dark fermentative biohydrogen production using various carbon sources

Substrate	Inoculum	Mode of operation	Maximum hydrogen yield	References
Glucose	Anaerobic sludge	Batch	1.10 mol/mol-hexose	Fan (2004)
Glucose	*Cl. acetobutylicum* ATCC 824	Continuous	1.08 mol/mol glucose	Zhang et al. (2006)
Glucose	*E. coli*	Continuous	2.00 mol/mol glucose	Bisaillon et al. (2006)
Sucrose	Wastewater treatment plant sludge	Batch	3.06 mol/mol glucose	Zhu and Beland (2006)
Sucrose	Anaerobic sludge	Batch	2.00 mol/mol glucose	Mu et al. (2007)
Sucrose	Cow dung compost	Batch	1.12 mol/mol glucose	Hu and Chen (2007)
Starch	*En. aerogenes*	Batch	1.09 mol/mol glycerol	Jo et al. (2008)
Glucose	*Ci. amalonaticus* Y19	Batch	2.49 mol/mol glucose	Oh et al. (2008)
Glucose	Upflow anaerobic sludge blanket (UASB) digester sludge	Batch	1.02 mol/mol glucose	Wang and Wan (2008)
Glucose	*Citrobacter* CDN-1	Batch	2.10 mol/mol glucose	Pandey et al. (2009)
Glucose	Digested sludge	Batch	1.78 mol/mol glucose	Xiao and Liu (2009)
Distillery effluent	*En. cloacae* IIT-BT 08	Batch	7.38 mol/kg$COD_{reduced}$	Mishra and Das (2014)
Glucose	Anaerobic sludge	Batch	1.86 mol/mol glucose	Sivagurunathan et al. (2014)
Distillery effluent	Coculture of *K. pneumoniae* IIT-BT 08 and *Ci. freundii* IIT-BT L139	Batch	8.76 mol/kg$COD_{reduced}$	Mishra et al. (2015)
Distillery effluent	Acidogenic mixed consortia	Batch	9.17 mol/kg$COD_{reduced}$	Mishra et al. (2015)
Glucose	Cow dung	Batch	2.60 mol/mol glucose	Kumari and Das (2016)
Glucose	*Cl. saccharoperbutylacetonicum*	Batch	2.48 mol/mol glucose	Mitra et al. (2017)

the microorganism inside the reactor. In the case of mixed culture, pH is one of the most important factors to determine the population of a specific group of bacteria inside the reactor (Antonopoulou et al., 2008). Khanal et al. (2003) reported that hydrogen production occurs in the exponential phase. When the microorganisms enter a stationary phase (pH below 4.5), the reaction shifts from acidogenesis to the solventogenic phase (Khanal, 2003). Grupe and Gottschalk (1992) performed the detailed analysis of the change in acidogenic and solventogenic phase due to pH change. They observed that intracellular acids concentration below 440 mM induce solventogenesis. This acid concentration can be maintained by maintaining the medium pH. Khanna et al. (2011) studied the effect of controlled pH on hydrogen production from glucose using *En. cloacae* IIT-BT 08. Significant improvement in hydrogen production was observed as compared to uncontrolled medium pH (Khanna et al., 2011).

The drop in the extracellular medium pH is due to the accumulation of VFAs (acetic acid, propionic acids, and butyric acids) in the fermentative medium. VFAs are concomitantly produced during hydrogen production. The concentration of each VFA depends on medium pH. At higher pH, more propionate and ethanol are produced, whereas low pH favors butyrate production.

The detailed study on the mechanism of medium pH changes was performed by Ginkel et al. (2001). He observed that accumulation of nonpolar undissociated acids leads to an increase in the ionic strength of the solution. The undissociated acids can easily penetrate the cells and released protons. This caused changes in the intercellular pH environment. This leads to denaturation of the intracellular enzyme system including the hydrogenase enzyme. Therefore, the hydrogen production decreased (Ginkel et al., 2001). In a recent study, Kumari and Das (2016) reported that by increasing the medium pH from 5 to 7, hydrogen production increased. The maximum hydrogen yield of 2.4 mol H_2/1 mol glucose was obtained at pH 6.5 using mixed microbial consortia (Figure 9.4a). Further increases in pH decreased the hydrogen production.

9.4.2 *Temperature*

The temperature of the medium is an important parameter for biohydrogen production. It is responsible for the rate of all cellular reaction. It affects the growth and metabolic pattern of the microorganism. The optimum fermentation temperature depends on the choice of microorganism used. On the basis of reaction temperature range, microorganisms can be categorized into four groups: psychrophiles (0–20°C), mesophiles (20–42°C), thermophiles (42–75°C), and extremophiles (75–90°C). Any change in temperature above or below the optimum value may affect the

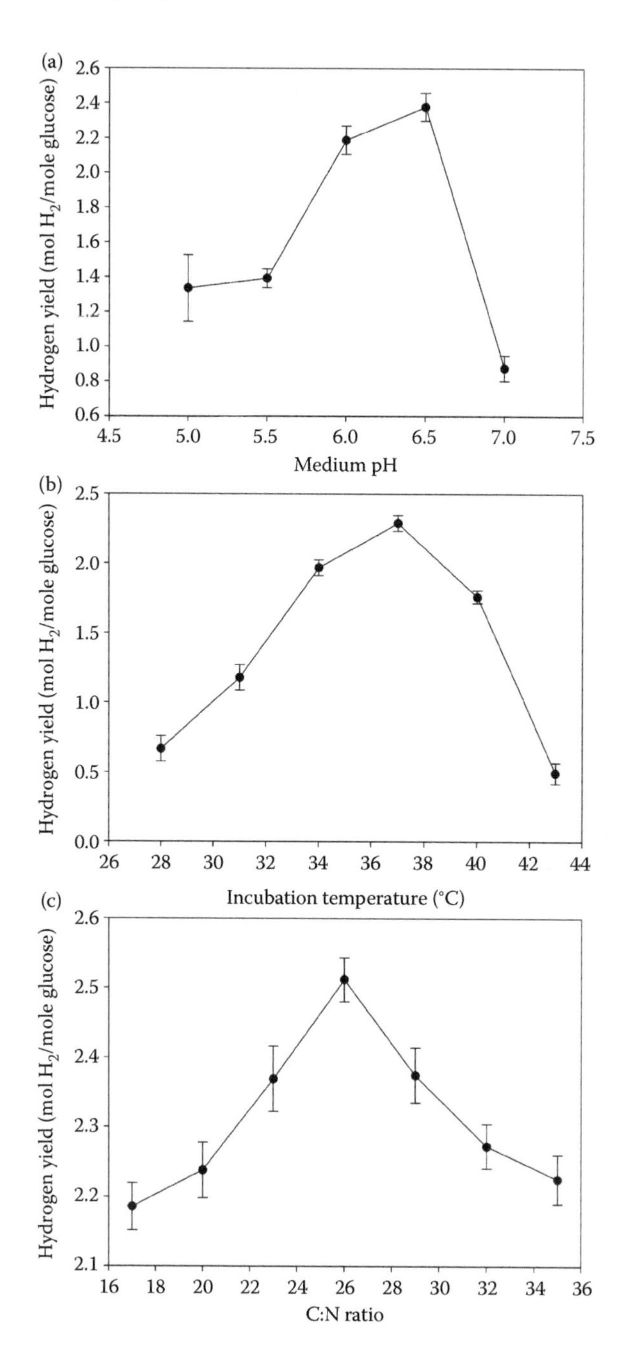

Figure 9.4 Effect of (a) medium pH, (b) incubation temperature, and (c) C:N ratio on hydrogen yield. (From Kumari, S., and Das, D. 2016. *Bioresource Technology*, 218, 1090–1097.)

growth of the microorganism. This optimum value can be determined by plotting the graph between numbers of cells generated per hour against temperature. At optimum temperature, the activities of bacteria are better. Nath et al. (2006) studied the effect of temperature variation on hydrogen production using *En. cloacae* IIT-BT 08. The temperature was varied from 25 to 40°C. The maximum hydrogen production was observed at 37°C. Kumari and Das (2016) also studied the effect of incubation temperature (28–43°C) on hydrogen yield. They observed the maximum hydrogen yield of 2.3 mol H_2/1 mole glucose at 37°C (Figure 9.4b). Hydrogen production decreased at higher temperature. This may be due to changes in the membrane structure. Therefore, a controlled bioreactor is always preferred to maintain the desired temperature for improved hydrogen production (Nath et al., 2006).

9.4.3 Nutrients

Supplementation of nitrogen, phosphate, and other inorganic trace minerals is essential to maximize the hydrogen yield using carbohydrates as a substrate for hydrogen production. Nitrogen is an essential component of amino acids and is required for optimal growth of the microorganism. Studies have shown that with respect to hydrogen production, organic nitrogen appears to be more suitable than inorganic nitrogen. The hydrogen yield of 2.4 mol H_2/1 mol glucose was obtained from starch by supplementing it with 0.1% (w/v) polypeptone. However, no significant improvement in hydrogen production was obtained when urea or other inorganic salts were used as a nitrogen source (Yokoi et al., 2001). However, although supplementation of the nitrogen source increases the overall hydrogen yield, it also adds to the overall cost of production. Thus, research is presently underway to find less expensive alternatives that can substitute as the nitrogen source in hydrogen production media. In this regard, corn steep liquor, which is a by-product of the corn starch manufacturing process, appears promising and like the best alternative supplementation for peptone as the nitrogen source (Yokoi et al., 2002). Additionally, the C:N ratio also plays a significant role in stabilizing the dark fermentation process and affecting the hydrogen productivity and specific hydrogen production rate (Lin, 2004). Kumari and Das (2016) studied the effect of C:N on hydrogen yield using mixed microbial consortia. The C:N ratio of 26 was found to be suitable for fermentative hydrogen production (Figure 9.4c) (Kumari and Das, 2016). Optimal phosphate concentrations are also desired to enhance the overall performance of the process. Phosphate acts as an important inorganic nutrient for optimal hydrogen production. Excess amounts of phosphate may lead to the production of more volatile fatty acids. However, higher VFA production is not desirable because it diverts the cellular reductants away from hydrogen production.

9.4.4 *Metal ions*

For any fermentative process, supplementation of suitable metal ions in media is essential. These metal ions act as enzyme cofactors and are also involved in the cellular transport processes. Hydrogenase, the key enzyme involved in hydrogen production, contains a bimetallic Fe-Fe center. It is also surrounded by FeS protein clusters (Nicolet et al., 2002). Therefore, several researchers have studied the effect of supplementation of iron on biohydrogen production. For example, Lee et al. (2001) studied the effect of iron concentration on hydrogen fermentation and found that higher Fe-ion concentration influences the system. The maximum rate of hydrogen production of 24 mL/g volatile suspended solids (VSS) h was obtained when the media were supplemented with 4,000 mg/L $FeCl_2$ (Lee et al., 2001). In addition, during the glycolysis process, the magnesium ion acts as an important cofactor for 10 different enzymes, including hexokinase, phosphofructokinase, and phosphoglycerate kinase. Lin and Lay (2004) also studied the effect of various trace elements (Mg, Na, Zn, Fe, K, I, Co, Mn, Ni, Cu, Mo, and Ca) on hydrogen production using *Cl. pasteurianum*. The study showed that suitable concentrations of Mg, Na, Zn, and Fe were necessary for a higher hydrogen yield (Lin and Lay, 2004).

9.4.5 *Fermentation time*

Fermentation time is the average time duration a microorganism can active inside the reactor. It significantly influences the enrichment of hydrogen-producing bacteria by restricting the growth of methanogenic bacteria. The maximum hydrogen production from distillery waste, dairy waste, or lignocellulosic waste was achieved within 14 h under batch mode without any methane production. As the fermentation time increased, the metabolic pathway shifts from acidogenic phase to methanogenic phase, which is unfavorable for hydrogen production. However, in the case of a continuous system, the two microbial communities can be separated by controlling the hydraulic retention time (HRT). This is a cost-effective way to separate the microorganism at a higher scale as compared to other pretreatment techniques. Kim et al. (2006) developed an undefined hydrogen-producing consortia by terminating the growth of methanogens using a HRT of 3 days. Moreover, the HRT also influences the type and concentration of VFAs produced after fermentation. This is due to the change in the bacterial communities with varying medium pH. The fermentation time mostly depends on the type of substrate used, the activity of the microorganism, the medium pH, and the reaction temperature (Kim and Lee, 2010).

9.4.6 Alkalinity

The alkalinity (buffering capacity) of the fermentative media influences the performance and yield of hydrogen production. This is governed by volatile fatty acids formed after dark fermentation. The higher concentration of VFA leads to a decrease in the buffering capacity of the system due to a decrease in medium pH. This problem can be overcome by codigesting with alkali-rich substrate. This approach can improve the hydrogen production without controlling the medium pH. Tenca et al. (2011) reported the improvement in hydrogen production by maintaining the endogenous pH using swine manure supplemented with alkali-rich substrate (Tenca et al., 2011). In another study, Shi et al. (2010) studied the effect of alkaline concentration on hydrogen production in a continuous stirrer tank reactor (CSTR). As the alkali concentration increased from 1.8 to 2.4×10^{-3} g/L, the hydrogen production also increased from 3 to 5 L/day (Shi et al., 2010). Hence, to improve the hydrogen production rate, it is important to regulate the alkalinity concentration inside the reactor.

9.5 Theoretical considerations

9.5.1 Determination of kinetic parameters using Monod model

Biohydrogen production in a batch system is affected by the limiting substrate concentration present in the fermentative media. Therefore, it is important to determine the kinetic parameters of cell growth and substrate degradation to understand the biological hydrogen production processes.

The unstructured and unsegregated Monod model is usually used to determine the cell growth kinetics of hydrogen-producing bacteria. The Monod equation (Equation 9.16) for the microbial growth is given as follows:

$$\mu = \frac{\mu_{maxS}}{K_s + S} \tag{9.16}$$

where

μ = specific growth rate (h^{-1}),
μ_{max} = maximum specific growth rate (h^{-1}),
K_s = saturation constant (g/L), and
S = limiting substrate concentration (g/L).

Kumar et al. (2000) studied the substrate degradation and cell growth kinetics of *En. cloacae* IIT-BT 08 using the Monod model. They observed that the values of experimental and simulated data are significantly different from each other due to substrate and product inhibition (Kumar et al., 2000).

In the case of gaseous products, inhibition takes place only due to the substrate because the products formed during the fermentation process are simultaneously collected in the gas collector. The classic Monod model could not explain the effect of substrate inhibition that occurs during biohydrogen production. It could not satisfactory explain the effect of pH, substrate diffusion, cell density, or other metal ions.

Andrews model is commonly used to study the effect of substrate inhibition in a biological system (Equation 9.17)

$$\mu = \frac{(\mu_{max} S)}{K_s + S + K_i S^2} \tag{9.17}$$

where K_i is the inhibition constant (g/L).

However, the Andrews model was unable to explain the effect of substrate inhibition in fermentative hydrogen production using *En. cloacae* IIT-BT 08. Therefore, a modified Andrews model was proposed to explain the effect of substrate inhibition on biological hydrogen production (Equation 9.18) (Kumar et al., 2000)

$$\mu = \frac{(\mu_{max} S)}{K_s + S - K_i S^2} \tag{9.18}$$

The Andrews model explain the nonlinear relationship between the specific cell growth rate (μ) and substrate concentration (S). Kumar et al. (2000) observed that the cell mass concentration profile determined from the modified Andrews model fit with the experimental value (Kumar et al., 2000). Nath et al. (2008) also compared the classic Monod model and modified Andrews model for progress of glucose degradation using *En. cloacae* DM11 and observed that the latter was the most suitable.

Several research works have been carried out to determine the kinetic parameters of microbial growth using different models. Kumar and Das (2000) determined the μ_{max} (1.12 h^{-1}) and K_s (8.89 g/L) of *En. cloacae* IIT-BT 08 using the Monod model. Chittibabu et al. (2006) reported the μ_{max} (0.4 h^{-1}) and K_s (11.11 g/L) values of recombinant *E. coli* BL21 using the classic Monod model. Further, Sharma and Li (2009) also used the classic Monod model to determine the μ_{max} (0.72 h^{-1}) and K_s (11.11 g/L) of mixed microflora developed from organic farm soil. Nath et al. (2008) used the Andrews model to determine the μ_{max} (0.4 h^{-1}) and K_s (5.51 g/L) of *En. cloacae* DM11.

9.5.2 Product formation kinetics using modified Gompertz model

During the dark fermentation process, two types of products are formed, liquid products and gaseous products. Hydrogen and carbon dioxides were the gaseous products and volatile fatty acids and solvents present in the spent medium were the liquid products.

The modified Gompertz equation (Equation 9.19) is a very well-known mathematical model for simulating the products formation

$$H(t) = P\exp\left\{-\exp\left[\frac{R_m e}{P}(\lambda - t) + 1\right]\right\} \tag{9.19}$$

where

$H(t)$ = cumulative hydrogen production (mL),
P = total gas production potential (mL),
R_m = maximum gas production rate (mL/h),
λ = lag time (h),
t = incubation time (h), and
The value of e is 2.718.

Mu et al. (2007) and Wang and Wan (2008) reported that the Gompertz equation is the most suitable model for direct fitting of experimental data to describe the progress of hydrogen production in a batch system. Several studies have been performed to evaluate the progress of hydrogen production in a batch system using the modified Gompertz equation (Lin et al., 2008; Kumari and Das, 2015, 2016, 2017; Mishra et al., 2015; Mitra et al., 2017).

Dark fermentative biohydrogen production is considered a growth-associated product. This could be explained by using the Luedeking–Piret model (Equation 9.20)

$$\vartheta = \alpha\mu + \beta \tag{9.20}$$

where

υ = specific hydrogen production rate (h^{-1}),
μ = specific growth rate (h^{-1}),
α = growth-associated coefficient (dimensionless), and
β = non growth-associated coefficient (h^{-1})

The plot of specific hydrogen production rate (υ) versus specific growth rate (μ) indicates that hydrogen is a growth-associated product. Kumar et al. (2000) determined the specific hydrogen production rate of *En. cloacae* IIT-BT 08 using the Luedeking–Piret model. Further, Lo et al. (2008) also determined the hydrogen production rate of *Cl. pasteurianum* CH_4 using the same model.

9.6 Recent development of biogasification

In the case of the dark fermentation process, the maximum hydrogen yield of 4 mol H_2/1 mole glucose can be achieved when acetate is the only end metabolite. This corresponds to the maximum gaseous energy recovery of 34.1% (considering the lower heating value of hydrogen). This yield is quite low for making the process economically viable compared to the existing chemical or electrochemical hydrogen generation processes (Hallenbeck and Ghosh, 2009). In the dark fermentation process, carbohydrates are usually converted to hydrogen and organic acids (acetic, malic, butyric, or lactic acid). Further, conversion of organic acids is not thermodynamically feasible mainly due to positive free-energy change (Perera et al., 2012; Ruggeri et al., 2010). Moreover, organic acids produced during dark fermentation can be further used as a substrate for photofermentation or microbial electrolysis cell to increase the hydrogen yield or for biomethane production to increase the overall gaseous energy recovery as well as chemical oxygen demand (COD) removal efficiency. The details of photofermentation and microbial electrolysis cell are presented in Sections 9.2.2 and 9.2.4, respectively.

In case of the dark fermentation–photofermentation integration process, there are a number of challenges, including the presence of uptake hydrogenase enzyme, requirement of light energy, sensitivity to fixed nitrogen, and also the need for shading effect due the pigments and hydrogen-impermeable photobioreactor. Similarly, in the case of MEC, additional voltage is required to convert dark fermentative effluent into hydrogen. Both processes require more research and development before they can be employed for practical applications.

On the other hand, anaerobic digestion (AD) is a simpler and well-established process compared to others. The two-stage process of biohydrogen followed by biomethane is known as biohythane (Kumari and Das, 2015). It has the advantages of enhancing gaseous energy recovery as well as COD reduction. Biomethane production in the second stage would be advantageous because it could be channelized to the existing infrastructure of compressed natural gas (CNG). In the biohythane concept, the second-stage biomethane could be used separately as a fuel or could be mixed with biohydrogen in a certain ratio to make it suitable for internal combustion (IC) engines.

9.7 Comparison of biohythane process with conventional anaerobic digestion process

The AD process is a decomposition of organic materials by a series of bacteria in an oxygen-free environment to produce biogas (a mixture of 50–75% v/v CH_4 and 25–50% v/v CO_2). The production of biogas from lignocellulosic biomass mainly involved four steps: (1) hydrolysis, (2) acidogenesis, (3) acetogenesis, and (4) methanogenesis. In the first hydrolysis step, the complex substrate will break down to a simple substrate, that is, polymeric carbohydrates are hydrolyzed to simple sugars, proteins are hydrolyzed to amino acids, and lipids are hydrolyzed to long-chain fatty acids. The second stage is the conversion of simple carbohydrates to short-chain VFAs (butyric acids, propionic acids) by acidogenic bacteria. Further, acetogenic bacteria convert these to short-chain VFAs like acetic acids along with carbon dioxide and hydrogen. The last step is the production of methane using this substrate by means of methanogenic bacteria (Figure 9.5). Among all the microbes involved in the four steps, methanogens are the slowest growing organism and are also very sensitive to environmental changes. The methanogens can be broadly classified into two groups: hydrogenotrophic methanogens and acetoclastic methanogens. The CO_2 and H_2 dissolved in the spent medium are being used as an energy source by hydrogenotrophic bacteria to produce methane, whereas acetoclastic methanogens convert acetate to methane and carbon dioxide. Due to the presence of hydrogenotrophic bacteria, the overall percentage of hydrogen yield decreased in the AD process. Two-stage biohythane production processes can overcome the problem of hydrogen consumption by separating acetogenic (hydrogen production stage) and methanogenic phase (methane production stage). The schematic representation of biohythane production is shown in Figure 9.6.

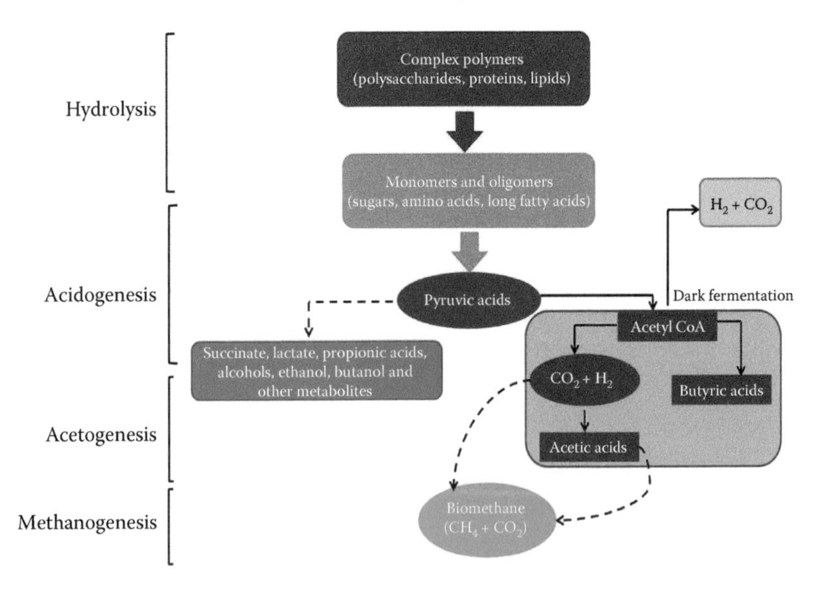

Figure 9.5 Schematic representation of conventional anaerobic digestion process.

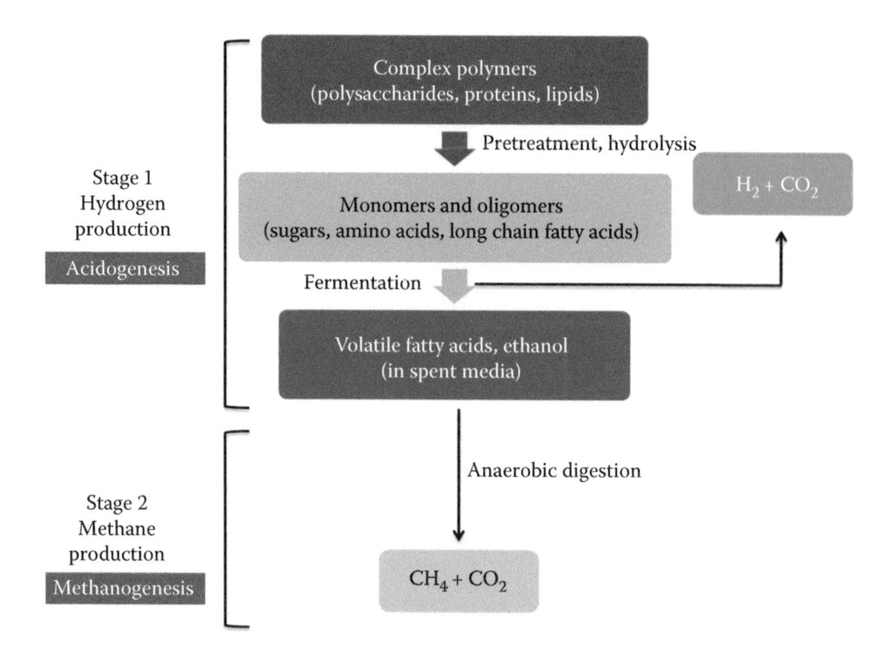

Figure 9.6 Schematic representation of biohythane process.

9.7.1 *Comparison of biohythane production process with AD process in terms of energy recovery*

The total theoretical gaseous energy recovery after the AD process and biohythane production process was calculated considering glucose as the model substrate. The calorific value of glucose is 2,827 KJ/mol and the lower heating values of hydrogen and methane are 241 and 801 KJ/mol, respectively. In the conventional AD process, 1 mol of glucose produces 3 mol of methane (Equation 9.13). Therefore, the maximum percent energy recovery efficiency is calculated as 3 × 801/2,826.6 = 85%.

In the case of the biohythane production process, 1 mol of glucose can produce a maximum 4 mol of hydrogen and 2 mol of methane (Equations 9.21 and 9.22). This corresponds to the maximum percent gaseous energy recovery of 91% (4 × 242 + 2 × 801)/2,826.6 = 91%).

$$\text{AD process: } C_6H_{12}O_6 \rightarrow 3CH_4 + 3CO_2 \tag{9.21}$$

$$\begin{matrix}\text{Biohythane process} \\ \text{(First stage)}\end{matrix} : C_6H_{12}O_6 + 2H_2O \rightarrow 2CH_3COOH + 2CO_2 + 4H_2 \tag{9.22}$$

$$\text{Second stage: } 2CH_3COOH + 2H_2O \rightarrow 2CH_4 + 4CO_2 + 2H_2O \tag{9.23}$$

The burning of hydrogen and hydrogen-methane mixture is much cleaner compared to solely methane. Moreover, biohythane production from complex biomass provides greater substrate consumption with high gaseous energy recovery compared to AD alone. Hence, it is advantageous to use biohythane production.

In recent years, some studies have been performed on biohythane production using lignocellulosic wastes as the initial feedstock (Monlau et al., 2015; Nathao et al., 2013; Pakarinen et al., 2011; Yang et al., 2011; Cheng et al., 2010; Pakarinen et al., 2009). Cheng and Liu (2012) reported the maximum gaseous energy recovery of 38% using untreated cornstalk. A further study was performed by Lin et al. (2015). They observed the maximum gaseous energy of 40% using alkali-pretreated water hyacinth. Kumari and Das (2015, 2016) also reported the maximum gaseous energy recovery up to 44.4% and 37.7% using alkali-pretreated sugarcane bagasse and fungal pretreated sugarcane top, respectively. Recently, Mishra et al. (2017) reported that using distillery effluent cosupplemented with groundnut deoiled cake (GDOC) can achieve the maximum gaseous energy up to 24 Kcal/L.

9.8 Challenges and future prospects

The major challenge of biohythane production is the pH balance in the acidogenic reactor (dark fermentation stage). Due to the formation of reduced end metabolites during the fermentation process, the pH dropped. This causes the inhibition of hydrogen-producing bacteria (as discussed in Section 9.4.1). On the other hand, in the methanogenic reactor the drop in pH due to VFAs is controlled by ammonia formation during methanogenesis. One way to overcome the pH in the acidogenic reactor is by recirculating the alkaline effluent of the methanogenic reactor to the acidogenic reactor. But the major problem of using this technique is the presence of methanogens in the effluent. It may hamper the purity of acidogenic bacteria. The other way to overcome the pH drop is by using high-pH substrate in the acidogenic reactor. Thus, further study to assess the effect of methanogenic effluent on the dark fermentation (DF) reactor in the biohythane process could provide a better technique for sustainability of the biohythane process.

Life-cycle analysis (LCA) of the biohythane process is needed to analyze the environmental effect of the overall process. Choi and Ahn (2014). studied the effect of high-pH-containing substrate on the DF process to replace the use of any buffer during pH drop. After LCA, it was observed that a phosphate buffer used during the process had a greater impact on the environment (Choi and Ahn, 2014).

Another challenge is the modeling and simulation of the integration process. It is necessary to find the kinetic parameters of biohydrogen production to design the reactor for fermentation study. It is also important to model the end products for further design of reactors for downstream processing. The modified Gompertz equation for product formation, logistic equation for biomass growth, Arrhenius equation for temperature effect, Monod equation for substrate utilization, and Luedeking–Piret model for the formation of by-products have been studied widely (Batstone et al., 2002; Mu et al., 2006; Wang and Wan, 2009). Upgrading these models is necessary, including complex biochemical processes including hydrolysis, acidogenesis, acetogenesis, and hydrogen production, when working with complex substrate.

9.9 Conclusion

Hydrogen has a tremendous potential as a clean and renewable energy source. As a result, biological hydrogen production has been the subject of basic and applied research for several decades. Among the various hydrogen production processes, dark fermentation technology has an excellent future potential for biohydrogen production. It has the ability to utilize a variety of feedstocks. Combining dark fermentation with the biomethanation process, that is, the biohythane process, is currently the most promising approach that can be used to increase the gaseous energy recovery as well as COD removal. Compared with the AD process, the substrate degradation and energy recovery efficiency is high. Therefore, hydrogen production via fermentation is more suitable to produce cleaner energy and to treat organic waste more efficiently as compared with the production with the process of photosynthesis. It may be concluded that the biohythane production technique from organic wastes must be optimized to meet the demand of large-scale gaseous energy production facilities.

Acknowledgments

Financial assistance obtained from the Department of Biotechnology (DBT) and Ministry of New and Renewable Energy (MNRE), Government of India, is acknowledged.

References

Antonopoulou, G., Stamatelatou, K., Venetsaneas, N., Kornaros, M., and Lyberatos, G. 2008. Biohydrogen and methane production from cheese whey in a two-stage anaerobic process. *Industrial & Engineering Chemistry Research*, 47, 5227–5233.

Azwar, M. Y., Hussain, M. A., and Abdul-Wahab, A. K. 2014. Development of biohydrogen production by photobiological, fermentation and electrochemical processes: A review. *Renewable & Sustainable Energy Reviews*, 31, 158–173.

Batstone, D. J., Keller, J., and Angelidaki, I. et al. 2002. The IWA Anaerobic Digestion Model No 1 (ADM1). *Water Science and Technology*, 45, 65–73.
Benemann, J. 1997. Feasibility analysis of photobiological hydrogen production. *International Journal of Hydrogen Energy*, 22, 979–987.
Bisaillon, A., Turcot, J., and Hallenbeck, P. C. 2006. The effect of nutrient limitation on hydrogen production by batch cultures of *Escherichia coli*. *International Journal of Hydrogen Energy*, 31, 1504–1508.
Bockris, J. O. M. 1981. The economics of hydrogen as a fuel. *International Journal of Hydrogen Energy*, 6, 223–241.
Cheng, X.-Y., and Liu, C.-Z. 2012. Enhanced coproduction of hydrogen and methane from cornstalks by a three-stage anaerobic fermentation process integrated with alkaline hydrolysis. *Bioresource Technology*, 104, 373–379.
Cheng, J., Xie, B., Zhou, J., Song, W., and Cen, K. 2010. Cogeneration of H_2 and CH_4 from water hyacinth by two-step anaerobic fermentation. *International Journal of Hydrogen Energy*, 35, 3029–3035.
Chin, H.-L., Chen, Z.-S., and Chou, C. P. 2003. Fedbatch operation using *Clostridium acetobutylicum* suspension culture as biocatalyst for enhancing hydrogen production. *Biotechnology Progress*, 19, 383–388.
Chittibabu, G., Nath, K., and Das, D. 2006. Feasibility studies on the fermentative hydrogen production by recombinant *Escherichia coli* BL-21. *Process Biochemistry*, 41, 682–688.
Choi, J., and Ahn, Y. 2014. Characteristics of biohydrogen fermentation from various substrates. *International Journal of Hydrogen Energy*, 39, 3152–3159.
Collect, C. 2004. Hydrogen production by *Clostridium thermolacticum* during continuous fermentation of lactose. *International Journal of Hydrogen Energy*, 29, 1479–1485.
Croese, E., Pereira, M. A., Euverink, G.-J. W., Stams, A. J. M., and Geelhoed, J. S. 2011. Analysis of the microbial community of the biocathode of a hydrogen-producing microbial electrolysis cell. *Applied Microbiology and Biotechnology*, 92, 1083–1093.
Das, D., and Veziroglu, T. N. 2001. Hydrogen production by biological processes: A survey of literature. *International Journal of Hydrogen Energy*, 26, 13–28.
Dasgupta, S., Laplante, B., Murray, S., and Wheeler, D. 2009. Climate change and the future impacts of storm-surge disasters in developing countries. *Center for Global Development Working Papers*, 182, 1–28.
Fan, Y. 2004. Optimization of initial substrate and pH levels for germination of sporing hydrogen-producing anaerobes in cow dung compost. *Bioresource Technology*, 91, 189–193.
Fascetti, E. D., Addario, E., Todini, O., and Robertiello, A. 1998. Photosynthetic hydrogen evolution with volatile organic acids derived from the fermentation of source selected municipal solid wastes. *International Journal of Hydrogen Energy*, 23, 753–760.
Furdui, C., and Ragsdale, S. W. 2000. The role of pyruvate ferredoxin oxidoreductase in pyruvate synthesis during autotrophic growth by the Wood-Ljungdahl pathway. *Journal of Biological Chemistry*, 275, 28494–28499.
Gaffron, H. 1942. Fermentative and photochemical production of hydrogen in algae. *Journal of General Physiology*, 26, 219–240.
Gaffron, H. and Rubin, J. 1942. Fermentative and photochemical production of hydrogen in algae. *Journal of General Physiology*, 26, 219–240.

Ginkel, S. V., Sung, S., and Lay, J.-J. 2001. Biohydrogen production as a function of pH and substrate concentration. *Environmental Science and Technology*, 35, 4726–4730.

Giordano, A., Sarli, V., Lavagnolo, M. C., and Spagni, A. 2014. Evaluation of aeration pretreatment to prepare an inoculum for the two-stage hydrogen and methane production process. *Bioresource Technology*, 166, 211–218.

Grupe, H. and Gottschalk, G. 1992. Physiological events in *Clostridium acetobutylicum* during the shift from acidogenesis to solventogenesis in continuous culture and presentation of a model for shift induction. *Applied and Environmental Microbiology*, 58, 3896–3902.

Guwy, A. J., Hawkes, F. R., Hawkes, D. L., and Rozzi, A. G. 1997. Hydrogen production in a high rate fluidised bed anaerobic digester. *Water Research*, 31, 1291–1298.

Hallenbeck, P. 2002. Biological hydrogen production. Fundamentals and limiting processes. *International Journal of Hydrogen Energy*, 27, 1185–1193.

Hallenbeck, P. C., and Ghosh, D. 2009. Advances in fermentative biohydrogen production: The way forward? *Recent Trends in Biotechnology*, 27, 287–297.

Happe, T., Schutz, K., and Bohme, H. 2000. Transcriptional and mutational analysis of the uptake hydrogenase of the filamentous cyanobacterium *Anabaena variabilis* ATCC 29413. *Journal of Bacteriology*, 182, 1624–1631.

Hawkes, F. R., Dinsdale, R., Hawkes, D. L., and Hussy, I. 2002. Sustainable fermentative hydrogen production: Challenges for process optimisation. *International Journal of Hydrogen Energy*, 27, 1339–1347.

Hillmer, P., and Gest, H. 1977. H_2 metabolism in photosynthetic bacterium *Rhodopseudomonas capsulata*: H_2 production by growing culture. *Journal of Bacteriology*, 129, 724–728.

Hu, B., and Chen, S. 2007. Pretreatment of methanogenic granules for immobilized hydrogen fermentation. *International Journal of Hydrogen Energy*, 32, 3266–3273.

Jo, J. H., Lee, D. S., Park, D., Choe, W.-S., and Park, J. M. 2008. Optimization of key process variables for enhanced hydrogen production by *Enterobacter aerogenes* using statistical methods. *Bioresource Technology*, 99, 2061–2066.

Jordan, K., Chodock, R., and Hand, A. R. et al. 2001. The origin of annular junctions: A mechanism of gap junction internalization. *Journal of Cell Science*, 114, 763–773.

Khanal, S. 2003. Biological hydrogen production: Effects of pH and intermediate products. *International Journal of Hydrogen Energy*, 29, 1123–1131.

Khanna, N., Kotay, S. M., Gilbert, J. J., and Das, D. 2011. Improvement of biohydrogen production by *Enterobacter cloacae* IIT-BT 08 under regulated pH. *Journal of Biotechnology*, 152, 9–15.

Kiely, P. D., Cusick, R., and Call, D. F. et al. 2011. Anode microbial communities produced by changing from microbial fuel cell to microbial electrolysis cell operation using two different wastewaters. *Bioresource Technology*, 102, 388–394.

Kim, S. H., Han, S. K., and Shin, H. S. 2006. Effect of substrate concentration on hydrogen production and 16S rDNA based analysis of the microbial community in a fermenter. *Process Biochemistry*, 41, 199–207.

Kim, M.-S., and Lee, D.-Y. 2010. Fermentative hydrogen production from tofu-processing waste and anaerobic digester sludge using microbial consortium. *Bioresource Technology*, 101, S48–S52.

Kirakosyan, A., and Kaufman, P. B. 2009. *Recent Advances in Plant Biotechnology*. Springer, Boston.

Kosourov, S., Tsygankov, A., Seibert, M., and Ghirardi, M. L. 2002. Sustained hydrogen photoproduction by *Chlamydomonas reinhardtii*: Effects of culture parameters. *Biotechnology Bioengineering*, 78, 731–740.

Kumar, N., and Das, D. 2000. Enhancement of hydrogen production by *Enterobacter cloacae* IIT-BT 08. *Process Biochemistry*, 35, 589–593.

Kumar, N., Monga, P. S., Biswas, A. K., and Das, D. 2000. Modeling and simulation of clean fuel production by *Enterobacter cloacae* IIT-BT 08. *International Journal of Hydrogen Energy*, 25, 945–952.

Kumari, S., and Das, D. 2015. Improvement of gaseous energy recovery from sugarcane bagasse by dark fermentation followed by biomethanation process. *Bioresource Technology*, 194, 354–363.

Kumari, S., and Das, D. 2016. Biologically pretreated sugarcane top as a potential raw material for the enhancement of gaseous energy recovery by two stage biohythane process. *Bioresource Technology*, 218, 1090–1097.

Kumari, S., and Das, D. 2017. Improvement of biohydrogen production using acidogenic culture. *International Journal of Hydrogen Energy*, 42, 4083–4094.

Lee, Y. J., Miyahara, T., and Noike, T. 2001. Effect of iron concentration on hydrogen fermentation. *Bioresource Technology*, 80, 227–231.

Levin, D. 2004. Biohydrogen production: Prospects and limitations to practical application. *International Journal of Hydrogen Energy*, 29, 173–185.

Lin, C. 2004. Carbon/nitrogen-ratio effect on fermentative hydrogen production by mixed microflora. *International Journal of Hydrogen Energy*, 29, 41–45.

Lin, C. Y., Chang, C. C., and Hung, C. H. 2008. Fermentative hydrogen production from starch using natural mixed cultures. *International Journal of Hydrogen Energy*, 33, 2445–2453.

Lin, R., Cheng, J., and Song, W. et al. 2015. Characterisation of water hyacinth with microwave-heated alkali pretreatment for enhanced enzymatic digestibility and hydrogen/methane fermentation. *Bioresource Technology*, 182, 1–7.

Lin, C. Y, and Lay, C. H. 2004. A nutrient formulation for fermentative hydrogen production using anaerobic sewage sludge microflora. *International Journal of Hydrogen Energy*, 30, 285–292.

Liu, H., Grot, S., and Logan, B. E. 2005. Electrochemically assisted microbial production of hydrogen from acetate. *Environmental Science and Technology*, 39, 4317–4320.

Lo, Y. C., Chen, W. M., Hung, C. H., Chen, S. D., and Chang, J. S. 2008. Dark H_2 fermentation from sucrose and xylose using H_2-producing indigenous bacteria: Feasibility and kinetic studies. *Water Research*, 42, 827–842.

Logan, B. E., Call, D., and Cheng, S. et al. 2008. Microbial electrolysis cells for high yield hydrogen gas production from organic matter. *Environmental Science and Technology*, 42, 8630–8640.

Markov, S. A., Thomas, A. D., Bazin, M. J., and Hall, D. O. 1997. Photo-production of hydrogen by cyanobacteria under partial vacuum in batch culture or in photobioreactor. *International Journal of Hydrogen Energy*, 22, 521–524.

Maza-Marquez, P., Vilchez-Vargas, R., and Kerckhof, F. M. et al. 2016. Community structure, population dynamics and diversity of fungi in a full-scale membrane bioreactor (MBR) for urban wastewater treatment. *Water Research*, 105, 507–519.

Melis, A., Zhang, L., Forestier, M., Ghirardi, M. L., and Seibert, M. 2000. Sustained photobiological hydrogen gas production upon reversible inactivation of oxygen evolution in the green alga *Chlamydomonas reinhardtii*. *Plant Physiology*, 122, 127–136.

Mishra, P., Balachandar, G., and Das, D. 2017. Improvement in biohythane production using organic solid waste and distillery effluent. *Waste Management*, 66, 70–78.

Mishra, P., and Das, D. 2014. Biohydrogen production from *Enterobacter cloacae* IIT-BT 08 using distillery effluent. *International Journal of Hydrogen Energy*, 39, 7496–7507.

Mishra, P., Roy, S., and Das, D. 2015. Comparative evaluation of the hydrogen production by mixed consortium, synthetic co-culture and pure culture using distillery effluent. *Bioresource Technology*, 198, 593–602.

Mitra, R., Balachandar, G., Singh, V., Sinha, P., and Das, D. 2017. Improvement in energy recovery by dark fermentative biohydrogen production followed by biobutanol production process using obligate anaerobes. *International Journal of Hydrogen Energy*, 42, 4880–4892.

Mohammadi, P., Ibrahim, S., Mohamad Annuar, M. S., and Law, S. 2011. Effects of different pretreatment methods on anaerobic mixed microflora for hydrogen production and COD reduction from palm oil mill effluent. *Journal of Cleaner Production*, 19, 1654–1658.

Monlau, F., Kaparaju, P., Trably, E., Steyer, J. P., and Carrere, H. 2015. Alkaline pretreatment to enhance one-stage CH_4 and two-stage H_2/CH_4 production from sunflower stalks: Mass, energy and economical balances. *Chemical Engineering Journal*, 260, 377–385.

Mu, Y., Wang, G., and Yu, H. Q. 2006. Kinetic modeling of batch hydrogen production process by mixed anaerobic cultures. *Bioresource Technology*, 97, 1302–1307.

Mu, Y., Yu, H.-Q., and Wang, G. 2007. Evaluation of three methods for enriching H_2-producing cultures from anaerobic sludge. *Enzyme and Microbial Technology*, 40, 947–953.

Nakashimada, Y., Rachman, M. A., Kakizono, T., and Nishio, N. 2002. Hydrogen production of *Enterobacter aerogenes* altered by extracellular and intracellular redox states. *International Journal of Hydrogen Energy*, 27, 1399–1405.

Nath, K., Kumar, A., and Das, D. 2006. Effect of some environmental parameters on fermentative hydrogen production by *Enterobacter cloacae* DM11. *Canadian Journal of Microbiology*, 52, 525–532.

Nath, K., Muthukumar, M., Kumar, A., and Das, D. 2008. Kinetics of two-stage fermentation process for the production of hydrogen. *International Journal of Hydrogen Energy*, 33, 1195–1203.

Nathao, C., Sirisukpoka, U., and Pisutpaisal, N. 2013. Production of hydrogen and methane by one and two stage fermentation of food waste. *International Journal of Hydrogen Energy*, 38, 15764–15769.

Nicolet, Y., Cavazza, C., and Fontecilla-Camps, J. C. 2002. Fe-only hydrogenases: Structure, function and evolution. *Journal of Inorganic Biochemistry*, 91, 1–8.

Oh, Y.-K., Kim, H.-J., Park, S., Kim, M.-S., and Ryu, D. D. Y. 2008. Metabolic-flux analysis of hydrogen production pathway in *Citrobacter amalonaticus* Y19. *Internationa Journal of Hydrogen Energy*, 33, 1471–1482.

O-Thong, S., Prasertsan, P., and Intrasungkha, N. et al. 2008. Optimization of simultaneous thermophilic fermentative hydrogen production and COD reduction from palm oil mill effluent by *Thermoanaerobacterium*-rich sludge. *Internal Journal of Hydrogen Energy*, 33, 1221–1231.

Pakarinen, O. M., Kaparaju, P. L. N., and Rintala, J. A. 2011. Hydrogen and methane yields of untreated, water-extracted and acid (HCl) treated maize in one- and two-stage batch assays. *International Journal of Hydrogen Energy*, 36, 14401–14407.

Pakarinen, O. M., Tahti, H. P., and Rintala, J. A. 2009. One-stage H_2 and CH_4 and two-stage H_2+CH_4 production from grass silage and from solid and liquid fractions of NaOH pre-treated grass silage. *Biomass and Bioenergy*, 33, 1419–1427.

Pandey, A., Sinha, P., and Kotay, S. M. 2009. Isolation and evaluation of a high H_2-producing lab isolate from cow dung. *International Journal of Hydrogen Energy*, 34, 7483–7488.

Perera, K. R. J., Ketheesan, B., Arudchelvam, Y., and Nirmalakhandan, N. 2012. Fermentative biohydrogen production II: Net energy gain from organic wastes. *International Journal of Hydrogen Energy*, 37, 167–178.

Ruggeri, B., Tommasi, T., and Sassi, G. 2010. Energy balance of dark anaerobic fermentation as a tool for sustainability analysis. *International Journal of Hydrogen Energy*, 35, 10202–10211.

Sasikala, K., Ramana, C. V., and Raghuveer, P. R. 1992. Photoproduction of hydrogen from the waste water of a distillery by *Rhodobacter sphaeroides* O.U.001. *International Journal of Hydrogen Energy*, 17, 23–27.

Sasikala, K., Ramana, C. V., and Subrahmanyam, M. 1991. Photo-production of hydrogen from wastewater of a lactic acid fermentation plant by a purple non-sulfur photosynthetic bacterium *Rhodobacter sphaeroides* O.U. 001. *Indian Journal of Experimental Biology*, 29, 74–75.

Serebryakova, L. T., Sheremetieva, M. E., and Lindblad, P. 2000. H_2-uptake and evolution in the unicellular cyanobacterium *Chroococcidiopsis thermalis* CALU 758. *Plant Physiology Biochemisty*, 38, 525–530.

Sharma, Y. and Li, B. 2009. Optimizing hydrogen production from organic wastewater treatment in batch reactors through experimental and kinetic analysis. *International Journal of Hydrogen Energy*, 34, 6171–6180.

Shaw, A. J., Jenney, F. E., Adams, M. W. W., and Lynd, L. R. 2008. End-product pathways in the xylose fermenting bacterium, *Thermoanaerobacterium saccharolyticum*. *Enzyme and Microbial Technology*, 42, 453–458.

Shi, X.-Y., Jin, D.-W., Sun, Q.-Y., and Li, W.-W. 2010. Optimization of conditions for hydrogen production from brewery wastewater by anaerobic sludge using desirability function approach. *Renewable Energy*, 35, 1493–1498.

Singh, S. P., Srivastava, S. C., and Pandey, K. D. 1994. Hydrogen production by *Rhodopseudomonas* at the expense of vegetable starch, sugarcane juice and whey. *International Journal of Hydrogen Energy*, 19, 437–440.

Sivagurunathan, P., Sen, B., and Lin, C. Y. 2014. Batch fermentative hydrogen production by enriched mixed culture: Combination strategy and their microbial composition. *Journal of Bioscience and Bioengineering*, 117, 222–228.

Sveshnikov, D. A., Sveshnikova, N. V., Rao, K. K., and Hall, D. O. 1997. Hydrogen metabolism of mutant forms of *Anabaena variabilis* in continuous cultures and under nutritional stress. *FEBS Microbiology Letters*, 147, 297–301.

Tanawade, S. S., Bapat, B. A., and Naikwade, N. S. 2011. Biofuels: Use of biotechnology to meet energy challenges. *International Journal of Biomedical Research*, 2, 25–31.

Tanisho, S. and Ishiwata, Y. 1994. Continuous hydrogen production from molasses by the bacterium *Enterobacter aerogenes*. *International Journal of Hydrogen Energy*, 19, 807–812.

Tenca, A., Schievano, A., Perazzolo, F., Adani, F., and Oberti, R. 2011. Biohydrogen from thermophilic co-fermentation of swine manure with fruit and vegetable waste: Maximizing stable production without pH control. *Bioresource Technology*, 102, 8582–8588.

Tsygankov, A. A., Serebryakova, L. T., Rao, K. K., and Hall, D. O. 1998. Acetylene reduction and hydrogen photoproduction by wild type and mutant strains of *Anabaena* at different CO_2 and O_2 concentrations. *FEMS Microbiology Letters*, 167, 13–17.

Ueno, Y., Haruta, S., Ishii, M., and Igarashi, Y. 2001. Characterization of a microorganism isolated from the effluent of hydrogen fermentation by microflora. *Journal of Bioscience Bioengineering*, 92, 397–400.

van de Werken, H. J. G., Verhaart, M. R. A., van Fossen, A. L. et al. 2008. Hydrogenomics of the extremely thermophilic bacterium *Caldicellulosiruptor saccharolyticus*. *Applied Environmental Microbiology*, 74, 6720–6729.

van Niel, E. W. J., Claassen, P. A. M., and Stams, A. J. M. 2003. Substrate and product inhibition of hydrogen production by the extreme thermophile, *Caldicellulosiruptor saccharolyticus*. *Biotechnology Bioengineering*, 81, 255–262.

Vrati, S. and Verma, J. 1983. Production of molecular hydrogen and single cell protein by *Pseudomonas capsulata* from cow dung. *Journal of Fermentation Technology*, 61, 157–162.

Wang, J. and Wan, W. 2008. Comparison of different pretreatment methods for enriching hydrogen-producing bacteria from digested sludge. *International Journal of Hydrogen Energy*, 33, 2934–2941.

Wang, J. and Wan, W. 2009. Kinetic models for fermentative hydrogen production: A review. *International Journal of Hydrogen Energy*, 34, 3313–3323.

Weissmen, J. C. and Bonemann, J. R. 1977. Hydrogen production by nitrogen starved culture of *Anabaena cylindrica*. *Applied Environmental Microbiology*, 33, 123–129.

Xiao, B. and Liu, J. 2009. Effects of various pretreatments on biohydrogen production from sewage sludge. *Chinese Science Bulletin*, 54, 2038–2044.

Yang, Z., Guo, R., Xu, X., Fan, X., and Luo, S. 2011. Hydrogen and methane production from lipid-extracted microalgal biomass residues. *International Journal of Hydrogen Energy*, 36, 3465–3470.

Yokoi, H., Maki, R., Hirose, J., and Hayashi, S. 2002. Microbial production of hydrogen from starch-manufacturing wastes. *Biomass and Bioenergy*, 22, 389–395.

Yokoi, H., Saitsu, A. Uchida, H. et al. 2001. Microbial hydrogen production from sweet potato starch residue. *Journal of Bioscience Bioengineering*, 91, 58–63.

Zhang, H., Bruns, M. A., and Logan, B. E. 2006. Biological hydrogen production by *Clostridium acetobutylicum* in an unsaturated flow reactor. *Water Research*, 40, 728–734.

Zhang, T., Liu, H., and Fang, H. H. P. 2003. Biohydrogen production from starch in wastewater under thermophilic condition. *Journal of Environmental Management*, 69, 149–156.

Zhu, H. and Beland, M. 2006. Evaluation of alternative methods of preparing hydrogen producing seeds from digested wastewater sludge. *International Journal of Hydrogen Energy*, 31, 1980–1988.

chapter ten

Fermentative hydrogen production

Current prospects and challenges

Anjana Pandey and Saumya Srivastava

Contents

10.1 Introduction

Hydrogen is the lightest and most ample element in the universe and has huge potential to be used as a future energy. The conversion efficiency of hydrogen is very high, which, along with its recyclability and nonpolluting nature, makes it a very promising element to be used as fuel of future (Sinha and Pandey, 2011).

Currently, primary production of molecular H_2 involves use of fossil fuels via steam reforming of natural gas or methane (CH_4). The total production of H_2 that exceeds 1 billion m^3/day is 48% from natural gas, 18% from coal, 30% from oil, and the rest, 4%, is from H_2O-splitting electrolysis (de Jong, 2008; Mohan and Pandey, 2013). For the production of pure H_2, another reaction, H_2O-gas shift reaction, is also used in combination with steam reforming, which is specifically used for ammonia production. Thermal decomposition, steam gasification, and catalytic oxidation are other thermochemical methods that are used for the production of H_2 (Chen and Syu, 2010; Mohan and Pandey, 2013). However, the fossil fuel method of H_2 production is responsible for the emission of greenhouse gases (GHGs). On the other hand, the biological pathways method for the production of H_2 from biomass is declared an emerging technology due to being eco-friendly and having a high-sustainability nature.

Several microorganisms have been discovered by using different biological organic substrates for improvement at the genetic as well as biochemical pathway level of an organism to produce hydrogen. This will also help to invent a better pathway in comparison with other methods (Sinha and Pandey, 2011).

A variety of organisms are available in nature from archaea to bacteria, cyanobacteria, and lower eukaryotes (i.e., green algae and protists) that take part in H_2 production (Boichenko et al., 2004; Hallenbeck, 2012). They can function either as a single culture or as a consortium of similar or mixed cultures. The heterotrophs are considered as major H_2-producing biocatalysts in the fermentation process. All the organisms do not require solar energy as an energy source and can survive in O_2-deficient conditions like some dark fermentative bacteria, which are obligate anaerobes.

Different natural biological processes are classified into four primary classes based on the systems that are responsible for evolving H_2: (1) H_2O splitting photosynthesis, (2) photofermentation, (3) dark fermentation, and (4) microbial electrolysis processing (Chandrasekhar et al., 2015).

There are two main enzymes, hydrogenase and nitrogenase, on which all the biohydrogen production technologies depend for hydrogen production. They can derive energy either directly from light energy or by indirect means via photosynthetically derived carbon compounds. Fermentative H_2 production is advantageous over photosynthetic H_2 production due to having a rapid H_2 production rate and simpler operation. In addition, the fermentative route of hydrogen production is able to produce very clean energy along with treating organic wastes, which made it a very significant method. Therefore, much attention has been given to fermentative H_2 production in recent years (Sinha and Pandey, 2011).

Fermentative H_2 production is very common under anaerobic conditions. The degradation of microorganisms leads to production of electrons from organic substrates to maintain electrical neutrality. In anaerobic environments, in the presence of hydrogenase enzyme, protons can act as an electron acceptor for production of molecular H_2 (Sinha and Pandey, 2011).

10.2 Fermentative hydrogen production

10.2.1 Anoxygenic photofermentation

Photofermentation comprises the transformation of light energy to biomass by production of H_2 and carbon dioxide (CO_2), the relation being stoichiometric. For photofermentation, purple nonsulfur (PNS) photosynthetic bacteria, comprising the *Rhodobacter* species, are useful in conversion of organic acids including butyrate, lactate, and acetate to H_2 and CO_2 in anaerobic conditions. Furthermore, these bacteria trap energy from the Sun to convert organic acids into H_2 by utilizing nitrogenases in the absence of ammonium ($NH_4{}^+$) ions (Azwar et al., 2014; Basak and Das, 2007; Mohan and Pandey, 2013). In particular, O_2-sensitive nitrogenase does not cause any problem for the process because the purple bacteria being used in the process undergo nonoxygenic photosynthesis (Basak and Das, 2007). However, these nitrogenases also retain several imperfections that affect the generation of H_2, such as suppression of their expression by NH_4^+, low catalytic activity, and lower photochemical efficacy (Brentner et al., 2010; Koku et al., 2002). Recently, a relatively high-yield conversion efficacy for production H_2 has been achieved (Ghosh et al., 2012). Because these processes are light-dependent, captured light energy regulates the electron stream in the photosynthetic system, resulting in generation of a proton gradient. This generated proton gradient is then utilized to attain both requirements for nitrogenase activity: adenosine triphosphate (ATP) production and high-energy electrons.

10.2.2 Dark fermentation

Dark fermentation is known to generate H_2. To the present, numerous studies on the production of H_2 by means of dark fermentation have been performed that utilize facultative (e.g., *Enterobacter aerogenes, E. cloacae, Escherichia coli, Citrobacter ntermedius,* etc.) and obligate anaerobes (including *Clostridium beijerinckii, C. paraputrificum, Ruminococcus albus,* etc.) (Chandrasekhar and Mohan, 2014a,b; Hallenbeck and Ghosh, 2009). The rate of dark fermentation is higher compared to photofermentation and photolysis. The lower yield of H_2 on substrates, due to production of different by-products, becomes the primary disadvantage of the process. The fermentation process results in generation of energy-rich reducing compounds [i.e., NAD(P)H and FADH] from different metabolic pathways, which are then reoxidized in a respiratory system, resulting in generation of ATP. In aerobic respiration, O_2 acts as the terminal electron accepter (TEA), which results in generation of ATP.

Glycolysis is the main metabolic pathway in which a substrate can be converted into pyruvate, which further acts as a key metabolic intermediate. In anaerobic conditions, the pyruvate undergoes acidogenic pathway alongside production of H_2, leading to synthesis of volatile fatty acids (VFAs).

10.2.3 Microbial electrolysis cells: Electro-fermentation

Microbial electrolysis cells (MECs) produce H_2 from numerous substrates and are being developed rapidly. These represent a class of microbial fuel cells (MFCs), which has been investigated for years. The MEC technology is also known as electro-fermentation or biocatalyzed electrolysis cells (Cheng and Logan, 2007; Rozendal et al., 2007). The MEC technology bears a resemblance to an MFC, with the primary difference is being the need of a small external voltage. On the thermodynamics basis, a potential greater than 0.110 V, in addition to the potential generated by microorganisms (−0.300 V), produces H_2 (Cheng and Logan, 2007). The normal redox potential for the reduction process is so observed to be −0.414 V for generation of H_2; hence, the potential requirement is quite low compared with the theoretically calculated value of required voltage of 1.230 V in electrolysis of H_2O (Rozendal et al., 2007).

10.3. Production technologies

10.3.1 Photobioreactors

Microbiological processes decide the design of a photobioreactor related to cyanobacteria and microalgae, among others. The photochemical efficiency, size, and absorption coefficient of these photoheterotrophic bacteria differ (Akkerman et al., 2002). The aim of designing photobioreactors is to achieve inexpensive and rapid growth with high density of the microalgae culture (Evens et al., 2000). Flat plate, tubular, pond, and pool type are the different photobioreactor designs that have been studied (Akkerman et al., 2002). Light environment, the climate and land space, construction materials of the reactor, culture mixing mechanism, maintenance of the reactor, and long-term operational stability with a high gas production are the decisive parameters in the cost analyses.

10.3.2 Dark fermentation bioreactors

Different studies on batch, semicontinuous, and continuous hydrogen-producing bioreactors have shown that batch hydrogen fermentation normally leads to lower hydrogen production rates in comparison with its counterpart. Continuous hydrogen production has been preferred from engineering's stance. Various biohydrogen bioreactor processes have also been developed with a high yield and production rate such as an anaerobic sequencing batch reactor (ASBR), fixed-bed bioreactor, membrane bioreactor (MBR), upflow anaerobic sludge blanket (UASB) bioreactor, and fluidized bed bioreactor in addition to the extensively studied continuous stirred-tank reactor (CSTR) (Show et al., 2011).

10.3.3 Continuous stirred-tank reactor

For continuous hydrogen production, the most commonly used reactor is CSTR (Ding et al., 2010; Younesi et al., 2008). Hydrogen-producing microbes are mixed completely and then suspended in the reactor liquor in a CSTR. A good substrate-microbes interaction leads to accomplishing mass transfer under such hydrodynamics. Because of the quickly mixed operating pattern, this system is not able to maintain high levels of fermentative biomass. Due to short hydraulic retention times (HRTs), biomass washout may restrict the rate of hydrogen production. Volatile suspended solids (VSS) that are retained in the bioreactor have normally ranged between 1 and 4 g VSS l^{-1} depending on the HRT (Show et al., 2007, 2010; Zhang et al., 2006). The highest rate of CSTR hydrogen production, 1.12 l h^{-1} l^{-1}, has been reported in fermentation of sucrose with a mixed hydrogen-producing culture.

10.3.4 Multistage bioreactors

Multistage bioreactors involve three or even four stages. They have been suggested to maximize the production of hydrogen from the substrate (U.S. Department of Energy, 2007; Wang et al., 2011). In the first stage, a direct photolysis reactor first filters the sunlight where the blue-green algae utilize the visible light, and in the second stage in a photofermentative reactor, photosynthetic microbes use the unfiltered infrared light. The effluent from the second-stage photofermentation and the biomass feedstock together is provided into a third-stage dark fermentation reactor in which the substrate is converted into hydrogen and organic acids by the bacteria. The fourth stage is to produce hydrogen by MECs. MECs can be operated during the night or low-light condition because they use the organic acids produced from the dark fermentation in a light-independent process (Show et al., 2011).

10.4 Factors affecting biohydrogen production

H_2 production by fermentation is a complex phenomenon, which is influenced by numerous factors including substrate, inoculum, reactor type, temperature, pH, and metal ion concentration. Due to the dependence of the hydrogen production phenomenon, it is always advantageous to have a suitable experimental design to improve the efficiency of the process (Das, 2009).

10.4.1 Effect of inoculum on fermentative hydrogen production

Microorganisms adept in production of H_2 are present in natural habitats including wastewater, sludge, soil, and compost. Hence, these sources act as a potential source of inoculums for production of H_2 by fermentation. Many research projects are presently using mixed cultures as inoculums for these sources. Mixed cultures are more useful than pure culture for fermentative H_2 production due to mixed cultures' simplicity to operate and ease of control. On treatment of mixed cultures under harsh conditions, H_2-producing bacteria have exhibited better survival compared to other microbes (Li and Fang, 2007; Nath et al., 2006).

10.4.2 Effect of pH on fermentative hydrogen production

pH is a vital parameter influencing the fermentative H_2 production because it alters the hydrogenase activity along with the metabolic pathways. It has been observed that in a specific range, an increase in pH resulted in an increase of H_2 production by bacteria, but at very high pH levels the hydrogen production rate reduces. Other experiments have shown that H_2 production via a fermentative pathway is feasible in lesser acidic conditions. Initial pH effects the duration of the log phase in batch production of H_2, though the H_2 production yield lowered at high starting pH values. It was observed that at higher initial pH, H_2 was produced swiftly to inhibitory levels along with acid, thus depleting the buffering capacity. As a result, the bacteria do not adapt to the changes in the environment and are depleted. Nonetheless, at lower initial pH values, the initiating conditions might not be advantageous for the bacteria, though due to adaptation at this pH, microbes produce H_2 at a reasonable rate for a longer time (Ferchichi et al., 2005; Khanal et al., 2004).

10.4.3 *Effect of temperature on fermentative hydrogen production*

Temperature is one of the significant factors controlling fermentative H_2 production. Temperature directly affects the growth and metabolic activity of a microorganism. Fermentative conversions can be accomplished at mesophilic (25–40°C), thermophilic (40–65°C), or hyperthermophilic (>80°C) temperatures. In many experimental setups, H_2 gas generation was favored at mesophilic conditions and in some thermophilic condition (Sompong et al., 2008a; Wu and Lin 2004).

10.4.4 *Effect of metal ion on fermentative hydrogen production*

Biohydrogen production necessitates vital micronutrients for bacterial growth and metabolism during the process of fermentation. Mg, Na, Fe, and Zn are significant trace elements that affect H_2 metabolism because of the fact that these metals are required as cofactors for enzymatic activity in bacteria (Alshiyab et al., 2008; Wang and Wan, 2008; Yang and Shen, 2006).

10.5 *Advanced approaches*

10.5.1 *The use of mixed microbial consortia*

Most of the reactors that have been described above depend on the formation of macroscopic aggregates of microbial cells, which are flocs or granules. The ability to form these particles is a common trait for mixed culture inoculum, not of pure cultures. There are also various advantages of using microbial consortia in the place of pure cultures. Industrial hydrogen fermentations need to undergo minimal pretreatment, which is carried out under nonsterile conditions using freely available complex feedstocks. These issues have been solved by microbial consortia because they require nonsterile conditions for growth and dominance. Due to complex community, they contain a group of the necessary hydrolytic activities, which are potentially more robust to different environmental conditions (Kleerebezem and van Loosdrecht, 2007). However, several issues are associated with microbial consortia. With changes in process parameters, their composition can vary over time and from reactor to reactor (Koskinen et al., 2007; Lin et al., 2006, 2008; Maintinguer et al., 2008). Construction of designer consortia containing a community of diverse members, each contributing a unique and essential metabolic capacity, could be a possible way to solve this issue (Brenner et al., 2008). The total community metabolic range in this system would be more than any individual member. However, the complex interactions that occur in natural consortia are not much known (Weibel, 2008).

10.5.2 Improving biohydrogen production by metabolic engineering of existing pathways

To increase biohydrogen production, much attention has been given to metabolic engineering (Hallenbeck and Ghosh, 2009). By increased flux via gene knockouts of competing pathways or by increasing homologous expression of enzymes associated with the hydrogen-generating pathways, improvements in hydrogen production can take place. *E. coli* has been used in a majority of studies because it is a laboratory workhorse for metabolic engineering. The basis behind using *E. coli* is (1) its genome size can be easily manipulated, (2) its metabolism is the best understood of all bacteria, and (3) it readily degrades different types of sugars. Therefore, it has been proven that *E. coli* can be used for metabolic engineering to achieve the maximal yields of hydrogen. The inactivation of lactate dehydrogenase (ldhA) (Bisaillon et al., 2006; Maeda et al., 2007; Turcot et al., 2008; Yoshida et al., 2006) or fumarate reductase (frd) (Maeda et al., 2007; Yoshida et al., 2006) leads to a modest increase in H_2 yields. In *E. coli,* formate, formed from pyruvate, is converted into H_2 and CO_2 by the formate–hydrogen lyase (FHL) complex, and manipulations for the increased expression of formate dehydrogenase H (FdhH) and hydrogenase 3 (Hyd3) enzymes, which are involved in this pathway increase the H_2 yield (Bisaillon et al., 2006; Maeda et al., 2007; Turcot et al., 2008; Yoshida et al., 2006). Inactivation of the hydrogenases Hyd1 and Hyd2 also leads to increased hydrogen production (Hallenbeck, 2005).

10.5.3 Other attempts

Different optimizations and modeling have been carried out for improvement of biohydrogen production. Several variables including pH, temperature, substrate concentration, and nutrient availability are responsible for rate and yield of hydrogen production. A variety of modeling methods including engineering, biology, environmental science, food processing, and industrial processing and their application in biohydrogen production have been developed (Wang and Wan, 2009). Artificial neural networks (ANNs) and design of experiments (DOE) are the two main types of widely used modeling methods that are relevant to biohydrogen production. ANNs have been successfully used to model the results of hydrogen production (Mu and Yu, 2007). DOE is based on statistical modeling for analysis of the relationship between a set of experimental factors for proposing an appropriate design for their experimental verification (Ilzarbe et al., 2008). Factorial design (FD) combined with response surface methodology (RSM) is a common and powerful DOE approach (Hanrahan and Lu, 2006). FD examines the interdependence of multiple factor responses and then gives permission for the identification of the most significant parameters responsible for controlling the process and the degree of interaction between them.

Many experiments on hydrogen production by dark fermentation have used FD and RSM with the aim of hydrogen production optimization. Their results showed that the analysis can be useful in optimization of different aspects of a biohydrogen fermentation process, such as conditions for spore germination (Fan et al., 2004), pretreatment strategies (Espinoza-Escalante et al., 2008; Wang et al., 2007), micronutrient formulations (Guo et al., 2009; Lin and Lay, 2005; Sompong et al., 2008b), and substrate composition (Turcot et al., 2008). Though most of the mesophilic fermentations require similar conditions, sufficient differences exist involving the need to justify optimizing the process. Additionally, these methods might be beneficial in instituting the process parameters for use of novel consortia and complex substrates. Also, these methods can rapidly optimize a specific fermentation process, yet these cannot improve the yield compared with the yield obtained by classical methods of optimization. Reverse micelles are also observed to be a method that elevates hydrogen production by increasing enzyme stability, improving enzyme kinetics, and by better distribution of the substrates and products among organic and aqueous phases (Biasutti et al., 2008; Monnard et al., 2008).

10.6 Future prospects

In spite of a great amount of research going on in the field of biohydrogen production, key difficulties must continue to be overcome before a viable and efficient process can be established for biological hydrogen production. The latest developments in both yield and volumetric synthesis rates of hydrogen fermentation processes are considerable. However, yields should significantly extend past the current metabolic limitation of glucose utilization. Thermodynamics (Thauer et al., 1977) and metabolic limitations propose that it is nearly possible to search for a microbe capable of complete transformation of substrates to hydrogen through fermentation. There are many opportunities, including metabolic engineering or multistage systems, that can further enhance fermentative hydrogen production.

References

Akkerman, I., Janssen, M., Rocha, J., and Wijffels, R. H., 2002. Photobiological hydrogen production: Photochemical efficiency and bioreactor design. *International Journal of Hydrogen Energy*, 27(11), 1195–1208.

Alshiyab, H., Kalil, M. S., Hamid, A. A., and Yusoff, W. M. W., 2008. Trace metal effect on hydrogen production using *C. acetobutylicum*. *OnLine Journal of Biological Sciences*, 8(1), 1–9.

Azwar, M. Y., Hussain, M. A., and Abdul-Wahab, A. K., 2014. Development of biohydrogen production by photobiological, fermentation and electrochemical processes: A review. *Renewable and Sustainable Energy Reviews*, 31, 158–173.

Basak, N. and Das, D., 2007. The prospect of purple non-sulfur (PNS) photosynthetic bacteria for hydrogen production: The present state of the art. *World Journal of Microbiology and Biotechnology*, 23(1), 31–42.

Biasutti, M. A., Abuin, E. B., Silber, J. J., Correa, N. M., and Lissi, E. A., 2008. Kinetics of reactions catalyzed by enzymes in solutions of surfactants. *Advances in Colloid and Interface Science*, 136(1), 1–24.

Bisaillon, A., Turcot, J., and Hallenbeck, P. C., 2006. The effect of nutrient limitation on hydrogen production by batch cultures of *Escherichia coli*. *International Journal of Hydrogen Energy*, 31(11), 1504–1508.

Boichenko, V. A., Greenbaum, E., and Seibert, M. 2004. Hydrogen production by photosynthetic microorganisms. *Photoconversion of Solar Energy, Molecular to Global Photosynthesis*, 2, 397–452.

Brenner, K., You, L., and Arnold, F. H., 2008. Engineering microbial consortia: A new frontier in synthetic biology. *Trends in Biotechnology*, 26(9), 483–489.

Brentner, L. B., Peccia, J., and Zimmerman, J. B., 2010. Challenges in developing biohydrogen as a sustainable energy source: Implications for a research agenda. *Environmental Science & Technology*, 44(7), 2243–2254.

Chandrasekhar, K., Lee, Y. J., and Lee, D. W., 2015. Biohydrogen production: Strategies to improve process efficiency through microbial routes. *International Journal of Molecular Sciences*, 16(4), 8266–8293.

Chandrasekhar, K. and Mohan, S. V., 2014a. Bio-electrohydrolysis as a pretreatment strategy to catabolize complex food waste in closed circuitry: Function of electron flux to enhance acidogenic biohydrogen production. *International Journal of Hydrogen Energy*, 39(22), 11411–11422.

Chandrasekhar, K. and Mohan, S. V., 2014b. Induced catabolic bio-electrohydrolysis of complex food waste by regulating external resistance for enhancing acidogenic biohydrogen production. *Bioresource Technology*, 165, 372–382.

Chen, W. H. and Syu, Y. J., 2010. Hydrogen production from water gas shift reaction in a high gravity (Higee) environment using a rotating packed bed. *International Journal of Hydrogen Energy*, 35(19), 10179–10189.

Cheng, S. and Logan, B. E., 2007. Sustainable and efficient biohydrogen production via electrohydrogenesis. *Proceedings of the National Academy of Sciences*, 104(47), 18871–18873.

Das, D., 2009. Advances in biohydrogen production processes: An approach towards commercialization. *International Journal of Hydrogen Energy*, 34(17), 7349–7357.

De Jong, W., 2008. Sustainable hydrogen production by thermochemical biomass processing. In: *Hydrogen Fuel: Production, Transport, and Storage*, Gupta, R. B. (Ed.), 185–225. Boca Raton, CRC Press.

Ding, J., Wang, X., Zhou, X. F., Ren, N. Q., and Guo, W. Q., 2010. CFD optimization of continuous stirred-tank (CSTR) reactor for biohydrogen production. *Bioresource Technology*, 101(18), 7005–7013.

Espinoza-Escalante, F. M., Pelayo-Ortiz, C., Gutierrez-Pulido, H. et al. 2008. Multiple response optimization analysis for pretreatments of tequila's stillages for VFAs and hydrogen production. *Bioresource Technology*, 99(13), 5822–5829.

Evens, T. J., Chapman, D. J., Robbins, R. A., and D'Asaro, E. A., 2000. An analytical flat-plate photobioreactor with a spectrally attenuated light source for the incubation of phytoplankton under dynamic light regimes. *Hydrobiologia*, 434(1), 55–62.

Fan, Y., Li, C., Lay, J. J., Hou, H., and Zhang, G., 2004. Optimization of initial substrate and pH levels for germination of sporing hydrogen-producing anaerobes in cow dung compost. *Bioresource Technology*, 91(2), 189–193.

Ferchichi, M., Crabbe, E., Gil, G. H., Hintz, W., and Almadidy, A., 2005. Influence of initial pH on hydrogen production from cheese whey. *Journal of Biotechnology*, 120(4), 402–409.

Ghosh, D., Sobro, I. F., and Hallenbeck, P. C., 2012. Optimization of the hydrogen yield from single-stage photofermentation of glucose by *Rhodobacter capsulatus* JP91 using response surface methodology. *Bioresource Technology*, 123, 199–206.

Guo, W. Q., Ren, N. Q., Wang, X. J. et al. 2009. Optimization of culture conditions for hydrogen production by *Ethanoligenens harbinense* B49 using response surface methodology. *Bioresource Technology*, 100(3), 1192–1196.

Hallenbeck, P. C., 2005. Fundamentals of the fermentative production of hydrogen. *Water Science and Technology*, 52(1–2), 21–29.

Hallenbeck, P. C., 2012. Hydrogen production by cyanobacteria. In: *Microbial Technologies in Advanced Biofuels Production*, 15–28. Springer, New York.

Hallenbeck, P. C. and Ghosh, D., 2009. Advances in fermentative biohydrogen production: The way forward? *Trends in Biotechnology*, 27(5), 287–297.

Hanrahan, G. and Lu, K., 2006. Application of factorial and response surface methodology in modern experimental design and optimization. *Critical Reviews in Analytical Chemistry*, 36(3–4), 141–151.

Ilzarbe, L., Álvarez, M. J., Viles, E., and Tanco, M., 2008. Practical applications of design of experiments in the field of engineering: A bibliographical review. *Quality and Reliability Engineering International*, 24(4), 417–428.

Khanal, S. K., Chen, W. H., Li, L., and Sung, S., 2004. Biological hydrogen production: Effects of pH and intermediate products. *International Journal of Hydrogen Energy*, 29(11), 1123–1131.

Kleerebezem, R. and van Loosdrecht, M. C., 2007. Mixed culture biotechnology for bioenergy production. *Current Opinion in Biotechnology*, 18(3), 207–212.

Koku, H., Eroglu, I., Gunduz, U., Yucel, M., and Turker, L., 2002. Aspects of the metabolism of hydrogen production by *Rhodobacter sphaeroides*. *International Journal of Hydrogen Energy*, 27(11), 1315–1329.

Koskinen, P. E., Kaksonen, A. H., and Puhakka, J. A., 2007. The relationship between instability of H_2 production and compositions of bacterial communities within a dark fermentation fluidised-bed bioreactor. *Biotechnology and Bioengineering*, 97(4), 742–758.

Li, C. and Fang, H. H., 2007. Fermentative hydrogen production from wastewater and solid wastes by mixed cultures. *Critical Reviews in Environmental Science and Technology*, 37(1), 1–39.

Lin, C. Y., Chang, C. C., and Hung, C. H., 2008. Fermentative hydrogen production from starch using natural mixed cultures. *International Journal of Hydrogen Energy*, 33(10), 2445–2453.

Lin, C. Y. and Lay, C. H., 2005. A nutrient formulation for fermentative hydrogen production using anaerobic sewage sludge microflora. *International Journal of Hydrogen Energy*, 30(3), 285–292.

Lin, C. Y., Lee, C. Y., Tseng, I. C., and Shiao, I. Z., 2006. Biohydrogen production from sucrose using base-enriched anaerobic mixed microflora. *Process Biochemistry*, 41(4), 915–919.

Maeda, T., Sanchez-Torres, V., and Wood, T. K., 2007. Enhanced hydrogen production from glucose by metabolically engineered *Escherichia coli*. *Applied Microbiology and Biotechnology*, 77(4), 879–890.

Maintinguer, S. I., Fernandes, B. S., Duarte, I. C. et al. 2008. Fermentative hydrogen production by microbial consortium. *International Journal of Hydrogen Energy*, 33(16), 4309–4317.

Mohan, S. and Pandey, A. 2013. Biohydrogen production: An introduction. In: *Biohydrogen*, Pandey, A., Chang, J.-S., Hallenbeck, P. C., and Larroche, C. (Eds.), pp. 1–24. Amsterdam, Elsevier.

Monnard, P. A., DeClue, M. S., and Ziock, H. J., 2008. Organic nano-compartments as biomimetic reactors and protocells. *Current Nanoscience*, 4(1), 71–87.

Mu, Y. and Yu, H. Q., 2007. Simulation of biological hydrogen production in a UASB reactor using neural network and genetic algorithm. *International Journal of Hydrogen Energy*, 32(15), 3308–3314.

Nath, K., Kumar, A., and Das, D., 2006. Effect of some environmental parameters on fermentative hydrogen production by *Enterobacter cloacae* DM11. *Canadian Journal of Microbiology*, 52(6), 525–532.

Rozendal, R. A., Hamelers, H. V., Molenkamp, R. J., and Buisman, C. J., 2007. Performance of single chamber biocatalyzed electrolysis with different types of ion exchange membranes. *Water Research*, 41(9), 1984–1994.

Show, K. Y., Lee, D. J., and Chang, J. S., 2011. Bioreactor and process design for biohydrogen production. *Bioresource Technology*, 102(18), 8524–8533.

Show, K. Y., Zhang, Z. P., Tay, J. H. et al. 2007. Production of hydrogen in a granular sludge-based anaerobic continuous stirred tank reactor. *International Journal of Hydrogen Energy*, 32(18), 4744–4753.

Show, K. Y., Zhang, Z. P., Tay, J. H. et al. 2010. Critical assessment of anaerobic processes for continuous biohydrogen production from organic wastewater. *International Journal of Hydrogen Energy*, 35(24), 13350–13355.

Sinha, P. and Pandey, A., 2011. An evaluative report and challenges for fermentative biohydrogen production. *International Journal of Hydrogen Energy*, 36(13), 7460–7478.

Sompong, O., Prasertsan, P., Intrasungkha, N., Dhamwichukorn, S., and Birkeland, N. K., 2008a. Optimization of simultaneous thermophilic fermentative hydrogen production and COD reduction from palm oil mill effluent by *Thermoanaerobacterium*-rich sludge. *International Journal of Hydrogen Energy*, 33(4), 1221–1231.

Sompong, O., Prasertsan, P., Karakashev, D., and Angelidaki, I., 2008b. Thermophilic fermentative hydrogen production by the newly isolated *Thermoanaerobacterium thermosaccharolyticum* PSU-2. *International Journal of Hydrogen Energy*, 33(4), 1204–1214.

Thauer, R. K., Jungermann, K., and Decker, K., 1977. Energy conservation in chemotrophic anaerobic bacteria. *Bacteriological Reviews*, 41(1), 100–180.

Turcot, J., Bisaillon, A., and Hallenbeck, P. C., 2008. Hydrogen production by continuous cultures of *Escherchia coli* under different nutrient regimes. *International Journal of Hydrogen Energy*, 33(5), 1465–1470.

U.S. Department of Energy, 2007. *Hydrogen, Fuel Cells and Infrastructure Technologies Program: Multi-Year Research, Development and Demonstration Plan.*

Wang, C. H., Lu, W. B., and Chang, J. S., 2007. Feasibility study on fermentative conversion of raw and hydrolyzed starch to hydrogen using anaerobic mixed microflora. *International Journal of Hydrogen Energy*, 32(16), 3849–3859.

Wang, A., Sun, D., Cao, G. et al. 2011. Integrated hydrogen production process from cellulose by combining dark fermentation, microbial fuel cells, and a microbial electrolysis cell. *Bioresource Technology*, 102(5), 4137–4143.

Wang, J. and Wan, W., 2008. Influence of Ni^{2+} concentration on biohydrogen production. *Bioresource Technology*, 99(18), 8864–8868.

Wang, J. and Wan, W., 2009. Experimental design methods for fermentative hydrogen production: A review. *International Journal of Hydrogen Energy*, 34(1), 235–244.

Weibel, D. B., 2008. Building communities one bacterium at a time. *Proceedings of the National Academy of Sciences*, 105(47), 18075–18076.

Wu, J. H. and Lin, C. Y., 2004. Biohydrogen production by mesophilic fermentation of food wastewater. *Water Science and Technology*, 49(5–6), 223–228.

Yang, H. and Shen, J., 2006. Effect of ferrous iron concentration on anaerobic biohydrogen production from soluble starch. *International Journal of Hydrogen Energy*, 31(15), 2137–2146.

Yoshida, A., Nishimura, T., Kawaguchi, H., Inui, M., and Yukawa, H., 2006. Enhanced hydrogen production from glucose using ldh- and frd-inactivated *Escherichia coli* strains. *Applied Microbiology and Biotechnology*, 73(1), 67–72.

Younesi, H., Najafpour, G., Ismail, K. S. K., Mohamed, A. R., and Kamaruddin, A. H., 2008. Biohydrogen production in a continuous stirred tank bioreactor from synthesis gas by anaerobic photosynthetic bacterium: *Rhodopirillum rubrum*. *Bioresource Technology*, 99(7), 2612–2619.

Zhang, Z. P., Show, K. Y., Tay, J. H. et al. 2006. Effect of hydraulic retention time on biohydrogen production and anaerobic microbial community. *Process Biochemistry*, 41(10), 2118–2123.

chapter eleven

Algal photobiohydrogen production

Archana Tiwari, Thomas Kiran, and Anjana Pandey

Contents

11.1 Introduction

Hydrogen is an element that is widespread in our environment, but the level of pure hydrogen in the atmosphere is very low. Utilization of hydrogen as a biofuel needs hydrogen-rich and efficient sources; biologically produced hydrogen can be evolved from microorganisms. Gaffron (1939) reported for the first time the photosynthetic molecular hydrogen production from microalgae. Although hydrogen is one of the most abundant elements on this planet, its pure form (H_2) exists at extremely low levels (<1 ppm) in the atmosphere. Fuel hydrogen must therefore be produced from hydrogen-rich substances. Biologically produced hydrogen or biohydrogen is considered a renewable, CO_2-neutral energy form (Figure 11.1 and Table 11.1).

A large number of microorganisms evolve hydrogen when growing on renewable feedstocks under special anaerobic conditions with low hydrogen pressure. Depending on carbon and energy sources, three different mechanisms are involved in microbial hydrogen evolution, presenting unique technical challenges for hydrogen production. Heterotrophic obligate or facultative anaerobes like *Clostridium* obtain both carbon and energy from carbohydrates such as glucose and deposit the excess reducing power in fermentative products and hydrogen. Photosynthetic bacteria like *Rhodobacter* can use broad organic substrates including lactic and acetic acids as the energy and carbon source under light irradiation. Light energy is essential to hydrogen evolution by photosynthetic cells. Photoautotrophic green microalgae and cyanobacteria use sunlight and CO_2 as the sole sources for energy and carbon. The production of hydrogen through biological means can be through following routes:

1. Direct photolysis of water (in eukaryotic algae).
2. During nitrogen fixation (in photosynthetic bacteria including cyanobacteria).

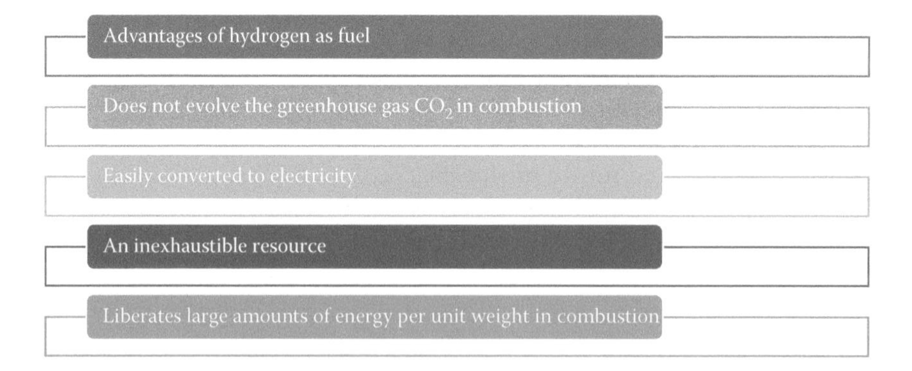

Figure 11.1 Advantages of hydrogen as fuel.

3. Nonphotosynthetic hydrogen production from organic compounds (in obligate anaerobic bacteria).
4. Fermentation (in fermentative bacteria).

11.2 Algae: Unique reservoir

Algae have a unique potential to make hydrogen directly from sunlight and water, although only in the complete absence of oxygen. As an energy carrier, hydrogen offers great promise as a fuel of the future because it can be applied in mobile applications with only water as an exhaust product and no NO_x emissions when used in a fuel cell. One major bottleneck for the full-scale implementation of hydrogen-based technology is the absence of a large-scale sustainable production method for hydrogen. Currently, hydrogen gas is produced by the process of steam reformation of fossil fuels. Large-scale electrolysis of water is also possible, but this production method costs more electricity than can be generated from the hydrogen it yields. Hydrogen can be produced biologically by a variety of means, including the steam reformation of bio-oils (Wang et. al, 2007), dark fermentation and photofermentation of organic materials (Kapdan and Kargi, 2006), and photolysis of water catalyzed by special microalgal species (Kapdan and Kargi, 2006; Ran et al., 2006). Several bacteria can extract hydrogen from carbohydrates in the dark and a group called purple nonsulfur bacteria can use energy from light to extract more hydrogen gas (H_2) from a wider range of substrates, while green sulfur bacteria can make H_2 from H_2S. These options are only interesting if a wastewater with these compounds is available (Rupprecht et al., 2006). Other algae can make hydrogen directly from sunlight and water, although only in the complete absence of oxygen. In practice, this means that hydrogen formation is only possible under conditions that either cost a great deal of energy, or prevent storing solar energy (Kapdan and Kargi, 2006), and a closed culture system is required. At the moment, it is only possible to produce a fraction of the theoretical maximum of 20 g $H_2/m^2/day$, making bulk-scale hydrogen production by algae not yet viable.

More knowledge of the organisms that can produce hydrogen (only a few have been investigated) and the required conditions as well as optimization of the biological route of solar energy to hydrogen through genetic modification is necessary. If these improvements prove to be possible, this would constitute a profitable and renewable hydrogen production (Melis and Happe, 2001). While many algal species show different interesting characteristics for biofuel production, there is still plenty of potential for modern biotechnology tools to increase their effectiveness. Innovative approaches are required so that hydrogen can be a successor of fossil fuels (Figure 11.2).

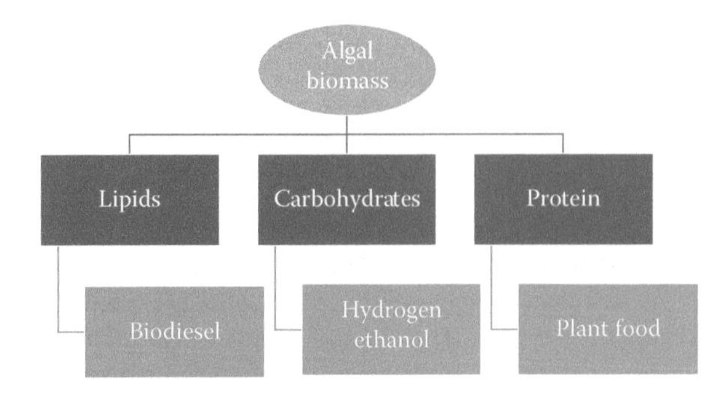

Figure 11.2 The different products generated from an algal biomass.

11.3 Algal photosynthesis and hydrogen production

Algae are photosynthetic organisms with unique properties. There are many microalgae reported to yield biofuels. Algae are photosynthetic autotrophs and inhabit a variety of habitats. They are tiny biological factories that use photosynthesis to transform carbon dioxide and sunlight into energy so efficiently that they can double their weight several times a day. As part of the photosynthesis process, algae produce oil and can generate 15 times more oil per acre than other plants used for biofuels, such as corn and switchgrass. Algae can grow in saltwater, freshwater, or even contaminated water, at sea or in ponds, and on land not suitable for food production (www.Sciencedaily.com). The photosynthesis process occurs in almost all algae, and in fact much of what is known about photosynthesis was first discovered by studying the green alga *Chlorella* (Figure 11.3) (Tiwari, 2014).

Photosynthesis comprises both light reactions and dark reactions (or Calvin cycle). During the dark reactions, carbon dioxide is bound to ribulose bisphosphate, a 5-carbon sugar with two attached phosphate groups, by the enzyme ribulose bisphosphate carboxylase. This is the initial step of a complex process leading to the formation of sugars. During the light reactions, light energy is converted into the chemical energy needed for the dark reactions. Chloroplasts are the sites of photosynthesis, the complex set of biochemical reactions that use the energy of light to convert carbon dioxide and water into sugars. Each chloroplast contains flattened, membranous sacs, called thylakoids, that contain the photosynthetic light-harvesting pigments, the chlorophylls, carotenoids, and phycobiliproteins (Tiwari and Pandey, 2013).

11.4 Factors affecting hydrogen production

Many factors play a significant role in hydrogen production by algae. Algae require light and nutrients to synthesize lipids, proteins, and

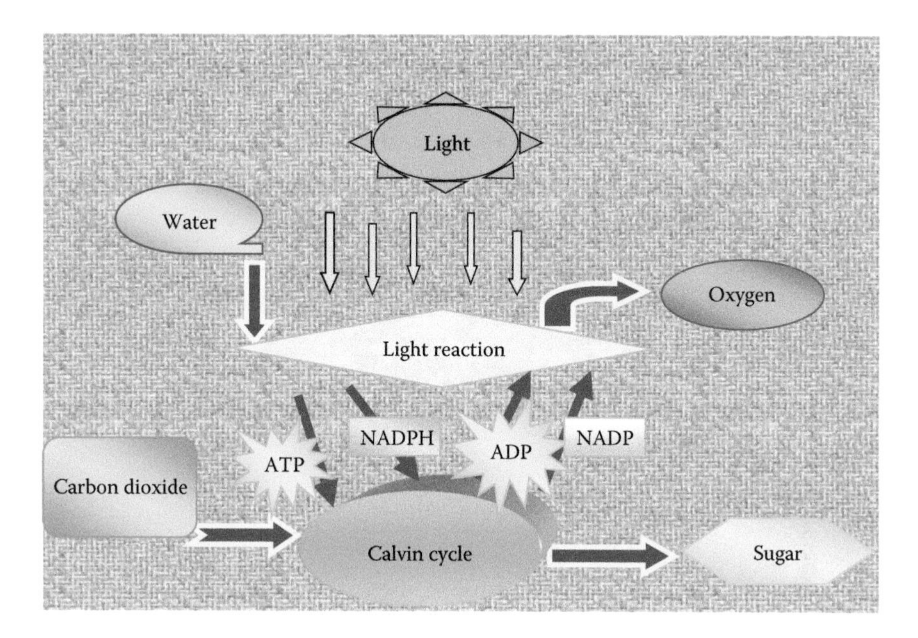

Figure 11.3 Photosynthesis in algae.

carbohydrates through photosynthesis. The yield of the metabolic product is directly proportional to diverse factors. These factors may be categorized as intrinsic or environmental (Figure 11.4).

The algal growth also requires macronutrients like nitrogen, phosphorus, sulfur, potassium, and magnesium. Micronutrients like iron and manganese are needed in small amounts. The growth and biochemical composition of algae are greatly influenced by diverse environmental factors along with the presence of macronutrients, micronutrients, and nonmineral nutrients. The plethora of these factors influence photosynthesis, and thereby fixation of carbon dioxide and the macromolecular composition yielding biofuels (Tiwari and Pandey, 2012).

11.4.1 Light

Photosynthetic algae require light as a source of energy to convert carbon dioxide to organic compounds like sugars. The intensity of light has a profound effect on photosynthesis and thus growth of algae. The intensity of light greater than saturation intensity or less than it severely lower the growth rate of algae. The photoadaptation leads to changes in various properties of the algal cell like pigments, rate of growth, respiratory rate, essential fatty acids, density of thylakoid membrane, and desaturation of chloroplast membranes. In algae, photoinhibition is caused due

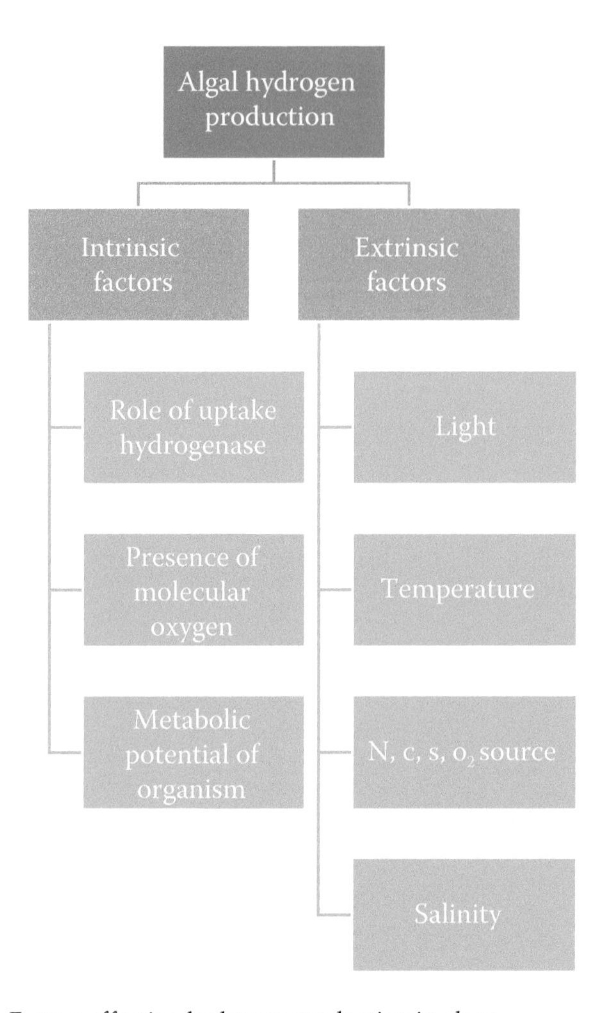

Figure 11.4 Factors affecting hydrogen production in algae.

to an increase in light intensity beyond saturation limits. It is reported that there is disruption of chloroplast lamellae owing to high light intensity and inactivation of enzymes involved in carbon dioxide fixation. In *Dunaliella viridis,* the growth rate decreased to 63% with increase in light intensity from 700 to 1500 μ mol $m^{-2} s^{-1}$ (Gordillo et al., 1998).

There is variation in light requirement by algal species for optimum hydrogen production. Some algae produce hydrogen in the presence of light, while some produce in both light and dark conditions. Nitrogen-fixing cells of *Anabaena variabilis* SPU 003 have the capacity to produce hydrogen mainly in darkness. *Spirulina platensis* is reported to produce hydrogen under anaerobic conditions, both in the dark and in the light, but several other species produce hydrogen only in the presence of light. In

Synechococcus elongatus PCC 7942, hydrogen production mediated by native hydrogenases occurs under in the dark under anaerobic conditions. *S. platensis* can produce hydrogen optimally at 32°C in complete anaerobic and dark conditions. In *A. variabilis* PK 84, hydrogen production was stimulated by light (Tsygankov et al., 1999). Hydrogen production in *Nostoc muscorum* is catalyzed by nitrogenase; more hydrogen is produced in this strain in the light than in the dark. *A. cylindrica* produces hydrogen under an argon atmosphere for 30 days in limited light (luminous intensity 6.0 W/m^{-2}) and 18 days under elevated light (luminous intensity 32 W/m^{-2}). Continuous hydrogen production by *A. cylindrica* for a prolonged period under light-limited conditions occurs in the absence of exogenous nitrogen. The effect of light on nitrogenase-mediated hydrogen production by most cyanobacteria is well studied. Nitrogenase function is saturated only at much higher light intensities than required for optimal growth. The rate of hydrogen production rates can be doubled if the luminous intensity exposure to cultures is changed from 20 to 60 W/m^2 (Hallenbeck et al., 1978).

11.4.2 Temperature

Temperature is an important environmental factor that has a profound impact on algal growth rate, cell size, biochemical composition, and nutrient requirements. The optimum temperature for hydrogen production by most microalgae ranges from 30–40°C and varies from species to species. In *Nostoc* cultured at 22°C, higher rates of hydrogen production were observed than at 32°C, while *N. muscorum* SPU 004 showed optimum hydrogen production at 40°C. In *A. variabilis* SPU 003, optimum hydrogen production was reported at 30°C (Dutta et al, 2005).

11.4.3 Salinity

Salinity affects the hydrogen production by microalgae. In freshwater, increasing salinity lowers hydrogen production because of the diversion of energy reductants for extrusion of sodium ions from within the cells or prevention of sodium ions influx. The production of hydrogen in marine cyanobacteria is unaffected by salinity to some extent (Rai and Abraham, 1995).

11.4.4 Carbon source

Carbon sources influence the hydrogen production considerably by influencing nitrogenase activity. The presence of different carbon sources causes variation in electron donation capabilities by the cofactor compounds to nitrogenase, thus influencing hydrogen production. The presence of simple organic compounds increases hydrogen production because electron donation by cofactor compounds enhances nitrogenase activity.

The addition of various sugars stimulated the production of hydrogen, with mannose giving the highest rate, 5.58 nmol of hydrogen produced per 1 mg dry weight per hour (Datta et al., 2000).

11.4.5 *Nitrogen source*

A number of nitrogenous compounds affect hydrogen production. Nitrate, nitrite, and ammonium inhibit nitrogenase activity in *A. variabilis* SPU 003 and *A. cylindrica*. In general, the nitrogenase activity is inhibited by nitrogenase sources (Datta et al., 2000).

11.4.6 *Molecular nitrogen*

The production of hydrogen is inhibited by the presence of molecular nitrogen because molecular nitrogen is a competitive inhibitor for hydrogen production.

11.4.7 *Oxygen*

Hydrogen-producing enzymes nitrogenase and hydrogenase are very sensitive to oxygen, so these enzymes function in anaerobic ambience. This problem can be encountered by the aid of biotechnological tools. Under aerobic outdoor conditions operated for 4 months, a maximum rate of 80 mL of hydrogen per reactor (4.35 L) per hour was obtained from a batch culture on a bright day (Fedorov et al., 2001).

11.4.8 *Sulfur*

The starvation of sulfur enhances hydrogen production as reported in *Gloeocapsa alpicola* and *Synechocystis* PCC 6803 (Antal and Lindbad, 2005). It is possible to inhibit the oxygenic photosynthesis and enhance hydrogen production by incubating in nutrients that lack sulfur. Sulfur is a very important component in the Photosystem II (PSII) repair cycle, and without sulfur the protein biosynthesis is greatly impaired and production of either cysteine or methionine becomes impossible. This results in a lack of the D1 protein (32-kDa reaction center protein), which is essential for Photosystem II and needs to be constantly replaced. For these reasons, during sulfur deprivation photosynthesis and respiration decrease, even in the presence of light. Because photosynthesis declines much quicker than respiration, an equilibrium point is reached after a while (usually after 22 h) and after that the amount of oxygen that is used in respiration is greater than the oxygen released by photosynthesis and the cell become anaerobic and at this point hydrogen production occurs in higher amounts, reaching peak production.

11.4.9 *Methane*

The presence of methane enhances hydrogen production in *G. alpicola* and *Synechocystis* PCC 6803 during dark anoxic conditions (Antal and Lindbad, 2005). Increased hydrogen production (up to four times) is observed in *G. alpicola* and *Synechocystis* PCC 6803 during dark anoxic incubation when methane is present and the medium pH is between 5.0 and 5.5. The effect of methane on the hydrogen evolution was maximal during the first hour of the incubation, followed by gradual declination.

11.4.10 *Micronutrients*

Many trace elements like Cu, Co, Mo, Zn, and Ni show pronounced enhancement in hydrogen production due to their involvement in the enzyme nitrogenase. Nickel adaptation permits normal biomass accumulation while significantly increasing the rate of fermentative hydrogen production in *Arthrospira maxima.* Reports show that the hydrogenase activities in cell extracts (*in vitro*) and whole cells (*in vivo*) correlates with the amount of Ni2þ in the growth medium (saturating activity at 1:5 mM Ni2þ). This and higher levels of nickel in the medium during photoautotrophic growth cause stress, leading to chlorophyll degradation and a retarded growth rate that is severe at ambient solar flux. Relative to nickel-free media (only extraneous Ni2þ), the average hydrogenase activity in cell extracts (*in vitro*) increases by 18-fold, while the average rate of intracellular H_2 production within intact cells increases sixfold. Nickel is inferred to be a limiting cofactor for hydrogenase activity in many cyanobacteria grown using photoautotrophic conditions, particularly those lacking a high-affinity Ni2þ transport system (Datta, 2000).

11.4.11 *Role of uptake hydrogenase*

The production of hydrogen is affected by the consumption of hydrogen produced by uptake hydrogenase. The produced hydrogen is usually lost due to the activity of the uptake hydrogenase. Higher hydrogen production can be achieved through knocking out the genes coding uptake hydrogenase. Deletion of this gene by mutation is reported to enhance production (Pandey et al., 2007). Furthermore, overexpression of the genes for bidirectional hydrogenase is also required.

11.4.12 *Metabolic potential of microorganisms*

On the basis of type of microalgal species selected for production, the efficiency of hydrogen production varies. The acumen of microbes to convert light energy to chemical energy of hydrogen by photosynthetic microorganisms is reported to be low (Hall et al., 1995).

11.5 Challenges in algal hydrogen production

There are several challenges to be encountered before exploiting the economic potential of producing hydrogen photobiologically. In addition to *Chlamydomonas*, several other algal species, including species of *Chlorella* (Rashid et al., 2011), *Scenedesmus* (Schulz et al., 1998), and *Tetraselmis* (D'Adamo et al., 2014), have also been reported to produce H_2 but at lower levels. The *Chlamydomonas reinhardtii* system has the advantage that all three genomes (nuclear, chloroplast, and mitochondrial) have been sequenced and transformed, and much more detailed information is available on the organization of photosynthetic complexes during H_2 production (Tolleter et al., 2011). Transformation techniques are well established and include particle bombardment, glass bead or silicon-carbide whisker methods, electroporation, and *Agrobacterium tumefaciens*-mediated gene transfer (Kumar et al., 2004). *Chlamydomonas* therefore remains the best characterized model organism for microalgal H_2 production, associated genetic engineering, and the identification of molecular bottlenecks.

11.5.1 Oxygen sensitivity of the hydrogenase

One of the major bottlenecks is overcoming the oxygen sensitivity of the hydrogenase enzymes, addressing the competition between hydrogenases and other enzymes for photosynthetic reductants, preventing the downregulation of photosynthesis caused by the nondissipation of the proton gradient across the photosynthetic membrane, and ensuring adequate efficiency when capturing and converting solar energy. Because both algal and cyanobacterial hydrogenases are sensitive to oxygen, an essential product of photosynthesis, hydrogen is produced only transiently when such cells are exposed to light. There are several strategies for extending the catalytic lifetime of such hydrogenases, including:

1. Computational simulation of pathways for oxygen gas diffusion into the catalytic site of hydrogenases and molecular engineering of these pathways, perhaps by narrowing the gas channels, to block O_2 from reaching the catalytic site.
2. Mutagenizing the hydrogenase gene and then screening for an oxygen-tolerant version of this enzyme.
3. Employing a metabolic switch such as sulfate deprivation to downregulate PSII-catalyzed oxygen evolution, inducing an anaerobic environment to sustain hydrogenase activity in the light.

4. Searching for more oxygen-tolerant hydrogenases from nature and then transferring such genes into green algae and cyanobacteria. Except for the case in which oxygen production would be suppressed, it will be necessary to separate the hydrogen and oxygen being produced to avoid accumulating flammable or explosive mixtures.

11.5.2 *Alteration of the thylakoid proton gradient*

Proton supply to the hydrogenase is another potential bottleneck. The transport of e^- from water to ferredoxin (PETF) via the photosynthetic electron transport chain is involved in establishing a proton gradient across the thylakoid membrane, which drives adenosine triphosphate (ATP) synthesis. While ATP production is essential during CO_2 fixation, the ATP requirement drops during H_2 production (Das et al., 2014) and electron transport is reduced at the point of cytochrome b_6f, leading to an impaired dissipation of the proton gradient and therefore decreased proton availability for the hydrogenase and reduced H_2 production. One strategy to improve H_2 production is to artificially dissipate the proton gradient to increase H_2 production transiently in the presence of the chemical uncoupler carbonyl cyanide *m*-chlorophenyl hydrazine (CCCP), which causes an efflux of H^+ from the thylakoid lumen into the stroma (Lee, 2013). This suggests that the integration of a proton channel into the thylakoid membrane could more permanently restore proton and electron flow to the hydrogenase. Such a proton channel, however, would need to be inducibly expressed because the addition of the uncoupler prior to anaerobiosis was found to abolish hydrogenase activity, suggesting that the proton gradient is important for initial hydrogenase expression (Lee, 2013) and aerobic growth. A similar strategy involves the development of a leaky ATPase to increase proton flow and reduce ATP production (Das et al., 2014). Reduced ATP production caused by the introduction of a proton channel or mutated ATPase may additionally reduce reactions competing for reducing equivalents and therefore increase electron supply to the hydrogenase (Kumar and Das, 2013).

11.5.3 *Optimizing light-capture efficiency*

The light-harvesting complex (LHC) antennae systems have a dual role: to capture photons and dissipate excess light energy to provide photoprotection (Takahashi et al., 2006). Biomass production efficiency, at least in the laboratory, can be improved through LHC antenna reduction because this enhances light distribution through the bioreactors, enables the use of increased operational cell concentrations, and can yield improved overall photosynthetic efficiencies of these systems (Beckmann et al., 2009).

Furthermore, *C. reinhardtii* antenna mutants have also been reported to exhibit an earlier onset of H_2 production under S-deprivation conditions (Oey et al., 2013), which is likely due to three factors:

1. Improved light distribution leading to higher photon conversion efficiencies of the overall culture.
2. The ability to lower the dissolved O_2 concentration through the use of higher cell densities, which balance O_2 production with metabolic load.
3. Altered photoinhibition and stabilization of PSII required for subsequent H_2 production (Volgusheva et al., 2013). Collectively, these properties resulted in cultures exhibiting higher H_2 production rates under the conditions tested (Oey et al., 2013).

The precise engineering of the LHC antenna requires a detailed understanding of the structural complexity and dynamic response of algae to light in larger-scale production systems (de Mooij et al., 2015). To date, the engineering of the nuclear-encoded antenna genes has utilized chemical or random insertion mutagenesis (Polle et al., 2002, 2003), the manipulation of antenna regulation proteins (e.g., NAB1 [Beckmann et al., 2009]), or RNA interference (RNAi) knockdown approaches (Oey et al., 2013, 2016).

The least-specific method of engineering antenna mutants involves the introduction of foreign DNA into the nuclear genome via random insertion (Zhang et al., 2014). As clear phenotypes are needed for the rapid screening of mutants, the expected light-green phenotypes of putative antenna mutants seems to be an attractive selection criteria. However, while this approach identifies mutants impaired in antenna or chlorophyll synthesis, it does not select for specific multiple mutations likely required for optimal H_2 production.

A more targeted approach to engineer antenna cell lines is via the indirect route of manipulating antenna regulation proteins. An example of this is the overexpression of the translational repressor NAB1, which results in the specific downregulation of specific LHC proteins (Beckmann et al., 2009). This approach relies on cellular regulatory mechanisms and is therefore limited in its target scope. Another approach is RNAi-mediated knockdown (De Riso et al., 2009) of specific target genes such as LHC genes, or genes involved in chlorophyll biosynthesis. A challenge for precise antenna engineering is that the coding regions of the LHC genes are highly homologous (Natali and Croce, 2015), which complicates the specific downregulation of target LHC genes. Additionally, the need to maintain the RNAi-expressing mutants, typically through ongoing selective pressure, makes this approach less than ideal for industrial scale-up. RNAi is therefore best suited for proof-of-principle studies to identify target genes.

The most elegant technologies facilitate precise and permanent genome editing enabling the fine-tuning of antenna genes and adjustments to the pigment content, expanding the available solar spectrum (Blankenship and Chen, 2013). Several genome editing systems have been developed in recent years, including zinc-finger nucleases (ZFNs) (Sizova et al., 2013), transcription activator-like effector nucleases (TALENs) (Gao et al., 2014), and the CRISPR/Cas systems (Cho et al., 2013; Mali et al., 2013), all of which rely on nuclease-induced DNA strand breaks and endogenous cell repair mechanisms to obtain mutants with specifically edited genomes. This allows processes such as photosynthesis to be fine-tuned for biotechnological applications. Another attraction of these techniques is the potential to develop mutants that are nongenetically modified organisms due to the transient introduction of the required nucleases. This would be a significant advance for large-scale applications such as biofuel production, particularly given current legislative limitations on the use of genetically modified organism (GMO) strains.

11.5.4 *Electron supply to the hydrogenase and availability of reduced PETF*

Limited electron flow to the hydrogenase is another potential bottleneck for sustainable H_2 production (Hallenbeck and Benemann, 2002). This can occur due to the limited availability of reduced PETF as a result of other competing pathways (Winkler et al., 2011) (e.g., ferredoxin-NADP+ reductase [FNR], sulfite reductase, nitrate reductase, glutamate synthase, and fatty acid desaturases). The hydrogenase (HYDA) can accept e^- via a direct (PSII-dependent; two photons per electron to HYDA) or indirect (PSII-independent from starch; three photons per electron to HYDA) route (Chochois et al., 2009). To improve electron flow, PETF, FNR, and the hydrogenase have all been engineered (Wittenberg et al., 2013) with a particular focus on the improvement of the affinity of the hydrogenase to PETF, the reduction of the affinity of PETF for FNR, and the fusion of PETF and PSI with the hydrogenase. However, the identified candidates have so far only been tested *in vitro* and are yet to be assessed for their performance *in vivo*. Photosynthetic, H_2-producing algal cells supply an excellent scaffold to carry out such proof-of-principle studies. While most engineering efforts have focused on the chloroplast-localized nuclear-encoded genes, recent efforts have been made toward the *in situ* overexpression of the hydrogenase in the plastid to uncouple it from its native control system (Reifschneider-Wegner et al., 2014). Engineering has also focused on indirect targets, including electron competitors such as ribulose-1,5-bisphosphate carboxylase/oxygenase (RuBisCo) (Pinto et al., 2013), cyclic electron flow (Tolleter et al., 2011), starch degradation (Chochois et al., 2009), and respiration (Ruehle et al., 2008), with mutants in all of these pathways reportedly yielding increased H_2 production levels. Finally, additional components have been added to the electron transfer pathway, including a plastid-expressed NAD(P)H dehydrogenase (Baltz et al., 2014), a hexose transporter (Doebbe et al., 2007), and exogenous hydrogenases (Chien et al., 2012).

11.6 Conclusions and future prospects

Hydrogen production mediated by microbes has many advantages and the requirements are relatively simple, such as photobioreactor and optimum conditions. Algal hydrogen production is a simple photosynthetic process in the source of hydrogen, which is nothing but water.

11.6.1 Engineering high-efficiency strains

The selection and engineering of high-efficiency microalgae cell lines is central to this process. Lessons learned from the optimization of microalgal H_2 production suggest that the highest efficiency strains will require no single mutations, but significant re-engineering of the H_2 production process. For example, downregulating the light-harvesting proteins has benefits because light-green cell lines enable higher cell densities to be used, resulting in more rapid O_2 scavenging, the early onset of H_2 production, and an overall higher H_2 yield (Oey et al., 2013).

11.6.2 Microbial electrolysis cells

The microbial electrolysis cells are designed to produce H_2 from electrogenic microorganisms which consume and release e^-, H^+, and CO_2. An externally supplied voltage allows the formation of H_2 gas from the released H^+. Because the H_2 production process itself does not occur inside the cell, microbial electrolysis cells are, strictly speaking, not biohydrogen production platforms. Additionally, due to the requirement of the external supply of an electric current and carbon-based feedstocks, the economic viability of scaling up this H_2 production strategy must be carefully considered (Heidrich et al., 2014).

11.6.3 Innovative photobioreactors

The scale-up of microalgae systems from the laboratory to commercial facilities theoretically offers the potential to couple H_2 fuel production to CO_2 sequestration (e.g., the residual biomass to biochar) or other coproducts. To establish the viability of a given process, rigorous techno-economic analyses (TEAs) and life-cycle analyses (LCAs), as well as pilot- and demonstration-scale trials, are required to confirm positive economic returns and energy returned on energy invested (ERoEI), as well as reduced greenhouse gas emissions. Achieving these targets will not only involve the engineering of next-generation cell lines as described above, but the design of photobioreactors able to deliver optimized production conditions for the chosen production strains and specific geographic locations (Slegers, 2014).

With the existing knowledge of bioengineering, it is possible to obtain a sufficient amount of hydrogen that on combustion will liberate energy and therefore could act as a substitute of coal in several operations.

Wide-scale usage of hydrogen as a fuel will bring about technological developments in many areas like agriculture, power generation, and the automotive industry. Moreover, many avenues will be encountered, such as the greenhouse effect being mitigated and leading to environmental conservation, boosting the economy of nations, and finding sustainable approaches for a better future.

References

Antal, T. K. and Lindblad, P. 2005. Production of H_2 by sulphur-deprived cells of the unicellular cyanobacteria *Gloeocapsa alpicola* and *Synechocystis* sp. PCC 6803 during dark incubation with methane or at various extracellular pH. *J Appl Microbiol*, 98: 114–120.

Baltz, A., Dang, K. V., Beyly, A. et al. 2014. Plastidial expression of Type II NAD(P) H dehydrogenase increases the reducing state of plastoquinones and hydrogen photoproduction rate by the indirect pathway in *Chlamydomonas reinhardtii*. *Plant Physiol*, 165: 1344–1352.

Beckmann, J., Lehr, F., Finazzi, G. et al. 2009. Improvement of light to biomass conversion by de regulation of light harvesting protein translation in *Chlamydomonas reinhardtii*. *J Biotechnol*, 142: 70–77.

Blankenship, R. E. and Chen, M. 2013. Spectral expansion and antenna reduction can enhance photosynthesis for energy production. *Curr Opin Chem Biol*, 17: 457–461.

Chien, L. F., Kuo, T. T., Liu, B. H. et al. 2012. Solar to bioH_2 production enhanced by homologous overexpression of hydrogenase in green alga *Chlorella* sp. DT. *Int J Hydrogen Energy*, 37:17738–17748.

Cho, S. W., Kim, S., Kim, J. M., and Kim, J. S. 2013. Targeted genome engineering in human cells with the Cas9 RNA guided endonuclease. *Nat Biotech*, 31: 230–232.

Chochois, V., Dauvillee, D., Beyly, A. et al. 2009. Hydrogen production in *Chlamydomonas*: Photosystem II dependent and independent pathways differ in their requirement for starch metabolism. *Plant Physiol*, 151: 631–640.

D'Adamo, S., Jinkerson, R. E., Boyd, E. S. et al. 2014. Evolutionary and biotechnological implications of robust hydrogenase activity in halophilic strains of *Tetraselmis*. *PLoS One*, 9: e85812

Das, D., Khanna, N., and Dasgupta, C. N. 2014. *Biohydrogen Production: Fundamentals and Technology Advances*, Boca Raton, FL: CRC Press.

Datta, M., Nikki, G., and Shah, V. 2000. Cyanobacterial hydrogen production. *World J Microbiol Biotechnol*, 16: 8–9.

De Mooij, T., Janssen, M., Cerezo Chinarro, O. et al. 2015. Antenna size reduction as a strategy to increase biomass productivity: A great potential not yet realized. *J Appl Phycol*, 27: 1063–1077.

De Riso, V., Raniello, R., Maumus, F. et al. 2009. Gene silencing in the marine diatom *Phaeodactylum tricornutum*. *Nucleic Acids Res*, 37: e96.

Doebbe, A., Rupprecht, J., Beckmann, J. et al. 2007. Functional integration of the HUP1 hexose symporter gene into the genome of *C. reinhardtii*: Impacts on biological H_2 production. *J Biotechnol*, 131: 27–33.

Dutta, D., De, D., Chaudhuri, S., and Bhattacharya, S. K. 2005. Hydrogen production by cyanobacteria. *Microb Cell Fact*, 4: 36–36.

Fedorov, A. S., Tsygankov, A. A., Rao, K. K., and Hall, D. O. 2001. Production of hydrogen by an *Anabaena variabilis* mutant in photobioreactor under aerobic outdoor conditions. In: Miyake, J., Matsunaga, T., and San Pietro, A. (Eds.), *BioHydrogen II*, pp. 223–228, Amsterdam: Elsevier.

Gaffron, H. 1939. Reduction of CO_2 with H_2 in green plants. *Nature*, 143: 204–205.

Gao, H., Wright, D. A., Li, T. et al. 2014. TALE activation of endogenous genes in *Chlamydomonas reinhardtii*. *Algal Res*, 1: 52–60.

Gordillo, F. J. L., Goutx, M., Figueroa, F. L., and Niell, F. X. 1998. Effects of light intensity, CO_2 and nitrogen supply on lipid class composition of *Dunaliella viridis*. *J Appl Phycol*, 10: 135–144.

Hall, D. O., Markov, S. A., Watanbe, Y., and Rao, K. K. 1995. The potential applications of cyanobacterial photosynthesis for clean technologies. *Photosynth Res*, 46: 159–167.

Hallenbeck, P. C. and Benemann, J. R. 2002. Biological hydrogen production: Fundamentals and limiting processes. *Int J Hydrogen Energy*, 27: 1185–1193.

Hallenbeck, P. C., Kochian, L. V., Weissmann, J. C., and Benemann, J. R. 1978. Solar energy conversion with hydrogen producing cultures of the blue green alga, *Anabaena cylindrica*. *Biotechnol Bioeng Symp*, 8: 283–297.

Heidrich, E. S., Edwards, S. R., Dolfing, J., Cotterill, S. E., and Curtis, T. P. 2014. Performance of a pilot scale microbial electrolysis cell fed on domestic wastewater at ambient temperatures for a 12 month period. *Bioresour Technol*, 173: 87–95.

Kapdan, I. K. and Kargi, F. 2006. Bio-hydrogen production from waste materials. *Enzyme Microb Technol*, 38, 569–582.

Kumar, K. and Das, D. 2013. CO_2 sequestration and hydrogen production using cyanobacteria and green algae. In: *Natural and Artificial Photosynthesis*, Reza, R. (Ed.), pp. 173–215. Hoboken: John Wiley & Sons Inc.

Kumar, S. V., Misquitta, R. W., Reddy, V. S., Rao, B. J., and Rajam, M. V. 2004. Genetic transformation of the green alga *Chlamydomonas reinhardtii* by *Agrobacterium tumefaciens*. *Plant Sci*, 166: 731–738.

Lee, J. 2013. Designer transgenic algae for photobiological production of hydrogen from water. In: *Advanced Biofuels and Bioproducts*, Lee, J. W. (Ed.), pp. 371–404. New York: Springer US.

Mali, P., Yang, L., Esvelt, K. M. et al. 2013. RNA guided human genome engineering via Cas9. *Science*, 339: 823–826.

Melis, A. and Happe, T. 2001. Hydrogen production. Green algae as a source of energy. *Plant Physiol*, 127: 740–748.

Natali, A. and Croce, R. 2015. Characterization of the major light harvesting complexes (LHCBM) of the green alga *Chlamydomonas reinhardtii*. *PLoS One*, 10: e0119211.

Oey, M., Ross, I. L., Stephens, E. et al. 2013. RNAi knock down of LHCBM1, 2 and 3 increases photosynthetic H_2 production efficiency of the green alga *Chlamydomonas reinhardtii*. *PLoS One*, 8: e61375.

Oey, M., Sawyer, A. L, Ross, I. L, and Hankamer, B. 2016. Challenges and opportunities for hydrogen production from microalgae. *Plant Biotech J*, 14: 1487–1499.

Pandey, A., Pandey, A., Srivastava, P., and Pandey, A. 2007. Using reverse micelle as microreactor for hydrogen production by coupled system of *Nostoc*/P4 and *Anabaena*/P4. *World J Microbiol Biotechnol*, 23: 269–274.

Pinto, T. S., Malcata, F. X., Arrabaca, J. D. et al. 2013. Rubisco mutants of *Chlamydomonas reinhardtii* enhance photosynthetic hydrogen production. *Appl Microbiol Biotechnol*, 97: 5635–5643.

Polle, J. E. W., Kanakagiri, S., Jin, E., Masuda, T., and Melis, A. 2002. Truncated chlorophyll antenna size of the photosystems—A practical method to improve microalgal productivity and hydrogen production in mass culture. *Int J Hydrogen Energy*, 27: 1257–1264.

Polle, J. E. W., Kanakagiri, S. D., and Melis, A. 2003. Tla1, a DNA insertional transformant of the green alga *Chlamydomonas reinhardtii* with a truncated light harvesting chlorophyll antenna size. *Planta*, 217: 49–59.

Rai, A. K. and Abraham, G. 1995. Relationship of combined nitrogen sources to salt tolerance in freshwater cyanobacterium *Anabena doliolum*. *J App Bact*, 78: 501–506.

Ran, C. Q., Chen, Z. A., Zhang, W., Yu, X. J., and Jin, M. F. 2006. Characterization of photo-biological hydrogen production by several marine green algae. *Wuhan Ligong Daxue Xuebao*, 28(Suppl 2): 258–263.

Rashid, N., Lee, K., and Mahmood, Q. 2011. Bio-hydrogen production by *Chlorella vulgaris* under diverse photoperiods. *Bioresour Technol*, 102: 2101–2104.

Reifschneider-Wegner, K., Kanygin, A., and Redding, K. E. 2014. Expression of the [FeFe] hydrogenase in the chloroplast of *Chlamydomonas reinhardtii*. *Int. J. Hydrogen Energy*, 39: 3657–3665.

Ruehle, T., Hemschemeier, A., Melis, A., and Happe, T. 2008. A novel screening protocol for the isolation of hydrogen producing *Chlamydomonas reinhardtii* strains. *BMC Plant Biol*. 8: 107.

Rupprecht, J., Hankamer, B., Mussgnug, J. H. et al. 2006. Perspectives and advances of biological H_2 production in microorganisms. *Appl Microbiol Biotechnol*, 72: 442–449.

Schulz, R., Schnackenberg, J., Stangier, K. et al. 1998. Light dependent hydrogen production of the green alga *Scenedesmus obliquus*. In: *BioHydrogen*, Zaborsky, O., Benemann, J., Matsunaga, T., Miyake, J., and San Pietro, A. (Eds.), pp. 243–251. New York: Springer US.

Sizova, I., Greiner, A., Awasthi, M., Kateriya, S., and Hegemann, P. 2013. Nuclear gene targeting in *Chlamydomonas* using engineered zinc finger nucleases. *Plant J*, 73: 873–882.

Slegers, P. M. 2014. Scenario studies for algae production. PhD Thesis. Wageningen University, Wageningen, Netherlands.

Takahashi, H., Iwai, M., Takahashi, Y., and Minagawa, J. 2006. Identification of the mobile light harvesting complex II polypeptides for state transitions in *Chlamydomonas reinhardtii*. *Proc Natl Acad Sci USA*, 103: 477–482.

Tiwari, A. 2014. *Cyanobacteria: Nature, Potentials and Applications*. New Delhi: Astral International Publishing House.

Tiwari, A. and Pandey, A. 2012. Cyanobacterial hydrogen production—A step towards clean environment. *Int J Hydrogen Energy*, 37: 139–150.
Tiwari, A. and Pandey, A. 2013. Algae derived biofuels. In: *Natural and Artificial Photosynthesis: Solar Power as an Energy Source*, Razeghifard, R. (Ed.). USA: John Wiley & Sons, Inc.
Tolleter, D., Ghysels, B., Alric, J. et al. 2011. Control of hydrogen photoproduction by the proton gradient generated by cyclic electron flow in *Chlamydomonas reinhardtii*. *Plant Cell*, 23: 2619–2630.
Tsygankov, A. A., Borodin, V. B., Rao, K. K., and Hall, D. O. 1999. H_2 photoproduction by batch culture of *Anabaena variabilis* ATCC 29413 and its mutant PK84 in a photobioreactor. *Biotechnol Bioeng*, 64: 709–715.
Volgusheva, A., Styring, S., and Mamedov, F. 2013. Increased photosystem II stability promotes H_2 production in sulfur deprived *Chlamydomonas reinhardtii*. *Proc Natl Acad Sci USA*, 110: 7223–7228.
Wang, Z., Pan, Y., Dong, T. et al. 2007. Production of hydrogen from catalytic steam reforming of bio-oil using C12A7-O-based catalysts. *Appl Catal*, 320: 24–34.
Winkler, M., Kawelke, S., and Happe, T. 2011. Light driven hydrogen production in protein based semi artificial systems. *Bioresour Technol*, 102: 8493–8500.
Wittenberg, G., Sheffler, W., Darchi, D., Baker, D., and Noy, D. 2013. Accelerated electron transport from photosystem I to redox partners by covalently linked ferredoxin. *Phys Chem Chem Phys*, 15: 19608–19614.
Zhang, R., Patena, W., Armbruster, U. et al. 2014. High throughput genotyping of green algal mutants reveals random distribution of mutagenic insertion sites and endonucleolytic cleavage of transforming DNA. *Plant Cell*, 26: 398–1409.

chapter twelve

Biohydrogen production in microbial electrolysis cells from renewable resources

Abudukeremu Kadier, Yong Jiang, Bin Lai, Pankaj Kumar Rai, Kuppam Chandrasekhar, Azah Mohamed, and Mohd Sahaid Kalil

Contents

12.1 Introduction

There is a growing demand for new energy sources due to global climate change and fossil fuel depletions, and future energy sources must meet the requirements of being carbon-free and renewable (Nikolaidis and Poullikkas, 2017; Kadier et al., 2017a). Recently, H_2 has been experiencing a striking increase in interest, and it has been globally accepted as the best alternative to the carbonaceous and rapidly depleting fossil fuels (Parkhey and Gupta, 2017). H_2 offers many advantages compared to other alternative energy sources. First, H_2 is an environmentally safe combustion fuel that is almost free of CO_2 and other greenhouse gas emission because its only oxidation product is water (Das, 2009; Sinha and Pandey, 2011; Abdeshahian et al., 2016). Second, it has the highest energy density: the energy content for H_2 is 120–142 MJ/kg, while for other possible biofuels it is 50 MJ/kg (CH_4), 26.8 MJ/kg (ethanol), and 44 MJ/kg (gasoline) (Carmona-Martínez et al., 2015). Third, H_2 can be derived from a wide variety of renewable feedstock and domestic waste materials, and thus it can be cost-effective, clean, sustainable, and renewable (Liu et al., 2010). Fourth, H_2 could be produced at ambient temperature and pressure conditions (Das, 2009; Sinha and Pandey, 2011). Currently, some of the most common methods of hydrogen production at the commercial scale are steam reformation of hydrocarbons (gasification), pyrolysis, plasma reformation, and water electrolysis. However, such methods of hydrogen production are highly energy-intensive and expensive (Kotay and Das, 2008; Manish and Banerjee, 2008) and thus it is of utmost necessity to identify widely available and cheaper raw materials sources and develop clean and renewable methods of hydrogen production.

A wide variety of renewable options are available at the laboratory scale to produce biohydrogen such as direct and indirect biophotolysis, dark fermentation, photofermentation, and sequential dark fermentation and photofermentation (Nikolaidis and Poullikkas, 2017). Microbial electrolysis cells (MECs) are a new and promising approach for H_2 production from organic matter, including wastewater and other renewable resources (Kadier et al., 2014; Zhang and Angelidaki, 2014). In comparison with other hydrogen production technologies, the MEC offers some great advantages. First, the MECs potentially offer the possibility to produce H_2 with lower energy inputs of 0.6 kWh/m^3 H_2 (Steinbusch et al., 2009) and 1 kWh/m^3 H_2 (Rozendal et al., 2008a), which are significantly lower compared to the typical energy requirement of 4.5–50.6 kWh/m^3 H_2 for water electrolysis (Miyake et al., 1999). Furthermore, in MEC no precious metal anodes were needed because of the self-sustaining biocatalysts. Moreover, high hydrogen recovery or conversion efficiency is achievable in the MEC; the highest overall hydrogen yield obtained in MECs was much higher than that of any types of fermentative hydrogen production methods. For instance,

Cheng and Logan (2011) reported an efficiency of 8.55 mol H_2/1 mol glucose under 0.6 V compared with the typical 4 mol H_2/1 mol glucose achieved by dark fermentation (Hallenbeck and Ghosh, 2009). In addition, in MECs, relatively higher purity of hydrogen is produced, and thus expensive gas purification processes are not needed (Kadier et al., 2016a). What is more, MEC integrates both pollution treatment and energy production, and so it has the advantages of cleanliness, energy saving, and waste utilization. Finally, MECs produce other value-added bioproducts such as bio-CH_4, H_2O_2, and bio-C_2H_5OH using microbes as biocatalysts (Kadier et al., 2016a).

12.2 Operating principles of MEC

In MECs, electrochemically active bacteria (EAB) at the anode breakdown or oxidize organic substrate by a sequence of metabolic reactions, generating electrons, protons, and carbon dioxide (CO_2) (Kadier et al., 2015c; Modestra et al., 2015). The produced electrons are carried from the anode electrode to the cathode electrode through an external electric circuit while protons are directly diffused into the electrolyte solution. Even if not essential, a proton exchange membrane (PEM) can be employed to facilitate H^+ transfer as well as make the electrode reactions distinct. Subsequently, e^- and H^+ are chemically reduced at the cathode (cathode chamber) to produce H_2 (Figure 12.1) (Chandrasekhar et al., 2015; Kadier et al., 2015b). Therefore, H_2 in MEC reactors is produced by chemical reduction reactions. Nevertheless, this is not a spontaneous reaction: MEC requires an externally supplied voltage (greater than 200 mV) in order to produce H_2 at the cathode by the combination of these e^- and H^+ under biologically assisted conditions. Nonetheless, the MEC device utilizes relatively less energy input ranges between 200 and 800 mV compared to typical water electrolysis, which required input energy value varying between 1230 and 1800 mV. This process is capable of attaining high H_2 yields by consuming organic substrate and waste as a carbon-rich energy source (Kadier et al., 2014; Modestra et al., 2015).

12.3 Thermodynamics and bioelectrochemical evaluation of MEC

Inopportunely, the thermodynamic evaluation confirms that H_2 gas produced by MECs is not a spontaneous process. The normal reduction potential to produce H_2 gas is less than the available potential at the anode for the reduction reaction. In reality, this potential variance is even higher owing to an ohmic resistance through the MEC device and also due to the overpotentials at the electrode–electrolyte interface. As a result, additional energy must be provided from an external source to the

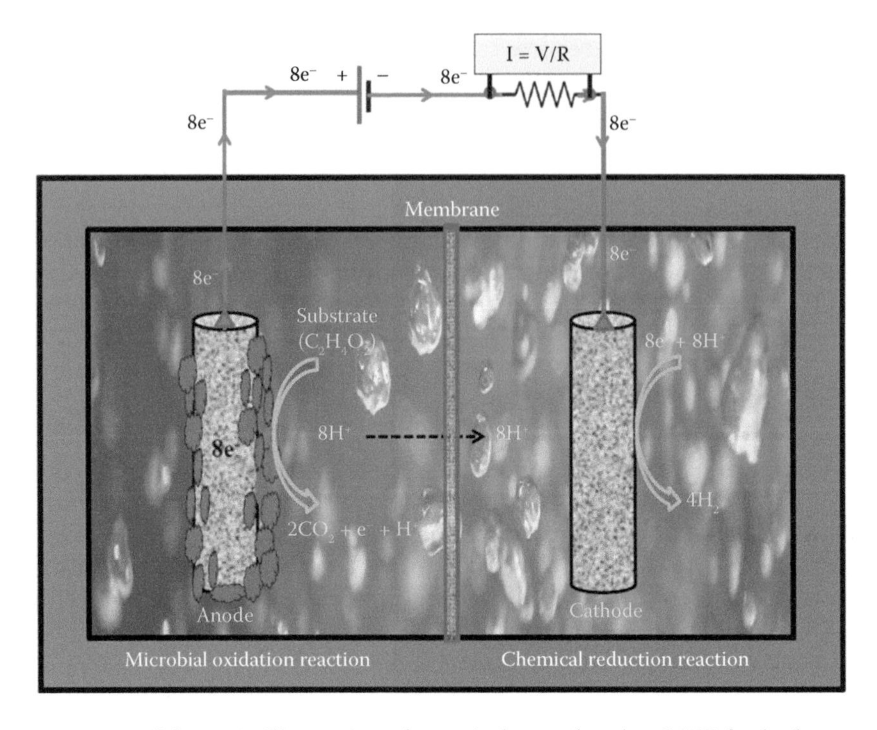

Figure 12.1 Schematic illustration of a typical two-chamber MEC for hydrogen production and its operation principles; the substrate gets oxidized at the anode compartment and generates e^- and H^+; the generated e^- and H^+ undergo a chemical reduction reaction to produce H_2 gas in the cathode compartment.

MEC system in order to drive e^- from the anode electrode to the cathode electrode. Hence, we can also term MECs as electrochemically assisted bioelectrochemical systems. The potential at the anode due to microbial oxidation of the substrate (acetate) is −0.28 V, whereas the potential at the cathode is −0.42 V for the chemical reduction reaction. The process can be described in the following electrochemical reactions, with acetate as an example substrate:

$$\text{Anode: } CH_3COO^- + 4H_2O \rightarrow 2HCO^{3-} + 9H^+ + 8e^- \quad (12.1)$$

$$\text{Cathode: } 8H^+ + 8e^- \rightarrow 4H_2 \ (E_{H+H_2} = -0.414\,\text{V versus normal hydrogen electrode}) \quad (12.2)$$

$$\text{Overall reaction in an MEC: } CH_3COO^- + 4H_2O \rightarrow 2HCO^{3-} + H^+ + 4H_2 \quad (12.3)$$

The reduction potential at the anode is much higher than at the cathode. Hence, only a little voltage from the external power source is required to drive this thermodynamically nonspontaneous reaction.

12.4 EAB and their electron transfer mechanisms

MECs are bioelectrochemical devices in which a chemical form of energy stored in the feedstock is transformed to H_2 through the metabolic activity of EAB (Chandrasekhar et al., 2015). Hence, the application of this novel technology in the wastewater treatment process must lead to green energy generation with simultaneous waste remediation. In recent years, the evidence of this process has growing interest in this field (Kadier et al., 2015a). The higher energy efficiency of MECs compared to traditional fermentation technology and adaptation to treatment of a wide range of substrates including wastewater has heightened academic interests (Guo et al., 2013b; Kadier et al., 2014). However, there are a few challenges in upscaling and commercialization of these new technologies for energy generation and/or waste remediation (Kadier et al., 2016a). In the direction of the commercialization of MECs, a wide-ranging investigation on EAB as biocatalysts in this bioelectrochemical process is useful. Understanding the principles of electron transfer between a microorganism and an electrode is essential in optimizing the current or hydrogen generation in MECs.

Until now, the exact mechanisms for electron transfer have not been completely understood, and three methods have been proposed to explain extracellular electron transfer (EET) to the anode in bioelectrochemical systems (BESs) as shown in Figure 12.2: (1) the direct electron transfer (DET) via outer-surface C-type cytochromes (Torres et al., 2010), (2) long-range or mediated electron transfer via electron shuttles (Newman and Kolter, 2000; von Canstein et al., 2008), and (3) electron

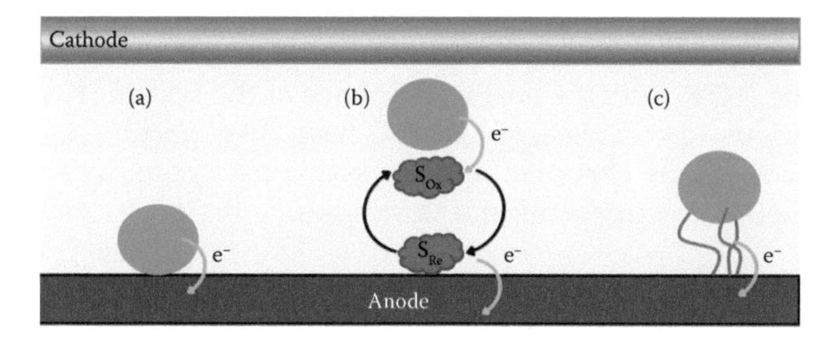

Figure 12.2 A schematic representation of possible extracellular electron transfer mechanisms from EAB to the anode surface: (a) direct transfer by contact; (b) indirect electron transfer by redox shuttles (S_{Ox} is oxidized electron shuttle and S_{Re} is reduced electron shuttle); and (c) electron transfer by conductive nanowire matrix.

transfer through conductive filamentous pili (i.e., nanowires) (Gorby et al., 2006; Reguera et al., 2005).

Direct electron transfer is related to membrane-bound or associated enzyme complexes and indirect electron transfer involves a reduction or oxidation of an organic or inorganic shuttle (pyocyanin and humic acids) and soluble compound shuttle (sulfur compounds and hydrogen). In an MEC, the composition of the microbial community will play a very important role because it impacts the e^- transport mechanisms to the electrode surface. In fact, few researchers have studied EAB communities for better understanding of biochemical and chemical interactions in the systems for example catalytic oxidation of substrate and competition among numerous EAB (Hasany et al., 2016). In addition, the type and nature of electrode materials as well as the applied potential range will also strongly influence the microbial community in MECs (Arunasri et al., 2016). Concurrently, the formations of biofilm and electrode oxidation processes are governed by the interaction of EAB with the type of electrode materials used. Furthermore, the chemical stability, surface charge, electrical conductivity, and electrical capacitance of the electrode used in MEC will greatly influence the interaction with EAB (Ambler and Logan, 2011; Zhang et al., 2010).

12.5 MEC reactor designs and the main operational modes

Numerous studies have demonstrated that to minimize the capital cost and to improve the hydrogen yield of MECs, the reactor designs and the operation parameters in different modes must be optimized (Kadier et al., 2016b). In this section, the reactor designs and the main operational modes of MECs are reviewed.

12.5.1 MEC reactor designs and configurations

The MEC reactor design directly affects the capital cost, the hydrogen production rate (HPR), and the energy efficiency of the system (Kadier et al., 2017b). Various reactor designs of MEC have been proposed, including two-chamber, single-chamber, and stacked construction by integrating MEC with other electrochemical technologies.

12.5.1.1 Two-chamber MECs

The use of two-chamber MECs, with the membrane as a separator, could reduce the crossover of organic matter and bacteria from the anode to the cathode chamber, and the crossover of hydrogen from cathode to anode, and it also helps to reduce the impurities such as carbon dioxide, methane, and hydrogen sulfide (Liu et al., 2010; Jiang et al., 2016). The use of a membrane also functions as a separator to avoid any short circuit, which is important because a short electrode distance is required in order to decrease the internal resistance.

The two-chamber configuration was first used as a concept proof of MECs by two groups almost at the same period (Liu et al., 2005; Rozendal et al., 2006b). However, in neutral condition, the transport rate of a proton and/or hydroxyl across the membrane is often limited, leading to a pH drop across the membrane (Rozendal et al., 2006a). The pH gradient across the membrane can inhibit the activity of EAB, increase the internal resistance, and lower the hydrogen production rate (Logan et al., 2008). The two-chamber MEC was proven to not be suitable to recover hydrogen from the nonbuffered wastewater due to the high pH drop (Ruiz et al., 2016). A novel operation mode of periodic polarity reversal was applied in two-chamber MECs for low buffered wastewater treatment (Jiang et al., 2016). The two-chamber MEC with flow through bioanode and flow through Ni foam cathode achieved a high HPR of 50 m^3 H_2/m^3 day (Jeremiasse et al., 2010).

The two-chamber MECs have been fabricated with different types, including bottle-type in Figure 12.3a (Liu et al., 2005), cylindrical type in Figure 12.3b (Cheng and Logan, 2007), disc-shaped (Rozendal et al., 2006b), and concentric tubular in Figure 12.3c (Kyazze et al., 2010). Two-chamber MECs with one liquid chamber and one gas chamber were also developed using a gas diffusion electrode (GDE) to stimulate the hydrogen gas evolution, as shown in Figure 12.3d (Tartakovsky et al., 2009). To scale up MEC, the packed bed electrode type in Figure 12.3e was used to increase the kinetic reaction rate per volume (Ditzig et al., 2007). A flat-type MEC in

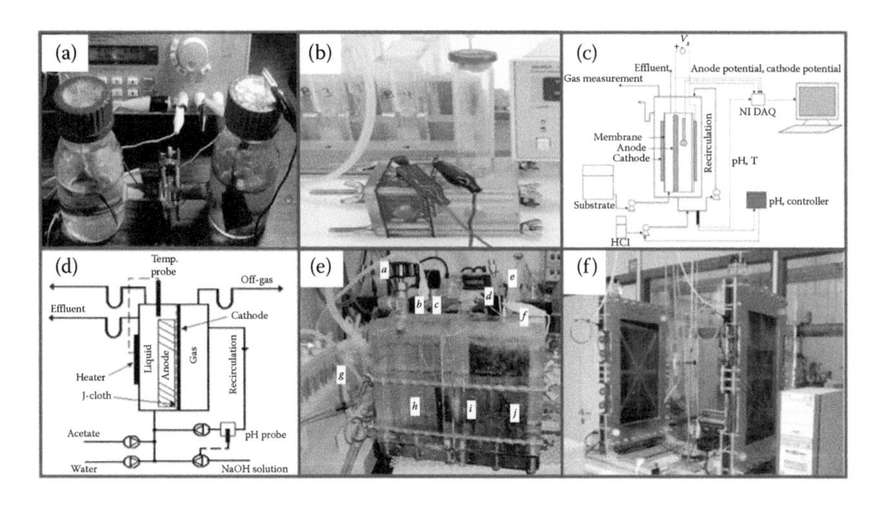

Figure 12.3 Two-chamber MECs used in previous studies.

Figure 12.3f was recently proposed to minimize the internal resistance by decreasing the distance of the electrodes and using the porous geotextile membrane as the separator (Escapa et al., 2015).

12.5.1.2 Single-chamber MECs

A single-chamber MEC was developed by removing membranes from two-chamber systems and soaking both the anode and the cathode in the same solution of one chamber. The use of single-chamber MECs can reduce the potential loss caused by the membrane resistance. It is easy to fabricate and to autoclave and it also avoids the problem related to membranes, such as fouling, degradation, and high cost (Liu et al., 2010). A continuous-flow single-chamber MEC with Ni-based gas diffusion cathode reached a HPR of 4.1 m^3 H_2/m^3 day (Manuel et al., 2010). The single-chamber MECs have been fabricated with different shapes with the main body made from various materials such as a glass bottle in Figure 12.4a (Hu et al., 2008) and a plastic cube in Figure 12.4b (Call and Logan, 2008). Small-scale MECs can be fabricated with glass serum bottles as in Figure 12.4c (Call and Logan, 2011), or glass serum tubes (Hu et al., 2009) for high-throughput

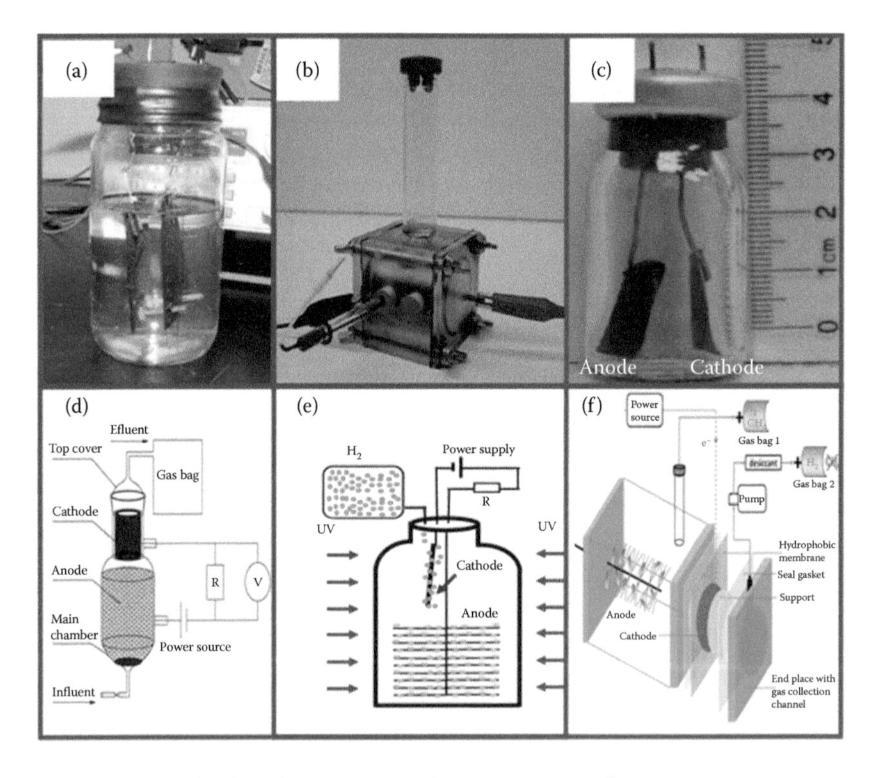

Figure 12.4 Single-chamber MECs used in previous studies.

bioelectrochemical research to select the EAB or the potential cathode materials. The greatest challenge in singlechamber MECs is the consumption of produced hydrogen by methanogens (Clauwaert and Verstraete, 2008) and the reoxidation of hydrogen by EAB (Lee and Rittmann, 2009). The decrease of hydrogen consumption in single-chamber MECs can be achieved by reactor construction and operation mode optimizing. For example, a cathode-on-top structure was proposed by putting the cathode above the anode for rapid recovery of hydrogen gas in Figure 12.4d (Guo et al., 2010; Lee et al., 2009). The ultraviolet irradiation was applied for methanogenesis inhibition in Figure 12.4e (Hou et al., 2014). Recently, the active gas harvesting method was applied to eliminate hydrogen consumption via rapid hydrogen extraction using a gas-permeable hydrophobic membrane and vacuum as in Figure 12.4f (Lu et al., 2016).

12.5.1.3 Stacked MECs

12.5.1.3.1 Multiple-electrode systems The use of multiple-electrode systems is a feasible way to increase the volumetric current densities of a scaled-up reactor because MECs must be scaled up to a certain extent for efficient electricity hydrogen production and wastewater treatment. A novel stacked stainless steel mesh cathode was designed to increase the cathode-to-anode ratio when using a carbon brush like the anode in Figure 12.5a (Guo et al., 2016). The electrode size and arrangement was optimized in a membraneless flat-plate MEC, and the best performance was achieved by using two gas diffusion cathodes sandwiched between the multiple anode and the hydrogen collection chamber in Figure 12.5b (Gil-Carrera et al., 2011). A 2.5-L MEC was designed with eight separate electrode pairs made of graphite brush and stainless steel mesh cathodes to examine the scalability of MEC in Figure 12.5c (Rader and Logan, 2010). The pilot-scale (1000 L) MEC constructed with the application of multiple-electrode systems contained 144 electrode pairs in 24 modules, and the current generation reached a maximum of 7.4 A/m^3, with the gas production reaching a maximum of 0.19 ± 0.04 m^3 H_2/m^3 day in Figure 12.5d (Cusick et al., 2011).

12.5.1.3.2 MEC connected to independent electrochemical reactor The MEC can be electrical and/or a water stream connected to an independent electrochemical reactor to provide a sustainable power supply, maximize the chemical recovery, and scale up the reactor. A microbial fuel cell (MFC) was used as the power supply and connected with a MEC in series in Figure 12.6a (Sun et al., 2008). HPR reached 0.002 m^3 H_2/m^3 day, with a R_{CAT} of 88%–96%. The number of MFC in the MFC–MEC coupled system was further increased and the electrical connection mode was also optimized (Jiang et al., 2014; Sun et al., 2009). Recently, the microbial desalination cell (MDC)–MEC coupled system was proposed to achieve

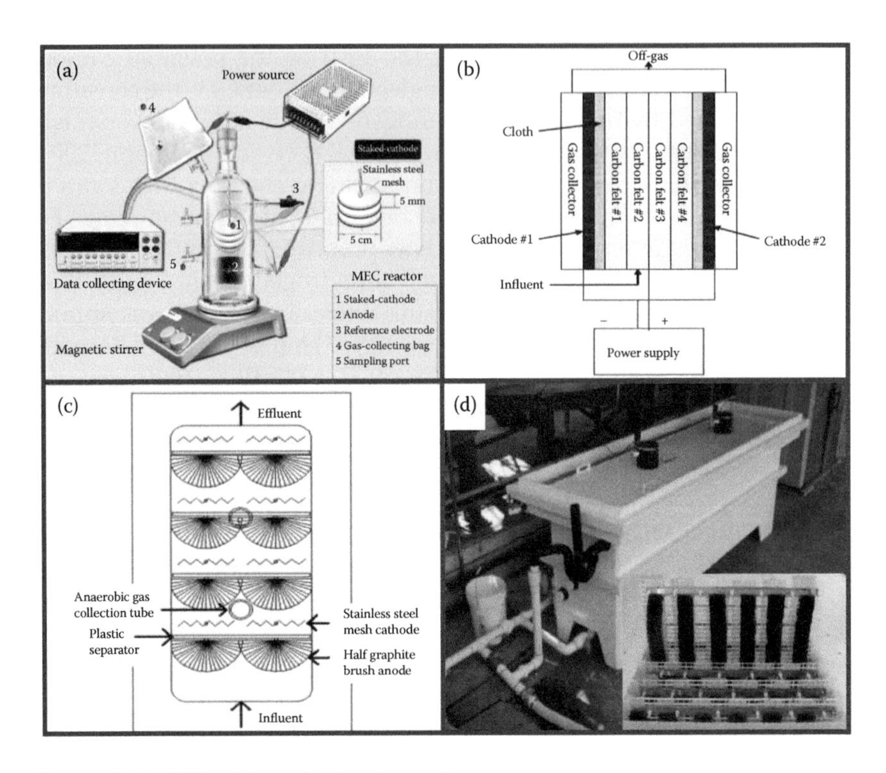

Figure 12.5 MEC with multiple-electrode systems.

the multiple goals of organic matter removal, desalination, and hydrogen production as in Figure 12.6b (Li et al., 2017). Other types of sustainable power supply devices, such as the thermoelectric microconverter (Chen et al., 2016) and dye-sensitized solar cell (DSSC) in Figure 12.6c (Chae et al., 2009), could also be used to provide the drive power for MEC. By connecting multiple MECs to a single DSSC, the HPR reached approximately 0.5 m^3 H_2/m^3 day, with substrate-to-hydrogen conversion efficiencies ranging from 42% to 65% (Ajayi et al., 2010). Recently, the use of a photocathode was also proposed, rather than using the independent photovoltaic cell, for solar energy conversion to provide power supply for MECs (Zang et al., 2014). A dark fermentation–MFC–MEC coupled system was proposed to maximize HPR by using fermentation effluent as a feed for MEC and using the MFC as a power supply as in Figure 12.6d, and a HPR of 0.48 m^3 H_2/m^3 day was achieved (Wang et al., 2011). When fermentable substrates are used, the MFC–MEC coupled system could be further used for multiple chemical productions, such as ethanol, hydrogen, and methane in Figure 12.6e (Sugnaux et al., 2016). The MFC–MEC coupled system could also be used for heavy metals recovery based on the difference of redox potential within the various heavy metals in the wastewater

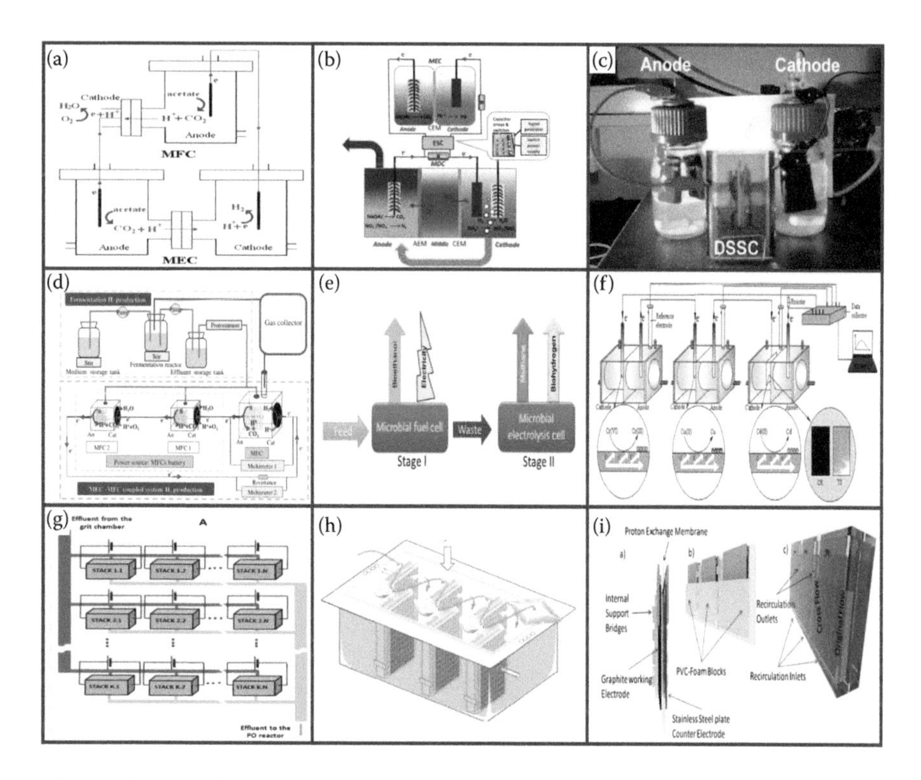

Figure 12.6 MECs connected to an independent electrochemical reactor.

in Figure 12.6e (Zhang et al., 2015). The scale up of MEC could be achieved by connecting multiple MEC models together, as shown in Figure 12.6g (Escapa et al., 2012), Figure 12.6i (Heidrich et al., 2013), and Figure 12.6h (Brown et al., 2014), rather than simply in a large-single MEC reactor.

12.5.1.3.3 Hybrid systems using MEC in situ integrated with other electrochemical technologies The MEC can be *in situ* integrated with other electrochemical technologies to construct various hybrid systems to extend the application area of MEC. For example, by combining the MEC with the MDC, the novel hybrid system termed a microbial electrolysis desalination cell (MEDC) was used as a novel device to desalinate salty water and to produce hydrogen gas as in Figure 12.7a (Mehanna et al., 2010). To solve the problem of pH differences in the anode and cathode during the desalination process, a microbial electrolysis desalination and chemical-production cell (MEDCC) with four chambers using a bipolar membrane was constructed (Chen et al., 2012). By exchanging the position of the anion exchange membrane (AEM) and the cation exchange membrane (CEM) in MEDC, the hybrid system was renamed a microbial saline-wastewater electrolysis cell (MSC) in Figure 12.7b (Kim and Logan, 2013). In MSC, the simultaneously

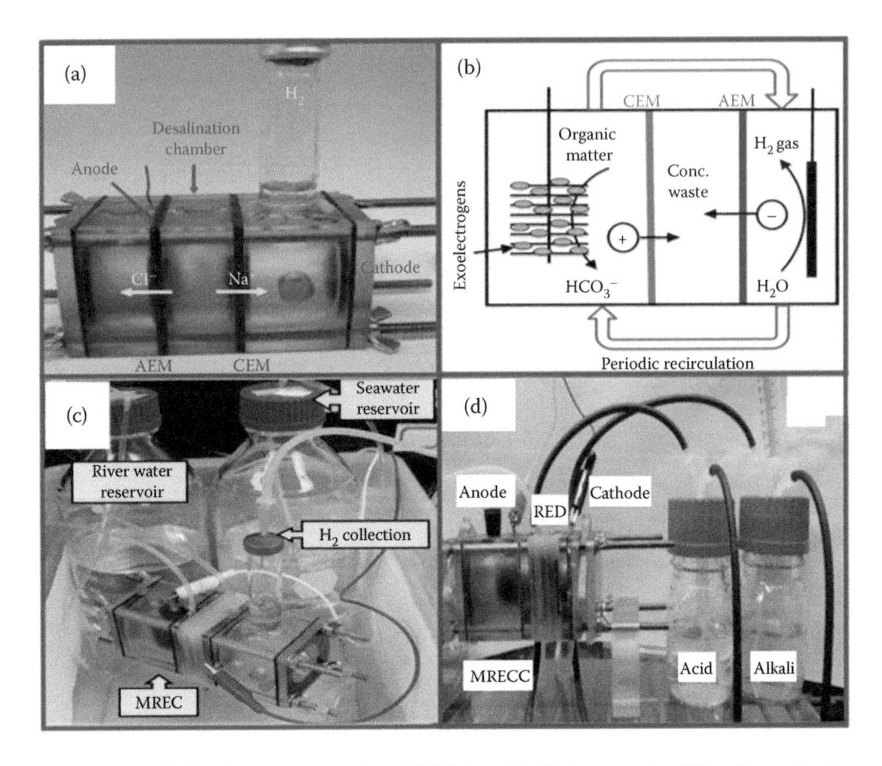

Figure 12.7 Hybrid systems using MEC *in situ* integrated with other electrochemical technologies.

organic matter removal and desalination within a single stream was achieved by using the bioanode to degrade organic matter and using the solution of the middle chamber to collect the concentrated wastewater. The structure of MSC could be further used for nutrient ions like ammonium and phosphate recovery from synthetic wastewater (Chen et al., 2015) and from urine (Ledezma et al., 2017). By combining a small reverse electrodialysis stack into an MEC, which is called a microbial reverse-electrodialysis electrolysis cell (MREC) in Figure 12.7c (Kim and Logan, 2011), both the energy from the organic matter oxidation and from the salinity gradient between seawater and river could be used for hydrogen production. The HPR increased from 0.8 to 1.6 m^3 H_2/m^3 day for seawater and river water flow rates ranging from 0.1 to 0.8 mL/min. The salinity gradient in MREC can be created by using the natural seawater and river water, and also the high and low concentrations of ammonium bicarbonate solution created by using low-grade waste heat sources (Nam et al., 2012). The productions of MREC have expanded beyond hydrogen with multiple chemicals such as methane, acetate, and hydrogen peroxide (Luo et al., 2014; Jiang et al., 2013). Moreover, the MRECs have been modified as the microbial reverse-electrodialysis electrolysis and chemical-production cell (MRECC) for hydrogen production and carbon dioxide sequestration by using the alkali produced in the cathode chamber in Figure 12.7d (Zhu et al., 2014).

12.5.2 *The main operational modes of MEC*

The operational modes could directly affect the performance of MEC in terms of hydrogen production rate, organic matter removal, and energy efficiency. In this section, the operational modes of MECs are reviewed.

12.5.2.1 *Batch modes*

The batch modes were frequently adopted in MEC studies as a concept of proof to evaluate the catalytic activity of novel cathode materials (Su et al., 2016), the feasibility of using new substrates in MEC (Selembo et al., 2009b; Liu et al., 2012), the efficiency of methanogen inhibition (Hu et al., 2008), and the possibility of producing other value-added chemicals (Cusick and Logan, 2012). For example, the exposure of cathodes to air was frequently adopted in fed-batch modes to suppress the growth of methanogens. A novel stainless steel fiber felt cathode was recently tested in MEC, and it was observed that the hydrogen production improved with the increase of the fed-batch circles due to a decrease in overpotential, which was caused by corrosion (Su et al., 2016). The batch mode was also applied in MEC for more accurately quantifying the hydrogen yield and electron flows (Lee et al., 2009). Moreover, the research experience obtained in fed-batch could be used as reference for continuous-flow modes.

12.5.2.2 *Fed-batch modes*

The fed-batch mode, in classical fermentation theory, was defined as "During fed-batch cultivation, one or more nutrients are supplied to the fermenter while cells and products remain in the fermenter until the end of operation" (Lee et al., 1999). According to this definition, the concentration of substrate in the fed-batch operated MEC, for example, acetate, should be added periodically without replacing the solution to control the substrate at a certain concentration and to avoid the potential inhibition of EAB caused by the high concentrations of substrates (Sharma and Li, 2010). However, in MEC studies, people are apt to confuse the two issues of fed-batch mode and batch mode. Most MECs claimed to be operated in fed-batch modes, where the MEC reactors were drained, refilled with substrate, and sparged with ultra-high-purity nitrogen after each fed-batch cycle (Call et al., 2009; Selembo et al., 2010; Wang et al., 2009b).

12.5.2.3 *Continuous-flow modes*

The performance of MECs operated in continuous-flow modes, where mixed cultures were generally used, were significantly affected by the following parameters: the hydraulic retention time (HRT), the organic loading rate, the cathode's catalysis ability, and the hydrogen gas collection method. For example, the HRT showed significant influences on hydrogen production and energy consumption in a tubular semipilot MEC fed with domestic wastewater, and it would require two MEC modules operating in series to achieve a relatively high-quality effluent when HRT was below 4 h (Gil-Carrera et al., 2013). A membraneless continuous-flow MEC with a gas-phase cathode was constructed, and a volumetric HPR of 6.3 m^3 H_2/m^3 day was achieved under substrate nonlimiting conditions (Tartakovsky et al., 2009). The biggest challenge of MECs in continuous-flow modes was the hydrogen consumption, caused by the reoxidation of the bioanode and the conversion to methane with carbon dioxide. For example, a single-chamber MEC using carbon fibers lacking metal catalysts as the cathode was proposed for hydrogen production. However, the hydrogen recycling between the cathode and the anode accounted for 62%–76% of observed current density, and it made the observed coulombic efficiency as high as 190%–310%, and the cathodic conversion efficiency was only 16%–24% (Lee and Rittmann, 2009). A pilot-scale (1000 L) continuous-flow MEC was constructed and tested for hydrogen generation from the treatment of winery wastewater. However, most of the product gas, with a maximum product rate of 0.19 ± 0.04 m^3 H_2/m^3 day, was converted to methane (86% $\pm$ 6%) (Cusick et al., 2011). The multielectrode-system-based MEC was constructed and the results show that it can be scaled up primarily based on cathode surface area, but that hydrogen can be completely consumed in a continuous-flow system unless methanogens can be completely eliminated (Rader and Logan, 2010).

12.6 Materials used in constructing MECs

The materials used to contract the MEC system are the key to determining the performance of the whole system and consequently its economic feasibility in industrial applications. While building a bioelectrochemical plant, the total cost of electrodes and membranes could account for more than 80% of the total investment in materials and 16% of the initial investment (Trapero et al., 2017). These numbers could be even more significant in the long-term operation of an MEC system in industry considering the relatively short life cycles of these materials compared to other components. In addition, the electrodes property could predominantly determine the formation of electroactive biofilm or electrochemical catalytic efficiency, which is the core of a bioelectrochemical system, and membranes could affect the system efficiency in terms of energy loss. Therefore, optimizing the usage of those materials is a critical factor that needs to be considered seriously (Kadier et al., 2017c). An overview of those principles is presented as Figure 12.8.

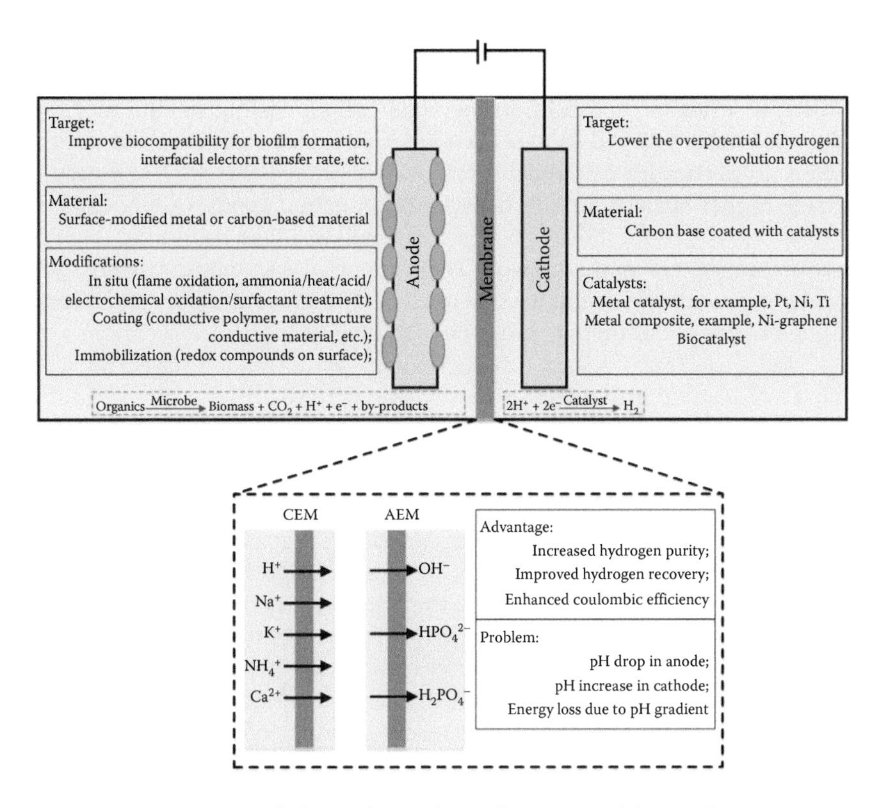

Figure 12.8 Overview of electrodes and membranes used for MEC construction.

12.6.1 Anode materials

When coupling hydrogen production with wastewater treatment, organics are degraded by the EAB in the anode and then protons and supplementary electric powers are provided for cathodic reduction. The abiotic–biotic interface between the anode and microorganisms plays a key role in this process, which is the same as for MFCs. Therefore, the same materials used for MFCs could also be used in MECs. Key parameters of anode material include biocompatibility, specific surface area, affinity to microbes, electric conductivity, and so on. To date, many materials have been investigated to be used as anodes, generally ranging from metallic to carbonous. The metallic materials generally show high electric conductivity and good mechanic strength, but they normally have weak cell attachment and have the risk of corrosion in long-term operation. In addition, some metal electrodes may be toxic to the microbial metabolism due to the contaminants in their composites. In contrast, the carbon-based materials are higher in specific surface area and always show good biocompatibility for biofilm formation, but their intrinsic ohmic resistance is a bit high, which will cause large ohmic energy loss, and the durability of carbon material is also an issue. Therefore, to further improve their capabilities to assist the abiotic–biotic interfacial electron transfer, a series of modification methods corresponding to different raw materials was developed in past decades. These are mainly to optimize the surface properties for cell attachment, such as increasing the roughness to provide higher surface area or hydrophilic function groups (C—N or C—O groups) on which microbes can attach, as well as other properties such as corrosion-resistance and longevity. These material and methods were thoroughly reviewed by a group (Guo et al., 2015; Kumar et al., 2013), and a summarized overview of them is presented in Table 12.1.

12.6.2 Cathode materials

In contrast to the biological anode oxidation, the reaction in the cathode is mostly an electrochemical reduction of protons to hydrogen gas. Therefore, reducing the overpotential (i.e., energy loss) of this reduction process is the core principle for choosing a cathode material and its corresponding modification methods. To date, platinum, which is a superior metal catalyst, is still the main selection for a cathode that reaches a good producing rate of hydrogen. Instead of using platinum wire or plate, the more common way is to coat nano or micro platinum particles on another cheaper conductive base (normally carbon material) using a chemical binder (Nafion or polytetrafluoroethylene [PTFE] solution), with the coating density of 0.25–1.0 mg Pt/cm^2. The coated platinum is an efficient catalyst to reduce the overpotential for the hydrogen evolution reaction and thus facilitate the production rate of hydrogen gas. This procedure has become the standard method to prepare cathodes for the MEC process. Due to the high cost of platinum and its potential of being poisoned by the buffer in electrolytes, other cheaper metal catalysts, such as nickel and titanium (Farhangi et al., 2014; Kadier et al., 2015b), and metal composites, like nickel foam-graphene (Cai et al., 2016), were also investigated for use in preparation of the cathode for MECs, following a similar procedure as for platinum. They generally gave lower performance than platinum, but can possibly be a future solution considering their lower cost. In addition to the chemical catalyst, studies also indicate the feasibility of using biocatalyst to facilitate hydrogen production (Rozendal et al., 2008b; Wang et al., 2014), which would potentially enhance the economic feasibility of MECs in the future. The underlying mechanism of biocathodes in hydrogen production is still not quite clear, which needs to be further investigated. The low cost and self-regeneration of biocathodes are the advantages, while issues like stability and biofouling can be problematic.

12.6.3 Membranes

In most MEC studies, membranes to separate the anodic and cathodic chambers is essential, although some membraneless system were developed.

Table 12.1 List of anode materials and their modification methods

Anode	Pros	Cons	Shape	Representative modification methods
Metallic: stainless steel, titanium, platinum, gold, silver, and nickel	High intrinsic electric conductivity; high durability	Corrosion; toxicity to microorganism; low specific surface area; high cost	Wire, plate, felt, mesh	*In situ* surface modification: Improve surface roughness and biocompatibility for biofilm formation, for example, flame oxidation (Guo et al., 2014) Surface coating: Improve the catalytic activity and protect electrode from being poisoned by coating with a polymer or composite layer, for example, polyaniline (Schroder et al., 2003) and its derivatives (Niessen et al., 2004)
Carbon-based: carbon, graphite	High specific surface area; low cost; high corrosion resistance; high biocompatibility	Low conductivity; low mechanism strength; short life cycle	Plate, rod, paper, cloth, granules, mesh, three-dimensional porous	*In situ* surface modification: Enhance biofilm formation by improving surface area, charges, and functional C—N or C—O groups, for example, ammonia treatment (Cheng and Logan, 2007), heat treatment (Wang et al., 2009a), acid treatment (Feng et al., 2010), electrochemical oxidation (Tang et al., 2011), and surfactant treatment (Guo et al., 2014) Redox compounds modification: Facilitate interfacial electron by immobilizing redox compounds on anode, for example, neutral red (Guo et al., 2013a) and methylene blue (Popov et al., 2012) Polymer and composite coating: Improve electron transfer rate and biocompatibility for biofilm by coating the surface with an organic conductive polymer layer or its composite, for example, polyaniline (Lai et al., 2011), polyaniline-metal (Prasad et al., 2007), polyaniline-carbon (Qiao et al., 2007), and polypyrrole-redox chemical (Feng et al., 2010)

The existence of a membrane could (1) improve the purity of hydrogen by avoiding mixing it with carbon dioxide produced in the anode, (2) enhance the hydrogen recovery rate by preventing hydrogen from being diffused and consumed by the anode microbial community, and (3) minimize the energy loss in the anode due to the growth of methanogens induced by hydrogen. Many different types of membranes have been investigated in the MEC field, including CEM, AEM, bipolar membrane, and charge mosaic membrane. A Nafion membrane, a kind of CEM, is most commonly used. However, a common issue with using this membrane (as well as other CEMs) is the pH gradient. The CEM membrane not only allows protons to immigrate but is also permeable to other positively charged ions like Na^+ and K^+. The concentrations of these positively charged ion species are normally much higher than protons in the electrolyte and they will dominate the charge balance between the two chambers. This will cause the increase of pH in the cathode chamber and decrease of pH in the anode, which consequently will cause significant energy loss according to the Nernst equation. AEM was indicated to be better to eliminate this pH gradient than CEM. It allows the negatively charged phosphate species to transfer through the membrane. These could help to buffer the pH drop in the anode, which would cause significant interference to cells' capability for extracellular electron transfer. The pH gradient is smaller while using AEM in the MEC system, but unfortunately it is still not negligible and causes energy loss and reduces HPR. Similarly, situations occurred for the MEC system using the other type of membranes or multimembranes as well. Therefore, additional control of pH will be required for a membrane MEC system to improve its performance.

12.7 Challenges and future prospects

Biohydrogen as a clean and sustainable fuel of the future is gaining attention worldwide. Biohydrogen production through MECs seems to be a potent alternative method due to high H_2 production, low energy input, and utilization of a wide range of organic waste materials. Despite the development of various new approaches and technologies, there are still hurdles to overcome for commercial application of MECs for hydrogen production. For any technology to be used commercially, scale-up studies are necessary in order to evaluate the commercial potential. There are only a few scale-up studies related to performance of MECs for H_2 production. Electrode material and its cost, reactor design, membrane cost, and methane production are a matter of concern during scale-up studies of MECs.

Platinum is considered to be best for the electrode (cathode) due to its high catalysis activity and has been widely used in MEC studies. However, it is reported that Pt is not a good candidate for scale up due to its high cost and negative environmental impacts (Kadier et al., 2016b) and poisoning by chemicals such as sulfide and thereby losing catalytic activities (Zhang and Angelidaki, 2014). Considerable research has been conducted to explore alternatives to Pt. The most promising material investigated so far is stainless steel and nickel alloys (Selembo et al., 2009a). In recent years, extensive studies have been carried out on nanostructured materials of tungsten (Harnisch et al., 2009) and palladium (Huang et al., 2011). The use of Pd nanoparticles reduced the cost of the cathode while maintaining the system performance.

Development of a biocathode (Rozendal et al., 2008b; Wang et al., 2014) provides new dimensions to research in the area of a low-cost cathode for MECs. Compared to chemical ones, the biocathode is low cost and self-generating without producing pollution (Kadier et al., 2016a). Biocathodes are currently a hot topic of research to implement MECs at the commercial scale due to their potential cost saving and sustainable nature.

In general, MECs use a membrane for the purity of hydrogen and to prevent microbial H_2 consumption. The most common membrane used is PEM (Ajayi et al., 2010; Chae et al., 2009; Selembo et al., 2009a). The drawbacks of using a membrane in MECs include the formation of pH gradient across the membrane and its cost (Hu et al., 2008).

Methane production is the most common reason for low H_2 yield in membraneless single-chambered MECs. Most studies reported hydrogenotrophic methanogens responsible for H_2 consumption and methane generation, making MECs inefficient for H_2 production. Several strategies have been employed to inhibit methanogens in MECs.

The applications of chemical inhibitors are effective but they increase the operational cost of MECs. The exposure of electrodes to oxygen has some positive effect, but methanogens are not completely removed. The development of an effective strategy to remove methanogens from MECs is a major hurdle to pass for application of MECs for efficient hydrogen production.

MEC designs also play an important role in efficient performance of the system. In order to achieve better H_2 yield, the reactor design should be such that it suppresses the growth of methanogens, thereby maintaining the performance of the MECs. MECs should be configured and designed in such a way that both the construction cost and energy loses were kept to a minimum. During design of MECs, a decrease in the activity of the anodic biofilm and cathode performance, and fouling and clogging the membrane or separator by biomass should be kept in mind. Other possible steps to make the H_2 production through MECs economically feasible is the integration of the system to other technologies like MFCs (Sun et al., 2008) and dark fermentation (Wang et al., 2011). There are various reports that integrated systems including MECs give better results in comparison to a single one (Ajayi et al., 2010; Sun et al., 2008; Wang et al., 2011).

References

Abdeshahian, P., Al-Shorgani, N. K. N., Salih, N. K. M., Shukor, H., Kadier, A., Hamid, A. A., Kalil, M. S. 2016. The Production of Biohydrogen by a Novel Strain Clostridium sp. YM1 in Dark Fermentation Process. *Int. J. Hydrogen Energy*. 39: 12524–12531.

Ajayi, F. F., Kim, K. Y., Chae, K. J. et al. 2010. Optimization studies of bio-hydrogen production in a coupled microbial electrolysis–dye sensitized solar cell system. *Photochem. Photobiol. Sci.* 9: 349–356.

Ambler, J. R. and Logan, B. E. 2011. Evaluation of stainless steel cathodes and a bicarbonate buffer for hydrogen production in microbial electrolysis cells using a new method for measuring gas production. *Int. J. Hydrogen Energy* 36: 160–166.

Arunasri, K., Modestra, J. A, Yeruva, D. K., Krishna, K. V., and Mohan, S. V. 2016. Polarized potential and electrode materials implication on electro-fermentative di-hydrogen production: Microbial assemblages and hydrogenase gene copy variation. *Bioresour. Technol.* 200: 691–698.

Brown, R. K., Harnisch, F., Wirth, S. et al. 2014. Evaluating the effects of scaling up on the performance of bioelectrochemical systems using a technical scale microbial electrolysis cell. *Bioresour. Technol.* 163: 206–213.

Cai, W., Liu, W., Han, J., and Wang, A. 2016. Enhanced hydrogen production in microbial electrolysis cell with 3D self-assembly nickel foam-graphene cathode. *Biosens. Bioelectron.* 80: 118–122.

Call, D. F. and Logan, B. E. 2008. Hydrogen production in a single chamber microbial electrolysis cell lacking a membrane. *Environ. Sci. Technol.* 42: 3401–3406.

Call, D. F. and Logan, B. E. 2011. A method for high throughput bioelectrochemical research based on small scale microbial electrolysis cells. *Biosens. Bioelectron.* 26: 4526–4531.

Call, D. F., Wagner, R. C., and Logan, B. E. 2009. Hydrogen production by *Geobacter* species and a mixed consortium in a microbial electrolysis cell. *Appl. Environ. Microbial.* 75: 7579–7587.

Carmona-Martínez, A. A., Trably, E., Milferstedt, K. et al. 2015. Long-term continuous production of H_2 in a microbial electrolysis cell (MEC) treating saline wastewater. *Water Res.* 15: 149–156.

Chae, K. J., Choi, M. J., Kim, K. Y. et al. 2009. A solar-powered microbial electrolysis cell with a platinum catalyst-free cathode to produce hydrogen. *Environ. Sci. Technol.* 43: 9525–9530.

Chandrasekhar, K., Lee, Y. J., and Lee, D. W. 2015. Biohydrogen production: Strategies to improve process efficiency through microbial routes. *Int. J. Mol. Sci.* 16: 8266–8293.

Chen, Y., Chen, M., Shen, N., and Zeng, R. J. 2016. H_2 production by the thermoelectric microconverter coupled with microbial electrolysis cell. *Int. J. Hydrogen Energy* 41: 22760–22768.

Chen, S., Liu, G., Zhang, R., Qin, B., and Luo, Y. 2012. Development of the microbial electrolysis desalination and chemical-production cell for desalination as well as acid and alkali productions. *Environ. Sci. Technol.* 46: 2467–2472.

Chen, X., Sun, D., Zhang, X., Liang, P., and Huang, X. 2015. Novel self-driven microbial nutrient recovery cell with simultaneous wastewater purification. *Sci. Rep.* 5: 15744.

Cheng, S. and Logan, B. E. 2007. Sustainable and efficient biohydrogen production via electrohydrogenesis. *Proc. Natl. Acad. Sci.* 104: 18871–18873.

Cheng, S. and Logan, B. E. 2011. High hydrogen production rate of microbial electrolysis cell (MEC) with reduced electrode spacing. *Bioresour. Technol.* 102: 3571–3574.

Clauwaert, P. and Verstraete, W. 2008. Methanogenesis in membraneless microbial electrolysis cells. *Appl. Microbiol. Biotechnol.* 82: 829–836.

Cusick, R. D., Bryan, B., Parker, D. S. et al. 2011. Performance of a pilot-scale continuous flow microbial electrolysis cell fed winery wastewater. *Appl. Microbiol. Biotechnol.* 89: 2053–2063.

Cusick, R. D. and Logan, B. E. 2012. Phosphate recovery as struvite within a single chamber microbial electrolysis cell. *Bioresour. Technol.* 107: 110–115.

Das, D. 2009. Advances in biohydrogen production processes: An approach towards commercialization. *Int. J. Hydrogen Energy* 34: 7349–7357.

Ditzig, J., Liu, H., and Logan, B. E. 2007. Production of hydrogen from domestic wastewater using a bioelectrochemically assisted microbial reactor (BEAMR). *Int. J. Hydrogen Energy* 32: 2296–2304.

Escapa, A., Gomez, X., Tartakovsky, B., and Moran, A. 2012. Estimating microbial electrolysis cell (MEC) investment costs in wastewater treatment plants: Case study. *Int. J. Hydrogen Energy* 37: 18641–18653.

Escapa, A., San-Martín, M. I., Mateos, R., and Moran, A. 2015. Scaling-up of membraneless microbial electrolysis cells (MECs) for domestic wastewater treatment: Bottlenecks and limitations. *Bioresour. Technol.* 180: 72–78.

Farhangi, S., Ebrahimi, S., and Niasar, M. S. 2014. Commercial materials as cathode for hydrogen production in microbial electrolysis cell. *Biotechnol. Lett.* 36: 1987–1992.

Feng, Y., Yang, Q., Wang, X., and Logan, B. E. 2010. Treatment of carbon fiber brush anodes for improving power generation in air–cathode microbial fuel cells. *J. Power Sources* 195: 1841–1844.

Gil-Carrera, L., Escapa, A., Carracedo, B., Moran, A., and Gomez, X. 2013. Performance of a semi-pilot tubular microbial electrolysis cell (MEC) under several hydraulic retention times and applied voltages. *Bioresour. Technol.* 146: 63–69.

Gil-Carrera, L., Mehta, P., Escapa, A. et al. 2011. Optimizing the electrode size and arrangement in a microbial electrolysis cell. *Bioresour. Technol.* 102: 9593–9598.

Godfrey, K., Popov, A., Dinsdale, R. et al. 2010. Influence of catholyte pH and temperature on hydrogen production from acetate using a two chamber concentric tubular microbial electrolysis cell. *Int. J. Hydrogen Energy* 35: 7716–7722.

Gorby, Y. A., Yamina, S., Mclean, J. S. et al. 2006. Electrically conductive bacterial nanowires produced by *Shewanella oneidensis* strain MR-1 and other microorganisms. *Proc. Natl. Acad. Sci. USA* 103: 11358–11363.

Guo, K., Tang, X., Du, Zh., and Li, H. 2010. Hydrogen production from acetate in a cathode-on-top single-chamber microbial electrolysis cell with a mipor cathode. *Biochem. Eng. J.* 51: 48–52.

Guo, K., Chen, X., Freguia, S., and Donose, B. C. 2013a. Spontaneous modification of carbon surface with neutral red from its diazonium salts for bioelectrochemical systems. *Biosens. Bioelectron.* 47: 184–189.

Guo, K., Donose, B. C., Soeriyadi, A. H. et al. 2014. Flame oxidation of stainless steel felt enhances anodic biofilm formation and current output in bioelectrochemical systems. *Environ. Sci. Technol.* 48: 7151–7156.

Guo, X., Liu, J., and Xiao, B. 2013b. Bioelectrochemical enhancement of hydrogen and methane production from the anaerobic digestion of sewage sludge in single-chamber membrane free microbial electrolysis cells. *Int. J. Hydrogen Energy* 38: 1342–1347.

Guo, K., Prevoteau, A., Patil, S. A., and Rabaey, K. 2015. Engineering electrodes for microbial electrocatalysis. *Curr. Opin. Biotechnol.* 33: 149–156.
Guo, Z., Thangavel, S., Wang, L. et al. 2016. Efficient methane production from beer wastewater in a membraneless MEC with stacked cathode: The effect of cathode-anode ratio on bioenergy recovery. *Energy Fuels* 31: 615–620.
Hallenbeck, P. C. and Ghosh, D. 2009. Advances in fermentative biohydrogen production: The way forward? *Trends Biotechnol.* 27: 287–297.
Harnisch, F., Sievers, G., and Schroder, U. 2009. Tungsten carbide as electro catalyst for the hydrogen evolution reaction in pH neutral electrolyte solutions. *Appl. Catal. B: Environ.* 89: 455–458.
Hasany, M., Mardanpour, M. M., and Yaghmaei, S. 2016. Biocatalysts in microbial electrolysis cells: A review. *Int. J. Hydrogen Energy* 41: 1477–1493.
Heidrich, E. S., Dolfing, J., Scott, K. et al. 2013. Production of hydrogen from domestic wastewater in a pilot-scale microbial electrolysis cell. *Appl. Microbial. Biotechnol.* 97: 6979–6989.
Hou, Y., Luo, H., Liu, G. et al. 2014. Improved hydrogen production in the microbial electrolysis cell by inhibiting methanogenesis using ultraviolet irradiation. *Environ. Sci. Technol.* 48: 10482–10488.
Hu, H. Q., Fan, Y., and Liu, H. 2008. Hydrogen production using single-chamber membrane-free microbial electrolysis cells. *Water Res.* 42: 4172–4178.
Hu, H. Q., Fan, Y., and Liu, H. 2009. Hydrogen production in single-chamber tubular microbial electrolysis cells using non-precious-metal catalysts. *Int. J. Hydrogen Energy* 34: 8535–8542.
Huang, Y. X., Liu, X. W., Sun, X. F. et al. 2011. A new cathodic electrode deposit with palladium nano-particles for cost-effective hydrogen production in a microbial electrolysis cell. *Int. J. Hydrogen Energy* 36: 2773–2776.
Jeremiasse, A. W., Hamelers, H. V. M., Saakes, M., and Buisman, C. J. M. 2010. Ni foam cathode enables high volumetric H_2 production in a microbial electrolysis cell. *Int. J. Hydrogen Energy* 35: 12716–12723.
Jiang, Y., Liang, P., Zhang, C. et al. 2016. Periodic polarity reversal for stabilizing the pH in two-chamber microbial electrolysis cells. *Appl. Energy* 165: 670–675.
Jiang, Y., Su, M., and Li, D. 2014. Removal of sulfide and production of methane from carbon dioxide in microbial fuel cells–microbial electrolysis cell (MFCs–MEC) coupled system. *Appl. Biochem. Biotechnol.* 172: 2720–2731.
Jiang, Y., Su, M., Zhang, Y. et al. 2013. Bioelectrochemical systems for simultaneously production of methane and acetate from carbon dioxide at relatively high rate. *Int. J. Hydrogen Energy* 38: 3497–3502.
Kadier, A., Kalil, M. S., Mohamed, A., Hasan, H. A., Abdeshahian, P., Fooladi, T., and Hamid, A. A. 2017a. Microbial electrolysis cells (MECs) as innovative technology for sustainable hydrogen production: Fundamentals and perspective applications. In: Hydrogen Production Technologies. Sankir, M. and Sankir N. D. (eds.). Wiley-Scrivener Publishing LLC, USA. 407–458.
Kadier, A., Kalil, M. S., Mohamed, A., Hamid, A. A. 2017b. A new design enhances hydrogen production by G. Sulfurreducens PCA strain in a single-chamber microbial electrolysis cell (MEC). *J. Teknol.* 79 (5–3); 71–79.
Kadier, A., Logroño, W., Rai, P. K., Kalil, M. S., Mohamed, A., Hasan H. A. and Hamid, A. A. 2017c. None-platinum electrode catalysts and membranes for highly efficient and inexpensive H2 production in microbial electrolysis cells (MECs): A review. *Iran. J. Catal.* 7 (2); 89–102.

Kadier, A., Abdeshahian, P., Simayi, Y. et al. 2015a. Grey relational analysis for comparative assessment of different cathode materials in microbial electrolysis cells. *Energy* 90: 1556–1562.

Kadier, A., Kalil, M. S., Abdeshahian, P. et al. 2016a. Recent advances and emerging challenges in microbial electrolysis cells (MECs) for microbial production of hydrogen and value-added chemicals. *Renew. Sustain. Energy Rev.* 61: 501–525.

Kadier, A., Simayi, Y., Abdeshahian, P. et al. 2016b. A comprehensive review of microbial electrolysis cells (MEC) reactor designs and configurations for sustainable hydrogen gas production. *Alexandria Eng. J.* 55: 427–443.

Kadier, A., Simayi, Y., Chandrasekhar, K., Ismail, M., and Kalil, M. S. 2015b. Hydrogen gas production with an electroformed Ni mesh cathode catalysts in a single-chamber microbial electrolysis cell (MEC). *Int. J. Hydrogen Energy* 40: 14095–14103.

Kadier, A., Simayi, Y., Kalil, M. S., Abdeshahian, P., and Hamid, A. A. 2014. A review of the substrates used in microbial electrolysis cells (MECs) for producing sustainable and clean hydrogen gas. *Renew. Energy* 71: 466–472.

Kadier, A., Simayi, Y., Logrono, W., and Kalil, M. S. 2015c. The significance of key operational variables to the enhancement of hydrogen production in a single-chamber microbial electrolysis cell (MEC). *Iran. J. Hydrog. Fuel Cell* 2: 85–97.

Kim, Y. and Logan, B. E. 2011. Hydrogen production from inexhaustible supplies of fresh and salt water using microbial reverse-electrodialysis electrolysis cells. *Proc. Natl. Acad. Sci.* 108: 16176–16181.

Kim, Y. and Logan, B. E. 2013. Simultaneous removal of organic matter and salt ions from saline wastewater in bioelectrochemical systems. *Desalination* 308: 115–121.

Kotay, S. M. and Das, D. 2008. Biohydrogen as a renewable energy resource—Prospects and potentials. *Int. J. Hydrogen Energy* 33: 258–263.

Kumar, G. G., Sarathi, V. G. S., and Nahm, K. S. 2013. Recent advances and challenges in the anode architecture and their modifications for the applications of microbial fuel cells. *Biosens. Bioelectron.* 43: 461–475.

Kyazze, G., Popov, A., Dinsdale, R., Esteves, S., Hawkes, F., Premier, G., and Guwy, A. 2010. Influence of catholyte pH and temperature on hydrogen production from acetate using a two chamber concentric tubular microbial electrolysis cell. *Int. J. Hydrogen Energy* 35: 7716–7722.

Lai, B., Tang, X., Li, H. et al. (2011). Power production enhancement with a polyaniline modified anode in microbial fuel cells. *Biosens. Bioelectron.* 28: 373–377.

Ledezma, P., Jermakka, J., Keller, J., and Freguia, S. 2017. Recovering nitrogen as a solid without chemical dosing: Bio-electroconcentration for nutrient recovery from urine. *Environ. Sci. Technol. Lett.* 4: 119–124.

Lee, J., Lee, S. Y., Park, S., and Middelberg, A. P. J. 1999. Control of fed-batch fermentations. *Biotechnol. Adv.* 17: 29–48.

Lee, H. S. and Rittmann, B. E. 2009. Significance of biological hydrogen oxidation in a continuous single-chamber microbial electrolysis cell. *Environ. Sci. Technol.* 44: 948–954.

Lee, H. S., Torres, C. I., Parameswaran, P., and Rittmann, B. E. 2009. Fate of H_2 in an upflow single-chamber microbial electrolysis cell using a metal-catalyst-free cathode. *Environ. Sci. Technol.* 43: 7971–7976.

Li, Y., Jordyn, S., Yuankai, H. et al. 2017. Energy-positive wastewater treatment and desalination in an integrated microbial desalination cell (MDC)-microbial electrolysis cell (MEC). *J. Power Sources* 356: 529–538.

Liu, H., Grot, S., and Logan, B. E. 2005. Electrochemically assisted microbial production of hydrogen from acetate. *Environ. Sci. Technol.* 39: 4317–4320.

Liu, H., Hu, H., Chignell, J., and Fan, Y. 2010. Microbial electrolysis: Novel technology for hydrogen production from biomass. *Biofuels* 1: 129–142.

Liu, W., Huang, S. H., Zhou, A. et al. 2012. Hydrogen generation in microbial electrolysis cell feeding with fermentation liquid of waste activated sludge. *Int. J. Hydrogen Energy* 37: 13859–13864.

Logan, B. E., Call, D., Cheng, S. et al. 2008. Microbial electrolysis cells for high yield hydrogen gas production from organic matter. *Environ. Sci. Technol.* 42: 8630–8640.

Lu, L., Hou, D., Wang, X., Jassby, D., and Ren, Z. J. 2016. Active H_2 harvesting prevents methanogenesis in microbial electrolysis cells. *Environ. Sci. Technol. Lett.* 3: 286–290.

Luo, X., Zhang, F., Liu, J. et al. 2014. Methane production in microbial reverse-electrodialysis methanogenesis cells (MRMCs) using thermolytic solutions. *Environ. Sci. Technol.* 48: 8911–8918.

Manish, S. and Banerjee, R. 2008. Comparison of biohydrogen production processes. *Int. J. Hydrogen Energy* 33: 279–286.

Manuel, M. F., Neburchilov, V., Wang, H., Guiot, S. R., and Tartakovsky, B. 2010. Hydrogen production in a microbial electrolysis cell with nickel-based gas diffusion cathodes. *J. Power Sources* 195: 5514–5519.

Mehanna, M., Kiely, P. D., Call, D. F., and Logan, B. E. 2010. Microbial electrodialysis cell for simultaneous water desalination and hydrogen gas production. *Environ. Sci. Technol.* 44: 9578–9583.

Miyake, J., Miyake, M., and Asada, Y. 1999. Biotechnological hydrogen production: Research for efficient light energy conversion. *J. Biotechnol.* 70: 89–101.

Modestra, J. A., Babu, M. L., and Mohan, S. V. 2015. Electro-fermentation of real-field acidogenic spent wash effluents for additional biohydrogen production with simultaneous treatment in a microbial electrolysis cell. *Sep. Purif. Technol.* 150: 308–315.

Nam, J. Y., Cusick, R. D., Kim, Y., and Logan, B. E. 2012. Hydrogen generation in microbial reverse-electrodialysis electrolysis cells using a heat-regenerated salt solution. *Environ. Sci. Technol.* 46: 5240–5246.

Newman, D. K. and Kolter, R. 2000. A role for excreted quinones in extracecullar electron transfer. *Nature* 405: 94–97.

Niessen, J., Schroder, U., Rosenbaum, M., and Scholz, F. 2004. Fluorinated polyanilines as superior materials for electrocatalytic anodes in bacterial fuel cells. *Electrochem. Commun.* 6: 571–575.

Nikolaidis, P. and Poullikkas, A. 2017. A comparative overview of hydrogen production processes. *Renew. Sustain. Energy Rev.* 67: 597–611.

Parkhey, P. and Gupta, P. 2017. Improvisations in structural features of microbial electrolytic cell and process parameters of electrohydrogenesis for efficient biohydrogen production: A review. *Renew. Sustain. Energy Rev.* 69: 1085–1099.

Popov, A. L., Kim, J. R., Dinsdale, R. M. et al. 2012. The effect of physico-chemically immobilized methylene blue and neutral red on the anode of microbial fuel cell. *Biotechnol. Bioprocess. Eng.* 17: 361–370.

Prasad, D., Arun, S., Murugesan, M. et al. 2007. Direct electron transfer with yeast cells and construction of a mediatorless microbial fuel cell. *Biosens. Bioelectron.* 22: 2604–2610.

Qiao, Y., Li, C. M., Bao, S. J., and Bao, Q. L. 2007. Carbon nanotube/polyaniline composite as anode material for microbial fuel cells. *J. Power Sources* 170: 79–84.

Rader, G. K. and Logan, B. E. 2010. Multi-electrode continuous flow microbial electrolysis cell for biogas production from acetate. *Int. J. Hydrogen Energy* 35: 8848–8854.

Reguera, G., McCarthy, K. D., Mehta, T. et al. 2005. Extracellular electron transfer via microbial nanowires. *Nature* 435: 1098–1101.

Rozendal, R. A., Hamelers, H. V. M., and Buisman, C. J. N. 2006a. Effects of membrane cation transport on pH and microbial fuel cell performance. *Environ. Sci. Technol.* 40: 5206–5211.

Rozendal, R. A, Hamelers, H. V. M., Euverink, G. J. W., Metz, S. J., and Buisman, C. J. N. 2006b. Principle and perspectives of hydrogen production through biocatalyzed electrolysis. *Int. J. Hydrogen Energy* 31: 1632–1640.

Rozendal, R. A, Hamelers, H. V. M., Rabaey, K., Keller, J., and Buisman, C. J. N. 2008a. Towards practical implementation of bioelectrochemical wastewater treatment. *Trends Biotechnol.* 26: 450–459.

Rozendal, R. A., Jeremiasse, A. W., Hamelers, H. V. M., and Buisman, C. J. N. 2008b. Hydrogen production with a microbial biocathode. *Environ. Sci. Technol.* 42: 629–634.

Ruiz, Y., Baeza, J. A., and Guisasola, A. 2016. Microbial electrolysis cell performance using non-buffered and low conductivity wastewaters. *Chem. Eng. J.* 289: 341–348.

Schroder, U., Niessen, J., and Scholz, F. 2003. A generation of microbial fuel cells with current outputs boosted by more than one order of magnitude. *Angew. Chem. Int. Ed.* 42: 2880–2883.

Selembo, P. A., Merrill, M. D., and Logan, B. E. 2009a. The use of stainless steel and nickel alloys as low-cost cathodes in microbial electrolysis cells. *J. Power Sources* 190: 271–278.

Selembo, P. A., Merrill, M. D., and Logan, B. E. 2010. Hydrogen production with nickel powder cathode catalysts in microbial electrolysis cells. *Int. J. Hydrogen Energy* 35: 428–437.

Selembo, P. A., Perez, J. M., Lloyd, W. A., and Logan, B. E. 2009b. High hydrogen production from glycerol or glucose by electrohydrogenesis using microbial electrolysis cells. *Int. J. Hydrogen Energy* 34: 5373–5381.

Sharma, Y. and Li, B. 2010. The variation of power generation with organic substrates in single-chamber microbial fuel cells (SCMFCs). *Bioresour. Technol.* 101: 1844–1850.

Sinha, P. and Pandey, A. 2011. An evaluative report and challenges for fermentative biohydrogen production. *Int. J. Hydrogen Energy* 36: 7460–7478.

Steinbusch, K. J. J., Arvaniti, E., Hamelers, H. V. M., and Buisman, C. J. N. 2009. Selective inhibition of methanogenesis to enhance ethanol and n-butyrate production through acetate reduction in mixed culture fermentation. *Bioresour. Technol.* 100: 3261–3267.

Su, M., Wei, L., Qiu, Z., Wang, G., and Shen, J. 2016. Hydrogen production in single chamber microbial electrolysis cells with stainless steel fiber felt cathodes. *J. Power Sources* 301: 29–34.

Sugnaux, M., Happe, M., Cachelin, C. P. et al. 2016. Two stage bioethanol refining with multi litre stacked microbial fuel cell and microbial electrolysis cell. *Bioresour. Technol.* 221: 61–69.
Sun, M., Sheng, G. P., Mu, Z. X. et al. 2009. Manipulating the hydrogen production from acetate in a microbial electrolysis cell–microbial fuel cell-coupled system. *J. Power Sources* 191: 338–343.
Sun, M., Sheng, G. P., Zhang, L. et al. 2008. An MEC-MFC-coupled system for biohydrogen production from acetate. *Environ. Sci. Technol.* 42: 8095–8100.
Tang, X., Guo, K., Li, H., Du, Z., and Tian, J. 2011. Electrochemical treatment of graphite to enhance electron transfer from bacteria to electrodes. *Bioresour. Technol.* 102: 3558–3560.
Tartakovsky, B., Manuel, M. F., Wang, H., and Guiot, S. R. 2009. High rate membrane-less microbial electrolysis cell for continuous hydrogen production. *Int. J. Hydrogen Energy* 34: 672–677.
Torres, C. I., Marcus, A. K., Lee, H. S. et al. 2010. A kinetic perspective on extracellular electron transfer by anode-respiring bacteria. *FEMS Microbiol. Rev* 34: 3–17.
Trapero, J. R., Horcajada, R., Linares, J. J., and Lobato, J. 2017. Is microbial fuel cell technology ready? An economic answer towards industrial commercialization. *Appl. Energy* 185: 698–707.
von Canstein, H., Ogawa, J., Shimizu, S., and Lloyd, J. 2008. Secretion of flavins by *Shewanella* species and their role in extracellular electron transfer. *Appl. Environ. Microb.* 74: 615–623.
Wang, X., Cheng, S., Feng, Y. et al. 2009a. Use of carbon mesh anodes and the effect of different pretreatment methods on power production in microbial fuel cells. *Environ. Sci. Technol.* 43: 6870–6874.
Wang, Y., Guo, W. O., Xing, D. F., Chang, G. S., and Ren, N. Q. 2014. Hydrogen production using biocathode single-chamber microbial electrolysis cells fed by molasses wastewater at low temperature. *Int. J. Hydrogen Energy* 39: 19369–19375.
Wang, A., Liu, W., Cheng, S. et al. 2009b. Source of methane and methods to control its formation in single chamber microbial electrolysis cells. *Int. J. Hydrogen Energy* 34: 3653–3658.
Wang, A., Sun, D., Cao, G. et al. 2011. Integrated hydrogen production process from cellulose by combining dark fermentation, microbial fuel cells, and a microbial electrolysis cell. *Bioresour. Technol.* 102: 4137–4143.
Zang, G. L., Sheng, G. P., Shi, C. et al. 2014. A bio-photoelectrochemical cell with a MoS_3-modified silicon nanowire photocathode for hydrogen and electricity production. *Energy Environ. Sci.* 7: 3033–3039.
Zhang, Y. and Angelidaki, I. 2014. Microbial electrolysis cells turning to be versatile technology: Recent advances and future challenges. *Water Res.* 56: 11–25.
Zhang, Y., Merrill, M. D, and Logan, B. E. 2010. The use and optimization of stainless steel mesh cathodes in microbial electrolysis cells. *Int. J. Hydrogen Energy* 35: 12020–12028.
Zhang, Y., Yu, L., Wu, D. et al. 2015. Dependency of simultaneous Cr(VI), Cu(II) and Cd(II) reduction on the cathodes of microbial electrolysis cells self-driven by microbial fuel cells. *J. Power Sources* 273: 1103–1113.
Zhu, X., Hatzell, M. C., and Logan, B. E. 2014. Microbial reverse-electrodialysis electrolysis and chemical-production cell for H_2 production and CO_2 sequestration. *Environ. Sci. Technol. Lett.* 1: 231–235.

chapter thirteen

Biogas production and quality control

Lijun Wang, Bo Zhang, and Gail Joseph

Contents

13.1 Introduction

Biogas generally refers to the gas generated by anaerobic digestion (AD), which mainly consists of methane (CH_4) and carbon dioxide (CO_2). The CH_4 in the biogas could be as high as 50%–60% by volume, and the remaining 40%–50% is CO_2 (Gebauer, 2004). AD is a biological process in which organic matters are degraded to a gaseous mixture of biogas in the absence of oxygen. Production of biogas by the AD of biomass has three major advantages: (1) the energy with the biomass can be converted to biofuels; (2) energy, nutrients, and water in the biomass can be recycled and reused; and (3) it provides an efficient waste treatment approach to reduce the harmful effect of the organic wastes on the environment (Ziemiński and Frąc, 2012).

Biogas has become an important form of bioenergy. AD of sewage sludge has already been common, and the number of on-farm digesters has been significantly increased around the world in the past decade (Edwards et al., 2015). There were more than 10,000 on-farm biogas plants in the European Union in 2013, which produced 9.2 million tons of oil equivalent of biogas (Lijó et al., 2017). The biogas accounted for approximately 3% of natural gas consumption in the European Union (EU) in 2013. It is estimated that biogas can be accounted as 10% of EU natural consumption in 2020 (Grando et al., 2017). In Germany, the total number of farm-based biogas plants has increased from 1050 in 2000 to 7850 in 2013 with the increase of electrical power capacity from 100 to 3543 MW (Blumenstein et al., 2016). In Italy, there were 994 biogas plants in 2012 with total electrical power capacity of 756 MW (Agostini et al., 2016). It is estimated that the biogas produced at more than 8000 large dairy and hog operations in the United States could potentially generate more than 13 million MWh of energy each year (1485 MW). In addition to more than 1200 sewage sludge digesters, there were 242 operating anaerobic digesters at livestock farms in the United States in 2015 (Edwards et al., 2015; U.S. EPA, 2017). China has been promoting the use of small household anaerobic digesters to process rural organic wastes since the 1970s, and there are approximately 5 million households using anaerobic digesters (Henderson, 2009). Meanwhile, the Chinese government is seeking to increase the production capacity to 44 billion m^3 of biogas by 2020 (NDRC, 2007).

AD is a very complicated biological process involving numerous microorganisms for hydrolysis, acidogenesis, acetogenesis, and methanogenesis with various operational environmental conditions. Various aspects of AD technology for biogas production have been reviewed in the literature, which include feedstock pretreatment (Carlsson et al., 2012; Carrere et al., 2016), various feedstocks of algae (Jankowska et al., 2017; McKennedy and Sherlock, 2015; Ward et al., 2014), aquatic plants (Zhang et al., 2015), food wastes (De Clercq et al., 2017; Kondusamy and Kalamdhad,

2014; Zhang et al., 2014), animal wastes (Salminen and Rintala, 2002), municipal solid waste (MSW) (Jain et al., 2015), sludge (Mirzoyan et al., 2010) and wastewater (Harris and McCabe, 2015), process control (Gaida et al., 2017; Nguyen et al., 2015), mathematical modeling (Donoso-Bravo et al., 2011; Madsen et al., 2011; Xie et al., 2016), and effects of additives and inhibitors (Rajagopal et al., 2013; Romero-Guiza et al., 2016; Yenigun and Demirel, 2013). This chapter is focused on the review on the biogas production principles and processes, and the latest technologies for the enhancement of biogas production and upgrading. The economics and sustainability of biogas production are discussed.

13.2 Biogas production principles and processes

13.2.1 Working principles

AD is a biological process of hydrolyzing insoluble organic compounds and subsequently gasifying intermediate compounds (Ziemiński and Frąc, 2012). AD of organic materials basically follows four stages: hydrolysis, acidogenesis, acetogenesis, and methanogenesis, as shown in Figure 13.1 (Wang, 2014). Although each of these stages could be a rate-limiting step,

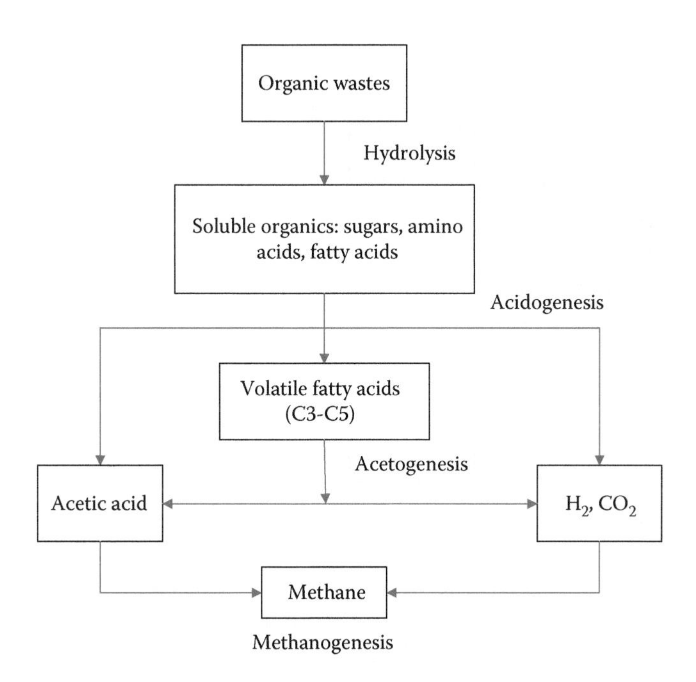

Figure 13.1 Successive steps in the anaerobic digestion process. (From Wang, L. 2014. *Sustainable Bioenergy Production*, CRC Press, Boca Raton.)

hydrolysis is generally considered as rate-limiting. The hydrolysis step degrades both insoluble organic material and high-molecular-weight compounds such as lipids, polysaccharides, proteins, and nucleic acids into soluble organic substances (e.g., sugars, amino acids, and fatty acids). The compounds formed during hydrolysis are further digested during the second step, acidogenesis. Volatile fatty acids (VFAs) are produced by acidogenic (or fermentative) bacteria along with ammonia (NH_3), CO_2, H_2S, and other by-products. The third AD step is acetogenesis, in which the higher organic acids and alcohols produced by acidogenesis are further split by acetogens to produce mainly acetic acid as well as CO_2 and H_2. This conversion is controlled to a large extent by the partial pressure of H_2 in the mixture. The last step of AD is methanogenesis, which produces methane by two groups of methanogenic microorganisms: the first group splits acetate into methane and carbon dioxide and the second group uses hydrogen as an electron donor and carbon dioxide as an acceptor to produce methane.

13.2.2 Operating parameters

13.2.2.1 Solids content

Solids concentration in a feedstock will significantly affect the digester design and operation cost. In a continuously stirred tank digester, an influent substrate concentration of 3%–8% total solids (TS) is added daily and an equal amount of effluent is withdrawn. Solid-state anaerobic digestion (SS-AD) generally can use the TS concentration of higher than 15% (Li et al., 2011). Animal manure, sewage sludge, and food waste are usually treated by regular liquid AD, while organic fractions of municipal solid waste (OFMSW) and lignocellulosic biomass such as crop residues and energy crops can be processed through SS-AD.

13.2.2.2 C:N ratio

The C:N ratio determines the nutrient levels of a digestion substrate. A high C:N ratio induces a low protein solubilization rate, which may avoid ammonia inhibition. However, a too high C:N ratio will provide insufficient nitrogen to maintain microbial cell biomass, resulting in low biogas production. A too low C:N ratio increases the risk of ammonia inhibition. The optimal C:N ratio for AD is between 20 and 30, with a ratio of 25 as the most commonly used (Mao et al., 2015). Table 13.1 gives the C:N ratios of various substrates (Hagos et al., 2017). Codigestion of lignocellulosic biomass and animal manure is widely used to obtain an optimum C:N for AD. However, it is difficult to determine the optimal C:N ratio for an AD process because the optimum value may be affected by different factors such as the substrate type, composition of trace elements, chemical composition, and biodegradability (Hagos et al., 2017). The maximum

Table 13.1 The C:N ratios of different AD feedstocks

Relatively lower C:N value materials		Relatively higher C:N value materials	
Substrates and materials	**C:N ratio**	**Substrates and materials**	**C:N ratio**
Cow dung	16–25	Rice straw	51–67
Poultry manure	5–15	Wheat straw	50–150
Pig manure	6–14	Sugar cane bagasse	140–150
Sheep dung	30–33	Com stalks and straw	50–56
Horse manure	20–25	Oat straw	48–50
Kitchen waste	25–29	Sugar beet and sugar foliage	35–40
Fruits and vegetable waste	7–35	Fallen leaves	50–53
Food waste	3–17	Seaweed	70–79
Peanut shoots and hulls	20–31	Algae	75–100
Waste cereals	16–40	Sawdust	200–500
Grass and grass trimmings	12–16	Potatoes	35–60
Alfalfa	12–17		
Slaughterhouse waste	22–37		
Goat manure	10–17		
Mixed food wastes	15–32		

Source: Hagos, K. et al. 2017. *Renewable and Sustainable Energy Reviews*, 76: 1485–1496.

methane yields were 330.1 and 340.1 L/kg volatile solids (VS) for anaerobic codigestion of 70% poultry droppings and 30% wheat straw at a C:N of 32, and 50% poultry droppings and 50% of meadow grass at a C:N of 31.52, respectively, compared to 254 L/kg VS for poultry droppings only at a C:N ratio of 6 (Rahman et al., 2017). It was also reported that the decrease of C:N increases the N_2O emission (Bauer et al., 2013).

13.2.2.3 *Temperature*

AD of organic wastes is usually operated at mesophilic temperature at approximately 35°C and thermophilic temperature at approximately 55°C. Relatively few studies have been conducted in the digestion of organic wastes at psychrophilic temperature at approximately 20°C (Jain et al., 2015). The rate of thermophilic digestion is normally faster than that of mesophilic digestion. Thermophilic digestion was usually more effective than mesophilic digestion for destroying pathogens. However, the application of high thermophilic temperatures increases the fraction of free ammonia and the logarithm of acid dissociation constant (pKa) of the VFAs, which play inhibiting roles to the microorganisms. Acidification may occur during thermophilic AD, inhibiting biogas production. Furthermore, thermophilic AD is more sensitive to environmental changes than the mesophilic AD. Although mesophilic AD exhibits better process stability and higher richness in bacteria, it has lower methane yield and suffers from poor biodegradability and nutrient imbalance (Mao et al., 2015). The optimal conditions for AD may be thermophilic hydrolysis and acidogenesis, and mesophilic methanogenesis in a two-stage process (Mao et al., 2015). Ambient or seasonal temperature AD has also been used to treat organic wastes because it does not require an extra heat supply, but it exhibits lower methane production and lower stability than the mesophilic AD. The thermophilic digestion has not been widely used because of the requirement of heating digesters to a high temperature (Jain et al., 2015). However, the thermophilic operation has been established as a reliable mode of SS-AD because the SS-AD under mesophilic conditions shows a poor start-up performance (Jain et al., 2015). Psychrophilic digestion requires less energy input for heating influents and maintaining the temperature of digesters, but has a much lower biogas production rate. Because temperature is a very important factor that affects the bacterial activity, any deviation of temperature from an optimum value may result in the unsatisfactory performance of the digesters. AD process failure can occur at a temperature change in excess of 1°C/day, and change in temperature of more than 0.6°C/day should be avoided (Turovskiy and Mathai, 2006). The AD of organic wastes at a high temperature may be economically sustainable if inexpensive external heat is available to warm the waste feedstock to the processing temperature. Digesters should be designed to ensure more uniform digester temperatures. Obtaining this uniform temperature profile will require a better understanding of the modes of heat transfer that occurs within and around the digester, and the fluid dynamics in the digester.

13.2.2.4 pH value

The pH value of AD slurry changes at various states of digestion because it is an indicator of the amount of VFAs generated during AD. Major intermediates in the AD reactions are VFAs. The optimum pH of acidogenesis is 5.5–6.5. If a high concentration of VFAs is formed, pH will decrease to the level at which the methanogenic bacteria are severely inhibited. On the other hand, the pH increases as the volatile acids are digested to produce CH_4 via methanogenesis. Many researchers have suggested the use of VFAs as a control parameter of AD because VFAs are indicative of the activity of the methanogenic consortia (Pind et al., 2003). Methanogenic bacteria are extremely sensitive to pH changes. A digester usually runs at a pH of 6.5–7.2, which is optimum for methanogenic bacteria. Therefore, it is important to have a buffering capacity in the system, and a buffering capacity of 70 mEq $CaCO_3$/L or a molar ratio of at least 1.4:1 of bicarbonate:VFA should be maintained for a stable and well-buffered AD process (Jain et al., 2015). Because there are different ranges of optimum pH for acidogenesis and methanogenesis, a two-stage AD may be a preferred operating mode (Mao et al., 2015).

13.2.2.5 Solids and hydraulic retention time

The solids retention time (SRT) is the average time that the organic wastes solids are in the system, whereas the hydraulic retention time (HRT) is the average time the sludge liquid is held in the digester. The subsequent steps of the digestion process are directly related to the SRT. The SRT is an important digester design and operation factor for the AD process, and is usually expressed in days. Each time solids are removed, a fraction of the bacterial population is also lost, thus suggesting that the cell growth must at least balance the cell removal to ensure a steady AD process. For a continuous AD process, the rate of feeding will depend on the retention period and type of feedstocks. Methane making microorganisms double in 2–4 days. Retention time should not be less than 2–4 days to avoid a large loss of bacteria with the slurry (Jain et al., 2015).

13.2.3 Anaerobic digesters

There are five typical types of anaerobic digesters: completely mixed digesters (i.e., continuously stirred tank digesters [CSTDs]), fixed film digesters, plug-flow digesters, anaerobic filters, and fluidized bed digesters (Nishio and Nakashimada, 2007). A CSTD is the most popular design. In the CSTD, the sludge is continuously fed for maximum efficiency. For practical reasons, the digesters are commonly fed once a day. In a CSTD, an influent substrate concentration of 3%–8% total solids is added daily and an equal amount of effluent is withdrawn (Gunaseelan, 1997). In a plug-flow digester, a volume of the medium with a suitable inoculum enters at one end of the tube and the digestion is completed when the medium reaches the other end. Unlike CSTDs, where the substrate concentration and temperature distribution are forced to uniform values with mechanical mixing, the performance of the plug-flow digester is strongly dependent on how well the fluid dynamics and fermentation kinetics are integrated. The anaerobic filter is primarily used for the digestion of wastewaters produced in large quantities. Microbes are attached to a solid support such as stones packed inside a tank. Wastewater flows upward through the tank and the retention time is only a few hours. Fluidized bed digester is a modified form of anaerobic filter. In the fluidized bed digester, the bacteria are attached to small glass spheres that are freely suspended in the up-flowing feed.

Among the 242 operating anaerobic digesters at livestock farms in the United States in 2015, 102 operations, or 42% of the total, were plug-flow digesters, and 90 operations, or 37% of the total, were CSTDs (U.S. EPA, 2017). An AD process involves the four main steps of hydrolysis, acidogenesis, acetogenesis, and methanogenesis with different microorganisms. Because various microorganisms have different growth rates and environmental responses, a two-stage or multiple-stage anaerobic digestion system can split the various AD steps to increase the stability, loading rate, and biogas production yield (Hagos et al., 2017).

13.2.4 *Feedstocks for biogas production*

AD is a versatile technology to treat a variety of organic wastes and produce fuel. Essentially all organic materials can be digested, except for stable woody materials because the anaerobic microorganisms are unable to degrade lignin. An excessive amount of lignin in the AD reactor may cause process failure. The potential sources of biogas on Earth are 75% in agricultural crops, by-products, and manure, 17% in municipal and industrial organic wastes, and 8% in sewage wastewater treatment facilities (Grando et al., 2017). Various renewable feedstocks (algae, aquatic plants, agricultural residues, and woody biomass), industrial wastewater, food processing wastes, and MSWs have been anaerobically digested for biogas production.

13.2.4.1 *Algae and aquatic biomass*

Algae, especially waste-grown algae, are viewed as next-generation biofuel feedstocks because of their superior photosynthetic efficiencies and higher carbon-capturing capabilities compared to terrestrial plants. Algae are usually composed of lipids, proteins, nucleic acids, and carbohydrates. The typical water content of plants is between 80% and 90%. Due to the growth environment of algae, the water content of algae is even higher, ~90%–95%. The energy consumed for drying algae is more than 75% of the total energy consumption. Using AD technology, wet algae can be directly employed to reduce the need for energy-intensive drying operations, which is almost inevitable in other biorefinery technologies. However, algae typically yield less methane than wastewater sludge (~0.3 versus 0.4 L CH_4/g volatile solids introduced) mainly because of ammonia toxicity and recalcitrant cell walls. Microalgae have a low C:N ratio of 6–9 due to the high protein content, which is much lower than the optimal C:N of 20–30 for AD (McKennedy and Sherlock, 2015). Ammonia toxicity might be counteracted by codigesting algae with high-carbon organic wastes such as crop residues and waste paper as well as feedstock pretreatment (Jankowska et al., 2017; Ras et al., 2011; Yen and Brune, 2007). Similar to algae, aquatic plants such as water hyacinth, duckweed, and cattails can be good candidates as feedstocks for anaerobic digestion due to the high yields of biomass, low fertilization and pesticide requirements, broad adaptability, greater ability to sequester carbon in the oil than most other plants, and high water content. There is a large variation in biogas production rate and yield among different aquatic plants (Alvarez and Lidén, 2008; Verma et al., 2007).

13.2.4.2 *Crop residues*

Crop residues such as corn stover, wheat straw, and rice straw are cellulosic materials, which are the structural portion of plants, including complex sugars that cannot directly be used for fermentation substrates. Most recent AD studies using cellulosic materials suggest or focus on pretreatment. Pretreatment is a key step for efficient utilization of biomass for biogas or biofuel production. The goal of the pretreatment is to increase the surface accessibility of carbohydrate polymers, pre-extract sugars, disrupt the lignin seal, and liberate the cellulose from the plant cell wall matrix. Pretreatment techniques have generally been divided into three categories: physical, chemical, and biological. Each pretreatment method has its advantages and disadvantages, and none of pretreatment approaches is suitable for all biomass species (Carrere et al., 2016).

13.2.4.3 *Animal wastes*

According to the U.S. Department of Agriculture, confined food animals produced roughly 335 million tons of wastes (dry weight) in 2005, representing almost a third of the total municipal and industrial waste produced every year (Wang, 2014). The amount of manure produced by confined food animals each year is about 100 times more than that of human sewage sludge processed in U.S. municipal wastewater plants. For instance, one dairy farm with 2500 cows produces as much waste as a city with approximately 411,000 residents. The amounts of animal wastes produced in the United States are given in Table 13.2 (Font-Palma, 2012; Wang, 2014).

AD breaks down complex organic fractions in animal wastes and produces biogas. AD can provide several benefits for the treatment of animal manure, which include (1) odor control, (2) reduction of gas emissions,

Table 13.2 Annual energy available in the United States from manure, sorted by animal category

Animal type	Animal units (millions)	Biogas energy per animal unit/day (thousand BTU)	Biogas energy/year (trillion BTU)
Fattened cattle	9.6	25.7	89.9
Milk cows	12.3	20.6	92.4
Other beef and dairy cattle	58.8	23.2	497
Swine	8.5	39.8	124
Poultry	6.1	56.0	125

Source: Font-Palma, C. 2012. *Energy Conversion and Management*, 53: 92–98; Wang, L. 2014. *Sustainable Bioenergy Production*, CRC Press, Boca Raton.

(3) potentially kills pathogens, (4) reduction of wastewater strength (oxygen demand), (5) conversion of organic nitrogen into plant-available ammonia nitrogen, (6) preservation of plant nutrients (e.g., N, P, K), and (7) production of biofuel such as biogas (Beddoes et al., 2007; Wilkie et al., 2004). The liquid portion in the digestate can be separated and utilized as a fertilizer. The solid portion can be composted to stabilize them and produce a more useful product (Beddoes et al., 2007).

13.2.4.4 *Food wastes*

Food wastes such as fruit and vegetable processing wastes (Bouallagui et al., 2005), meat processing wastes and slaughterhouse wastes (Salminen and Rintala, 2002), and solid fish wastes (Gebauer, 2004) can be treated either aerobically or anaerobically to reduce pollutant and pathogen risk and recover materials from the wastes. Food processing wastes have high moisture content, which makes them difficult to convert in a thermochemical conversion process without predrying. AD is a commercially demonstrated technology and is widely used for treating organic wastes with a high moisture content (i.e., >80%–90% moisture). Gebauer (2004) used a CSTD for the mesophilic anaerobic treatment of sludge from saline fish farm effluents with total solids of 8.2%–10.2% by weight, chemical oxygen demand (COD) of 60–74 g/L, and sodium of 10–10.5 g/L. After AD treatment, the COD content of the sludge decreased from 60% to 36% and methane yields were between 0.114 and 0.184 L/g COD loaded. However, the process was strongly inhibited, presumably by sodium, and unstable, with propionic acid being the main compound of the VFAs. When diluting the sludge 1:1 with tap water (Na content = 5.3 g/L), the inhibition could be overcome and a stable process with low VFA concentrations was achieved (Gebauer, 2004). In another study, codigestion of fish wastes and sisal pulp has been shown to improve the digestibility of both materials and biogas yields (Mshandete et al., 2004). Hejnfelt and Angelidaki (2009) investigated the methane potential from pig meat and bone flour, fat, blood, hair, meat, ribs, raw waste in batch, and semicontinuously fed reactor experiments. The methane potential measured by batch assays is between 225 and 619 $dm^3\ kg^{-1}$, corresponding to 50%–100% of the calculated theoretical methane potential. Dilution of the by-products had a positive effect on the specific methane yield, and high concentrations of long-chain fatty acids and ammonia in the by-products inhibit the biogas process at concentrations higher than 5 g lipids dm^{-3} and 7 g N dm^{-3}, respectively.

Codigestion of 5% pork by-products mixed with pig manure at 37°C showed 40% higher methane production compared to digestion of manure alone (Hejnfelt and Angelidaki, 2009). Fruit and vegetable processing wastes (FVWs) contain 8%–18% TS, with a total VS content of 86%–92%. The organic fraction includes approximately 75% easy biodegradable matter (sugars and hemicellulose), 9% cellulose, and 5% lignin. AD of FVWs has been studied under different operating conditions using different types of bioreactors (Bouallagui et al., 2005). Overall, AD can achieve the conversion of 70%–95% of organic matter to methane, with a volumetric organic loading rate (OLR) of 1–6.8 g versatile solids/L day. The major limitation of this process is the rapid acidification, resulting a low pH value in the reactor, and a higher VFAs content. VFAs further stress and inhibit the activity of methanogenic bacteria. Continuous two-phase systems appear as more highly efficient technologies for AD of FVWs. The greatest advantage lies in the buffering of the organic loading rate taking place in the first stage, allowing a more constant feeding rate of the methanogenic bacteria in the second stage. Using a two-stage system consisting of a thermophilic liquefaction reactor and a mesophilic anaerobic filter, more than 95% volatile solids were converted to methane at a volumetric loading rate of 5.65 g VS/L day. The average methane production yield was about 420 L/kg added VS.

13.2.4.5 Municipal solid wastes

The organic fraction of MSW could be anaerobically digested to biogas in the digestion treatment system. Initializing AD of MSW normally requires inoculation with sewage sludge or codigestion with manure, following a successively higher concentration of MSW (Hartmann and Ahring, 2006; Macias-Corral et al., 2008). According to EU Regulation EC 1772/2002, MSW needs to be pasteurized or sterilized before and/or after AD (Ariunbaatar et al., 2014). Thus, there is a growing interest in studying the pretreatment methods to enhance AD of the organic fraction of MSW and eliminate the cost of sterilization (Jain et al., 2015). So far, physical pretreatments are widely applied (Cesaro and Belgiorno, 2014). Technologies including the use of rotary drum (Zhu et al., 2009), shredder with magnetic separation (Hansen et al., 2007), electroporation, and sonication (Cesaro and Belgiorno, 2013) were found to be effective for MSW separation and pretreatment prior to AD, and enhanced the biogas production. But size reduction showed no significant effect on AD performance (Zhang and Banks, 2013).

The variables that govern AD of the organic fraction of MSW include solids content, processing temperature, the reactor configuration (single-stage versus multistage processes), and feedstock characteristics such as composition of biodegradable fractions, C:N ratio, and particle size. In general, a higher organic fraction can result in a higher biogas yield, and thermophilic processes are more efficient than mesophilic processes in terms of higher biogas yields at different OLRs. A high biogas yield was achieved by means of wet thermophilic processes at OLRs lower than 6 kg-VS m^{-3} day^{-1}, while high-solids processes appeared to be relatively more efficient when OLRs applied are higher than 6 kg-VS m^{-3} day^{-1}

(De Baere, 2006). Multistage systems showed a higher reduction of recalcitrant organic matter compared to single-stage systems, but they were rarely applied in a full scale. Larger particle radius resulted in lower COD degradation, which further results in a lower methane production rate (Esposito et al., 2011). In addition, reduction of mixing levels may be used as an operational tool to stabilize unstable digesters (Stroot et al., 2001). An extended cost-benefit calculation shows that the highest overall benefit of the process is achieved at an OLR that is lower and a HRT that is longer than those values of OLR and HRT at which the highest biogas production is achieved (Hartmann and Ahring, 2006).

13.2.4.6 Industrial wastewater

AD is viewed as the most suitable option for the treatment of high-strength organic effluents (Poh et al., 2015). The presence of biodegradable components in the effluents coupled with the advantages of anaerobic process over other treatment methods makes it an attractive option. For this purpose, a number of anaerobic reactors for the digestion of organic effluents from sugar and distillery, pulp and paper, slaughterhouse, and dairy units, among others, were well developed (Lettinga, 1995). Rajeshwari et al. (2000) provided a comprehensive review of anaerobic reactors for industrial wastewater treatment. In summary, although most of the high-rate reactors have proved their applicability for different high-strength wastewaters over a range of organic loading rates, there exist certain differences in the preference of a particular type of digester over others in terms of various factors such as the requirement of pretreatment, dilution, and control of operating conditions. There are no governing factors that dictate the suitability of any particular reactor design for a specific effluent. By suitable modifications in the reactor designs and by altering the effluent characteristics, the existing high-rate digesters can be accommodated for treatment of organic effluents. More recently, Ersahin et al. (2011) reviewed applications of the AD technology for treatment of various types of industrial wastewaters and showed that high-rate anaerobic reactors enable treatment of industrial effluents at high organic loading rates with a considerably lower ecological footprint.

13.3 Enhancement of biogas production

13.3.1 Biomass pretreatment

Lignocellulosic biomasses such as crop and forest residues are abundant nonfood renewable sources for biogas production. However, the rigid structure of lignocellulosic materials limits their digestibility. Pretreatment of lignocellulosic materials is necessary to make them more amendable for AD. Pretreatment of feedstocks may also be able to obtain suitable by-products.

Pretreatment technologies include (1) chemical pretreatment, (2) physical pretreatment, and (3) biological pretreatment (Sun and Cheng, 2002). Example physical pretreatment methods are microwave treatment and milling. Example chemical pretreatment methods use alkali (NaOH, ammonia) and acid (H_2SO_4, H_2O_2). Alkalis are more efficient than acids in altering the structure of lignin. Alkali-assisted extrusion and enzymatic hydrolysis of lignocellulosic biomass have been used to improve the biogas production rate and yield. The combination of alkali extrusion and enzymatic hydrolysis of a corn cob could increase the biogas yield by 22.3% (Pérez-Rodríguez et al., 2017). Pretreatment with ammonia has been shown to improve digestibility and biogas yield. Ammonia pretreatment has also been found to increase the nitrogen content and conditioning C:N ratio, making corn stover more degradable (Angelidaki et al., 1999). Acid pretreatment involves the use of concentrated or dilute acids to break the rigid structure of lignocellulosic material. The most commonly used acid is sulfuric acid (H_2SO_4), which has been commercially used to pretreat a wide variety of biomass types like wheat straw (Delgenes et al., 1990; Saha et al., 2005) and corn stover (Cheng, 2010; Fernandez et al., 1999). Acid treatment can be used in fractionating the components of the lignocellulosic biomass to remove hemicellulose. However, a thorough washing or detoxification step is necessary to remove acids before fermentation. Another drawback is the accumulation of fermentation inhibitors like furfural and hydroxymethyl furfural (HMF), which reduce the effectiveness of the pretreatment method and further process (Hasan et al., 2014).

Although most of the chemical and physical pretreatment methods have been shown to be very effective, they are not widely used in industries due to some of their disadvantages such as specialized instrument requirement, inefficient and high energy usage, and sometimes production of acid or alkaline wastewater as a by-product (Wang et al., 2012). Biological pretreatment has become a more attractive approach to enhance the bioconversion of lignocellulose to bioenergy, mainly due to its low chemical and energy usage and environmental friendliness. The biodegradation of cellulosic materials utilizing microbial cocultures or complex communities has been proposed as a highly efficient approach for biotechnological applications because it avoids the problems of feedback regulation and metabolite repression posed by isolated single strains (Galbe and Zacchi, 2012). Fungal and bacterial pretreatment has been the two most popular microbial pretreatment options used for bioconversion of lignocellulose to simple sugars. A majority of the industrial pretreatment methods have mainly focused on the action of fungi that produce an enzyme to be able to degrade hemicellulose, lignin, and polyaromatic phenols. White-rot and soft-rot fungi have been found to successfully degrade lignocellulosic material (Anderson and Akin, 2008; Sun and Cheng, 2002). White-rot fungi such as *Phanerochaete chrysosporium, Phlebia radiata, Dichomitus squalens, Rigidoporus lignosus,* and *Jungua separabilima* delignify wood and wheat straw (Itoh et al., 2003). Wheat straw when pretreated with *Pleurotus ostreatus* for 60 days converted 33% of cellulose to glucose (Taniguchi et al., 2005). *Cyathus stercoreus* is known to be highly proficient in degrading corn stover by up to 36% cellulose conversion (Keller et al., 2003). White-rot fungi have high selectivity for lignin degradation over cellulose. Most white-rot fungi, such as *P. chrysosporium,* simultaneously degrade cellulose and hemicellulose and lignin, which results in low cellulose recovery (Eriksson et al., 1990). *P. chrysosporium* when acted upon cotton stalk preferentially degrade lignin and part of the hemicellulose to produce a cellulose-rich residue (Anderson and Akin, 2008). *Ceriporiopsis subvermispora* is one of the most effective wood-decaying fungi and can selectively degrade lignin with very low cellulose loss (Itoh et al., 2003). Few aerobic microorganisms selectively degrade lignin or hemicellulose (Zhong et al., 2011). An increase of 33% in methane yield was obtained when orange processing waste was treated with selected fungi strains (Srilatha et al., 1995). Yeasts such as *Saccharomyces cerevisiae* sp., *Coccidioides immitis* sp., and *Hansenula anomala* sp. consume the sugars present in the corn straw under anaerobic condition for the biomass growth. They expose the crystalline and noncrystalline cellulose materials in the straw by destroying the crystal texture of the corn straw.

Office paper treated with selected aerobic bacteria enhanced the enzymatic hydrolysis, leading to 94% sugar recovery (Taherzadeh and Karimi, 2008). Aerobic and facultative anaerobic bacteria such as β-*proteobacterium* HMD444 and strictly anaerobic bacteria (*Thermoanaerobacterium thermosaccharolyticum* strain M18, *Thermanaerovibrio acidaminovorans* DSM 6589, and *Clostridium* sp. strains LDC-8-c12, 5-8, CO6-72, and others, can be utilized as the microbial consortium for biological pretreatment of lignocellulose. Cellulolytic bacteria (*Bacillus licheniformis* sp., *Pseudomonas* sp., *B. subtilis* sp., and *Pl. florida* sp.) can degrade cellulose and hemicellulose, while *Pl. florida* sp. can also degrade lignin. Lactic acid bacteria of *Lactobacillus deiliehii* sp. can be added to adjust the pH value in the physiological processes of various types of microbial populations (Zhong et al., 2011). Pretreating dry ground corn straw powder with complex microbial agents has been reported to reduce total lignin, cellulose, and hemicellulose by 6%–25% and yield 33% more biogas and 76% more methane (Zhong et al., 2011).

13.3.2 Anaerobic codigestion

Anaerobic codigestion has been widely used to enhance biogas production of digesters. Codigestion is a main factor like pretreatment and digester configuration that affects biogas production. There are several important advantages for codigestion, which include (1) improvement of the process stabilization, (2) dilution of inhibitory substances, (3) nutrient balance, (4) accomplishment of the requirement of moisture content in a digester, (5) reduction in greenhouse gas (GHG) emissions to the atmosphere, (6) synergetic effects of microorganisms, (6) increase of the load of biodegradable organic matter, and (7) improvement of the economics of an AD plant (Hagos et al., 2017). Among the 994 biogas plants in Italy in 2012, codigestion accounted 74.2%, compared to 22.4% using only energy crops and 3.2% using only livestock manure (Agostini et al., 2016).

Mshandete et al. (2004) used a batchwise digester to codigest sisal pulp and fish wastes for improving the digestibility of the materials and biogas yield through adjusting the ratio between carbon and nitrogen. The methane yields from sisal pulp and fish waste alone were 0.32 and 0.39 m^3 CH_4/kg VS, respectively, at TS of 5%; codigestion with 33% fish waste and 67% sisal pulp representing 16.6% of TS gave a methane yield of 0.62 m^3 CH_4/kg VS added (Mshandete et al., 2004). Fruit and vegetable wastes and chicken manure are another promising combination to be used as codigestion feedstock. Callaghan et al. (2002) investigated the codigestion of fruit and vegetable wastes and chicken manure in a continuously stirred tank reactor at 35°C, HRT of 21 days, and loading rate in the range of 3.19–5.01 kg VS/m^3 day. They found that by increasing the fruit and vegetable wastes from 20% to 50%, the methane yield increased from 0.23 to 0.45 m^3/kg VS added. The increase of fruit and vegetable wastes in the feedstock can decrease the ammonia inhibition from chicken manure (Callaghan et al., 2002).

13.3.3 Microbial management

AD is a synergistic process of a consortium of microbes for hydrolysis, acidogenesis, acetogenesis, and methanogenesis, which are achieved by specific groups of bacteria. Hydrolytic bacteria first hydrolyze complex particulate compounds to soluble monomeric substrates that can be fermented to volatile organic acids by acidogenic (fermentative) bacteria. The volatile organic acids are further converted to acetate, hydrogen, and carbon dioxide by acetogenic bacteria, which are eventually processed into biogas through methanogenesis by methanogenic bacteria. Cellulolytic microorganisms such as *Cellulomonas, Clostridium,* and *Bacillus* produce cellulases that hydrolyze lignocellulosic biomass (Jain et al., 2015).

The stability of biogas production is highly dependent on the microbial community composition of the reactor. The composition is determined by nature of biomass substrate and physical-chemical parameters of the digestion process. The AD process is complex due to various factors. It becomes difficult to understand when and how a shift in a microbial community occurs during the AD process. The information regarding the microbial behavior will be beneficial to optimize the anaerobic digestion process. The microbial communities in AD are not well understood because of their complexity and metabolic pathways (St-Pierre and Wright, 2013). However, it is necessary to analyze the microbial community in an AD process to improve its efficiency. Restricting the study to just culture-dependent techniques would give an incomplete picture of the microbial ecology and physiology associated with an AD process. It also creates a biased analysis because the environmental factors that influence microbial activity will not have been considered (Manyi-Loh et al., 2013). Understanding the microbial community structure in the AD process will help identify the driving forces (composition of microbial consortium) of biogas-producing microbial communities. The findings will allow rational design of the microbial communities to promote higher efficacy of an AD system.

Genomic characterization along with biochemical analysis has advanced the understanding of feedstock properties, reactor configuration, and the influence of operation conditions on a microbial community structure and dynamics and its overall influence on efficiency of biogas production (Ito et al., 2012; Werner et al., 2011). The use of next-generation sequencing (NGS) has revealed a large phylogenetic and metabolic diversity in the digester reactions. Understanding the metabolic capabilities of these organisms, the influence of environmental factors on the community structure, functional redundancy, and interspecies interactions is key to increasing the efficacy of the digestion process and product formation. For a successful NGS project, expertise in both wet laboratories and bioinformatics is crucial to ensure high-quality data and data interpretation (Buermans and den Dunnen, 2014). Good sequence data are important because of their high quality, robustness and low noise. The sequence itself is hard evidence of its correctness. A sequencing system usually does not allow random sequences and will take countersteps to reduce and discard any random sequences, if formed, during data analysis (Pavan et al., 2000).

Operational temperature is a major AD factor. In a case study (Pap et al., 2015), the AD process was carried out under three different temperatures and samples were drawn for metagenomics analyses. The NGS approach was used to monitor the change of microbial community structure in the biogas reactor under the change of temperature. There was a striking difference observed between the microbial communities at mesophilic and at thermophilic temperatures as shown in Figure 13.2. Under mesophilic conditions, *Bacteriodes* (45.4% of total bacteria), Firmicutes (24.7% of total bacteria), and Proteobacteria (10.4% of total bacteria) were the most dominant bacterial populations. The thermophilic bacterial population was found to dominate the digestion process because the high temperature stabilized toward the end of the cycle with phylum Firmicutes (66.5%

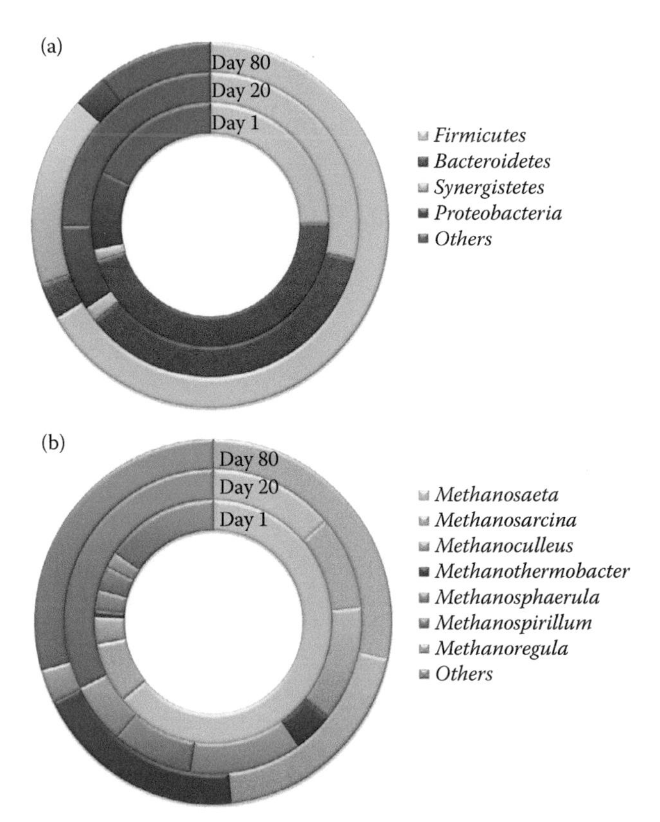

Figure 13.2 Composition of the phylogenetic diversity of the microbial communities from the biogas reactors: (a) relative abundances of the bacterial and (b) archaeal populations at various sampling times (Day 1, 20, and 80). (From Pap, B., Gyorkei, A., Boboescu, I.Z. et al. 2015. Temperature-dependent transformation of biogas-producing microbial communities points to the increased importance of hydrogenotrophic methanogenesis under thermophilic operation. *Bioresource Technology*, 177: 375–380.)

of total bacteria) dominating the reaction process. The next-generation sequencing mediated metagenomics study showed an increase from 0.2% of the total bacteria population under mesophilic conditions to 11.3% of the total bacteria under stable thermophilic operation. While studying the archaeal population in the anaerobic digester, genus *Methanosaeta* (63.9% of the total archaeal community) was the most abundant under mesophilic operation. Under thermophilic conditions, *Methanosarcina* was identified as the most dominant genus (28.3% of the total archaeal community).

Another study demonstrated the effect of increase of TS on microbial population in AD. In this study, food waste was anaerobically digested at a mesophilic (37°C) temperature with three different TS contents (5%, 15%,

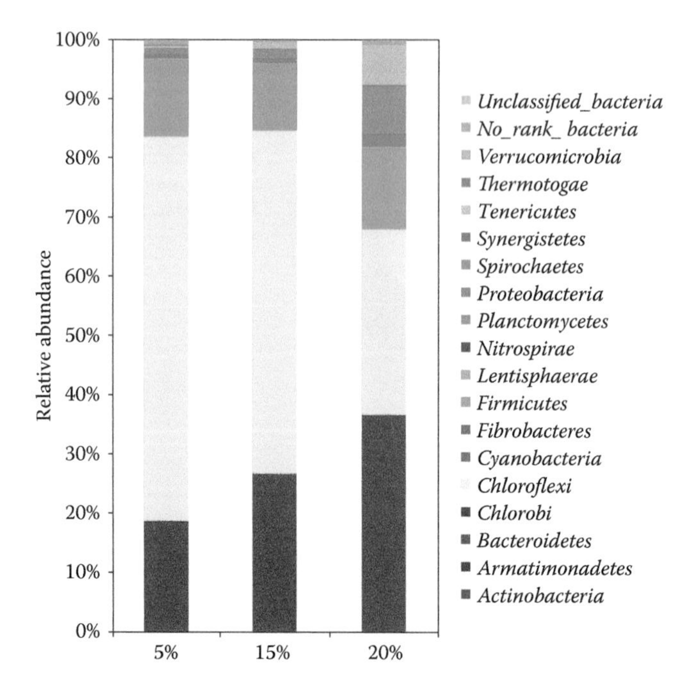

Figure 13.3 Bacterial community in digester with different TS concentration from pyrosequencing. (From Yi, J., Dong, B., Jin, J., and Dai, X. 2014. Effect of increasing total solids contents on anaerobic digestion of food waste under mesophilic conditions: Performance and microbial characteristics analysis. *PloS One*, 9: e102548.)

and 20%). This was the first documented use of pyrosequencing technology to characterize the microbial communities in anaerobic digesters with different TS contents (Yi et al., 2014). The highest bacterial community was observed in an anaerobic digester with TS content of 5% and the highest archaeal community was observed in a digester with TS content of 15% as shown in Figure 13.3.

13.3.4 *Additives and trace element supplementation*

AD is a multistage process of hydrolysis, acidogenesis, acetogenesis, and methanogenesis that depends on susceptible microorganisms. Trace elements such as Fe, Co, Ni, and S must be adequate for supporting the metabolism of microorganisms to maintain an effective digestion process. Fe has been identified as the most effective element for stabilizing food waste AD (Mao et al., 2015). Supplementation with a mixture of Co, Mo, Ni, Se, and W was reported to increase methane production to the range of 45%–65% for inoculums with low concentrations of trace metals (Facchin et al., 2013). Although supplementation of trace elements could be a simple way to achieve AD process stabilization and efficient biogas generation, the economic feasibility of trace elements depends on their costs (Mao et al., 2015).

Codigestion of various substrates is a good strategy to supplement trace elements in a digester because some feedstocks such as sewage sludge and livestock manure are rich in trace elements. However, published research has generally focused on the C and N balance and dilution of toxicants of anaerobic codigestion and less research has been done to analyze the effect of trace element bioavailability on the improvement of anaerobic codigestion (Choong et al., 2016).

Some inhibiting materials such as ammonium are present in a digester. Various organic and inorganic nanoparticles have been used as additives to improve biogas production (Yadvika et al., 2004). Zeolite has been used to adsorb heavy metals and ammonium that are toxic to microbial bacteria and carry microorganisms on the surface during AD (Hagos et al., 2017).

13.4 Economics of biogas production

Compared to direct aerobic composting, AD technology is complex and requires a large investment, but it can recover an amount of energy and reduce pollutant emissions from the wastes. The initial investment of an AD plant includes the capital costs of all the fixed assets such as the plant, machinery, and buildings. The investment costs for biogas plants reported in the literature vary broadly with variations of 20%–30% or higher depending on the technology and various equipment for pretreatment, storage, and handling of different feedstocks (Pantaleo et al., 2013). The AD revenues may come from sales of electricity and upgraded biogas. The economics of an AD technology thus depends on energy costs. The profitability of AD heavily depends on farm type and the availability of land to allow disposal of digestate. It was reported that AD could be economically viable on medium and large arable farms in England at 2009 prices (Jones and Salter, 2013). Agostini et al. (2016) reported that the generation of electricity by the biogas from the AD of energy crops such as maize and sorghum alone could lead to economic losses and provide no or very limited GHG savings in Italy. However, AD of dairy cattle manure and codigestion of manure with up to 30% sorghum cultivated on no tillage could generate profit and provide GHG savings. Sorghum is a better energy crop than maize as AD feedstock in terms of both economics and GHG savings. Codigestion of manure and energy crops can be economically profitable because the costs associated with mitigation of GHG emissions are very high. The use of animal manure and agricultural residues as feedstocks could make AD profitable (Agostini et al., 2016).

The recovered nutrients in the AD effluents can increase the net profitable margin of dairy and arable farms. However, AD effluents are generally not suitable to directly dispose of on the land because they are too wet, contain some phytotoxic volatile fatty acids, and are not hygienized if the digestion does not occur at a thermophilic temperature. Therefore, aerobic posttreatment after anaerobic digestion is needed (Mata-Alvarez et al., 2000). The future of AD should be sought in the context of an overall sustainable waste management perspective (Mata-Alvarez et al., 2000).

Furthermore, the methane produced from industrial anaerobic wastewater treatment is utilized on-site as fuel for the biogas engine to cogenerate heat and electricity. Heat might be used to keep the AD temperature. Additional profit could be earned if the waste treatment plants adopt an AD system by connecting the electricity generated to the national grid. The amount of electricity supplied to the national grid will then be paid to the plant according to feed-in-tariff (Chin et al., 2013). This makes the AD technology more economically attractive.

13.5 Environmental impact of biogas production

Life-cycle assessment (LCA) is an emerging tool to measure and compare the environmental impacts of human activities. LCA consists of two procedures: the selection of impact indicators and the analysis of inventory data for emissions. The impact indicators used in the LCA may include global warming potential, acidification, eutrophication, photochemical oxidation, and energy use. The energetic balance and the potential environmental impacts of AD are often selected as the indicators in comparative LCA case studies.

LCA was conducted to evaluate environmental impacts of two MSW-to-energy schemes currently practiced in Thailand: incineration and anaerobic digestion. The AD scheme was found to have more advantages: (1) potential impacts such as global warming, acidification, stratospheric ozone depletion, and photo-oxidant formation can be avoided due to net electricity production and fertilizer production and (2) the AD resulted in higher net energy output compared to the incineration scheme. But the incineration had less potential impact for nutrient enrichment. LCA analysis provides invaluable information to policy makers, and suggests an integrated method of management (Chaya and Gheewala, 2007).

Tilche and Galatola (2008) looked into the potential of AD for the reduction of GHG emissions in Europe. The LCA was performed for the 27 European countries on the basis of their 2005 Kyoto Protocol declarations. They considered two different possible biogas applications: electricity production from manure waste, and upgraded methane production for light goods vehicles. Their work shows that biogas may noticeably contribute to GHG reductions, and its use as a biofuel may allow for true negative GHG emissions, showing a net advantage with respect to other biofuels (Tilche and Galatola, 2008).

Lijó et al. (2017) conducted LCA of 15 farm-based biogas plants in Italy. Most of them adopt a codigestion using energy crops such as maize, ryegrass, and sorghum and animal wastes such as hog, cow, and chicken manure. They found that there was no significant difference in GHG emissions from the AD process itself for the 15 AD plants. However, the selection of feedstocks plays a key role in the overall eco-efficiency of an AD plant due to their different origins and compositions. This is because the cultivation of energy crops represented an important source of GHGs due to the use of diesel for agricultural machinery. Another source of GHGs was nitrous oxide specially produced during the storage and utilization of the solid fraction of the digestate. Digestate management was identified as an important issue to minimize GHG emissions from AD plants (Lijó et al., 2017).

13.6 Biogas quality control and upgrading

The main constituents of biogas are CH_4 and CO_2, but there are significant quantities of undesirable contaminants such as H_2S, NH_3, and siloxanes. Purification processes are needed to remove unwanted components such as H_2S. When the biogas is used as fuel for vehicles, injected into the natural gas network, or used in fuel cell chemical synthesis, high purification grade must be achieved and CO_2 must also be removed (Grando et al., 2017). There are several approaches that can be used to convert biogas into transportation fuels as shown in Figure 13.4. Methane in biogas

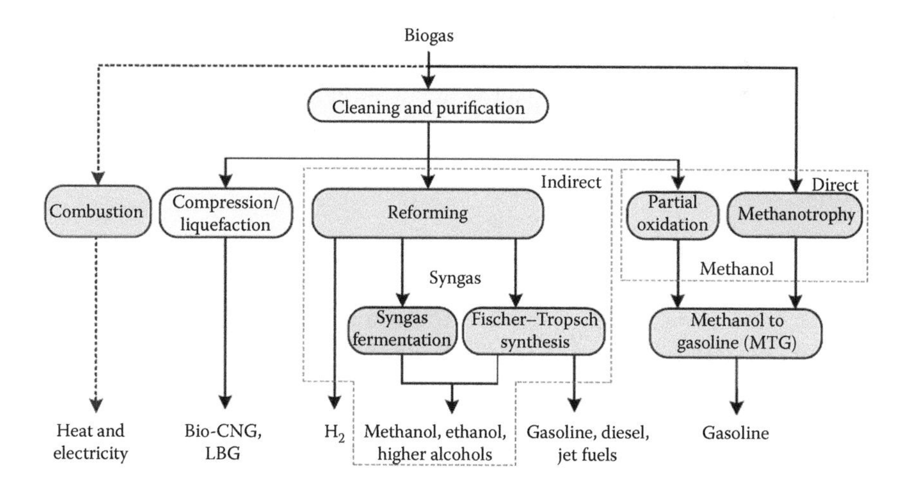

Figure 13.4 Route map of biogas cleaning and conversion to transportation fuels. (From Yang, L., Ge, X., Wan, C., Yu, F., and Li, Y. 2014. *Renewable and Sustainable Energy Reviews*, Progress and perspectives in converting biogas to transportation fuels. 40: 1133–1152.)

can be concentrated to the same standards as natural gas. Biogas can be reformed into cleaner fuels like syngas and hydrogen. The syngas can be used in a Fischer–Tropsch process to produce liquid fuels. Nickel-based catalysts have been widely studied for the catalytic reforming of biogas with steam into syngas (Bereketidou and Goula, 2012; Lino et al., 2017). Because biogas contains both CH_4 and CO_2, dry reforming of CH_4 with CO_2 is considered as one of the promising technologies for upgrading biogas. Mo-based catalysts have been studied for dry reforming of biogas (Gaillard et al., 2017).

For the downstream utilization, the impurities, particularly H_2S, in the raw biogas must be removed. There are various biogas purification methods available such as water scrubbing, chemical absorption, pressure swing absorption, membrane separation, and biofilters (Zhao et al., 2010). A packed column with an aqueous solution can be used to absorb CO_2 and H_2S in the biogas to generate CH_4-enriched fuel (Tippayawong and Thanompongchart, 2010). For water scrubbing, raw biogas is usually pressurized and introduced at the bottom of the scrubbing tower. Amine scrubbing can completely remove H_2S and achieve high efficiency and reaction rates for the removal of CO_2 as compared to water scrubbing and can be operated at a low pressure. Many different amines such as diethanolamine (DEA), monoethanolamine (MEA), methyldiethanolamine (MDEA), diisopropanolamine (DIPA), and aminoethoxy ethanol (DGA) are used. The disadvantage is the requirement of waste chemical treatment (Kadam and Panwar, 2017). Other biogas purification methods include pressure swing adsorption with an adsorbent such as zeolite and membrane permeation (Kadam and Panwar, 2017). In Europe, the most preferable technology for biogas cleaning is water scrubbing with a share of 40%, followed by pressure swing adsorption and chemical scrubbing. The chemical scrubbing and physical absorption consume the highest amount of energy, but water scrubbing requires less energy (Kadam and Panwar, 2017).

13.7 Conclusions

AD is a versatile technology that has the potential to recover energy from a variety of organic wastes. Almost all organic wastes, such as renewable feedstock (algae, aquatic plants, agricultural residues, and woody biomass), industrial wastewater, MSWs, and food processing wastes, can be digested using the AD process. Direct AD of organic matters such as lignocellulosic biomass and animal manure by microorganisms is difficult because of their unbalanced nutrients and low bioaccessibility. Various pretreatments and codigestions have been studied to improve biogas production. Proper microbial management and trace element supplementation can improve biogas production via AD. The use of animal manure and agricultural residues as feedstocks could make AD profitable. LCA shows that AD as a waste treatment technology has minor or negative impacts to human life, reduces GHG emissions, and generates high net energy output. Digestate management was identified as an important issue to minimize GHG emissions from AD plants. The economic and environmental impacts of AD should be sought in the context of an overall sustainable waste management perspective. The raw biogas has undesirable impurities such as H_2S and CO_2. These impurities can be removed using various commercially available cleaning technologies such as water scrubbing, pressure swing adsorption, and amine scrubbing to generate biomethane.

Acknowledgments

This publication was made possible by Grant Numbers NC.X2013-38821-21141 and NC.X-294-5-15-130-1 from the U.S. Department of Agriculture, National Institute of Food and Agriculture (USDA-NIFA). Its contents are solely the responsibility of the authors and do not necessarily represent the official views of the National Institute of Food and Agriculture.

References

Agostini, A., Battini, F., Padella, M. et al. 2016. Economics of GHG emissions mitigation via biogas production from sorghum, maize and dairy farm manure digestion in the Po valley. *Biomass and Bioenergy*, 89: 58–66.

Alvarez, R. and Lidén, G. 2008. Anaerobic co-digestion of aquatic flora and quinoa with manures from Bolivian Altiplano. *Waste Management*, 28: 1933–1940.

Anderson, W. F. and Akin, D. E. 2008. Structural and chemical properties of grass lignocelluloses related to conversion for biofuels. *Journal of Industrial Microbiology & Biotechnology*, 35(5), 355–366.

Angelidaki, I., Ellegaard, L., and Ahring, B. K. 1999. A comprehensive model of anaerobic bioconversion of complex substrates to biogas. *Biotechnology and Bioengineering*, 63: 363–372.

Ariunbaatar, J., Panico, A., Esposito, G., Pirozzi, F., and Lens, P. N. L. 2014. Pretreatment methods to enhance anaerobic digestion of organic solid waste. *Applied Energy*, 123: 143–156.

Bauer, F., Persson, T., Hulteberg, C., and Tamm, D. 2013. Biogas upgrading–Technology overview, comparison and perspectives for the future. *Biofuels, Bioproducts and Biorefining*, 7: 499–511.

Beddoes, J. C., Bracmort, K. S., Burns, R. T., and Lazarus, W. F. 2007. An analysis of energy production costs from anaerobic digestion systems on U.S. livestock production facilities. *Technical Note 1*, U.S. Department of Agriculture, Natural Resources Conservation Service, Washington, D.C.

Bereketidou, O. and Goula, M. 2012. Biogas reforming for syngas production over nickel supported on ceria–alumina catalysts. *Catalysis Today*, 195: 93–100.

Blumenstein, B., Siegmeier, T., and Moller, D. 2016. Economics of anaerobic digestion in organic agriculture: Between system constraints and policy regulations. *Biomass and Bioenergy*, 86: 105–119.

Bouallagui, H., Touhami, Y., Cheikh, R. B., and Hamdi, M. 2005. Bioreactor performance in anaerobic digestion of fruit and vegetable wastes. *Process Biochemistry*, 40: 989–995.

Buermans, H. and den Dunnen, J. 2014. Next generation sequencing technology: Advances and applications. *Biochimica et Biophysica Acta (BBA)-Molecular Basis of Disease*, 1842: 1932–1941.

Callaghan, F., Wase, D., Thayanithy, K., and Forster, C. 2002. Continuous co-digestion of cattle slurry with fruit and vegetable wastes and chicken manure. *Biomass and Bioenergy*, 22: 71–77.

Carlsson, M., Lagerkvist, A., and Morgan-Sagastume, F. 2012. The effects of substrate pre-treatment on anaerobic digestion systems: A review. *Waste Management*, 32: 1634–1650.

Carrere, H., Antonopoulou, G., Affes, R. et al. 2016. Review of feedstock pretreatment strategies for improved anaerobic digestion: From lab-scale research to full-scale application. *Bioresource Technology*, 199: 386–397.

Cesaro, A. and Belgiorno, V. 2013. Sonolysis and ozonation as pretreatment for anaerobic digestion of solid organic waste. *Ultrasonics Sonochemistry*, 20: 931–936.

Cesaro, A. and Belgiorno, V. 2014. Pretreatment methods to improve anaerobic biodegradability of organic municipal solid waste fractions. *Chemical Engineering Journal*, 240: 24–37.

Chaya, W. and Gheewala, S. H. 2007. Life cycle assessment of MSW-to-energy schemes in Thailand. *Journal of Cleaner Production*, 15: 1463–1468.

Cheng, J. 2010. *Biomass to Renewable Energy Processes*, CRC Press Inc, Boca Raton, FL, USA.

Chin, M. J., Poh, P. E., Tey, B. T., Chan, E. S., and Chin, K. L. 2013. Biogas from palm oil mill effluent (POME): Opportunities and challenges from Malaysia's perspective. *Renewable and Sustainable Energy Reviews*, 26: 717–726.

Choong, Y. Y., Norli, I., Abdullah, A. Z., and Yhaya, M. F. 2016. Impacts of trace element supplementation on the performance of anaerobic digestion process: A critical review. *Bioresource Technology*, 209: 369–379.

De Baere, L. 2006. Will anaerobic digestion of solid waste survive in the future? *Water Science & Technology*, 53: 187–94.

De Clercq, D., Wen, Z., Gottfried, O., Schmidt, F., and Fei, F. 2017. A review of global strategies promoting the conversion of food waste to bioenergy via anaerobic digestion. *Renewable and Sustainable Energy Reviews*, 79: 204–221.

Delgenes, J. P., Moletta, R., and Navarro, J. M. 1990. Acid-hydrolysis of wheat straw and process considerations for ethanol fermentation by *Pichia-stipitis* y7124. *Process Biochemistry*, 25: 132–135.

Donoso-Bravo, A., Mailier, J., Martin, C. et al. 2011. Model selection, identification and validation in anaerobic digestion: A review. *Water Research*, 45: 5347–5364.

Edwards, J., Othman, M., and Burn, S. 2015. A review of policy drivers and barriers for the use of anaerobic digestion in Europe, the United States and Australia. *Renewable and Sustainable Energy Reviews*, 52: 815–828.

Eriksson, K.-E., Blanchette, R. A., and Ander, P. 1990. *Microbial and Enzymatic Degradation of Wood and Wood Components*, Springer-Verlag, Berlin.

Ersahin, M. E., Ozgun, H., Dereli, R. K., and Ozturk, I. 2011. Anaerobic treatment of industrial effluents: An overview of applications. in: *Waste Water—Treatment and Reutilization*, F. S. G. Einschlag (Ed.), InTech, Rijeka, Croatia, pp. 3–28.

Esposito, G., Frunzo, L., Panico, A., and Pirozzi, F. 2011. Modelling the effect of the OLR and OFMSW particle size on the performances of an anaerobic co-digestion reactor. *Process Biochemistry*, 46: 557–565.
Facchin, V., Cavinato, C., Pavan, P., and Bolzonella, D. 2013. Batch and continuous mesophilicanaerobic digestion of food waste: Effect of trace elements supplementation. *Chemical Engineering Transactions*, 32: 457–462.
Fernandez, A., Huang, S., Seston, S. et al. 1999. How stable is stable? Function versus community composition. *Applied and Environmental Microbiology*, 65: 3697–3704.
Font-Palma, C. 2012. Characterisation, kinetics and modelling of gasification of poultry manure and litter: An overview. *Energy Conversion and Management*, 53: 92–98.
Gaida, D., Wolf, C., and Bongards, M. 2017. Feed control of anaerobic digestion processes for renewable energy production: A review. *Renewable and Sustainable Energy Reviews*, 68: 869–875.
Gaillard, M., Virginie, M., and Khodakov, A. Y. 2017. New molybdenum-based catalysts for dry reforming of methane in presence of sulfur: A promising way for biogas valorization. *Catalysis Today*, 289: 143–150.
Galbe, M. and Zacchi, G. 2012. Pretreatment: The key to efficient utilization of lignocellulosic materials. *Biomass and Bioenergy*, 46: 70–78.
Gebauer, R. 2004. Mesophilic anaerobic treatment of sludge from saline fish farm effluents with biogas production. *Bioresource Technology*, 93: 155–167.
Grando, R. L., de Souza Antune, A. M., da Fonseca, F. V. et al. 2017. Technology overview of biogas production in anaerobic digestion plants: A European evaluation of research and development. *Renewable and Sustainable Energy Reviews*, 80: 44–53.
Gunaseelan, V. N. 1997. Anaerobic digestion of biomass for methane production: A review. *Biomass and Bioenergy*, 13: 83–114.
Hagos, K., Zong, J., Li, D., Liu, C., and Lu, X. 2017. Anaerobic co-digestion process for biogas production: Progress, challenges and perspectives. *Renewable and Sustainable Energy Reviews*, 76: 1485–1496.
Hansen, T. L., Jansen, J. l. C., Davidsson, Å., and Christensen, T. H. 2007. Effects of pre-treatment technologies on quantity and quality of source-sorted municipal organic waste for biogas recovery. *Waste Management*, 27(3): 398–405.
Harris, P. W. and McCabe, B. K. 2015. Review of pre-treatments used in anaerobic digestion and their potential application in high-fat cattle slaughterhouse wastewater. *Applied Energy*, 155: 560–575.
Hartmann, H. and Ahring, B. K. 2006. Strategies for the anaerobic digestion of the organic fraction of municipal solid waste: An overview. *Water Science and Technology*, 53(8): 7–22.
Hasan, N. A., Young, B. A., Minard-Smith, A. T. et al. 2014. Microbial community profiling of human saliva using shotgun metagenomic sequencing. *PLoS One*, 9(5): e97699.
Hejnfelt, A. and Angelidaki, I. 2009. Anaerobic digestion of slaughterhouse by-products. *Biomass and Bioenergy*, 33(8): 1046–1054.
Henderson, J. P. 2009. Anaerobic digestion in rural China. *Biocycle*, 38(1): 79–81.
Ito, T., Yoshiguchi, K., Ariesyady, H. D., and Okabe, S. 2012. Identification and quantification of key microbial trophic groups of methanogenic glucose degradation in an anaerobic digester sludge. *Bioresource Technology*, 123: 599–607.

Itoh, H., Wada, M., Honda, Y., Kuwahara, M., and Watanabe, T. 2003. Bioorganosolve pretreatments for simultaneous saccharification and fermentation of beech wood by ethanolysis and white rot fungi. *Journal of Biotechnology*, 103(3): 273–280.

Jain, S., Jain, S., Wolf, I. T., Lee, J., and Tong, Y. W. 2015. A comprehensive review on operating parameters and different pretreatment methodologies for anaerobic digestion of municipal solid waste. *Renewable and Sustainable Energy Reviews*, 52: 142–154.

Jankowska, E., Sahu, A. K., and Oleskowicz-Popiel, P. 2017. Biogas from microalgae: Review on microalgae's cultivation, harvesting and pretreatment for anaerobic digestion. *Renewable and Sustainable Energy Reviews*, 75: 692–709.

Jones, P. and Salter, A. 2013. Modelling the economics of farm-based anaerobic digestion in a UK whole-farm context. *Energy Policy*, 62: 215–225.

Kadam, R. and Panwar, N. 2017. Recent advancement in biogas enrichment and its applications. *Renewable and Sustainable Energy Reviews*, 73: 892–903.

Keller, F. A., Hamilton, J. E., and Nguyen, Q. A. 2003. Microbial pretreatment of biomass. *Applied Biochemistry and Biotechnology*, 105: 27–41.

Kondusamy, D. and Kalamdhad, A. S. 2014. Pre-treatment and anaerobic digestion of food waste for high rate methane production—A review. *Journal of Environmental Chemical Engineering*, 2: 1821–1830.

Lettinga, G. 1995. Anaerobic digestion and wastewater treatment systems. *Antonie van Leeuwenhoek*, 67: 3–28.

Li, J., Jha, A. K., He, J. et al. 2011. Assessment of the effects of dry anaerobic co-digestion of cow dung with waste water sludge on biogas yield and biodegradability. *International Journal of Physical Sciences*, 6: 3723–3732.

Lijó, L., Lorenzo-Toja, Y., Gonzalez-Garcia, S. et al. 2017. Eco-efficiency assessment of farm-scaled biogas plants. *Bioresource Technology*, 237: 146–155.

Lino, A. V. P., Assaf, E. M., and Assaf, J. M. 2017. Hydrotalcites derived catalysts for syngas production from biogas reforming: Effect of nickel and cerium load. *Catalysis Today*, 289: 78–88.

Macias-Corral, M., Samani, Z., Hanson, A. et al. 2008. Anaerobic digestion of municipal solid waste and agricultural waste and the effect of co-digestion with dairy cow manure. *Bioresource Technology*, 99: 8288–8293.

Madsen, M., Holm-Nielsen, J. B., and Esbensen, K. H. 2011. Monitoring of anaerobic digestion processes: A review perspective. *Renewable and Sustainable Energy Reviews*, 15: 3141–3155.

Manyi-Loh, C. E., Mamphweli, S. N., Meyer, E. L. et al. 2013. Microbial anaerobic digestion (bio-digesters) as an approach to the decontamination of animal wastes in pollution control and the generation of renewable energy. *International Journal of Environmental Research and Public Health*, 10: 4390–4417.

Mao, C., Feng, Y., Wang, X., and Ren, G. 2015. Review on research achievements of biogas from anaerobic digestion. *Renewable and Sustainable Energy Reviews*, 45: 540–555.

Mata-Alvarez, J., Mace, S., and Llabres, P. 2000. Anaerobic digestion of organic solid wastes. An overview of research achievements and perspectives. *Bioresource Technology*, 74: 3–16.

McKennedy, J. and Sherlock, O. 2015. Anaerobic digestion of marine macroalgae: A review. *Renewable and Sustainable Energy Reviews*, 52: 1781–1790.

Mirzoyan, N., Tal, Y., and Gross, A. 2010. Anaerobic digestion of sludge from intensive recirculating aquaculture systems: Review. *Aquaculture*, 306: 41–46.

Mshandete, A., Kivaisi, A., Rubindamayugi, M., and Mattiasson, B. 2004. Anaerobic batch co-digestion of sisal pulp and fish wastes. *Bioresource Technology*, 95: 19–24.

National Development and Reform Commission (NDRC). 2007. *National Middle and Long Term Plan of Renewable Energy Development 2006–2020*. Beijing, People's Republic of China.

Nguyen, D., Gadhamshetty, V., Nitayavardhana, S., and Khanal, S. K. 2015. Automatic process control in anaerobic digestion technology: A critical review. *Bioresource Technology*, 193: 513–522.

Nishio, N. and Nakashimada, Y. 2007. Recent development of anaerobic digestion processes for energy recovery from wastes. *Journal of Bioscience and Bioengineering*, 103: 105–112.

Pantaleo, A., De Gennaro, B., and Shah, N. 2013. Assessment of optimal size of anaerobic co-digestion plants: An application to cattle farms in the province of Bari (Italy). *Renewable and Sustainable Energy Reviews*, 20: 57–70.

Pap, B., Gyorkei, A., Boboescu, I. Z. et al. 2015. Temperature-dependent transformation of biogas-producing microbial communities points to the increased importance of hydrogenotrophic methanogenesis under thermophilic operation. *Bioresource Technology*, 177: 375–380.

Pavan, P., Battistoni, P., Mata-Alvarez, J., and Cecchi, F. 2000. Performance of thermophilic semi-dry anaerobic digestion process changing the feed biodegradability. *Water Science and Technology*, 41: 75–81.

Pérez-Rodríguez, N., Garcia-Bernet, D., and Domínguez, J. M. 2017. Extrusion and enzymatic hydrolysis as pretreatments on corn cob for biogas production. *Renewable Energy*, 107: 597–603.

Pind, P. F., Angelidaki, I., and Ahring, B. K. 2003. Dynamics of the anaerobic process: Effects of volatile fatty acids. *Biotechnology and Bioengineering*, 82: 791–801.

Poh, P. E., Tan, D. T., Chan, E.-S., and Tey, B. T. 2015. Current advances of biogas production via anaerobic digestion of industrial wastewater. in: *Advances in Bioprocess Technology*, P. Ravindra (Ed.), Springer International Publishing, Cham, pp. 149–163.

Rahman, M. A., Møller, H. B., Saha, C. K. et al. 2017. Optimal ratio for anaerobic co-digestion of poultry droppings and lignocellulosic-rich substrates for enhanced biogas production. *Energy for Sustainable Development*, 39: 59–66.

Rajagopal, R., Masse, D. I., and Singh, G. 2013. A critical review on inhibition of anaerobic digestion process by excess ammonia. *Bioresource Technology*, 143: 632–641.

Rajeshwari, K., Balakrishnan, M., Kansal, A., Lata, K., and Kishore, V. 2000. State-of-the-art of anaerobic digestion technology for industrial wastewater treatment. *Renewable and Sustainable Energy Reviews*, 4: 135–156.

Ras, M., Lardon, L., Bruno, S., Bernet, N., and Steyer, J.-P. 2011. Experimental study on a coupled process of production and anaerobic digestion of *Chlorella vulgaris*. *Bioresource Technology*, 102: 200–206.

Romero-Guiza, M. S., Vila, J., Mata-Alvarez, J., Chimenos, J. M., and Astals, S. 2016. The role of additives on anaerobic digestion: A review. *Renewable and Sustainable Energy Reviews*, 58: 1486–1499.

Saha, B. C., Iten, L. B., Cotta, M. A., and Wu, Y. V. 2005. Dilute acid pretreatment, enzymatic saccharification and fermentation of wheat straw to ethanol. *Process Biochemistry*, 40: 3693–3700.

Salminen, E. and Rintala, J. 2002. Anaerobic digestion of organic solid poultry slaughterhouse waste—A review. *Bioresource Technology*, 83: 13–26.
Srilatha, H. R., Nand, K., Babu, K. S., and Madhukara, K. 1995. Fungal pretreatment of orange processing waste by solid-state fermentation for improved production of methane. *Process Biochemistry*, 30: 327–331.
St-Pierre, B. and Wright, A.-D. G. 2013. Metagenomic analysis of methanogen populations in three full-scale mesophilic anaerobic manure digesters operated on dairy farms in Vermont, USA. *Bioresource Technology*, 138: 277–284.
Stroot, P. G., McMahon, K. D., Mackie, R. I., and Raskin, L. 2001. Anaerobic codigestion of municipal solid waste and biosolids under various mixing conditions—I. Digester performance. *Water Research*, 35: 1804–1816.
Sun, Y. and Cheng, J. 2002. Hydrolysis of lignocellulosic materials for ethanol production: A review. *Bioresource Technology*, 83: 1–11.
Taherzadeh, M. J. and Karimi, K. 2008. Pretreatment of lignocellulosic wastes to improve ethanol and biogas production: A review. *International Journal of Molecular Sciences*, 9: 1621–1651.
Taniguchi, M., Suzuki, H., Watanabe, D. et al. 2005. Evaluation of pretreatment with *Pleurotus ostreatus* for enzymatic hydrolysis of rice straw. *Journal of Bioscience and Bioengineering*, 100(6): 637–643.
Tilche, A. and Galatola, M. 2008. The potential of bio-methane as bio-fuel/bio-energy for reducing greenhouse gas emissions: A qualitative assessment for Europe in a life cycle perspective. *Water Science and Technology*, 57: 1683–1692.
Tippayawong, N. and Thanompongchart, P. 2010. Biogas quality upgrade by simultaneous removal of CO_2 and H_2S in a packed column reactor. *Energy*, 35: 4531–4535.
Turovskiy, I. S. and Mathai, P. 2006. *Wastewater Sludge Processing*, John Wiley & Sons, Hoboken, New Jersey, USA.
U.S. Environmental Protection Agency (EPA). 2017. AgSTAR data and trends, https://www.epa.gov/agstar/agstar-data-and-trends. Accessed on October 6, 2017.
Verma, V., Singh, Y., and Rai, J. 2007. Biogas production from plant biomass used for phytoremediation of industrial wastes. *Bioresource Technology*, 98: 1664–1669.
Wang, L. 2014. *Sustainable Bioenergy Production*, CRC Press, Boca Raton.
Wang, L. H., Wang, Q. H., Cai, W. W., and Sun, X. H. 2012. Influence of mixing proportion on the solid-state anaerobic co-digestion of distiller's grains and food waste. *Biosystems Engineering*, 112: 130–137.
Ward, A. J., Lewis, D. M., and Green, F. B. 2014. Anaerobic digestion of algae biomass: A review. *Algal Research*, 5: 204–214.
Werner, J. J., Knights, D., Garcia, M. L. et al. 2011. Bacterial community structures are unique and resilient in full-scale bioenergy systems. *Proceedings of the National Academy of Sciences*, 108: 4158–4163.
Wilkie, A., Castro, H., Cubinski, K., Owens, J., and Yan, S. 2004. Fixed-film anaerobic digestion of flushed dairy manure after primary treatment: Wastewater production and characterisation. *Biosystems Engineering*, 89: 457–471.
Xie, S., Hai, F. I., Zhan, X. et al. 2016. Anaerobic co-digestion: A critical review of mathematical modelling for performance optimization. *Bioresource Technology*, 222: 498–512.

Yadvika, Santosh, Sreekrishnan, T. R., Kohli, S., and Rana, V. 2004. Enhancement of biogas production from solid substrates using different techniques—A review. *Bioresource Technology*, 95: 1–10.

Yang, L., Ge, X., Wan, C., Yu, F., and Li, Y. 2014. Progress and perspectives in converting biogas to transportation fuels. *Renewable and Sustainable Energy Reviews*, 40: 1133–1152.

Yen, H.-W. and Brune, D. E. 2007. Anaerobic co-digestion of algal sludge and waste paper to produce methane. *Bioresource Technology*, 98: 130–134.

Yenigun, O. and Demirel, B. 2013. Ammonia inhibition in anaerobic digestion: A review. *Process Biochemistry*, 48: 901–911.

Yi, J., Dong, B., Jin, J., and Dai, X. 2014. Effect of increasing total solids contents on anaerobic digestion of food waste under mesophilic conditions: Performance and microbial characteristics analysis. *PloS One*, 9: e102548.

Zhang, Y. and Banks, C. J. 2013. Impact of different particle size distributions on anaerobic digestion of the organic fraction of municipal solid waste. *Waste Management*, 33: 297–307.

Zhang, C., Su, H., Baeyens, J., and Tan, T. 2014. Reviewing the anaerobic digestion of food waste for biogas production. *Renewable and Sustainable Energy Reviews*, 38: 383–392.

Zhang, B., Wang, L. J., and Li, R. 2015. Production of biogas from aquatic plants. in: *Aquatic Plants: Composition, Nutrient Concentration and Environmental Impact*, C. E. Rodney (Ed.), Nova Science Publishers, Inc., New York, pp. 33–46.

Zhao, Q., Leonhardt, E., MacConnell, C., Frear, C., and Chen, S. 2010. Purification technologies for biogas generated by anaerobic digestion. *CSANR Research Report 2010-01* Washington State University, Pullman, Washington, USA. Online: http://csanr.wsu.edu/wp-content/uploads/2013/02/CSANR2010-001.Ch09.pdf. Accessed October 6, 2017.

Zhong, W., Zhang, Z., Luo, Y. et al. 2011. Effect of biological pretreatments in enhancing corn straw biogas production. *Bioresource Technology*, 102: 11177–11182.

Zhu, B., Gikas, P., Zhang, R. et al. 2009. Characteristics and biogas production potential of municipal solid wastes pretreated with a rotary drum reactor. *Bioresource Technology*, 100: 1122–1129.

Ziemiński, K. and Frąc, M. 2012. Methane fermentation process as anaerobic digestion of biomass: Transformations, stages and microorganisms. *African Journal of Biotechnology*, 11: 4127–4139.

section four

Solid biofuels

chapter fourteen

An appraisal on biochar functionality and utility in agronomy

Sonil Nanda, Ajay K. Dalai, Kamal K. Pant, Iskender Gökalp, and Janusz A. Kozinski

Contents

14.1 Introduction

Today, the exploitation of fossil fuels worldwide has led to rising greenhouse gas (GHG) emissions, especially CO_2, air pollution, and global warming. These environmental issues have recently increased global attention for alternative energy supply, especially from renewable resources. Any form of renewable carbon in waste plant biomass and animal manure, as well as organic matter in industrial effluents, municipal solid wastes, sewage sludge, and other waste materials can be transformed to hydrocarbon-based biofuels. Biofuels, especially those produced from agricultural crop residues, forestry biomass, algae, energy crops, and new plants for photosynthesis, use oil-seed plants and are considered carbon-neutral because of the CO_2 emitted from their combustion. In other words, the energy content in biofuels is derived from biological carbon fixation.

While biofuels are regarded as carbon-neutral, biochar is considered as carbon-negative. Biochar is a product of biomass pyrolysis and gasification that can capture the carbon in the environment for a long time, making the biomass conversion process carbon-negative. Biochar can capture CO_2 from the atmosphere, thus alleviating global warming through reduced GHG emissions. Biochar is densified with stable aromatic carbon and thus cannot be readily returned to the atmosphere in the form of CO_2 even under ambient environmental conditions (Lehmann, 2007a; Lehmann and Joseph, 2009). Therefore, it persists in the soil for a longer time as a carbon sink, preserving the recalcitrant carbon for hundreds to thousands of years.

The historical occurrence of biochar or carbon black in the soil can be evidenced to the discovery of Amazonian *terra preta* in South America (Erickson, 2003). *Terra preta,* also known as black earth, is a dark and fertile soil extremely rich in charred carbon from plant residues, bones, and animal manure attributed by anthropogenic activities by pre-Columbian inhabitants between 450 BC and 950 AD. The char in *terra preta* is believed to result from incomplete combustion of plant residues, animal feces, and bones by natural fires or controlled or deliberate burning as a management strategy to address soil fertility (Sohi et al., 2010). *Terra preta* soil is approximately 2 m deep and regenerates at a rate of 1 cm/year.

Although biochar is predominantly a carbonaceous material, it also inherits a certain proportion of inorganic elements from its precursor (biomass), which are otherwise present as the mineral matter in the latter (Azargohar et al., 2013, 2014; Mohanty et al., 2013; Nanda et al., 2014a). The applications and functionalities of biochar are manifold. Nanda et al. (2016a) comprehensively reviewed the multifarious applications of biochar in energy (e.g., cogasification, cofiring, and combustion), materials (e.g., activated carbon, adsorbents, catalyst, and specialty materials), environment (e.g., carbon sequestration and sorption of pollutants), agronomy (e.g., water retention and plant nutrition), and medicine (e.g., adsorption of drugs and toxins).

Due to the large carbon content in biochar, it can potentially substitute coal in cogasification, cofiring, and copyrolysis (Sjostrom et al., 1999; Zhu et al., 2008; Ellis et al., 2015; Farid et al., 2016). In all these processes, the involvement of char promotes the reactivity of biomass and coal mixtures, resulting in its abridged yield and promoted gas production. Biochar also acts as an effective catalyst for thermochemical cracking of bio-oil and tars during pyrolysis or gasification (Ekstrom et al., 1985; Freel and Graham, 1998). Syngas produced from gasification of biochar can be used to produce hydrogen, methanol, alkanes, and higher alcohols. Hydrogen can be generated through the water-gas shift reaction during biochar gasification, whereas alkanes and alcohols can be generated through Fischer–Tropsch catalysis (Dias et al., 2007; Nanda et al., 2014b). Activated carbon is an activated form of biochar having an increased surface area and small low-volume pores available for adsorption or chemical reactions. The uses of activated carbon are many, a few of which include adsorption of pollutants, air and water purification, battery electrodes, supercapacitors, catalyst support, decaffeination, gas storage, metal extraction, gold purification, and biomedical applications (Nanda et al., 2016a).

Carbon capture and sequestration and carbon credit approaches of biochar can be recognized owing to its high carbon content and thermal stability (Lira et al., 2013). The solidity of biochar evidenced from *terra preta* soil is due to its resistance to mineralization compared to any other organic matter. This suggests that biochar can offer many agronomic benefits to farmlands and degraded soils. Farmlands used for agriculture contain a relatively small fraction of the global soil carbon pool, but this proportion is significant when compared to the annual atmospheric flux (Sohi et al., 2010). Through many dynamic or reversible interactions with soil nutrients and minerals, biochar can enhance soil fertility and provide value to crop productivity.

The interest in biochar is growing owing to its utilization in energy production, agronomic practices, biomaterial manufacturing, and greenhouse gas mitigation. In addition, integrating pyrolysis with biochar application to the soil for carbon sequestration could also lead to a noteworthy strategy for mitigating climate change. This chapter aims to throw light on pyrolysis and gasification as the technologies to produce biochar from a broad range of feedstocks. A special emphasis of this chapter is on the agronomic applications of biochar, especially in carbon sequestration, soil amelioration, crop improvement, and heavy metal adsorption.

14.2 Biochar production

Heating rate and temperature play a significant role in the generation of biochar from different organic wastes. Biochar can be produced through several thermochemical technologies such as pyrolysis, gasification, combustion, and carbonization. The occurrence of biochar can also be felt by forest fires that are either set naturally (increase in diurnal temperature) or are prescribed (controlled to clear forest cover) (Busse et al., 2009). Pyrolysis is the most widely used technology to produce biochar. Pyrolysis operates in the absence of oxygen, resulting in bio-oil, biochar, and gases.

The yields of bio-oil, biochar, and producer gas from pyrolysis are greatly influenced by process parameters such as temperature, heating rate, residence time, gas flow rate, reactor geometry, feedstock type, and feedstock properties (particle size, moisture, minerals, fixed carbon, and volatiles) (Mohan et al., 2006). The tailoring of these operating conditions classifies pyrolysis into slow, fast, flash, vacuum, intermediate, and pressurized ultrapyrolysis (Huber et al., 2006; Samanya et al., 2012; Nanda et al., 2014b). Each type of pyrolysis has variable yields of bio-oil, biochar,

Table 14.1 Operating conditions and product yields from different pyrolysis

Pyrolysis	Temperature (°C)	Heating rate (°C/s)	Residence time	Product yield (weight %)		
				Bio-oil	Biochar	Producer gas
Fast pyrolysis	400–800	10–200	0.5–5 s	50	20	30
Flash pyrolysis	800–1000	1000	0.5 s	75	12	13
Intermediate pyrolysis	~500	–	10–20 s	50	25	25
Slow pyrolysis	300–700	0.1–1	0.1–1.5 h	30	35	35

and gases. For example, fast and flash pyrolysis involve high temperatures, fast heating rates, and short residence times, resulting in a higher bio-oil yield and lower biochar yield. Table 14.1 summarizes the operating conditions and product yields from different types of pyrolysis. Fast pyrolysis typically operates at 400–500°C with a heating rate of 10–200°C/s and vapor residence time of 30–1500 ms (Bridgwater, 1999). Flash pyrolysis requires a temperature range of 400–600°C with a heating rate greater than 1000°C/s and vapor residence time less than 100 ms (Samolada and Vasalos, 1991; Wagenaar et al., 1993). In contrast, slow pyrolysis involves moderate temperatures, slow heating rates, and longer residence time, leading to a higher yield of biochar and lower yield of bio-oil. Slow pyrolysis prefers 300–700°C with a slower heating rate of 0.1–1°C/s and residence time of 0.1–1.5 h (Maschio et al., 1992). Intermediate pyrolysis operates at approximately 500°C for 10–20 s (Ahmad et al., 2014). Nearly 50% of the products consist of bio-oil, whereas biochar and gas yields are almost equal.

Biomass pyrolysis leads to primary and secondary cracking reactions during the vapor release process (Bridgwater and Peacocke, 2000). With the increase in temperature, the primary cracking reaction causes the gases and volatile matter (including water vapor) in the biomass to evolve and then be rapidly quenched during the condensation process. Higher heating rate, short vapor residence time, rapid quenching, and temperatures around 500°C are preferable for suppressing secondary cracking reactions and maximizing bio-oil yield (Luo et al., 2005). On the other hand, slower heating rate and longer vapor residence time enhance gas production by favoring secondary cracking reactions (Bridgwater, 1999). Longer residence time (up to 1.5–2 h) leads to increased condensation rates and promotes secondary reactions to form heavy molecular weight compounds such as tars and biochar. The producer gas comprises the gaseous components that remain noncondensable during the entire process and include H_2, CO, CO_2, CH_4, C_2H_4, and C_{2+} (Kanaujia et al., 2014).

The operating parameters in pyrolysis require physical and chemical attributes to biochar such as carbon content, volatile matter, aromaticity, elemental composition, surface area, porosity, alkalinity, and ion-exchange capacity. These key properties of biochar are sensitive to the feedstock (precursor), pyrolysis temperature, heating rate, residence time, oxidation medium, and potential postprocessing treatments (e.g., activation, acid washing, solvent washing, etc.). As the pyrolysis temperature rises, there is a significant increase in biochar's aromatic carbon content, ash content, pH, surface area, and pore volume (Nanda et al., 2016a). On the other hand, in the low-temperature pyrolysis, there is a considerable reduction in biochar's yield, volatiles content, electrical conductivity, and cation exchange capacity. The applications of biochar also depend on its specific physicochemical and structural properties.

14.3 Carbon sequestration

Significant research has aimed to counterbalance greenhouse gas emissions (especially CO_2) through sequestration of the emitted carbon in the atmosphere. Nanda et al. (2016c) describe some key strategies that can potentially sequester carbon in terrestrial and geological systems. Certain physicochemical carbon-capturing technologies include absorption, adsorption, membrane separation, and cryogenic distillation. The biological routes for carbon capture include algal and bacterial systems, dedicated energy crops, and coalbed methanogenesis. Finally, the geological carbon-sequestering routes include biochar amendment and oceanic carbon storage. Terrestrial ecosystems are attractive sites for carbon storage because more than 80% of organic carbon is preserved in soil (Watson et al., 2000).

Carbon capture from the atmosphere using adsorption technologies at an industrial scale can cost up to US$500/ton carbon, whereas biomass- or biochar-based carbon sequestering can be achieved at roughly half the cost (Matovic, 2011). Furthermore, a significant investment of energy and finance is involved with compressing CO_2 and pumping it into the ground. In contrast, biochar production and application as a carbon-sequestering agent seems like an attractive option because it offers the benefits of low cost, soil fertility, and soil amelioration. Integrating biomass conversion processes (e.g., pyrolysis) with biochar application to soil can lead to a net withdrawal of CO_2 from the atmosphere. Biochar amendment to soil can also decrease N_2O and CH_4 emissions, which are 300 and 23 times more potent greenhouse gases than CO_2, respectively (Renner, 2007). Approximately 32 MJ/kg of heat energy is forfeited by storing biochar rather than burning it (Matovic, 2011). This exceeds the carbon capture and storage penalty, which is estimated at 10%–30% for large power plants.

Photosynthesis in plants transforms solar energy into chemical energy as carbohydrates. These carbohydrates in plants can be transformed to hydrocarbon biofuels such as alcohols, bio-oils, syngas, and biogas through thermochemical, hydrothermal, and biochemical conversion technologies. Biofuels upon combustion emit CO_2, which is again assimilated by plants through photosynthesis, completing the carbon cycle in nature. Because plants can fix CO_2 through photosynthesis, biochar can act as a carbon storehouse. Photosynthesis fixes 120 Gt CO_2 from the atmosphere into carbohydrates each year (Schlesinger, 1995). From this drawn proportion of CO_2, approximately 60 Gt CO_2 is returned to the atmosphere through plant respiration, but the remaining 60 Gt CO_2 per year is invested in fresh plant growth.

Around 50% of the carbon content in biomass is converted to bio-oils and producer gas through pyrolysis, while the remaining 50% of biomass carbon is stabilized as biochar (Lehmann et al., 2006). Noncharred organic materials such as plant residues and animal matter also add carbon to the soil through natural decomposition at a slower rate. Carbon continues to release from the decomposition of dead matter until most of it is lost from the soil within a decade. Unlike decomposition, biochar is relatively stable to preserve carbon for centuries and millenniums as evidenced through *terra preta* soil.

Carbon storage by biochar contrasts with carbon sequestration by afforestation, reforestation, and no-tillage agriculture (Nanda et al., 2016a). The farmlands converted to no-tillage lands may lose carbon-capturing efficiency within two decades. Dedicated energy crops also have the ability to produce net greenhouse gases through indirect land-use change. In addition, forests mature over decades and start releasing CO_2 (Lehmann, 2007b). This makes biochar application a different approach to acting as a long-standing sink for holding carbon and reducing CO_2 emissions upon oxidation. Biochars act as net carbon sinks, approximately 75% of which arise due to carbon sequestration (Ennis et al., 2012).

A long half-life of biochar can ensure an efficient carbon sequestration and its stability in the environment (Lehmann et al., 2006). The resistance of biochar against biotic and abiotic oxidation is as variable as its chemical properties and origin. Significant studies on *terra preta* soil indicate that biochar carbon is the oldest fraction of carbon in soil, even older than the carbon in soil aggregates and organo-mineral complexes (Lehmann et al., 2006). The longevity of biochar also depends on its production procedure. Temperature and residence time increase the extent of aromatic structure formation in biochar (Ennis et al., 2012). Schmidt and Noack (2000) suggested that atomic H:C ratios more than 0.2 for biochar derived at 800°C had partial continuous transformation from aromatic to graphitic structures. The aromatic structures can also occur in different forms such as turbostratic carbon (found in biochar produced through high-temperature pyrolysis) and amorphous carbon (found in biochar produced through low-temperature pyrolysis) (Lehmann and Joseph, 2009; Keiluweit et al., 2010; Nguyen et al., 2010). Biochar with low aromatic structures and small aromatic clusters have high surface functionality. Although better surface functionality can cause the biochar to exhibit higher cation exchange capacity and increased water retention capacity, it also leads to greater degradation. In contrast, higher aromatic carbon content in biochar results in greater recalcitrance in soil with protracted carbon sequestration potential.

The decomposition rate of biochar is substantially reduced when it migrates deep into the soil layer or in the bottom benthic regions of river, lake, or sea sediments (Lehmann et al., 2006). The transport mechanisms and leaching of biochar from the top soil surface to the bottom layers is less understood. Biochar buried in deep-sea sediments can be up to 13,900 years older than the age of other soil organic carbon such as humic substances (Masiello and Druffel, 1998). This is a clear indication that the time scale for carbon sequestration by biochar can be up to thousands of years, or it can be regarded as a permanent carbon pool.

14.4 Soil amelioration

Biochar application to soil can significantly improve its quality. It is known to positively enhance the bulk density of soil by enhancing the available water content, rooting pattern of plants, and soil fauna (Major et al., 2010). The micropores of biochar allow the flow of air and water, thereby reducing the overall bulk density of the soil-biochar mixture. The preexisting closed pores in biochar open up when it mixes with the available water in the soil. This increases the biochar particles' internal surface area for adsorption of soil particles and humic substances.

Biochar is efficient in adsorbing ammonia from inorganic fertilizers to act as a buffer in agricultural soil (Iyobe et al., 2004). The low nitrogen concentrations and high C:N ratios of biochar decrease ammonia concentration through biological immobilization. Apart from ammonia, biochar can also efficiently adsorb nitrate, phosphate, soil minerals, and hydrophobic organic pollutants (Lehmann et al., 2006).

Cation exchange capacity determines soil fertility, nutrient retention in soil, and ability to protect groundwater contamination from pollutant cations (Nanda et al., 2016a). As the organic matter increases, the soil gains its ability to retain cations in a bioavailable form. In general, biochar particles demonstrate large surface area, high charge density, and more negative surface charge than other organic soil matter (Liang et al., 2006).

This reflects on the higher cation exchange capacity of biochar when compared to any other soil organic matter to adsorb cations per unit carbon (Sombroek et al., 2003).

Elevated pyrolysis temperature imparts higher pH, surface area, and carbon yield in the resulting biochar. Azargohar et al. (2013) reported a rise in the pH of canola meal biochar from 8.5 (at 300°C) to 10.1 (at 700°C). The surface area for poultry litter biochar increased from 2.7 m^2/g at 300°C to 5.8 m^2/g at 600°C (Song and Guo, 2012). Mohanty et al. (2013) reported the surface areas of biochars produced at 450°C with a high heating rate of 450°C/min from pinewood, timothy grass, and wheat straw as 185, 203, and 184 m^2/g, respectively. However, at a slow heating rate of 2°C/min, biochars from pinewood, timothy grass, and wheat straw showed relatively lower surface areas of 166, 179, and 178 m^2/g, respectively.

The cation exchange capacities for biochar and other soil organic matter are less at lower pH and vice versa. Biochar tends to be alkaline in nature due to the presence of considerable amounts of alkali and alkaline earth metals (Mohanty et al., 2013; Nanda et al., 2014a). The alkaline nature of biochar suggests its application to acidic soils to neutralize the pH. As the soil pH rises, the cation exchange capacity increases with higher plant-available exchangeable minerals. Biochar also retains various macronutrients and micronutrients in the soil to make them available to plant roots through adsorption. Biochar particulates also show resistance to microbial activity and decomposition, which is a function of chemical stability (Schmidt et al., 2000).

Soil with a considerably high amount of biochar amendment can hold up to 40% of soil organic carbon (Lehmann et al., 2006). The high levels of organic carbon in soil can improve its fertility and promote crop productivity. It is sometimes questionable as to how much biochar a soil can hold without being lethal to the plants. However, studies indicate that very high biochar loadings of 14×10^4 kg biochar/ha have also improved crop yields (Lehmann and Rondon, 2006).

Biochar when amended into soil provides an excellent habitat for beneficial soil microorganisms that aid in enhancing compositing, soil fertility, and crop productivity (Singh et al., 2015). Studies have shown that biochar-amended soils had greater gene abundance of nitrogen-fixing bacteria (Ducey et al., 2013). Biochar incorporation into soil can enhance water-holding capacity and water permeability in soil. An increase in water-holding capacity can have greater benefits in sandy soils amended with biochar compared to loamy and clayey soils (Atkinson et al., 2010). This suggests soil with high clay fractions are less likely to restore from biochar applications. Soil water content induced by biochar tends to induce soil oxygen tension, which affects soil microorganisms and their capacity to oxidize soil organic matter (Van Gestel et al., 1993). Organic matter decomposition, nutrient availability, and microbial activity in soil are functions of soil saturation, which depends on soil wetting and drying cycles. The water content in soil influences the stability of biochar and its oxidative breakdown (Nguyen and Lehmann, 2009).

Biochar is often used to reclaim infertile and degraded soils due to its ability to improve physicochemical characteristics of the soil, enhance bioavailability of nutrients, increase crop productivity, and promote sorption of organic contaminants (Choppala et al., 2012). Neither do all soil types exhibit improvements and restoration nor do all crops behave positively with biochar amendment. Kuppusamy et al. (2016) highlighted some limitations associated with biochar-amended soil that include (1) binding and deactivation of chemical fertilizers, herbicides, and nutrients in soil; (2) surplus availability of nutrients leading to soil salinity; (3) increase in soil pH and conductivity; (4) impact on germination and other biogeochemical processes in soil; and (5) leaching of certain heavy metals and polycyclic aromatic hydrocarbons that may be present in biochar soil.

14.5 Improvement in crop productivity

The effects of biochar on soil biology are advantageous because it positively alters the soil microflora and composition. Most of the studies have shown that biochar amendment to soil results in enhancement of beneficial fungi and nitrogen-fixing microorganisms, which in turn improve crop productivity. The improved crop productivity ranges from 20% to 220% at biochar amendment rates of 0.4–8 tons C/ha (Lehmann and Rondon, 2006). Carbon and nitrogen sources are indispensable to soil microorganisms for their metabolism. Biochar is a good source of carbon and nitrogen, but its recalcitrance makes these elements exist in nonmineralizable forms to microorganisms. Microorganisms and their enzymes have access to the internal pores of biochar to colonize, immobilize, and gain protection against the predators or external environment (Lehmann et al., 2011).

The beneficial microbial population in soil amended with biochar increases enormously. Plant growth-promoting rhizobacteria and mycorrhizal fungi are the key microbial groups benefited by biochar amendment (Graber et al., 2010). Biochar-amended agricultural soils help to enhance the proliferation of arbuscular mycorrhiza and ectomycorrhiza, the two forms of mycorrhizal fungi. Mycorrhizal fungi symbiotically colonize plant roots either internally (by arbuscular mycorrhiza) or externally (by ectomycorrhiza) (Nanda et al., 2016a). Mycorrhizal fungi flourish in soil by developing interconnected hyphal networks. The internal pores of biochar serve as a suitable dais for building hyphal networks and expediting the exchange of nutrients from the soil to plant roots (Warnock et al., 2010). The fungal hyphae together with plant roots also prevent soil erosion by holding the biochar-soil particles in the rhizosphere (i.e. plant's rooting zone).

Biochar incorporation to soil improves the abundance of mycorrhizal fungi. The plants having a symbiotic relationship with mycorrhizal fungi serve important roles through phytoextraction and phytostabilization (Vassilev et al., 2004). Biochar serves as a source of reduced-carbon compounds and nutrients, as well as a refuge for soil microorganisms including mycorrhization helper bacteria and phosphate solubilizing bacteria (Warnock et al., 2007). Mycorrhizal fungi indirectly benefit from the increased populations including mycorrhization helper bacteria and phosphate solubilizing bacteria. Biochar also stimulates spore germination of arbuscular mycorrhiza, leading to their multiplication in soil (Rillig et al., 2010). Biochar also increases nitrification by proliferating free-living nitrogen-fixing soil bacterial such as *Azospirillum* and *Azotobacter* (Gundale and DeLuca, 2007; Berglund et al., 2004). Considering these stimulating properties, it is evident that biochar has many valuable roles in soil fertility, reclaiming marginal and degraded soils, and improving crop production.

As discussed earlier, an increase in the water-holding capacity of soil amended with biochar has many benefits to sandy soils where soil moisture is critical to establishing initial vegetation. Biochar amendment also reduces the requirement of irrigation in agricultural farms because it enhances the water-holding capacity of the soil. Water-holding capacity of the soil, particularly in the rhizosphere, can augment nutrient movement and leaching (Atkinson et al., 2010). Owing to the high surface area and porosity of biochar, it can maintain soil moisture, organic carbon, and essential plant nutrients for a prolonged period.

The sorptive properties of biochar, especially for hydrophobic substances, could serve as reservoirs for carbon, nutrients, metabolic compounds, and also for some inhibitory compounds such as allelochemicals (Warnock et al., 2007). When the nutrient retention ability and sorptive potential of biochar come together, it loses its sensitivity to beneficial and inhibitory compounds in the soil. Biochemical signaling between mycorrhizal fungi and the host plant is an important function in the rhizosphere. Plant roots secrete compounds such as flavonoids, sesquiterpenes, and strigolactones, which act as some signaling compounds (Warnock et al., 2007). Biochar amendment can exert both direct and indirect effects in the rhizosphere. The direct effects include adsorption or desorption of signaling compounds and inhibitory allelochemicals, whereas the indirect effects include changes in pH and electrical conductivity (Ennis et al., 2012). Although the provision of a sink for fungicidal molecules would promote growth and root colonization, the adsorption of chemical signals would be disadvantageous. Soil response to biochar amendment includes increased microbial population, respiration, and respiration efficiency (Ennis et al., 2012). Sorption by biochar leads to alterations in soil substrate concentration and availability of nutrients over the longer term (Wardle et al., 1998). Therefore, soil microflora in biochar-amended soil varies according to the type of biochar, amendment amount, amendment frequency, and age (Ennis et al., 2012).

Adoption of biochar in soil systems as an organic fertilizer is of added value if it is generated from biogenic wastes that otherwise pose management issues. Animal manure, one of such biogenic wastes, is available in surplus amounts as fecal waste matter from poultry and livestock farming. Owing to odor, pests, and CH_4 and N_2O emissions during composting, animal manure necessitates effective management and valorization approaches (Junior et al., 2015). Pyrolysis and gasification of animal manure can not only yield energy products (bio-oil, biochar, and syngas), but also reduce manure volume and recover inherent nutrients, as well as reduce odor, pests, and pollution. There are only a few studies reported on the thermochemical conversion of poultry litter (Das et al., 2008; Agblevor et al., 2010), chicken litter (Kim and Agblevor, 2007), horse manure (Nanda et al., 2016b), and swine manure (Ro et al., 2010). There is very little information available on the soil applications for biochar derived from animal manure, especially from poultry litter (Gaskin et al., 2008; Cantrell et al., 2012; Song and Guo, 2012; Azargohar et al., 2014), dairy manure (Cao and Harris, 2010), swine manure (Cantrell et al., 2012), and horse manure (Nanda et al., 2016b). In most of these studies, biochar from animal manure is reported to be alkaline with the presence of essential soil nutrients and rich carbon content, which are desirable features of an organic fertilizer. However, more research in this area is required to diversify the selection of animal manure for thermochemical conversion and use of its biochar as a resilient organic fertilizer.

Compared to other organic fertilizers, biochar from plant materials is generally low in nutrient content and nitrogen levels (Lehmann et al., 2003). Biochar produced from animal manure often has higher nutrient content, but its agronomic values as soil amendment have not been widely studied. When used directly, animal manure, especially poultry litter, has a higher potential risk of phosphorus contamination to surface waters (Chan et al., 2008). However, Zhao et al. (2016) reported that incorporation of biochar generated from chicken manure into Cd-spiked soil can significantly enhance the soil pH, cation exchange capacity, soil organic matter, and carbonate levels. An increased biochar application also decreases the bioavailability of heavy metals and increases the content of stable fractions. Therefore, conversion of animal manure and poultry litter to biochar through pyrolysis can be an effective alternative to utilize this biogenic resource in agriculture.

14.6 *Adsorption of heavy metals and other soil pollutants*

Biochar application to soil can reduce the mobility of heavy metals and other organic soil contaminants such as insecticides (Hilber et al., 2009;

Hua et al., 2009). The biochar surface can exhibit either positive or negative charge depending on the cations present in it. The positive surface charge has a tendency to decrease as the biochar oxidizes, thereby changing its absorption potential (Cheng et al., 2008). The concentration of soluble compounds such as phenols can also be reduced upon biochar addition to the soil (Gundale and DeLuca, 2007). At low soil pH, certain elements such as Al, Cu, and Mn can be toxic to plant growth, but their concentration can be significantly reduced by biochar amendment (Atkinson et al., 2010). At slightly higher soil pH, the availability of other beneficial elements such as N, P, Ca, Mg, and Mo can increase with biochar incorporation.

There are two main routes through which biochar can reduce soil or groundwater pollution (Lehmann, 2007a): (1) by retaining nutrients and minerals in the soil to lower their chances of leaching into groundwater and (2) by retaining nutrients in the soil to reduce the requirement of additional chemical fertilizers. Biochar slows the bioavailability of herbicide, which is dependent on mineralization. Therefore, herbicide phytotoxicity is reduced with the application of biochar, which also reduces leaching and runoff. Land contaminated with heavy metals may lead to their immobilization following biochar amendment (Ennis et al., 2012). However, it is also suggested that with biochar addition to soil, the metals or metalloids may be displaced as organic complexes (Cao et al., 2003).

The strong sorption ability of biochar is attributed to its surface properties originating from the precursor. The oxygen-containing functional groups can improve the sorption of aqueous Cr on biochar by reducing Cr(VI) to Cr(III) (Dong et al., 2011). The high levels of cationic nutrients such as Na, Mg, K, and Ca in biochar also increase its cation exchange capacity and enhance sorption of heavy metals through an ion exchange process under acidic pH conditions (Mohan et al., 2007).

Biochars have high sorption capacity for metals due to their surface heterogeneity (Kasozi et al., 2010). Biochar with a high surface area and pore volume can physically sorb metallic ions on its surface and retained within the pores (Kumar et al., 2011). The negatively charged surfaces of biochar can sorb positively charged metallic ions through electrostatic attraction. Inherited or induced ligands and functional groups on the biochar surface can also interact with metals, form complexes, and immobilize them within their pores or surface or form precipitates of their solid mineral phases (Inyang et al., 2016). Biochar aids in the removal of heavy metals from the soil through a variety of mechanisms such as complexation, chemisorption, ion exchange, physical sorption, and precipitation (Inyang et al., 2016). Table 14.2 summarizes a few studies on the sorption of heavy metals by biochar produced from diverse sources.

In complexation, multiatom structures or complexes are formed with specific metal–ligand interactions. Transition metals with partially filled d-orbitals have a high affinity for ligands and are effectively removed

Table 14.2 Summary of a few studies showing the sorption of heavy metals by biochar

Biochar precursor	Pyrolysis temperature (°C)	Heavy metals	Sorption mode	Reference
British oak, common ash, sycamore, and birch	–	Cu and Pb	Complexation	Karami et al. (2011)
Broiler litter	350–700	Cu	Electrostatic interaction	Uchimiya et al. (2010)
Broiler litter	350–800	Cu, Pb, Sb, and Zn	Stabilization	Uchimiya et al. (2012)
Cherry, common ash, and oak	400	Ar	Immobilization	Hartley et al. (2009)
Chicken manure and green wastes	550	Cd, Cu, and Pb	Immobilization	Park et al. (2011)
Cottonseed hulls	200–800	Cd, Cu, Ni, and Pb	Sequestration	Uchimiya et al. (2011)
Dairy manure	450	Pb	Immobilization	Cao et al. (2011)
Hardwood	–	Ar, Cd, Cu, and Zn	Immobilization	Beesley et al. (2010)
Oak wood	400	Pb	Immobilization	Ahmad et al. (2012)
Rice straw	300	Pb	Nonelectrostatic adsorption	Jiang et al. (2012)
Sewage sludge	550	Ar, Cd, Cr, Co, Cu, Ni, Pb, and Zn	Immobilization	Khan et al. (2013)

using this technique (Crabtree, 2009). Oxidation of biochar increases the oxygen content through the formation of carboxyl groups, thus promoting metal complexation (Harvey et al., 2011). Chemisorption occurs through electrostatic interaction between surface-charged biochar and metal ions with an opposite charge. This mechanism depends on the pH of the soil solution and point of zero charge (PZC) of biochar (Inyang et al., 2016). Biochar generated at high temperatures leads to the formation of graphene structures that favor chemisorption (Keiluweit and Kleber, 2009).

Adsorption of heavy metals on the biochar surface through the ion exchange process depends on the size of the metallic ion and surface functional group of biochar. Ion exchange occurs by a selective replacement of positively charged ions on biochar surfaces with a target metallic ion (Inyang et al., 2016). Transition metals have a strong binding ability to these ion exchange sites on the biochar surface.

On the other hand, physical sorption involves the removal of heavy metals by diffusional movement of ions into biochar pores without any chemical bonding. During precipitation, solid precipitates are formed either in solution or on the biochar surface. Precipitation is one of the most common mechanisms for heavy metal immobilization. Metals and rare earth elements such as Cu, Ni, Pb, and Zn are most prone to precipitate on the biochar surface (Inyang et al., 2016).

14.7 Conclusions and perspectives

Biochar is an attractive carbonaceous material that has miscellaneous applications in carbon sequestration, soil amelioration, crop improvement, air and water purification, pharmaceuticals, electronics, and smart materials, to name a few. These applications largely depend on the physical and chemical properties of biochar, which are contributed by its production temperature, heating rate, sweep gas flow rate, residence time, biomass particle size, and other proximate properties of biomass. Biochar generated at higher temperatures has large surface area and micropore volume, making it an effective sorption agent for organic contaminants. However, biochar produced at lower temperatures has an electrostatic affinity toward cationic nutrients in the soil.

Biochar improves the soil quality by increasing its water-holding capacity, nutrients concentration, cation exchange capacity, and the population of plant growth-promoting microorganisms. The alkaline nature of biochar and its surface charge also helps to neutralize acidic and degraded soils. The increased water-holding capacity induced by biochar is highly beneficial to sandy soils, which also increases their nitrogen and phosphorus contents. The crop improvement attribute of biochar is associated with the proliferation of nitrogen-fixing bacteria and mycorrhiza fungi, greater plant root development, amplified availability of essential nutrients, and high soil organic carbon and fertility. The stable carbon in biochar remains in the soil for centuries, making it less prone to mineralization. This also makes biochar a persistent soil conditioner and a colossal carbon sink.

There are several instances to rationalize biochar as an effective agent for soil restoration and crop improvement, but there remains a gap in understanding the ecological interactions between biochar, soil, plant roots, and soil microorganisms. Long-term geological studies and life-cycle analysis can help better the understanding of these interfaces. The constructive features of biochar together with the synergistic interactions of biochar–mycorrhizae–plant can not only demonstrate agronomic values but also sequester carbon in soils to contribute to climate change mitigation. This interaction can be potentially implemented for the reclamation of degraded lands, infertile soils, lands contaminated by industrial pollution and mine wastes, and acidic soils arising from collective fertilizer use. These efficiencies of biochar associated with energy and environment can help develop agronomic strategies for sustainable agricultural activities with reduced greenhouse gas emissions and improved soil quality and plant growth.

Acknowledgments

The authors would like to thank the Natural Sciences and Engineering Research Council of Canada (NSERC) for funding this environmental research.

References

Agblevor, F. A., Beis, S., Kim, S. S., Tarrant, R., and Mante, N. O. 2010. Biocrude oils from the fast pyrolysis of poultry litter and hardwood. *Waste Management* 30: 298–307.

Ahmad, M., Lee, S. S., Yang, J. E. et al. 2012. Effects of soil dilution and amendments (mussel shell, cow bone, and biochar) on Pb availability and phytotoxicity in military shooting range soil. *Ecotoxicology and Environmental Safety* 79: 225–31.

Ahmad, M., Rajapaksha, A. U., Lim, J. E. et al. 2014. Biochar as a sorbent for contaminant management in soil and water: A review. *Chemosphere* 99: 19–33.

Atkinson, C. J., Fitzgerald, J. D., and Hipps, N. A. 2010. Potential mechanisms for achieving agricultural benefits from biochar application to temperate soils: A review. *Plant and Soil* 337: 1–18.

Azargohar, R., Nanda, S., Kozinski, J. A., Dalai, A. K., and Sutarto, R. 2014. Effects of temperature on the physicochemical characteristics of fast pyrolysis biochars derived from Canadian waste biomass. *Fuel* 125: 90–100.

Azargohar, R., Nanda, S., Rao, B. V. S. K., and Dalai, A. K. 2013. Slow pyrolysis of deoiled canola meal: Product yields and characterization. *Energy and Fuels* 27: 5268–79.

Beesley, L., Jimenez, E. M., and Eyles, J. L. G. 2010. Effects of biochar and greenwaste compost amendments on mobility, bioavailability and toxicity of inorganic and organic contaminants in a multi-element polluted soil. *Environmental Pollution* 158: 2282–7.

Berglund, L. M., DeLuca, T. H., and Zackrisson, T. H. 2004. Activated carbon amendments of soil alters nitrification rates in Scots pine forests. *Soil Biology and Biochemistry* 36: 2067–73.

Bridgwater, A. V. 1999. Principles and practice of biomass fast pyrolysis processes for liquids. *Journal of Analytical and Applied Pyrolysis* 51: 3–22.

Bridgwater, A. V. and Peacocke, G. V. C. 2000. Fast pyrolysis processes for biomass. *Renewable and Sustainable Energy Reviews* 4: 1–73.

Busse, M. D., Cochran, P. H., Hopkins, W. E. et al. 2009. Developing resilient ponderosa pine forests with mechanical thinning and prescribed fire in central Oregon's pumice region. *Canadian Journal of Forest Research* 39: 1171–85.

Cantrell, K. B., Hunt, P. G., Uchimiya, M., Novak, J. M., and Ro, K. S. 2012. Impact of pyrolysis temperature and manure source on physicochemical characteristics of biochar. *Bioresource Technology* 107: 419–28.

Cao, X. and Harris, W. 2010. Properties of dairy-manure-derived biochar pertinent to its potential use in remediation. *Bioresource Technology* 101: 5222–8.

Cao, X., Ma, L., Liang, Y., Gao, B., and Harris, W. 2011. Simultaneous immobilization of lead and atrazine in contaminated soils using dairy-manure biochar. *Environmental Science and Technology* 45: 4884–9.

Cao, X., Ma, L. Q., and Shiralipour, A. 2003. Effects of compost and phosphate amendments on arsenic mobility in soils and arsenic uptake by the hyperaccumulator, *Pteris vittata* L. *Environmental Pollution* 126: 157–67.

Chan, K. Y., Van Zwieten, L., Meszaros, I., Downie, A., and Joseph, S. 2008. Using poultry litter biochars as soil amendments. *Australian Journal of Soil Research* 46: 437–44.

Cheng, C.-H., Lehmann, J., and Engelhard, M. H. 2008. Natural oxidation of black carbon in soils: Changes in molecular form and surface change along a climosequence. *Geochimica et Cosmochimica Acta* 72: 1598–610.

Choppala, G., Bolan, N., Megharaj, M., Chen, Z., and Naidu, R. 2012. The influence of biochar and black carbon on reduction and bioavailability of chromate in soils. *Journal of Environmental Quality* 41: 1175–84.

Crabtree, R. H. 2009. *The Organometallic Chemistry of the Transition Metals*, New York: Wiley.

Das, K. C., Garcia-Perez, M., Bibens, B., and Melear, N. 2008. Slow pyrolysis of poultry litter and pine woody biomass: Impact of chars and bio-oils on microbial growth. *Journal of Environmental Science and Health, Part A* 43: 714–24.

Dias, J. M., Alvim-Ferraz, M. C. M., Almeida, M. F., Rivera-Utrilla, J., and Sanchez-Polo, M. 2007. Waste materials for activated carbon preparation and its use in aqueous-phase treatment: A review. *Journal of Environmental Management* 85: 833–46.

Dong, X. L., Ma, L. N. Q., and Li, Y. C. 2011. Characteristics and mechanisms of hexavalent chromium removal by biochar from sugar beet tailing. *Journal of Hazardous Materials* 190: 909–15.

Ducey, T. F., Ippolito, J. A., Cantrell, K. B., Novak, J. M., and Lentz, R. D. 2013. Addition of activated switchgrass biochar to an aridic subsoil increases microbial nitrogen cycling gene abundances. *Applied Soil Ecology* 65: 65–72.

Ekstrom, C., Lindman, N., and Pettersson, R. 1985. Catalytic conversion of tars, carbon black and methane from pyrolysis/gasification of biomass. In: *Fundamentals of Thermochemical Biomass Conversion*, ed. R. P. Overend, T. A. Milne, and L. K. Mudge, pp. 601–18. Essex: Elsevier Applied Science Publishers.

Ellis, N., Masnadi, M. S., Roberts, D. G., Kochanek, M. A., and Ilyushechkin, A. Y. 2015. Mineral matter interactions during co-pyrolysis of coal and biomass and their impact on intrinsic char co-gasification reactivity. *Chemical Engineering Journal* 279: 402–8.

Ennis, C. J., Evans, A. G., Islam, M., Ralebitso-Senior, T. K., and Senior, E. 2012. Biochar: Carbon sequestration, land remediation, and impacts on soil microbiology. *Critical Reviews in Environmental Science and Technology* 42: 2311–64.

Erickson, C. 2003. Historical ecology and future explorations. In: *Amazonian Dark Earths: Origin, Properties, Management*, ed. J. Lehmann, D. C. Kern, B. Glaser, and W. I. Woods, pp. 455–93. Dordrecht, Netherlands: Kluwer Academic Publishers.

Farid, M. M., Jeong, H. J., and Hwang, J. 2016. Kinetic study on coal–biomass mixed char co-gasification with H_2O in the presence of H_2. *Fuel* 181: 1066–73.

Freel, B. A. and Graham, R. G. 1998. Method and apparatus for a circulating bed transport fast pyrolysis reactor system. U.S. Patent 5,792,340.

Gaskin, J. W., Steiner, C., Harris, K., Das, K. C., and Bibens, B. 2008. Effect of low-temperature pyrolysis conditions on biochar for agricultural use. *Transactions of the ASABE* 51: 2061–69.

Graber, E. R., Harel, Y. M., Kolton, M. et al. 2010. Biochar impact on development and productivity of pepper and tomato grown in fertigated soilless media. *Plant and Soil* 337: 481–96.

Gundale, M. J. and DeLuca, T. H. 2007. Charcoal effects on soil solution chemistry and growth of *Koeleria macrantha* in the ponderosa pine/Douglas-fir ecosystem. *Biology and Fertility of Soils* 43: 303–11.

Hartley, W., Dickinson, N. M., Riby, P., and Lepp, N. W. 2009. Arsenic mobility in brownfield soils amended with green waste compost or biochar and planted with *Miscanthus*. *Environmental Pollution* 157: 2654–62.

Harvey, O. R., Herbert, B. E., Rhue, R. D., and Kuo, L.-J. 2011. Metal interactions at the biochar-water interface: Energetics and structure-sorption relationships elucidated by flow adsorption microcalorimetry. *Environmental Science and Technology* 45: 5550–6.

Hilber, I., Wyss, G. S., Mader, P. et al. 2009. Influence of activated charcoal amendment to contaminated soil on dieldrin and nutrient uptake by cucumbers. *Environmental Pollution* 157: 2224–30.

Hua, L., Wu, W., Liu, Y., McBride, M. B., and Chen, Y. 2009. Reduction of nitrogen loss and Cu and Zn mobility during sludge composting with bamboo charcoal amendment. *Environmental Science and Pollution Research* 16: 1–9.

Huber, G. W., Iborra, S., and Corma, A. 2006. Synthesis of transportation fuels from biomass: Chemistry, catalysts, and engineering. *Chemical Reviews* 106: 4044–98.

Inyang, M. I., Gao, B., Yao, Y. et al. 2016. A review of biochar as a low-cost adsorbent for aqueous heavy metal removal. *Critical Reviews in Environmental Science and Technology* 46: 406–33.

Iyobe, T., Asada, T., Kawata, K., and Oikawa, K. 2004. Comparison of removal efficiencies for ammonia and amine gases between woody charcoal and activated carbon. *Journal of Health Science* 50: 148–53.

Jiang, T.-Y., Jiang, J., Xu, R.-K., and Li, Z. 2012. Adsorption of Pb(II) on variable charge soils amended with rice-straw derived biochar. *Chemosphere* 89: 249–56.

Junior, C. C., Cerri, C. E. P., Pires, A. V., and Cerri, C. C. 2015. Net greenhouse gas emissions from manure management using anaerobic digestion technology in a beef cattle feedlot in Brazil. *Science of the Total Environment* 505: 1018–25.

Kanaujia, P. K., Sharma, Y. K., Garg, M. O., Tripathi, D., and Singh, R. 2014. Review of analytical strategies in the production and upgrading of bio-oils derived from lignocellulosic biomass. *Journal of Analytical and Applied Pyrolysis* 105: 55–74.

Karami, M., Clemente, R., Jimenez, E. M., Lepp, N. W., and Beesley, L. 2011. Efficiency of green waste compost and biochar soil amendments for reducing lead and copper mobility and uptake to ryegrass. *Journal of Hazardous Materials* 191: 41–8.

Kasozi, G. N., Zimmerman, A. R., Nkedi-Kizza, P., and Gao, B. 2010. Catechol and humic acid sorption onto a range of laboratory-produced black carbons (biochars). *Environmental Science and Technology* 44: 6189–95.

Keiluweit, M. and Kleber, M. 2009. Molecular-level interactions in soils and sediments: The role of aromatic pi-systems. *Environmental Science and Technology* 43: 3421–9.

Keiluweit, M., Nico, P. S., Johnson, M. G., and Kleber, M. 2010. Dynamic molecular structure of plant-derived black carbon (biochar). *Environmental Science and Technology* 44: 1247–53.

Khan, S., Chao, C., Waqas, M., Arp, H. P. H., and Zhu, Y. G. 2013. Sewage sludge biochar influence upon rice (*Oryza sativa* L.) yield, metal bioaccumulation and greenhouse gas emissions from acidic paddy soil. *Environmental Science and Technology* 47: 8624–32.

Kim, S. S. and Agblevor, F. A. 2007. Pyrolysis characteristics and kinetics of chicken litter. *Waste Management* 27: 135–40.

Kumar, S., Loganathan, V. A., Gupta, R. B., and Barnett, M. O. 2011. An assessment of U(VI) removal from groundwater using biochar produced from hydrothermal carbonization. *Journal of Environmental Management* 92: 2504–12.

Kuppusamy, S., Thavamani, P., Megharaj, M., Venkateswarlu, K., and Naidub, R. 2016. Agronomic and remedial benefits and risks of applying biochar to soil: Current knowledge and future research directions. *Environment International* 87: 1–12.

Lehmann, J. 2007a. Bio-energy in the black. *Frontiers in Ecology and the Environment* 5: 381–87.

Lehmann, J. 2007b. A handful of carbon. *Nature* 447: 143–4.

Lehmann, J., de Silva, J. P., Jr., Steiner, C. et al. 2003. Nutrient availability and leaching in an archaeological Anthrosol and a Ferralsol of the Central Amazon basin: Fertilizer, manure and charcoal amendments. *Plant and Soil* 249: 343–57.

Lehmann, J., Gaunt, J., and Rondon, M. 2006. Biochar sequestration in terrestrial ecosystems—A review. *Mitigation and Adaptation Strategies for Global Change* 11: 403–27.

Lehmann, J. and Joseph, S. 2009. *Biochar for Environmental Management: Science and Technology*. London: Earthscan.

Lehmann, J., Rillig, M. C., Thies, J. et al. 2011. Biochar effects on soil biota—A review. *Soil Biology and Biochemistry* 43: 1812–36.

Lehmann, J. and Rondon, M. 2006. Biochar soil management on highly weathered soils in the humid tropics. In: *Biological Approaches to Sustainable Soil Systems*, ed. N. Uphoff. Boca Raton: CRC Press.

Liang, B., Lehmann, J., Solomon, D. et al. 2006. Black carbon increases cation exchange capacity in soils. *Soil Science Society of America Journal* 70: 1719–30.

Lira, C. S., Berruti, F. M., Palmisano, P., et al. 2013. Fast pyrolysis of Amazon tucuma (*Astrocaryum aculeatum*) seeds in a bubbling fluidized bed reactor. *Journal of Analytical and Applied Pyrolysis* 99: 23–31.

Luo, Z., Wang, S., and Cen, K. 2005. A model of wood flash pyrolysis in fluidized bed reactor. *Renewable Energy* 30: 377–92.

Major, J., Lehmann, J., Rondon, M., and Goodale, C. 2010. Fate of soil-applied black carbon: Downward migration, leaching and soil respiration. *Global Change Biology* 16: 1366–79.
Maschio, G., Koufopanos, C., and Lucchesi, A. 1992. Pyrolysis, a promising route for biomass utilization. *Bioresource Technology* 42: 219–31.
Masiello, C. A. and Druffel, E. R. M. 1998. Black carbon in deep-sea sediments. *Science* 280: 1911–3.
Matovic, D. 2011. Biochar as a viable carbon sequestration option: Global and Canadian perspective. *Energy* 36: 2011–6.
Mohan, D., Pittman, C. U., Jr., Bricka, M. et al. 2007. Sorption of arsenic, cadmium, and lead by chars produced from fast pyrolysis of wood and bark during bio-oil production. *Journal of Colloid and Interface Science* 310: 57–73.
Mohan, D., Pittman, C. U., Jr., and Steele, P. H. 2006. Pyrolysis of wood/biomass for bio-oil: A critical review. *Energy and Fuels* 20: 848–89.
Mohanty, P., Nanda, S., Pant, K. K. et al. 2013. Evaluation of the physiochemical development of biochars obtained from pyrolysis of wheat straw, timothy grass and pinewood: Effects of heating rate. *Journal of Analytical and Applied Pyrolysis* 104: 485–93.
Nanda, S., Azargohar, R., Kozinski, J. A., and Dalai, A. K. 2014a. Characteristic studies on the pyrolysis products from hydrolyzed Canadian lignocellulosic feedstocks. *Bioenergy Research* 7: 174–91.
Nanda, S., Dalai, A. K., Berruti, F., and Kozinski J. A. 2016a. Biochar as an exceptional bioresource for energy, agronomy, carbon sequestration, activated carbon and specialty materials. *Waste and Biomass Valorization* 7: 201–35.
Nanda, S., Dalai, A. K., Gökalp, I., and Kozinski J. A. 2016b. Valorization of horse manure through catalytic supercritical water gasification. *Waste Management* 52: 147–58.
Nanda, S., Mohammad, J., Reddy, S. N., Kozinski, J. A., and Dalai, A. K. 2014b. Pathways of lignocellulosic biomass conversion to renewable fuels. *Biomass Conversion and Biorefinery* 4: 157–91.
Nanda, S., Reddy, S. N., Mitra, S. K., and Kozinski, J. A. 2016c. The progressive routes for carbon capture and sequestration. *Energy Science and Engineering* 4: 99–122.
Nguyen, B. T. and Lehmann, J. 2009. Black carbon decomposition under varying water regimes. *Organic Geochemistry* 40: 846–53.
Nguyen, B., Lehmann, J., Hockaday, W. C., Joseph, S., and Masiello, C. A. 2010. Temperature sensitivity of black carbon decomposition and oxidation. *Environmental Science and Technology* 44: 3324–31.
Park, J. H., Choppala, G. K., Bolan, N. S., Chung, J. W., and Cuasavathi, T. 2011. Biochar reduces the bioavailability and phytotoxicity of heavy metals. *Plant and Soil* 348: 439–451.
Renner, R. 2007. Rethinking biochar. *Environmental Science and Technology* 41: 5932–3.
Rillig, M. C., Wagner, M., Salem, M. et al. 2010. Material derived from hydrothermal carbonization: Effects on plant growth and arbuscular mycorrhiza. *Applied Soil Ecology* 45: 238–42.
Ro, K. S., Cantrell, K. B., and Hunt, P. G. 2010. High-temperature pyrolysis of blended animal manures for producing renewable energy and value-added biochar. *Industrial Engineering and Chemistry Research* 49: 10125–31.

Samanya, J., Hornung, A., Apfelbacher, A., and Vale, P. 2012. Characteristics of the upper phase of bio-oil obtained from co-pyrolysis of sewage sludge with wood, rapeseed and straw. *Journal of Analytical and Applied Pyrolysis* 94: 120–25.

Samolada, M. C. and Vasalos, I. A. 1991. A kinetic approach to the flash pyrolysis of biomass in a fluidized bed reactor. *Fuel* 70: 883–9.

Schlesinger, W. H. 1995. An overview of the global carbon cycle. In: *Soils and Global Change*, ed. R. Lal, J. Kimble, E. Levine, and B. A. Stewart, pp. 9–25. Boca Raton: Lewis Publishers.

Schmidt, M. W. I. and Noack, A. G. 2000. Black carbon in soils and sediments: Analysis, distribution, implications, and current challenges. *Global Biogeochemical Cycles* 14: 777–93.

Singh, R., Babu, J. N., Kumar, R. et al. 2015. Multifaceted application of crop residue biochar as a tool for sustainable agriculture: An ecological perspective. *Ecological Engineering* 77: 324–47.

Sjostrom, K., Chen, G., Yu, Q., Brage, C., and Rosen, C. 1999. Promoted reactivity of char in co-gasification of biomass and coal: Synergies in the thermochemical process. *Fuel* 78: 1189–94.

Sohi, S. P., Krull, E., Lopez-Capel, E., and Bol, R. 2010. A review of biochar and its use and function in soil. In: *Advances in Agronomy*, ed. D. L. Sparks, pp. 47–82. Burlington: Academic Press.

Sombroek, W., Ruivo, M. D. L., Fearnside, P. M., Glaser, B., and Lehmann, J. 2003. Amazonian dark earths as carbon stores and sinks. In: *Amazonian Dark Earths: Origin, Properties, Management*, ed. J. Lehmann, D. C. Kern, B. Glaser, and W. I. Woods, pp. 125–39. Dordrecht, Netherlands: Kluwer Academic Publishers.

Song, W. and Guo, M. 2012. Quality variations of poultry litter biochar generated at different pyrolysis temperatures. *Journal of Analytical and Applied Pyrolysis* 94: 138–45.

Uchimiya, M., Bannon, D. I., Wartelle, L. H., Lima, I. M., and Klasson, K. T. 2012. Lead retention by broiler litter biochars in small arms range soil: Impact of pyrolysis temperature. *Journal of Agriculture and Food Chemistry* 60: 5035–44.

Uchimiya, M., Wartelle, L. H., Klasson, T., Fortier, C. A., and Lima, I. M. 2011. Influence of pyrolysis temperature on biochar property and function as a heavy metal sorbent in soil. *Journal of Agriculture and Food Chemistry* 59: 2501–10.

Uchimiya, M., Wartelle, L. H., Lima, I. M., and Klasson, K. T. 2010. Sorption of deisopropylatrazine on broiler litter biochars. *Journal of Agriculture and Food Chemistry* 58: 12350–6.

Van Gestel, M., Merckx, R., and Vlassak, K. 1993. Microbial biomass responses to soil drying and rewetting: the fate of fast- and slow-growing microorganisms in soils from different climates. *Soil Biology and Biochemistry* 25: 109–23.

Vassilev, A., Schwitzguebel, J. P., Thewys, T., van der Lelie, D., and Vangronsveld, J. 2004. The use of plants for remediation of metal-contaminated soils. *The Scientific World Journal* 4: 9–34.

Wagenaar, B. M., Prins, W., and van Swaaij, W. P. M. 1993. Flash pyrolysis kinetics of pine wood. *Fuel Processing Technology* 36: 291–8.

Wardle, D. A., Zackrisson, O., and Nilsson, M.-C. 1998. The charcoal effect in boreal forests: Mechanisms and ecological consequences. *Oecologia* 115: 419–26.

Warnock, D. D., Lehmann, J., Kuyper, T. W., and Rillig, M. C. 2007. Mycorrhizal responses to biochar in soil—Concepts and mechanisms. *Plant and Soil* 300: 9–20.

Warnock, D. D., Mummey, D. L., McBride, B. et al. 2010. Influences of non-herbaceous biochar on arbuscular mycorrhizal fungal abundances in roots and soils: Results from growth-chamber and field experiments. *Applied Soil Ecology* 46: 450–6.

Watson, R. T., Noble, R., Bolin, B. et al. 2000. Land use, land-use change, and forestry. *Intergovernmental Panel on Climatic Change Special Report*, Cambridge: Cambridge University Press.

Zhao, B., Xu, R., Ma, F., Li, Y., and Wang, L. 2016. Effects of biochars derived from chicken manure and rape straw on speciation and phytoavailability of Cd to maize in artificially contaminated loess soil. *Journal of Environmental Management* 184: 569–74.

Zhu, W., Song, W., and Lin, W. 2008. Catalytic gasification of char from co-pyrolysis of coal and biomass. *Fuel Processing Technology* 89: 890–69.

section five

Bioenergy

chapter fifteen

New materials and configurations for microbial fuel cells

V. M. Ortiz-Martínez, M. J. Salar-García, F. J. Hernández-Fernández, A. P. de los Ríos, F. Tomás-Alonso, and J. Quesada-Medina

Contents

15.1 Introduction

Microbial fuel cells (MFCs) can be considered as one of the most promising technologies to help address the challenge of climate change (Hernández-Fernández et al., 2015). These devices can convert the chemical energy present in a substrate, such as wastewater, into electricity, offering a twofold benefit, electricity generation and water treatment. Bacteria in the anode chamber degrade the substrate (oxidation) and transfer the electrons released to the anode through an external circuit. Protons pass through the proton exchange membrane (PEM) to the cathode, where they react with oxygen (reduction) and the electrons from the anode to form water. MFCs can be set up in double-chamber or single-chamber configurations. Double-chamber designs include both an anodic and a cathodic chamber, commonly separated by a selective separator for ion exchange, while single-chamber MFCs only comprise an anodic chamber, with the cathode electrode exposed to the air (Du et al., 2007; Oliveira et al., 2013). The main components (anode, cathode, and separator) and the final design in which they are assembled play a key role in MFC performance and efficiency. In recent years, studies focused on MFCs have grown exponentially because this technology still shows certain limitations associated with their practical implementation. The high cost of some materials, such as proton exchange membranes or precious catalysts, and the complexity of MFC configurations have limited their commercial use. This chapter describes the most significant advances made in terms of materials and designs for MFC construction. The information covered ranges from the simplest and lowest cost materials to those more complex options that can even outperform noble metals. Furthermore, this chapter also includes the evolution of the MFC assemblies developed over recent years, including the discussion of operational parameters.

15.2 Anode materials

Carbonaceous materials are the most commonly used options to prepare MFC anodes. Carbon-based materials show high conductivity and long-term stability, and are biocompatible and relatively cheap in comparison with other materials. For these reasons, carbon is the most widespread material to prepare this type of electrode. Anode types can be grouped into three categories depending on their structure: (1) plane, (2) packed, and (3) brush (Figure 15.1). The most common plain carbonaceous electrodes used are carbon paper, felt, cloth, or veil, and graphite plates or sheets (Wei et al., 2011; Zhou et al., 2011).

Graphite plates or sheets show higher strength than carbon paper, which usually is very thin and slightly brittle. Roughened carbon-based materials, such as carbon cloth, exhibit higher values of power output than carbon sheets due to their higher specific area. However, their higher cost limits their large-scale applications (ter Heijne et al., 2008; Zhang et al., 2010). Carbon or graphite felt or foam are thicker than carbon paper, sheet, or cloth. Their special texture encourages the growth of the biofilm around the anode due to their high specific surface. The performance of MFCs using electrodes made of graphite rod (GR), felt, and

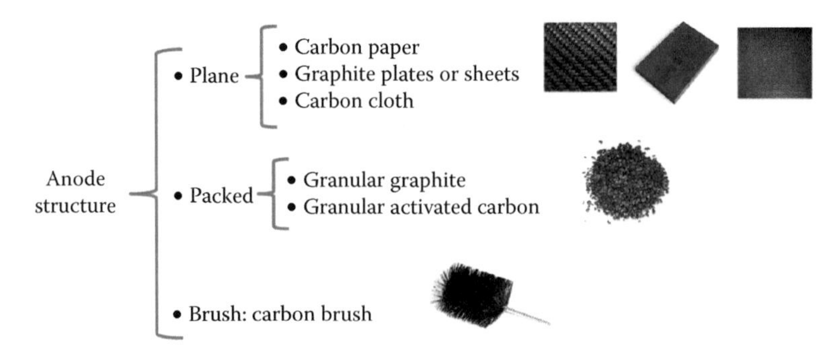

Figure 15.1 Classification of main MFC electrode assemblies.

foam was evaluated by Chaudhuri and Lovley (2003). Graphite rod and felt electrodes exhibit similar values of current and biomass activity, but electrodes based on graphite foam allow MFCs to reach up to 2.4 times higher values of current density.

Packed structures show higher specific surface available for biofilm growth when compared with plane materials. For that reason, the use of packed electrodes in microbial fuel cells has exponentially increased in recent years (di Lorenzo et al., 2010; Li et al., 2010). However, this type of material commonly needs an electron connector, such as a graphite rod, its porosity is reduced, and the material has to be tightly packed to avoid dead zones (Logan, 2007).

The use of brush structures overcomes the drawbacks of plane and packed materials because their specific surface and porosity are higher, and the current collection is more efficient than in the case of the materials described above. Logan et al. (2007) were pioneers in the use of anodes made of carbon brush. They reported that the larger the brush anode, the higher the power density generated, reaching a maximum value of 1430 $mW \cdot m^{-2}$, which is more than two times higher than the power density obtained with electrodes made of plain carbon paper.

Anode conductivity can be improved by using metals or metal oxides, whose conductivity is higher than that of carbonaceous materials. However, the application of some metals is limited due their corrosive characteristics. Moreover, their smooth surface often hinders biofilm growth. Because of this, these materials tend to reach lower power outputs when compared with carbon-based materials. Among the metallic materials investigated, stainless steel and titanium are the most common options for cathode construction (Wei et al., 2011). The performance of these materials can be enhanced by using rougher structures such as metallic mesh. For instance, anodes made of stainless steel mesh (SSM) exhibit higher values of current density when compared with anodes based on plain graphite (Erable and Bergel, 2009). Other metals such as gold are also suitable for the growth of electroactive bacteria, for example, *Geobacter sulfurreducens,* as demonstrated by Richter et al. (2008), although noble metals are not an affordable option.

The structure of the anode is a key factor for biofilm growth. It has been recently reported that bacterial adhesion and electron transfer to the anode can be improved by the treatment of the anode surface. These treatments can be grouped as (1) physical or chemical methods, (2) coating an anode surface with conductive or electroactive layers, and (3) preparation of metal–graphite composites (Figure 15.2).

Acid or heat treatments improve MFC performance by facilitating the adhesion of microorganisms because both specific anode area and the positive charges on the electrode surface increase by this method, and the interference of the contaminants is reduced (Wang et al., 2009).

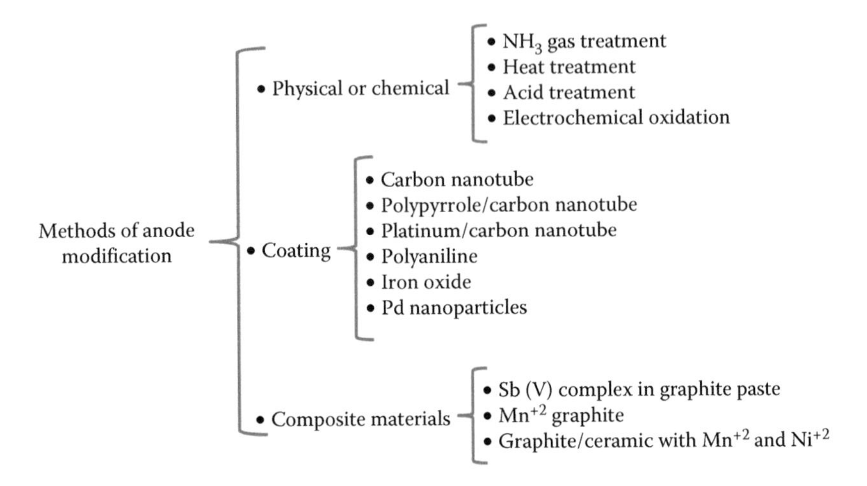

Figure 15.2 Methods of anode surface modification. (From Wei, J., Liang, P., and Huang, X. 2011. *Bioresource Technology*, 102: 9335–9344.)

Regarding the addition of a surface coating, the most commonly used materials are carbon nanotubes, conductive polymers, metals, or composites of such materials. Carbon nanotubes are the most widespread coating material because they highly improve the performance of MFC devices. However, although carbon nanotubes facilitate electron transfer from microorganisms to the anode, it has been demonstrated that nanomaterials can negatively affect bacteria growth. Thus, further research may be needed for a better understanding of the interaction mechanism between both bacteria and nanostructured materials (Liang et al., 2011; Sun et al., 2010). Other types of anode surface modification such as the addition of conductive polymers can also improve the values of current intensity because the specific area of the anode also increases (Zhao et al., 2010). Finally, it has been reported that composite materials can increase power output in MFCs. For example, when Mn^{+4}-graphite anodes are used, the energy produced can be up to 1000 times higher than in the case of using fine graphite (Park and Zeikus, 2002).

The nature of the anode material used greatly affects the composition of the anodic microbial community. Recently, Wang et al. (2016) analyzed the effect of several low-cost anode materials such as carbon fiber felt (CFF), SSM, GR, and foamed nickel (FN) on the biodiversity of the anodic microbial community and on the bioelectricity production and pollutant removal rates in a microbial fuel cell coupled with constructed wetland (CW-MFC). The results confirm that the relative abundance of Proteobacteria in an anode made of CFF or FN is higher than that found in anodes based on SSM or GR. Carbon fiber felt and foam nickel anodes allow CW-MFC to reach higher values of bioenergy versus stainless steel mesh and graphite rod. The results also show that foam nickel-based anodes improve the abundance of *Dechloromonas*, widely known for its ability for phosphate accumulation and denitrifying properties.

Titanium and titanium dioxide nanotubes (TNs) are another type of anode materials that have recently been investigated. Feng et al. (2016) reported the *in situ* growth of TNs on titanium surface, offering an effective option for anode construction. The synthesis of nanotubes over the titanium surface makes it rougher, more hydrophilic, and more conductive. All these properties enhance the development of the biofilm over the anode surface. MFCs using the TN-modified titanium anodes achieve 190 times higher current densities than those generated by bare titanium anodes and also higher compared with the values achieved with most of the anodes made of carbon felt. These results show that TN-modified titanium could be a promising anode material in MFCs.

Another promising material for MFC electrodes is graphene. However, this material exhibits two main drawbacks, its hydrophobicity and the isolation of the binder, which can limit the performance of this type of anodes, and thus it may need to be pretreated to serve as an effective anode. Xue et al. (2017) treated graphene powder with a cationic surfactant, cetyltrimethylammonium bromide (CTAB), to charge the powder positively, increasing its hydrophilicity. The treated powder was deposited onto the surface of stainless steel mesh by using vacuum filtration, obtaining a novel anode that was tested in a single-chamber MFC. This work reports that electrocatalytic activity and microbial variety are higher when MFCs contain CTAB-graphene modified anodes versus devices using bare graphite anodes. This work provides a strategy to synthesize novel binder-free anodes suitable for the growth of an electrocatalytic anodic biofilm.

All these research works show the interest of improving the performance of microbial fuel cells by developing novel anode materials, which facilitate the growth of electroactive biofilm, electron transfer to anode electrode, and bioenergy production.

15.3 Cathode materials

Cathode performance can be another possible bottleneck in MFC systems. Most of the materials previously described for anode construction can also be used as supporting material for the cathode. The main difference with the anode is that the cathode commonly needs a catalyst, which in turns varies depending on the type of electron acceptor. MFC cathodes can be classified into three main groups: (1) air cathodes, (2) aqueous air cathodes, and (3) biocathodes (Wei et al., 2011).

Oxygen is the most common electron acceptor because it is sustainable, is abundant in nature, and exhibits a high redox potential (Freguia et al., 2007). There is a wide variety of catalysts suitable for the oxygen reduction reaction (ORR); platinum is the most widespread material due to its high catalytic performance (He et al., 2015; Lu and Li, 2012). Logan et al. (2005) reported that MFCs using cathodes based on platinum produce up to five times higher power density versus plain carbon-based cathodes. However, the high price of this noble metal limits its large-scale applications. For that reason, more sustainable and cheaper options have been investigated over the last years. Table 15.1 summarizes some of the alternative catalysts reported for cathode construction in MFCs.

The two most widespread cathode assemblies used with catalysts are air cathodes and aqueous air cathodes. The structure of air cathodes usually is made of three different layers: (1) a diffusion layer exposed to air, (2) a conductive supporting material as intermediate layer, and (3) a final layer made of catalyst mixed with a binder. Aqueous air cathodes consist of a conductive supporting material such as carbon cloth or carbon paper coated with a layer of catalyst and binder (Wei et al., 2011). The

Table 15.1 Alternative catalysts used in microbial fuel cells

Catalyst	Electrode material	MFC configuration	Power ($mW \cdot m^{-2}{}_{cathode}$)	Reference
Lead dioxide	Titanium	Double chamber	77–78	Morris et al. (2007)
Fe/Fe_2O_3	Carbon felt	Double chamber	341.4	Zhuang et al. (2010)
Cobalt material (cobalt tetramethylphenylporphyrin [CoTMPP])	Carbon cloth	Air cathode single chamber	369	Cheng et al. (2006b)
Manganese dioxide	Carbon cloth	Air cathode single chamber	• Cobalt-doped: 180 • Copper-doped: 165	Lu and Li (2012)
Activated carbon and black acetylene	Stainless steel mesh	Air cathode single chamber	• 15% ratio cathodes: 1510 • 10% ratio ccathodes: 1560 • 5% ratio cathodes: 1510	Zhang et al. (2014)

performance of the aqueous air cathodes is limited by the concentration of dissolved oxygen in water, which is lower than in the air. Because of this, air cathodes reach higher values of power densities, being more suitable for the commercial applications of MFCs (Logan et al., 2015). In the case of air cathodes, supporting materials based on carbon such as carbon cloth or carbon paper are very common. However, in recent years, cheaper alternatives such as stainless steel have been investigated (Cheng et al., 2006a; You et al., 2011).

The two most common binders employed to fix the catalyst over the surface of the cathodic support are Nafion, a commercial solution of perfluorosulfonic acid, and polytetrafluorethylene (PTFE). Cheng et al. (2006b) found that Nafion allows MFCs to produce a maximum power density of 480 mW · m^{-2}, while cathodes containing PTFE reach lower values, 360 mW · m^{-2}. However, as in other cases, the most important drawback of Nafion is its high cost, which limits its large-scale applications.

PTFE has been also employed as a diffusion layer. Cheng et al. (2006a) reported that four PTFE diffusion layers allow MFCs to increase the coulombic efficiency by 171% and the maximum power output by 42%, in comparison with cathodes lacking the diffusion layer. Moreover, the use of PTFE avoids water losses in the system.

To reduce the final cost of MFCs, many researchers have proposed the possibility of employing carbonaceous materials with high specific surface to reduce the overpotential of the cathodic reaction instead of using a metal catalyst. Several authors have studied granular graphite as a suitable material for preparing aqueous air cathodes, while activated carbon is considered more adequate for air cathodes (Tran et al., 2010; Zhang et al., 2009a).

As in the case of the anode, an alternative option to improve the specific area of the cathode and thus MFC performance consists of modifying its surface. For aqueous air cathodes, nitric acid and thermal treatment of cathodes made of activated graphite granules increase the specific area of the electrode (Erable et al., 2009). On the other hand, the modification of carbonaceous materials has proven to be beneficial for enhancing the performance of air cathodes. For instance, the treatment of a graphite air cathode with ferric sulfate, fine graphite powder, kaolin, and nickel chloride to obtain Fe^{3+}-doped electrodes allows MFCs to reach better performances than with untreated electrodes (Park and Zeikus, 2003). Among the treatments investigated, carbon powder treated with nitric outperforms platinum-based cathodes in MFCs in terms of power output (Duteanu et al., 2010). These low-cost materials could encourage the large-scale implementation of microbial fuel cells by reducing their capital cost.

Biocathodes are a sustainable alternative to common cathodes coated with a metal or carbon catalyst. They are usually made of carbonaceous materials such as graphite plate or carbon felt covered with enzymes or microorganisms. Laccases or bilirubin oxidases are some of the main redox enzymes employed to prepare biocathodes (He and Angenent, 2006; Huang et al., 2011). Lacasse-based cathodes can increase the level of power output up to 10 times when compared with cathodes coated with platinum (Schaetzle et al., 2009). Despite the numerous advantages of using enzymes as catalysts, they still pose several limitations related to their denaturing under certain conditions, the difficulty in being immobilized, and their short lifetime (Erable et al., 2012). With the use of microorganisms, some of these limitations can be overcome. Both pure strain and mixed cultures have been used to catalyze the oxygen reduction reaction, although mixed populations can favor power output versus a pure strain (Rabaey et al., 2008). Although it remains unclear which are the best supporting material for biocathode construction, it has been demonstrated that the higher the specific surface of the supporting material employed, the better the performance of the biocathodes. Accordingly, packed or brush structures exhibit better results than plane structures (Wei et al., 2011).

Apart from oxygen, there are other inorganic compounds that can act as electron acceptors. The redox potential of NO_3^- competes with that of oxygen and therefore the denitrification process takes places simultaneously with energy production in MFCs (Clauwaert et al., 2007). Otherwise, autotrophic denitrification allows NO_2^- to be removed by collecting the electrons produced (Virdis et al., 2008). Some metal-containing ions can also be also used as electron acceptors. Ferricyanide is the most popular metal-containing ion compound used as an electron acceptor in MFCs. However, its application is limited versus oxygen because this compound must be chemically synthesized (Rabaey et al., 2004). Permanganate is a low-cost alternative that can display even better performance than ferricyanide (You et al., 2006). In addition to ferricyanide and permanganate, heavy metal and noble metals, usually contained in wastewater, have also been employed for this purpose. This fact improves the wastewater treatment capacity of MFCs. Cu^{+2}, VO_3^-, $Cr_2O_7^-$, Fe^{+3}, Ag^+, or Hg^+ are only a few examples of metal-containing ions that act as electron acceptors in MFCs (He et al., 2015).

Several organic compounds have also been studied as electron acceptors. Azo dyes are very common in industrial wastewater and its decolorization can take place simultaneously with the energy production. Acid orange 7, orange 1, methyl orange, or methylene blue are some of the azo dyes that have been successfully degraded at the cathode in MFCs. In addition, laccase is one of the most suitable biocatalysts for azo dye decolorization. Other pollutants such as nitrogenous aromatic compounds or chlorophenols are commonly found in wastewater and their removal is necessary due to their high level of toxicity (He et al., 2015). In addition to bioenergy production, microbial fuel cells have shown great potential for wastewater treatment by recovering a huge variety of organic and inorganic pollutants in the cathode. In this sense, both electrode material and electron acceptor are two key factors when it comes to the design of MFCs. This second benefit could strengthen the prospects of the commercial applications of these devices.

15.4 Separators

Another key component in microbial fuel cell devices is the separator. A good separator must meet the following requirements: (1) low cost, (2) minimum crossover of substrate and oxygen, and (3) high ionic conductivity. In addition to these factors, the performance of separators may also depend on the type of material and its properties, thickness, specific area, stability, and configuration. There is a huge variety of materials that have been used as separator in MFCs, from the simplest type of material, such as earthen pot, to complex composite membranes. Figure 15.3 shows the different types of materials tested as separators in microbial fuel cells (Daud et al., 2015; Khera and Chandra, 2012).

Thus far, ion exchange membranes have been the most commonly used separators in MFCs. They consist of a polymeric network containing charged functional groups that promote the transfer of the ions oppositely charged. Cation exchange membranes (CEMs) often contain sulfonate groups, while tertiary amines are the most common positively charged groups used in anion exchange membranes (AEMs). Figure 15.4 shows a schematic representation of ion and mass transfer across two types of ion exchange membranes.

Nafion membranes, which consist of a sulfonated tetrafluoroethylene (TFE) copolymer, are commonly employed in MFCs as cation exchange

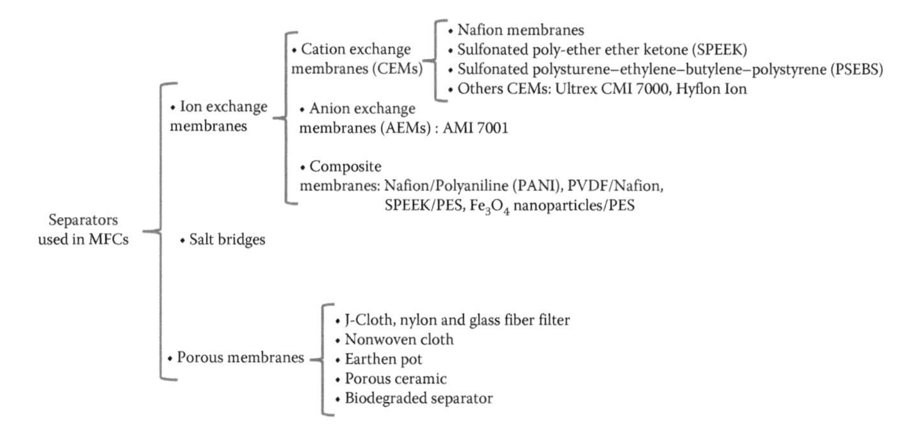

Figure 15.3 Separators used in MFCs. (From Daud, S. M., Kim, B. H., Ghasemi, M., and Daud, W. R. W. 2015. *Bioresource Technology*, 195: 170–179.)

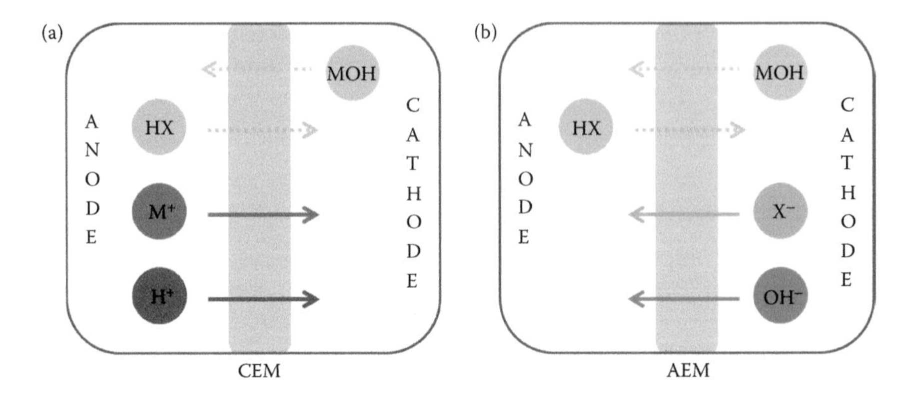

Figure 15.4 Scheme of ion and mass transfer through (a) cation exchange membrane and (b) anion exchange membrane (MOH and HX are undissociated bases and acids, respectively). (From Harnisch, F. and Schroder, U. 2009. *Chemical Society Reviews*, 39: 4433–4448.)

membranes, more specifically as PEMs, due to their high proton conductivity. Their structure contains hydrophilic sulfonate groups attached to a hydrophobic fluorocarbon backbone (Choi et al., 2006). This type of membrane exhibits numerous advantages such as high cation conductivity due to the sulfonate groups, and high chemical and long-term stability (Peighambardous et al., 2010). There are different types of Nafion membranes with different physical properties, thus exhibiting different MFC performance. For example, Nafion 112 allows MFCs to produce a higher power density (32 $mW \cdot m^{-2}$) when compared with Nafion 117. The first type is thinner than Nafion 117 and presents lower ohmic resistances (Rahimnejad et al., 2010).

A low-cost alternative to Nafion membranes consists of the use of membranes based on sulfonated poly-ether ketone (SPEEK). This polymer is thermally and mechanically stable and its sulfonation degree can be modified by varying the conditions of the sulfonation reaction (Yee et al., 2013). However, SPEEK-based membranes have higher internal resistance, lower conductivity, and higher activation loss than Nafion and therefore MFCs using SPEEK can offer lower levels of power than those containing Nafion-type separators (up to 20% less). Although the levels of power output are lower, the cost of SPEEK is up to three times cheaper than in the case of Nafion. On the other hand, SPEEK-based MFCs can achieve higher rates of chemical oxygen demand (COD) removals, up to 18% higher removal percentages (Ghasemi et al., 2013).

Another polymer employed as cation exchange membranes in MFCs is sulfonated polystyrene–ethylene–butylene–polystyrene (SPSEBS).

This thermoplastic elastomer consists of styrene blocks dispersed in an ethylene–butylene matrix. This nonfluorinated structure shows chemical, thermal, and mechanical stability. Due to its higher proton conductivity, SPSEBS can offer higher values of power output versus Nafion when used as separator in MFCs (Ayyaru et al., 2012).

In addition to the materials described above, there are other types of commercial membranes that are commonly employed as separators in MFCs. Ultrex CMI-7000 is an acid polymer membrane based on a polystyrene structure cross-linked with divinyl benzene-containing sulfonate groups. In terms of cation exchange and durability, this type of membrane exhibits similar performance to that of Nafion, although it generally displays higher ohmic resistances (Sotres et al., 2014). On the other hand, Hyflon ion, a copolymer based on TFE and perfluorosulfonylfluoridevinyl ether as side chain, have also been reported as an alternative separator. In this case, the results obtained by Ieropoulos et al. (2010) show that MFCs using Hyflon Ion can outperform MFCs working with Nafion membranes, achieving up to two times more power density.

Another class of exchange membranes are AEMs. This type of membrane contains a positively charged group bound to a polymeric structure, through which anions can be transferred. Although the most widespread membranes used in MFCs are cation exchange membranes, anion exchange membranes are also usually employed as separators in these devices (Varcoe et al., 2014). Zuo et al. (2008) reported a maximum power output of 610 $mW \cdot m^{-2}$ with a microbial fuel cell using an AMI-7001, a commercial anion exchange membrane. This value was 25% higher than that obtained employing a CEM in the same system. When AEMs are used, protons are consumed to neutralize the anions (OH^-) that cross the AEM, avoiding the pH reduction in the anode.

In recent years, other types of membranes more complex than CEMs and AEMs have been developed. They consist of a mixture of polymeric materials and organic or inorganic compounds. The result is a novel membrane with new properties that are capable of outperforming commercial membranes (Ghasemi et al., 2013). For instance, Nafion is commonly combined with other materials to improve its properties. Composite membranes based on Nafion and polyaniline (PANI) show better performance and less biofouling cover than plain Nafion when both are compared as separators in MFCs. PANI promotes proton conductivity and reduces the biofouling of the composite (Mokhtarian et al., 2013).

Nafion has also been combined with polyvinylidene fluoride (PVDF) (Shahgaldi et al., 2014). Alternative composite membranes can include cation polymers such as sulfonated poly(ether ether ketone) (SPEEK). Lim et al. (2012) reported that composite membranes based on SPEEK and polyethersulfone (PES) are cheaper than Nafion and show better performance than this type of commercial membrane. The mixture of a polymer and a conducting inorganic material allows an organic-inorganic composite membrane to be manufactured, for example, Fe_3O_4 nanoparticles and polyethersulfone. This type of material can offer up to 29% more power output than Nafion (Rahimnejad et al., 2012).

The second group of separators includes salt bridges. Their structure is simpler than ion exchange membranes and their price is lower. This type of separator consists of a glass tube filled with an electrolytic solution that conducts the ions between the two chambers. KCl and phosphate buffer solution are commonly used as inert electrolytes. In addition, agar is often added to avoid the mixing of the fluids and oxygen diffusion from the cathode to the anode through the separator. However, the main drawback of using a salt bridge as a separator is their high internal resistance, disfavoring power performance. This high internal resistance is probably caused by the long distance of proton diffusion (Min et al., 2005).

The third group of separators includes porous membranes, such as J-cloth, nonwoven cloth, glass fiber, earthen pot ceramic, biodegradable bag, and natural rubber. They are not ion-selective materials and can be classified according to pore size (Table 15.2) (Daud et al., 2015).

Nonwoven fabric (NWF) filter is also used in MFCs for wastewater treatment due its low cost and internal resistance in comparison with Nafion. Furthermore, the performance of NWF is better than that of Nafion due to its high water permeability. Compared with proton exchange membranes, NWF exhibits high long-term stability. The pore size in this material is larger than in the case of Nafion, which promotes proton exchange (Choi et al., 2013). J-cloth, nylon, and glass fiber filter show similar proton transport properties, although oxygen transfer rates are different in each of them (Zhang et al., 2009b).

Earthen pot is a low-cost material recently employed as a separator in MFCs. In 2010, Behera et al. (2010) used earthen pot as a separator in an air-cathode single-chamber MFC for the treatment of rice mill wastewater. They reached power densities up to four times higher than those obtained with Nafion. A few years later, Winfield et al. (2013) investigated porous ceramic and biodegradable bags (BioBags) as separators and compared the results with those obtained with cation exchange membranes. Their

Table 15.2 Classification of porous membranes

Microporous filtration membranes	Coarse-pore filter materials
• Microfiltration membranes (MFMs) • Ultrafiltration membranes (UFMs)	• Fabrics • Glass fiber • Nylon • Mesh • Cellulose filters

results confirm that porous separators can show higher proton conductivity than cation exchange membranes, and so are a promising alternative to expensive commercial membranes.

All these results confirm the large efforts made to develop environmentally friendly and low-cost alternative materials to commercial membranes. There is a wide variety of materials that offer chemical stability, low internal resistance, good long-term stability, and low biofouling tendency, among other advantages. They can be used in bioelectrochemical processes, offering promising prospects to replace traditional separators.

15.5 MFC configurations

In recent years, different types of laboratory-scale MFC configurations have been developed for research purposes. As commented above, one of the main drawbacks of this technology is its upscaling. Electrode spacing and separators are key factors that strongly affect the internal resistance of the system and thus the level of power output obtained in these devices. Consequently, the efficiency of MFCs is lower compared with other renewable energy technologies. To date, the pilot-scale attempts to scale up this technology have not been wholly satisfactory. There are some bottlenecks that need to be addressed, such as charge transfer and mass diffusion processes, ohmic losses, or the cathodic reaction kinetics, which usually require expensive catalysts. In this regard, several assembly options have been proposed for practical implementations of MFCs. They include alternative and low-cost materials, such as those that have been described in the above sections. As regards the upscaling, two approaches can be considered. On the one hand, new prototypes try to enhance the levels of power generated while increasing anode capacity for wastewater treatment (Zhou and Gu, 2013). Other authors propose the so-called miniaturization and multiplication approach as a possible viable strategy to scale up this technology. This implies the connection of a large number of miniaturized MFCs as a stack. Miniature MFCs offer large surface-area-to-volume ratio and short electrode spacing, reducing ohmic losses and increasing power performance (Chouler et al., 2016). Considering these two perspectives (macroscale and microscale), the following is a brief look at the different MFC designs recently reported.

As commented in the introduction, MFCs are typically set up in double-chamber or single-chamber assemblies. Double-chamber designs include both an anodic and a cathodic chamber, commonly separated by a selective separator for ion exchange, although configurations without a separator are also possible. Nevertheless, double-chamber designs are usually more complex and offer more limitations when it comes to scaling up this technology. Compared with double-chamber assemblies, single-chamber MFCs are a promising alternative that offers lower costs and simpler designs. In addition, they have proven to achieve higher efficiencies.

The MFC configurations reported in the literature include a wide range of designs, such as cylindrical, rectangular, and flat-plate shapes arranged in double- and single-chamber systems, including options without membranes. Figure 15.5 shows a schematic representation of some of these configurations.

Several continuously fed tubular configurations have been reported in recent years. Rabaey et al. (2005) designed an upflow tubular single-chamber MFC with an external cathode consisting of a woven graphite mat and an inner anode based on granular graphite. The cathode and the anodic chamber are separated by a conventional cation exchange membrane (Ultrex). This configuration was capable of producing levels of power outputs in the interval 60–90 $W \cdot m^{-3}$. Alternatively, He et al. (2005), developed another upflow design consisting of respective cylindrical Plexiglas-based anodic and cathodic chambers vertically assembled and separated by a proton exchange membrane. Anode and cathode electrodes consisted of reticulated vitreous carbon housed in their respective chambers. The power performance was lower in this case, with a maximum power density of 170 $mW \cdot m^{-2}$. More recently, Kim et al. (2009) developed a tubular reactor based on a polypropylene tube, which houses activated carbon inside as an anode electrode, and is folded with an exposed-to-air cathode folded around the tube. This configuration includes hydrogel between the membrane and the cathode, which increases cathode potential. The maximum power output achieved is of approximately 6 $W \cdot m^{-3}$, working with an anion exchange membrane.

Stack configurations allow voltage or current responses to be improved. Higher values of voltage output and currents can be obtained by connecting stacked MFCs in series or in parallel, respectively. However, the long-term impact of voltage reversal, which disfavors power generation by stacked MFCs, needs to be further investigated. Chouler et al. (2016) proposed innovative air cathode miniature MFCs connected as a stack. The individual cells are made of two square and flat plates made of polydimethylsiloxane between which cathode, separator, and anode are sandwiched. Both anode and cathode are manufactured with carbon cloth and the separator consists of a conventional proton exchange membrane (Nafion). By electrically connecting three individual units in parallel, the power output reached a maximum value of 1.2 $W \cdot m^{-3}$, which is more than 10 times higher compared to individual units. Tubular reactors can be also assembled in stack. Zhuang et al. (2012) reported an MFC consisting of five tubular air cathodes connected as a stack, employing manganese dioxide (MnO_2) as a catalyst on carbon fiber cloth (cathode) and graphite felt as an anode. They obtained a maximum power density of 175.7 $W \cdot m^{-2}$ with the parallel connection.

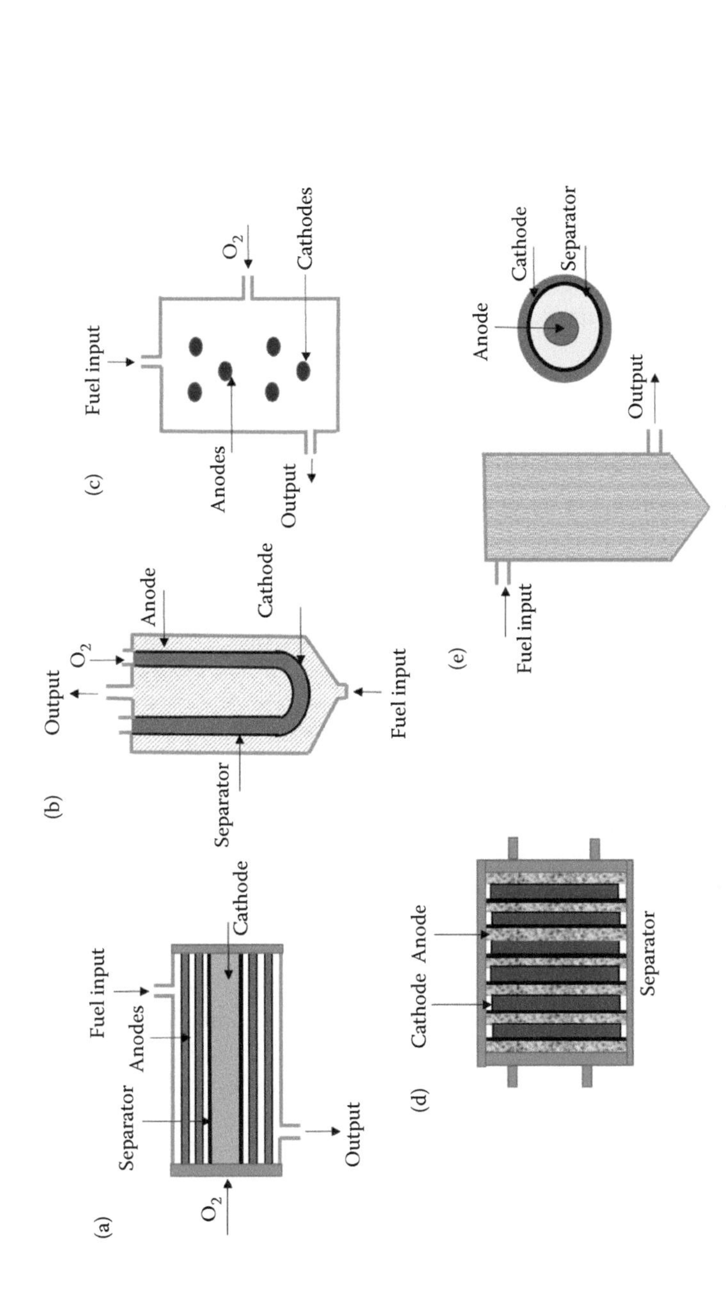

Figure 15.5 Several reported MFC configurations: (a) cylindrical single-chamber MFC; (b) cylindrical double-chamber MFC with U-shaped cathode; (c) membraneless MFC with rectangular shape; (d) stacked MFC; and (e) downflow single-chamber MFC. (Drawn with modifications after Du, Z., Li, H., and Gu, T. 2007. *Biotechnology Advances*, 25: 464–482.)

Other innovative designs include rotatable bioelectrochemical contactors, such as the configuration developed by Cheng et al. (2012). This type of MFC comprises a series of rotating disks acting as electrodes, with one half exposed to air (cathode) and the other submerged in water (anode). Each half comprises stainless steel mesh as a current collector, with an electrically conductive carbon fiber sheet placed over both sides of the mesh. The systems were capable of producing approximately 4 $mW \cdot m^{-2}$, with a COD removal rate of 0.91 kg COD $m^{-3}day^{-1}$. Compared with conventional activated sludge processes, these rotatable bioelectrochemical contactors can be considered more energy efficient because active aeration is not required.

Other innovative submersible MFC configurations have been recently reported. This is the case of the design presented by Zhang and Angelikadi (2012). In this setup, the cathode and the anode are pressed together on respective sides of a proton exchange membrane. This design includes a rectangular-shaped cathode chamber made of polycarbonate and two electrode-membrane units, the complete self-stack MFC being then submerged. This system is capable of producing a maximum power density of 294 $mW \cdot m^{-2}$. In addition, several reactors can be stacked together to enhance MFC performance. This design was developed in order to harvest electricity from submerged sediments. Moreover, this configuration offers operational flexibility because the system can be operated alternatively as two different cells or as a single self-stacked cell. Koroglu et al. (2014) developed a novel multitube cylindrical glass reactor that includes an array of carbon fiber tubes as anodes and tubular stainless steel electrodes as cathodes, separated by a proton exchange membrane (Nafion). This domestic wastewater-fed reactor can offer a maximum power output of 25.138 $mW \cdot m^{-2}$ with high COD removal ratios, up to 87%.

Other continuous cylindrical reactors include advanced materials such as ionic liquids as part of the membrane. Salar-Garcia et al. (2016) built a continuous scalable single-chamber MFC based on polymer inclusion membranes embedded into a carbon cloth cathode. The reactor consists of a plastic housing on which a window is made to place the cathode-membrane assembly. This way, one side of the membrane remains exposed to the air, while the other side faces the wastewater contained in the anode chamber. Inside the reactor, a perforated tube contains granulated graphite, which acts as an anode. Several types of ionic liquid were tested as membrane electrolytes, specifically triisobutyl(methyl)phosphoniumtosylate ($[P_{I4,I4,I4,1}]$[TOS]) and methyltrioctylammonium chloride ([MTOA][Cl]). The system offers a maximum power density of 12.3 $W \cdot m^{-3}$ (at a flow rate of 0.25 mL min^{-1}) when the ionic liquid triisobutyl(methyl) phosphoniumtosylate is employed as a membrane, and also allows a high percentage of chemical oxygen demand removal to be achieved (60%).

Finally, as commented in the section on separator materials, ceramic materials have been employed for MFC construction. Gajda et al. (2015) reported low-cost designs with low maintenance requirements. One of the possible configurations including this type of material consists of a cylindrical terracotta cave around which a carbon veil is externally folded to form the anode. The inner cavity houses the cathode electrode attached to the cylindrical wall that is made of a carbon veil layer on which a mixture of activated carbon and polytetrafluoroethylene is hot-pressed. The so-assembled ceramic cave is placed inside a plastic case, which acts as the anode chamber. This configuration allows a clean catholyte to be obtained in the inner cavity of the cylinder while the MFC is run under an external resistance load. The catholyte generated is a basic solution of salts coming from the anode. This is a representative case of alternative configurations that minimize the costs of the MFC construction due to the use of inexpensive materials.

15.6 Conclusions

Many scientific studies have demonstrated that MFC technology offers potentially promising prospects for power generation and wastewater treatment, and also for applications such as metal recovery and dye removal. New low-cost materials are constantly being developed and investigated in MFCs for electrode construction (anode and cathode) and to prepare novel separators in order to increase the efficiency of this technology. In this regard, ceramic materials, nanostructured materials, and nonplatinum catalysts are examples of current research lines that could improve the efficiency of these devices. This chapter summarizes the new materials and configurations recently reported with the purpose of increasing the possibilities of their practical implementation. These materials have been assembled in innovated single-chamber and double-chamber designs. All these research efforts contribute to development and enhancement of the MFC technology, which could become a preferred option among sustainable bioenergy processes.

References

Ayyaru, S., Letchoumanane, P., Dharmalingam, S., and Stanislaus, A. R. 2012. Performance of sulfonated polystyrene-ethylene-butylene-polystyrene membrane in microbial fuel cell for bioelectricity production. *Journal of Power Sources*, 217: 204–208.

Behera, M., Jana, P. S., More, T. T., and Grangrekar, M. M. 2010. Rice mill wastewater treatment in microbial fuel cell fabricated using proton exchange membrane and earthen pot at different pH. *Bioelectrochemistry*, 79: 228–233.

Chaudhuri, S. K. and Lovley, D. R. 2003. Electricity generation by direct oxidation of glucose in mediatorless microbial fuel cells. *Nature Biotechnology*, 21: 1229–1232.

Cheng, K. Y., Ho, G., and Cord-Ruwisch, R. 2012. Energy-efficient treatment of organic wastewater streams using a rotatable bioelectrochemical contactor (RBEC). *Bioresource Technology*, 126: 431–436.

Cheng, S. A., Liu, H., and Logan, B. E. 2006a. Increased performance of single-chamber microbial fuel cells using an improved cathode structure. *Electrochemistry Communication*, 8: 489–494.

Cheng, S. A., Liu, H., and Logan, B. E. 2006b. Power densities using different cathode catalysts (Pt and CoTMPP) and polymer binders (Nafion and PTFE) in single chamber microbial fuel cells. 2006. *Environmental Science and Technology*, 40: 364–369.

Choi, S., Kim, J. R., Cha, J. et al. 2013. Enhanced power production of a membrane electrode assembly microbial fuel cell (MFC) using a cost effective poly[2,5-benzimidazole] (ABPBI) impregnated non-woven fabric filter. *Bioresource Technology*, 128: 14–21.

Choi, Y., Wang, G., and Yau, S. T. 2006. Electronic composite of sulfonated tetrafluorethylene copolymer with potassium ferricyanide exhibiting room temperature negative differential resistance. *Applied Physics Letters*, 89: 233116.

Chouler, J., Padgett, G. A., Cameron, P. J. et al. (2016). Towards effective small scale microbial fuel cells for energy generation from urine. *Electrochimica Acta*, 192: 89–98.

Clauwaert, P., Rabaey, K., Aelterman, P. et al. 2007. Biological denitrification in microbial fuel cells. *Environmental Science and Technology*, 41: 3354–3360.

Daud, S. M., Kim, B. H., Ghasemi, M., and Daud, W. R. W. 2015. Separators used in microbial electrochemical technologies: Current status and future prospects. *Bioresource Technology*, 195: 170–179.

Di Lorenzo, M., Scott, K., Curtis, T. P., and Head, I. M. 2010. Effect of increasing anode surface area on the performance of a single chamber microbial fuel cell. *Chemical Engineering Journal*, 156: 40–48.

Du, Z., Li, H. and Gu, T. 2007. A state of the art review on microbial fuel cells: A promising technology for wastewater treatment and bioenergy. *Biotechnology Advances*, 25: 464–482.

Duteanu, N., Erable, B., Kumar, S. M. S., Ghangrekar, M. M., and Scott, K. 2010. Effect of chemically modified Vulcan XC-72R on the performance of air-breathing cathode in a single-chamber microbial fuel cell. *Bioresource Technology*, 101: 5250–5255.

Erable, B. and Bergel, A. 2009. First air-tolerant effective stainless steel microbial anode obtained from a natural marine biofilm. *Bioresource Technology*, 100: 3302–3307.

Erable, B., Duteanu, N., Kumar, S. M. S. et al. 2009. Nitric acid activation of graphite granules to increase the performance of the non-catalyzed oxygen reduction reaction (ORR) for MFC applications. *Electrochemistry Communications*, 11: 1547–1549.

Erable, B., Feron, D., and Bergel, A. 2012. Microbial catalysis of the oxygen reduction reaction for microbial fuel cells: A review. *ChemSusChem*, 5: 975–987.

Feng, H., Liang, Y., Guo, K. et al. 2016. TiO_2 nanotube arrays modified titanium: A stable, scalable and cost-effective bioanode for microbial fuel cells. *Environmental Science and Technology Letters*, 3: 420–424.

Freguia, S., Rabaey, K., Yuan, Z., and Keller, J. 2007. Non-catalyzed cathodic oxygen reduction at graphite granules in microbial fuel cells. *Electrochimica Acta*, 53: 598–603.

Gajda, I., Greenman, J., Melhuish, C., and Ieropoulos, I. 2015. Simultaneous electricity generation and microbially-assisted electrosynthesis in ceramic MFCs. *Bioelectrochemistry*, 104: 58–64.

Ghasemi, M., Daud, W. R. W., Ismail, A. F. et al. 2013. Simultaneous wastewater treatment and electricity generation by microbial fuel cell: Performance comparison and cost investigation of using Nafion 117 and SPEEK as separators. *Desalination*, 325: 1–6.

Harnisch, F. and Schroder, U. 2009. From MFC to MXC: Chemical and biological cathodes and their potential for microbial bioelectrochemical systems. *Chemical Society Reviews*, 39: 4433–4448.

He, Z. and Angenent, L. T. 2006. Application of bacterial biocathodes in microbial fuel cells. *Electroanalysis*, 18: 2009–2015.

He, Z., Minteer, S. D., and Angenent, L. T. 2005. Electricity generation from artificial wastewater using an upflow microbial fuel cell. *Environmental Science and Technology*, 39: 5262–5267.

He, C. S., Mu, Z. X., Yang, H. Y. et al. 2015. Electron acceptors for energy generation in microbial fuel cells fed with wastewaters: A mini-review. *Chemosphere*, 140: 12–17.

Hernández-Fernández, F. J., Pérez de los Rios, A., Salar-García, M. J. et al. 2015. Recent progress and perspectives in microbial fuel cells for bioenergy generation and wastewater treatment. *Fuel Processing Technology*, 138: 284–297.

Huang, L., Regan, J. M., and Quan, X. 2011. Electron transfer mechanisms, new applications, and performance of biocathode microbial fuel cells. *Bioresource Technology*, 102: 316–323.

Ieropoulos, I., Greenman, J., and Melhuish, C. 2010. Improved energy output levels from small-scale microbial fuel cells. *Bioelectrochemistry*, 78(1): 44–50.

Khera, J. and Chandra, A. 2012. Microbial fuel cells: Recent trends. *Proceedings of the National Academy of Sciences, India Section A: Physical Sciences*, 82: 31–41.

Kim, J. R., Premier, G. C., Hawkes, F. R., Dinsdale, R. M., and Guwy, A. J. 2009. Development of a tubular microbial fuel cell (MFC) employing a membrane electrode assembly cathode. *Journal of Power Sources*, 187: 393–399.

Koroglu, E. O., Baysoy, D. Y., Cetinkaya, A. Y., Ozkaya, B., and Mehmet, C. 2014. Novel design of a multitube microbial fuel cell (UM2FC) for energy recovery and treatment of membrane concentrates. *Biomass Bioenergy*, 69: 58–65.

Li, F. X., Sharma, Y., Lei, Y., Li, B. K., and Zhou, Q. X. 2010. Microbial fuel cells: The effects of configurations, electrolyte solutions, and electrode materials on power generation. *Applied Biochemistry and Biotechnology*, 160: 168–181.

Liang, P., Wang, H. Y., Xia, X. et al. 2011. Carbon nanotube powders as electrode modifier to enhance the activity of anodic biofilm in microbial fuel cells. *Biosensors and Bioelectrons*, 26: 3000–3004.

Lim, S. S., Daud, W. R. W., Jahim, J. M. et al. 2012. Sulfonated poly(ether ether ketone)/poly(ether sulfone) composite membranes as an alternative proton exchange membrane in microbial fuel cells. *International Journal of Hydrogen Energy*, 37: 11409–11424.

Logan, B. E. 2007. *Microbial Fuel Cells*, 1st Ed. John Wiley & Sons, Inc., Hoboken.

Logan, B. E., Cheng, S. A., Watson, V., and Estadt, G. 2007. Graphite fiber brush anodes for increased power production in air–cathode microbial fuel cells. *Environmental Science and Technology*, 41: 3341–3346.

Logan, B. E., Murano, C., Scott, K., Gray, N. D., and Head, I. M. 2005. Electricity generation from cysteine in a microbial fuel cell. *Water Research*, 39: 942–995.

Logan, B. E., Wallack, M. J., Kim, K. Y. et al. 2015. Assessment of microbial fuel cell configurations and power densities. *Environmental Science and Technology Letters*, 2: 206–214.

Lu, M. and Li, S. F. Y. 2012. Cathode reactions and applications in microbial fuel cells: A review. *Critical Reviews in Environmental Sciences and Technology*, 42: 2504–2525.

Min, B., Cheng, S., and Logan, B. E. 2005. Electricity generation using membrane and salt bridge microbial fuel cells. *Water Research*, 39: 1675–1686.

Mokhtarian, N., Ghasemi, M., Daud, W. R. W. et al. 2013. Improvement of microbial fuel cell performance by using Nafion polyaniline composite membranes as separator. *Journal of Fuel Cell Science and Technology*, 10: 1–6.

Morris, J. M., Jin, S., Wang, J., Zhu, C., and Urynowicz, M. A. 2007. Lead dioxide as an alternative catalyst to platinum in microbial fuel cells. *Electrochemistry Communication*, 9: 1730–1734.

Oliveira, V. B., Somoes, M., Melo, L. F., and Pinto, A. M. F. R. 2013. Overview on the developments of microbial fuel cells. *Biochemical Engineering Journal*, 73: 53–64.

Park, D. H. and Zeikus, J. G. 2002. Impact of electrode composition on electricity generation in a single-compartment fuel cell using *Shewanella putrefaciens*. *Applied Microbiology and Biotechnology*, 59: 58–61.

Park, D. H. and Zeikus, J. G. 2003. Improved fuel cell and electrode designs for producing electricity from microbial degradation. *Biotechnology and Bioengineering*, 81: 348–355.

Peighambardous, S. J., Rowshanzamir, S., and Amjadi, M. 2010. Review of the proton exchange membranes for fuel cells applications. *International Journal of Hydrogen Energy*, 35: 9349–9384.

Rabaey, K., Boon, N., Siciliano, S. D., Verhaege, M., and Verstraete, W. 2004. Biofuel cells select for microbial consortia that self-mediate electron transfer. *Applied Environmental Microbiology*, 70: 5373–5382.

Rabaey, K., Clauwaert, P., Aelterman, P., and Verstraete, W. 2005. Tubular microbial fuel cells for efficient energy generation. *Environmental Science and Technology*, 39: 8077–8082.

Rabaey, K., Read, S. T., Clauwaert, P. et al. 2008. Cathodic oxygen reduction catalyzed by bacteria in microbial fuel cells. *ISME Journal*, 2: 519–527.

Rahimnejad, M., Ghasemi, M., Najafpour, G. D. et al. 2012. Synthesis, characterization and application studies of self-made Fe_3O_4/PES nanocomposite membranes in microbial fuel cell. *Electrochimica Acta*, 85: 700–706.

Rahimnejad, M., Jafari, T., Haghparast, F., Najafpour, G. D., and Goreyshi, A. 2010. Nafion as a nanoproton conductor in microbial fuel cells. *Turkish Journal of Engineering and Environmental Sciences*, 34: 289–292.

Richter, H., McCarthy, K., Nevin, K. P. et al. 2008. Electricity generation by *Geobacter sulfurreducens* attached to gold electrodes. *Langmuir*, 24: 4376–4379.

Salar-Garcia, M. J., Ortiz-Martinez, V. M., Baicha, Z., de los Rios, A. P., and Hernandez-Fernandez, F. J. 2016. Up-flow microbial fuel cell based on novel embedded ionic liquid-type membrane-cathode assembly. *Energy*, 101: 113–120.

Schaetzle, O., Barriere, F., and Schroder, U. 2009. An improved microbial fuel cell with laccase as the oxygen reduction catalyst. *Energy Environmental Science*, 2: 96–99.

Shahgaldi, S., Ghasemi, M., Daud, W. R. W. et al. 2014. Performance enhancement of microbial fuel cell by PVDF/Nafion nanofibre composite proton exchange membrane. *Fuel Processing Technology Journal*, 124: 290–295.

Sotres, A., Diaz-Marcos, J., Guivernau, M. et al. 2014. Microbial community dynamics in two-chambered microbial fuel cells: Effect of different ion exchange membranes. *Journal of Chemical Technology and Biotechnology*, 90(8): 1497–1506.

Sun, J. J., Zhao, H. Z., Yang, Q. Z., Song, J., and Xue, A. 2010. A novel layer-by-layer self-assembled carbon nanotube-based anode, preparation, characterization, and application in microbial fuel cell. *Electrochimica Acta*, 55: 3041–3047.

ter Heijne, A., Hamelers, H. V. M., Saakes, M., and Buisman, C. J. N. 2008. Performance of non-porous graphite and titanium-based anodes in microbial fuel cells. *Electrochimica Acta*, 53: 5697–5703.

Tran, H. T., Ryu, J. H., Jia, Y. H. et al. 2010. Continuous bioelectricity production and sustainable wastewater treatment in a microbial fuel cell constructed with non-catalyzed granular graphite electrodes and permeable membrane. *Water Science and Technology*, 61: 1819–1827.

Varcoe, J. R., Atanassov, P., Dekel, D. R. et al. 2014. Anion-exchange membranes in electrochemical energy systems. *Energy and Environmental Science*, 7: 3135–3191.

Virdis, B., Rabaey, K., Yuan, Z., and Keller, J. 2008. Microbial fuel cells for simultaneous carbon and nitrogen removal. *Water Research*, 42: 3013–3024.

Wang, X., Cheng, S. A., Feng, Y. J. et al. 2009. Use of carbon mesh anodes and the effect of different pretreatment methods on power production in microbial fuel cells. *Environmental Science and Technology*, 43: 6870–6874.

Wang, J., Song, X., Wang, Y. et al. 2016. Microbial community structure of different electrode materials in constructed wetland incorporating microbial fuel cell. *Bioresource Technology*, 221: 697–702.

Wei, J., Liang, P., and Huang, X. 2011. Recent progress in electrodes for microbial fuel cells. *Bioresource Technology*, 102: 9335–9344.

Winfield, J., Chambers, L. D., Rossiter, J., and Ieropoulos, I. 2013. Comparing the short and long term stability of biodegradable, ceramic and cation exchange membranes in microbial fuel cells. *Bioresource Technology*, 148: 480–486.

Xue, L. X., Yang, N., Ren, Y. P. et al. 2017. Effect of binder-free graphene-cetyltrimethylammonium bromide anode on the performance of microbial fuel cells. *Journal of Chemical Technology and Biotechnology*, 92: 157–162.

Yee, R. S. L., Zhang, K., and Ladewing, B. P. 2013. The effect of sulfonated poly(ether eteher ketone) ion exchange preparation condition on membrane properties. *Membranes*, 3: 182–195.

You, S. J., Wang, X. H., Zhang, J. N. et al. 2011. Fabrication of stainless steel mesh gas diffusion electrode for power generation in microbial fuel cell. *Biosensors and Bioelectrons*, 26: 2142–2146.

You, S. J., Zhao, Q. L., Zhang, J. N., Jiang, J. Q., and Zhao, S. Q. 2006. A microbial fuel cell using permanganate as the cathodic electron acceptor. *Journal of Power Sources*, 162: 1409–1415.

Zhang, Y. and Angelidaki, I. 2012. Self-stacked submersible microbial fuel cell (SSMFC) for improved remote power generation from lake sediments. *Biosensors and Bioelectronics*, 35: 265–270.

Zhang, F., Cheng, S. A., Pant, D., Van Bogaert, G., and Logan, B. E. 2009a. Power generation using an activated carbon and metal mesh cathode in a microbial fuel cell. *Electrochemistry Communication*, 11: 2177–2179.

Zhang, X., Cheng, S., Wang, X., Huang, X., and Logan, B. E. 2009b. Separators characteristics for increasing performance of microbial fuel cells. *Environmental Science and Technology*, 43: 8456–8461.

Zhang, F., Saito, T., Cheng, S. A., Hickner, M. A., and Logan, B. E. 2010. Microbial fuel cell cathodes with poly(dimethylsiloxane) diffusion layers constructed around stainless steel mesh current collectors. *Environmental Science and Technology*, 44: 1490–1495.

Zhang, X., Xia, X., Ivanov, I., Huang, X., and Logan, B. E. 2014. Enhanced activated carbon cathode performance for microbial fuel cell by blending carbon black. *Environmental Science and Technology*, 48: 2075–2081.

Zhao, Y. K., Watanabe, K., Nakamura, R. et al. 2010. Three-dimensional conductive nanowire networks for maximizing anode performance in microbial fuel cells. *Chemistry A European Journal*, 16: 4982–4985.

Zhou, M., Chi, M., Luo, J., He, H., and Jin, T. 2011. An overview of electrode materials in microbial fuel cells. *Journal of Power Sources*, 196: 4427–4435.

Zhou, M. and Gu, T. 2013. The next breakthrough in microbial fuel cells and microbial electrolysis cells for bioenergy and bioproducts. *Journal of Microbial and Biochemical Technology*, S12: 003.

Zhuang, L., Zheng, Y., Zhow, S. et al. 2012. Scalable microbial fuel cell (MFC) stack for continuous real wastewater treatment. *Bioresource Technology*, 106: 82–88.

Zhuang, L., Zhou, S. G., Li, Y. T., Liu, T. L., and Huang, D. Y. 2010. *In situ* Fenton-enhanced cathodic reaction for sustainable increased electricity generation in microbial fuel cells. *Journal of Power Sources*, 95: 1379–1382.

Zuo, Y., Cheng, S., and Logan, B. E. 2008. Ion exchange membrane cathode for scalable microbial fuel cells. *Environmental Science and Technology*, 42: 6967–6972.

chapter sixteen

*Bioelectrochemical biorefining**

Abhijeet P. Borole

Contents

16.1 Definition of the concept: Bioelectrochemical biorefining

Bioelectrochemical biorefining (BER) is a new concept that has evolved from the field of microbial fuel cells (MFCs) and electrolysis cells. It targets integration of microbial electrochemical cells into the biorefinery (Figure 16.1). The idea is to use low-value resources to generate electrons and then use the electrons to produce value-added products. This can enable production of fuels and chemicals from waste and biomass using electrons as intermediates. Thus the name bioelectrochemical biorefining.

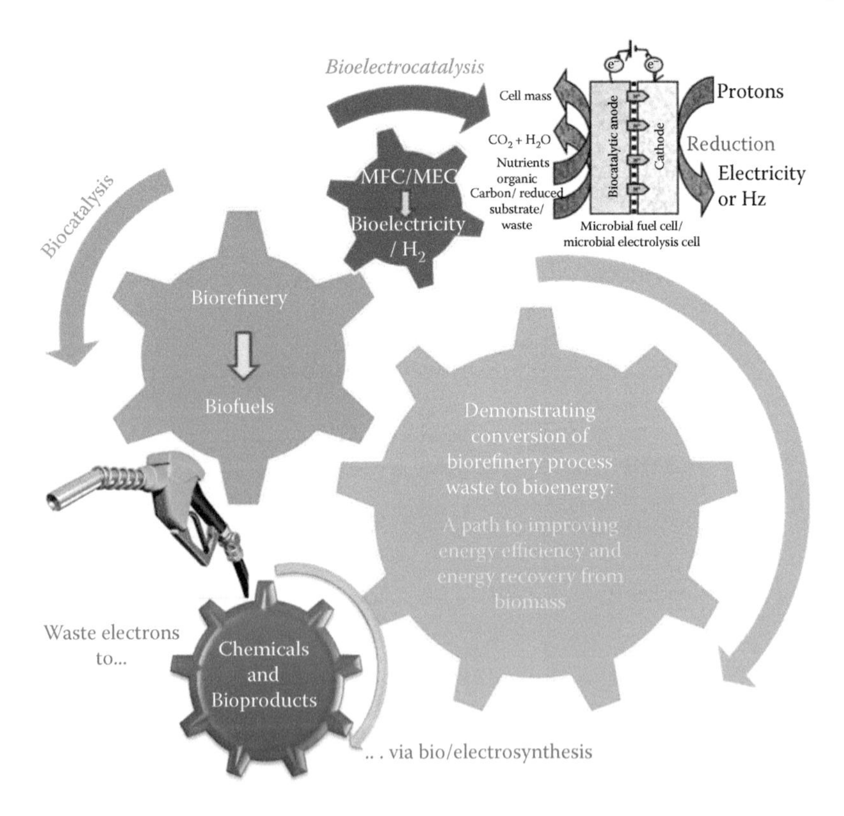

Figure 16.1 Integration of microbial electrochemical cells into biorefineries for production of fuels and chemicals.

Electrons are the simplest and most common energy carriers of many energy transfer systems and energetic molecules. Thus, using them as the vehicle for conversion of existing renewable resources into products needed for the 21st century is the primary goal of this marriage between microbial electrochemical technology and biorefineries.

The reactor system needed for bioelectrochemical biorefining can range from simple microbial electrochemical cells to more complicated designs that integrate biomass and waste processing. In its simplest form, it includes an anode, a cathode, and potentially a separator such as a membrane between the two electrodes. In the anode chamber, conversion of the substrates into electrons takes place, followed by their transfer to the electrode. In addition to the electrons, other products that are generated in the anode include protons, carbon dioxide, and intermediates or dead-end metabolites. The electrons and protons are utilized in the cathode to generate a range of products. Figure 16.2 shows the typical reactions that can occur in the anode and the cathode. The primary objective of the bioelectrochemical biorefining concept is to improve the utilization of the biomass resource resulting in higher process and energy conversion efficiency while generating a product that can also increase economic returns.

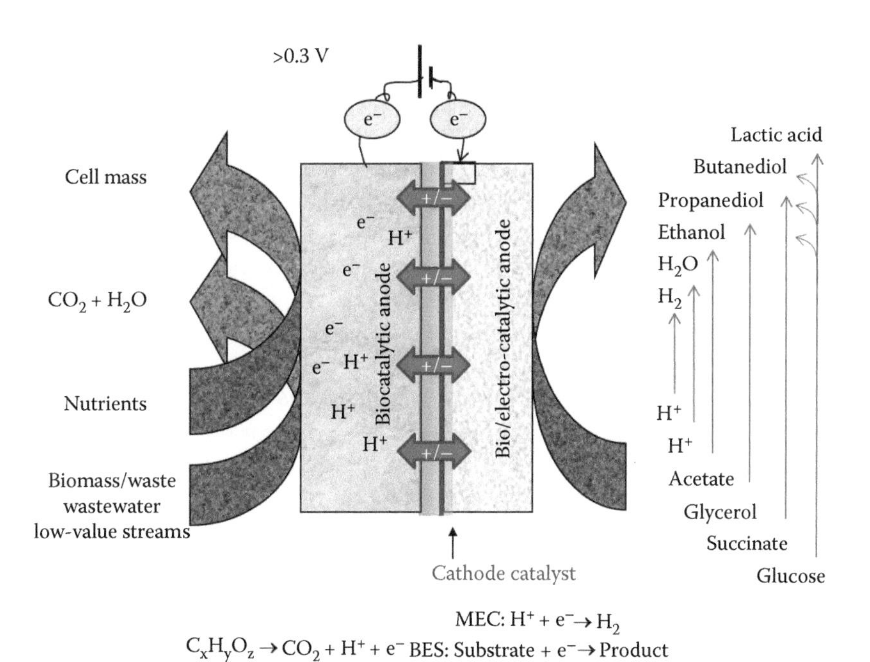

Figure 16.2 Concept of bioelectrochemical biorefining for conversion of biomass and waste into value-added products.

16.2 Anode processes

The substrates for the anode in a BER cell can include any type of molecule or atom from which electrons can be extracted. It has applications in all forms of industrial operations; however, the focus of this chapter is on those suitable for the bioenergy and biofuels industry. Thus, the substrates of most relevance are the organic and inorganic compounds generated in the biorefinery that have low or no value. This includes all forms of waste and wastewaters generated in the biorefinery. In addition to the waste, biomass can be directly used as the substrate as well. In a biorefinery, the waste products typically generated result from processes converting biomass into fuels and chemicals. This includes pretreatment, bioconversion, thermochemical conversion, and separations. The conversion operations are considered here and the bioelectrochemical processes employed to extract the electrons and their performance are described in the following subsections.

16.2.1 Bioconversion-derived substrates

Fermentation is a typical bioconversion process employed for conversion of biomass into biofuels. The most studied process is the generation of ethanol from a range of biomass resources. Besides the ethanol, many by-products are generated in the biorefinery. This includes organic acids, aldehydes, ketones, and hydrogen. Most of these by-products are produced at low enough concentration that it is not economically viable to recover them. Thus, a waste stream is generated after fuel recovery that is typically sent to the wastewater treatment facility. This stream, usually referred to as the stillage, can serve as a feedstock for the anode. In addition to the soluble by-products, conversion of lignocellulosic biomass also results in a lignin-rich stream. Typically, the lignin is separated and used as boiler fuel. However, it can also serve as a substrate for the BER cell if lignin can be solubilized. Further discussion on this alternative is included in Section 16.2.2.

16.2.1.1 Corn stover-derived waste streams

Corn stover has been a commonly used substrate for ethanol production over the last two decades. Significant information is available on the products and by-products of this process. A typical conversion process includes a biomass pretreatment, followed by fermentation and product separation by distillation or flash distillation. An example is presented here that targeted production of electrons from a dilute acid-pretreated corn stover, fermented to produce ethanol (Borole et al., 2013). The waste stream derived from corn stover fermentation after separation of the fuel was reported to consist of a total chemical oxygen demand (COD) of 83.3 g/L (Pannell et al., 2016). This stream contains a large amount of xylose due to incomplete fermentation (Table 16.1). In full-scale plants, this is expected

Table 16.1 Chemical characterization of stover process water

Sample	Cellobiose	Glucose	Xylose	Arabinose	Xylitol	Lactic acid	Glycerol	EtOH	2-furfural	HMF	Acetic acid	Other[a]
SPW1	0.8	0.2	15.5	6.6	0.4	0.6	0.8	0.4	0.0042	0.004	9.0	36
SPW2	0.94	0.36	20.28	6.45	NA	NA	NA	25.98	0.04	0.009	8.7	25.1

Source: Pannell, T. C., R. K. Goud, D. J. Schell, and A. P. Borole, Effect of fed-batch vs. continuous mode of operation on microbial fuel cell performance treating biorefinery wastewater. *Biochemical Engineering Journal*, 116(2016), 85–94.

Note: Concentrations given in g/L.

[a] The concentrations of other components in the samples were estimated based on maximum potential solubilization of oligomers derived from carbohydrates and other components of biomass. It includes up to 6 g/L phenolic compounds and up to 20 g/L soluble oligomers. The SPW2 sample contained 66.4 g/L total organic content.

to be lower; however, at the time of the study more efficient strains were likely not available. It also contains a large amount of arabinose and a range of fermentation by-products such as lactic acid and carboxylic acids. In addition, the pretreatment process used prior to fermentation generates various hydrolyzed products from biomass such as furanic compounds, primarily furfural and hydroxymethylfurfural (HMF). Thus, this is a complex stream with a range of molecules besides acetic acid. This stream derived from corn stover fermentation and postethanol recovery is termed as stover process water (SPW1). The SPW1 contains a small fraction of ethanol; however, in addition to SPW1, the study by Pannell et al. also reported using a sample collected prior to ethanol recovery, which was identified as SPW2 (Table 16.1). The total COD of SPW2 used in the study was 130.9 g/L.

In addition to the acids and alcohols, a range of phenolic compounds were present in this stream; however, they were not quantified individually. It is important to know about the presence of phenolic compounds because they can be toxic to many microorganisms (Palmqvist and Hahn-Hagerdal, 2000). The furanic compounds can also exhibit toxicity to microorganisms and therefore the mixture of furanic and phenolic compounds requires special attention. Their fate in the anode is discussed below; however, assessment of individual furanic and phenolic compounds in the bioanode is discussed further in Section 16.2.2.

16.2.1.2 *Development of bioanode for treating bioconversion by-products*

As described above, the waste streams generated from biorefinery processes employing fermentation can result in complex, mixed-substrate streams. In order to efficiently convert the organic matter present in these streams into electrons, it is necessary to understand the anodic processes. Work conducted using microbial fuel cells and electrolysis cells over the past decade has resulted in investigations of various substrates in the anode (Pant et al., 2010; Lu and Ren, 2016). The primary substrate that serves as a typical electron donor in the anode to produce electrons is acetate. Several microorganisms have been reported to use acetate as a substrate; however, the most common is *Geobacter sulfurreducens* (Bond and Lovley, 2003). The process of producing electrons from substrates and their subsequent transfer to electrodes, such that a current is generated, is called exoelectrogenesis. This phenomenon is the underlying process occurring in the bioanode that is used to generate electrons from wastewater and other process streams to enable the bioelectrochemical biorefining process to occur.

Development of the bioanode for utilization of acetate for production of electrons or current can simply be achieved by inoculating the anode with *G. sulfurreducens* and providing the substrate (Esteve-Nunez et al., 2005).

Alternately, a mixed community such as that from anaerobic digester can also be used to inoculate the anode, which typically is enriched in *G. sulfurreducens* when acetate is used as the substrate (Yates et al., 2012). This is the first step in development of the bioanode for research in the field of microbial fuel cells, that is, if the objective is to produce current from acetate. It is important to note the dominant role *Geobacter* spp. play in MFC bioanodes with acetate as the substrate because they appear to also be present in bioanodes developed using alternate substrates. However, it is possible that they may not play a dominant role in all bioelectrochemical systems.

It is relatively simple to develop a bioanode for single substrates while achieving high performance because growth of a single species is relatively easy, if the growth conditions are known. However, development of bioanodes for utilization of real wastewaters, which are usually complex and contain a mixture of substrates, requires specific attention to various factors as has been reported by Borole et al. (2011). Optimization of bioanode for conversion of complex substrates requires development of biofilms that are electroactive and capable of utilizing most if not all the components of the waste stream. This requires consideration of the design parameters of the anode as well as the overall bioelectrochemical cell, and attention to process parameters and biological parameters which pertain to the inoculum. Efficient conversion of a complex mixture of substrates into electrons usually involves much more than exoelectrogenesis. The biorefinery process water such as SPW1 and SPW2 have a range of sugars, carboxylic acids, alcohols, furanic, and phenolic compounds as well as their oligomers. Thus, in addition to exoelectrogenesis, hydrolysis, fermentation, or acidogenesis/acetogenesis are needed for the cell to function efficiently. Additionally, minimization of methanogenesis and most other anaerobic processes that use alternate electron acceptors is necessary (Borole et al., 2011). The presence of sulfate, nitrate, oxygen, and iron, among others, can negatively affect enrichment of exoelectrogens; thus it is important to eliminate these electron acceptors at least during the enrichment process. Figure 16.3 shows the range of parameters that need to be considered for electroactive biofilm optimization.

Using the enrichment strategy based on the optimization process discussed above (Figure 16.3), a consortium was reported that could efficiently utilize the corn stover process water. This process was patented including the application to ethanol biorefineries (Borole, 2010, 2012). Using such a bioanode, the mixture of more than 20 contaminant molecules present in the SPW1 and SPW2 samples were treated and the electrochemical performance of the anode was reported under various operating conditions. Several other questions were also investigated including to what extent the bioanode removes organics in the biorefinery stream and at what coulombic efficiencies. A previous report which did not use the optimization process indicated above had indicated that the conversion of sugars

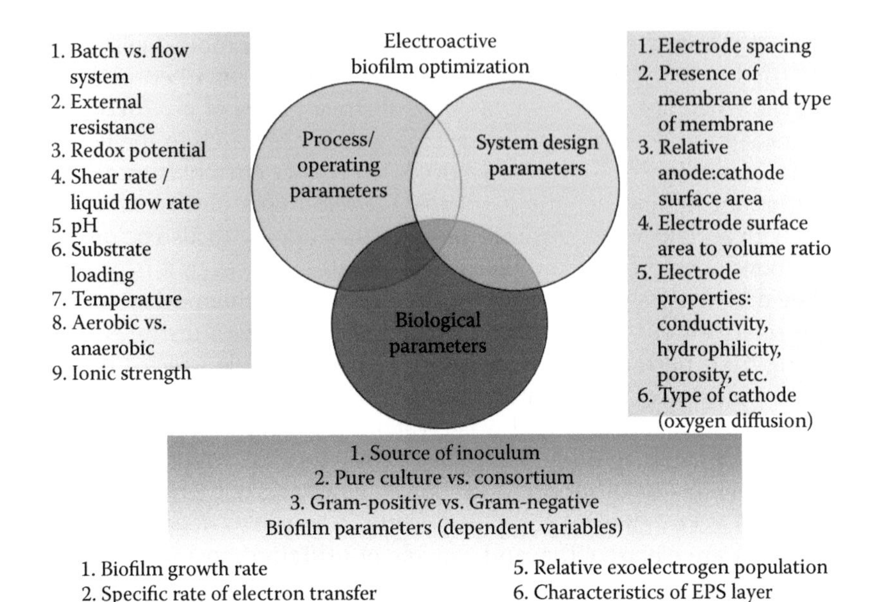

Figure 16.3 Parameters to be considered during optimization of electroactive biofilms. (From Borole, A. P., G. Reguera, B. Ringeisen, et al., *Energy & Environmental Science*, 4(2011) 4813–4834.)

could not be achieved in the presence of toxic or inhibitory molecules (Catal et al., 2008a). Therefore, investigations targeting study of biorefinery wastewater streams containing the inhibitor molecules are important.

16.2.1.3 Performance parameters for bioanode characterization

Assessment of the bioelectrochemical phenomenon requires new parameters to describe the conversion rate and efficiency, among others, because the process involves exchange or transfer of electrons from one entity to another. The primary parameters that describe anode performance include coulombic efficiency (CE), current density, potential efficiency, and COD removal efficiency.

16.2.1.3.1 Anode coulombic efficiency CE is defined as the ratio of electrons or charge (expressed in coulombs) produced as current to the maximum electrons that represent the amount of substrate consumed. It is calculated using the following formula:

$$\mathrm{CE}_{\mathrm{anode}} = \frac{I_{\mathrm{obs}}t}{n_{sT}n_eF}$$

where I_{obs} is the produced current; n_{sT} is the moles of substrate used; n_e is the number of electrons available per mole of substrate (based on complete conversion to CO_2); F is Faraday constant; and t is the duration of the experiment.

Depending on the operating mode of the experiment or the mode in which the substrate is delivered, the formula can be altered to represent batch mode, fed-batch mode, or continuous mode. The difference in the calculation arises from the denominator, which is calculated differently for each mode of operation. In addition to the coulombic efficiency, a related parameter known as anode conversion efficiency (ACE) has also been reported. When the exact amount of substrate removal is not known, this parameter can be used, which is based on total substrate added into the system. It is useful for waste streams that have low concentrations or for operations that are run continuously so the effluent concentrations cannot be determined accurately.

The COD for these samples can be below the detection limit, necessitating the use of the total substrate added into the system rather than the precise amount of substrate removed. The ACE has been reported to be calculated as follows:

$$\mathrm{ACE} = \frac{I_{\mathrm{obs}} t}{n_{sT} n_e F}$$

where n_{sT} is the total moles of substrate supplied during the experiment.

16.2.1.3.2 Current density Current density is defined as the rate of production of current per unit area or volume of the reactor. The latter is more specifically called volumetric current density. In most cases the area used in this calculation is the projected surface area of the anode or the membrane separating the anode and cathode. Because the counter ions have to transfer through the membrane, on which current depends, this is a relevant parameter in calculating current density.

16.2.1.3.3 Potential efficiency A potential difference exists between the theoretical voltage at which the current is expected and the actual voltage the anode reaches when it is left under open circuit conditions. The theoretical voltage is the redox potential at which the relevant reaction takes place and depends on the substrate being used as well as the microbes or proteins involved in the reaction. A different redox potential is reached when the cell is operating under closed circuit conditions and is producing current continuously. The potential efficiency is the ratio of the theoretical voltage required for reaction to the voltage of the reactor measured under load conditions.

16.2.1.3.4 COD removal efficiency For a complex mixture such as a waste or a biomass stream, COD is the most suitable parameter characterizing the energy content of the stream. Thus, removal of COD offers an easy way of tracking the conversion process. COD removal efficiency is the COD removed from the effluent of the reactor as a percentage of the feed introduced into the reactor.

16.2.1.4 Conversion of biorefinery streams in bioanode

Extraction of electrons from different substrates fed into the anode via the waste or biomass stream is important to achieve high efficiency. Electrochemical utilization of acetate is the most common reaction in the anode, therefore the rate of utilization of acetate is high. In conversion of the SPW2 stream, most of the acetate was reported to be removed (Borole et al., 2013). Conversion of the sugars was also reported (Figure 16.4). Xylose, galactose, and arabinose were the major sugars present in the SPW2, and the rate of removal is dependent on the concentration of the sugar present. Similar observations have been reported in other studies using pure sugars (Catal et al., 2008b). Conversion of xylose has been studied in detail by Huang et al., (2008), who determined the biochemical kinetics of its conversion in the anode. Conversion of xylose was also studied in continuously fed MFCs, which showed that its conversion varies with time and part of the energy is stored internally, which can be

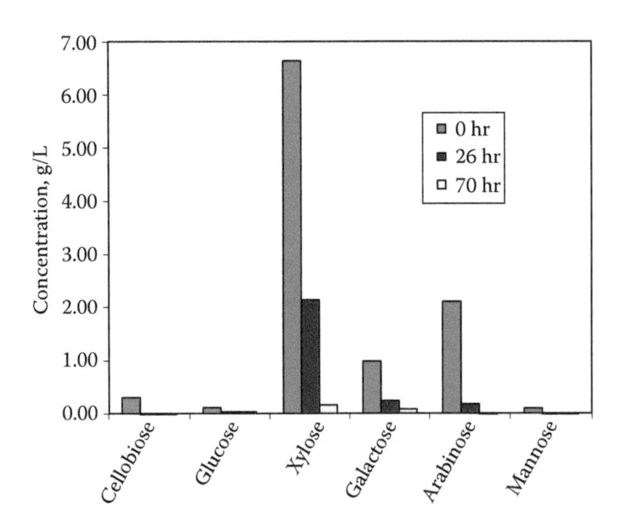

Figure 16.4 Removal of sugars from SPW2 at 32% loading; near complete removal of sugars from SPW2 sample in 70 h indicates high rate of conversion of the sugars in the bioanode. (From Borole, A. P., C. Hamilton, and D. Schell, *Environmental Science & Technology*, 47(2013) 642–648.)

recovered if the bioanode is operated over long periods of time (Huang and Logan, 2008).

In addition to the sugars, removal of the furanic and phenolic compounds was also reported (Borole et al., 2013). The components that were identified and monitored over time included furfural, HMF,

Table 16.2 Removal of furanic and phenolic compounds identified in the shredded pine wood (SPW) stream by the electroactive microbial consortium developed in bioanode

1%	2-furfural	5-HMF	4-Hydroxybenzaldehyde	Vanillic acid
1% initial	0.43	0.57	0.10	0.06
1% 30 h	0.03	0.03	0.02	0.00
1% 49 h	0.00	0.05	0.00	0.00
2%				
2% initial	0.40	0.22	0.05	0.11
2% 48 h	0.06	0.04	0.00	0.00
4%				
4% initial	1.02	1.59	0.17	0.29
4% 72 h	0.14	0.02	0.00	0.16
8%				
8% initial	2.03	3.18	0.35	0.58
8% 33 h	0.36	0.02	0.05	0.19
16%				
16% initial	4.23	5.92	0.69	1.43
16% 26 h	1.21	0.05	0.03	0.40
32%				
32% initial	7.75	12.60	1.38	2.31
32% 26 h	1.67	0.08	0.19	0.76
32% 70 h	0.52	0.04	0.12	0.75
64%				
64% initial	16.27	26.80	2.67	3.16
64% 20 h	3.31	0.15	0.46	0.64
100%				
100% initial	25.41	39.77	4.32	7.21
100% 3 h	7.36	20.37	5.40	7.32
100% 48 h	2.91	3.55	0.00	2.41

Source: Borole, A. P., C. Hamilton, and D. Schell, Conversion of residual organics in corn stover-derived biorefinery stream to bioenergy via microbial fuel cells, *Environmental Science & Technology*, 47(2013) 642–648.)

Note: The study reported batch experiments with 1%–100% of the SPW (by volume) present in the anode medium; the initial concentrations reported were measured during the first hour after addition of SPW2 in the anode recirculation solution.

4-hydroxybenzaldehyde, and vanillic acid. Table 16.2 shows the conversion of these molecules over time in the bioanode (Borole et al., 2013). Luo et al. (2010) reported utilization of furfural in bioanode as well.

In addition to the four individual furanic and phenolic molecules, removal of many other compounds was reported (Borole et al., 2013); however, they were not identified or quantified. Figure 16.5 shows a high-pressure liquid chromatogram (HPLC) of the SPW2 sample before and after the treatment. This shows that almost all the peaks observed by HPLC, which included compounds other than furanic and phenolic compounds, were removed as well. A few new peaks were observed, which indicated that some new intermediates were also produced. However, it was reported that all the intermediates generated gradually disappeared over time (Borole et al., 2013). This observation was based on comparison of the intermediates in runs conducted at various loadings of the SPW substrate in the bioanode (from 1% to 100%).

16.2.1.5 *Bioanode performance for biorefinery streams*

Utilization of the organic compounds present in the biorefinery waste streams in the bioanode indicates the potential removal of these compounds from the stream. However, in order for the process to be valuable, it is necessary to efficiently recover the electrons in the form of current.

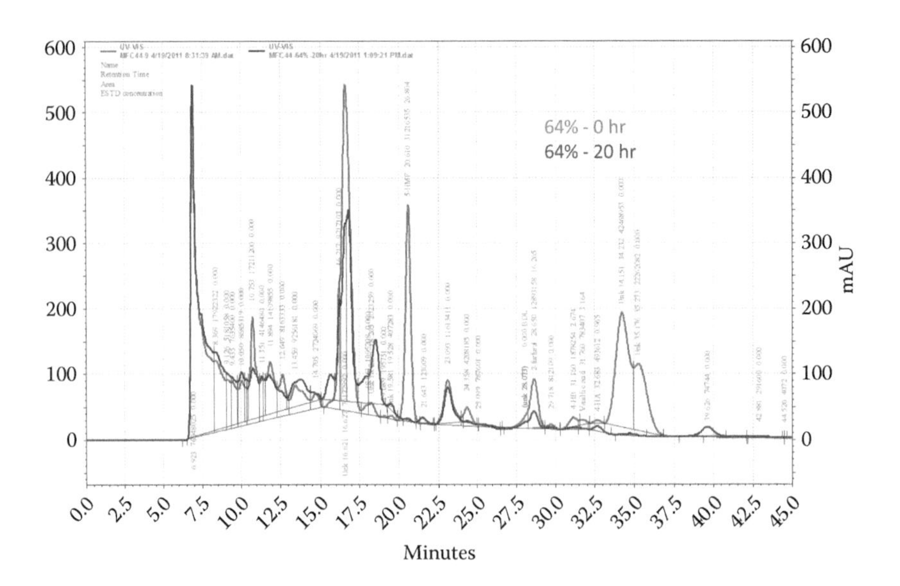

Figure 16.5 Chromatogram showing reduction in phenolics and furan aldehyde compounds present in 64% SPW sample after a 20-h treatment. (From Borole, A.P., C. Hamilton, and D. Schell, *Environmental Science & Technology*, 47(2013) 642–648.)

Current density associated with conversion of the SPW streams in the bioanode was reported and found to be a function of the organic loading as well as the mode of substrate addition (Pannell et al., 2016). Pannell et al. reported two types of bioanode operations, one in which the substrate was added only at the beginning of the run (batch mode), and another in which the substrate was continuously fed into the bioanode . Current densities reported under the two conditions are shown in Figure 16.6. These results are for operation of the bioanode as a microbial fuel cell; however, the current generated can also be used for other cathodic reactions as shown in Figure 16.2. The fed-batch versus continuous mode of operation was reported to have a significant effect on the current density over time. While the current density remained relatively high in the continuously fed bioanode, it dropped significantly in the fed-batch system. This was reported to be due to deterioration of the electroactive component of the microbial consortium in the fed-batch system as discussed further in the subsection on microbial characterization.

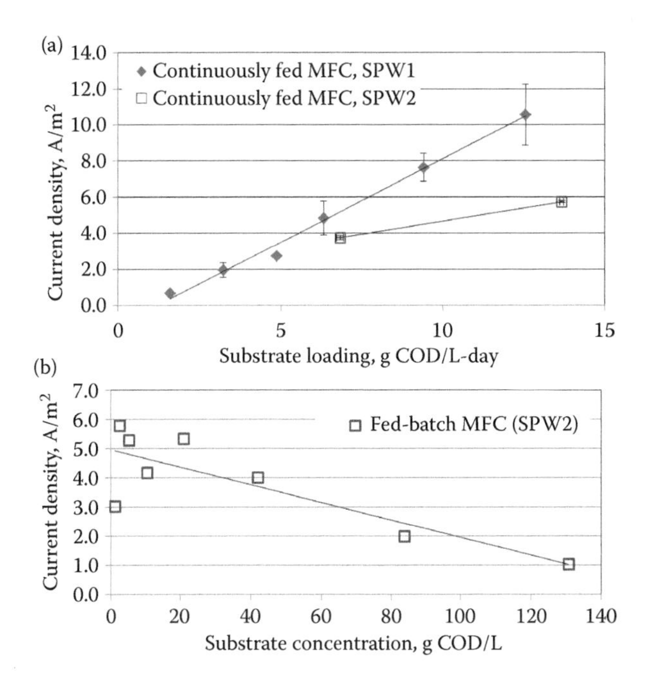

Figure 16.6 Current density as a function of substrate loading rate for (a) continuously fed bioanode and (b) fed-batch mode of operation; two different microbial fuel cells were reported using the two substrates SPW1 and SPW2; the blue diamonds represent current density obtained for SPW1 as the substrate, while the red squares represent the current density obtained using SPW2 as the substrate. (From Pannell, T. C., R. K. Goud, D. J. Schell, and A. P. Borole, *Biochemical Engineering Journal*, 116(2016) 85–94.)

In addition to the current density, the efficiency of capturing the electrons based on substrate utilized is also important. Pannell et al. reported anode conversion efficiency versus coulombic efficiency due to the inability to measure low concentrations present in the effluent of the continuously fed bioanode. This may also indicate that the effluent from the bioanode may be easier to recycle or dispose of into the environment owing to its low concentration of contaminants. This may imply lower wastewater treatment costs for the biorefinery. A high ACE ranging from 40% to 60% was reported for the continuously fed bioanode; however, the ACE dropped precipitously in the batch run (Figure 16.7). The cause of this is similar to that indicated for the drop in current density and may be linked to the change in the composition of the microbial composition. The COD removal was reported to range from 15% to 82% in the batch experiments (Pannell et al., 2016). This indicates that a substantial portion of the organic matter present in SPW2 can be removed. The high ACE reported in the continuously fed bioanode further supports this inference and that the substrate consumed can be efficiently recovered as current, with minimum recovery of 40% and a maximum of 60%.

16.2.1.6 *Alternate lignocellulosic feedstocks*

Several other studies have reported bioanode performance using lignocellulosic process-derived waste or substrate streams. Untreated corn stover was used in a bioanode of a microbial fuel cell; however, it yielded a lower

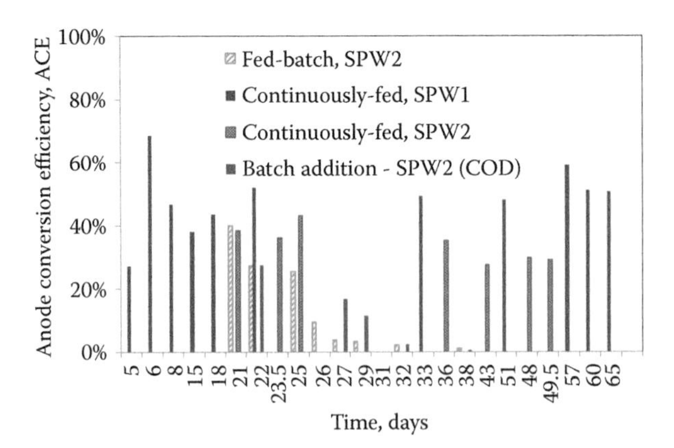

Figure 16.7 Changes in anode conversion efficiency of the bioanode reported over a 73-day period; the results are compared with ACE reported under batch conditions previously; the precipitous drop in ACE in batch experiment conditions shows the inefficient conversion of substrates to electricity, potentially due to growth of nonexoelectrogenic bacteria in the anode under initial high substrate concentrations representative of batch operation. (From Pannell, T. C., R. K. Goud, D. J. Schell, and A. P. Borole, *Biochemical Engineering Journal*, 116(2016) 85–94.)

current, reaching only 2 A/m^2 (Wang et al., 2009) compared to more than 10 A/m^2 reported by Pannell et al. (2016). Use of corn stover pretreated via steam explosion was reported to increase the current density to a maximum of 2.4 A/m^2 (Wang et al., 2009). Conversion of cellulose, one of the three biopolymers present in lignocellulosic biomass in a bioanode, was reported using rumen bacteria (Rismani-Yazdi et al., 2007). This study showed that electrons can be extracted directly from biopolymers if a suitable consortium of microorganisms is used. Use of externally provided cellulose-degrading enzymes followed by utilization of the hydrolysis product in a bioanode has also been reported (Rezaei et al., 2008). This strategy allowed improvement in current production compared to unhydrolyzed cellulose; however, the maximum current reported was 1 A/m^2. The use of the enzyme to degrade cellulose was reported to increase the CE from 51% to 73%. Conversion of cellulose to electrons in a bioanode was reported using a pure culture of *Enterobacter cloacae* at a maximum current density of 0.5 A/m^2 (Rezaei et al., 2009). Use of a defined coculture of *Clostridium cellulolyticum* and *G. sulfurreducens* has been reported to degrade cellulose as well, resulting in a current density of approximately 0.1 A/m^2 (Ren et al., 2008). These studies indicate that while conversion of the biomass polymers can be achieved by alternate methods, production of current is low. Thus, electroactive biofilm optimization strategies such as those identified in Figure 16.2 may be necessary to produce high current density when using complex, mixed substrate, or biomass-derived streams.

16.2.1.7 Characterization of anode consortia utilizing fermentation waste streams

A few studies have investigated identification of novel consortia capable of transforming biomass or the waste streams generated from fermentation of lignocellulosic biomass. Wang et al. (2009) used liquid and solid fraction of the steam exploded corn stover as substrates in MFC to produce electricity. The resulting community included *Rhodopseudomonas*, uncultured Verrucomicrobia, Clostridia, Betaproteobacteria, and several other uncultured microbes. Pannell et al. (2016) used dilute acid pretreated corn stover for fermentation and the fermentation effluent as a substrate for electricity production. The consortia included Clostridia as the dominant class, with Betaproteobacteria, Gammaproteobacteria, Bacteroidia, Deltaproteobacteria, and a few unassigned species. Figure 16.8 shows the major species present within these classes. Several microbes including *G. sulfurreducens* that have history of association with electroactivity were reported. The current state of knowledge is insufficient to associate the other microbial members with specific substrates present in the biorefinery waste streams. A better understanding of the structure–function relationships of these communities will enable development of

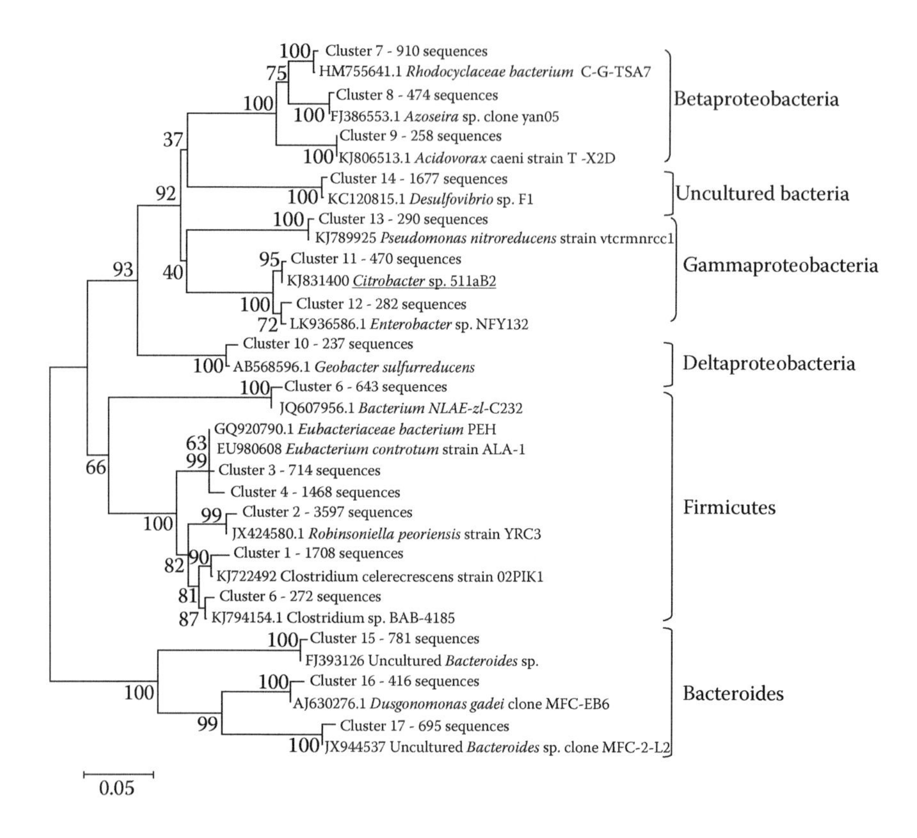

Figure 16.8 Phylogenetic tree of the anode microbial consortia enriched under continuously fed bioanode conditions in an MFC; underlined names indicate microbes that have been associated with exoelectrogenic activity. (From Pannell, T. C., R. K. Goud, D. J. Schell, and A. P. Borole, *Biochemical Engineering Journal,* 116(2016) 85–94.)

optimization methods leading to further improvements in performance of the bioanode to achieve current densities of commercial interest.

16.2.2 Thermoconversion-derived substrates

The alternate pathway for biofuels production via thermochemical processing also generates waste streams that may serve as suitable substrate streams for producing electrons in bioelectrochemical biorefining cells. Gasification and pyrolysis are the two major thermochemical pathways that are used to produce biofuels. Gasification of biomass generates syngas and char or coke as two major products. Biofuels are produced via syngas using thermoconversion or bioconversion pathways. Char is not suitable for use as a substrate for producing electrons. Thus, the gasification route does not produce streams that can be used for producing electrons. The char, however, may serve as a material to produce electrodes for bioelectrochemical systems. This has been reported by Huggins et al. (2014). The syngas stream is a potential candidate for electron generation, however, that would reduce the feedstock for biofuel production. Nevertheless, some researchers have reported the use of syngas as a feed for bioanode for generation of electrons and production of electricity, hydrogen, or other end products. Hussain et al. (2011) investigated mesophilic as well as thermophilic pathways for conversion of syngas to electricity. They reported mechanisms of syngas conversion and indicated the potential of this method to produce electricity. This is relevant for bioelectrochemical biorefining in the context of conversion of the syngas, which is generated via pyrolysis. Because the ultimate goal is to produce biofuels and chemicals, the use of syngas for hydrogen production or generation of electrons via a bioanode could be valuable in a biorefinery for enhancing efficiency and economics.

The pyrolysis process generates a bio-oil, which is then hydrotreated for production of biofuels. This process requires a considerable amount of hydrogen. Because hydrogen is one of the products of bioelectrochemical cells (Figure 16.2), producing hydrogen from any wastes generated in the pyrolysis process may form an important product of bioelectrochemical biorefining. Incidentally, the pyrolysis process does produce an aqueous waste, which is typically associated with bio-oil (Ren et al., 2015). A fast pyrolysis process generates a bio-oil with 40% or more water; however, the oil and water mixture forms a single phase. In catalytic pyrolysis, however, a separate aqueous phase is generated (Tomasini et al., 2014). Either way, the aqueous phase can be separated from fast pyrolysis bio-oil to be used as a substrate for electron generation in a bioanode (Park et al., 2016; Ren et al., 2017). Use of the aqueous phase for hydrogen production via aqueous phase reforming has also been reported; however, the presence of furans, acetic acid, and other carbonyl groups makes it difficult for the catalysts to survive a long period without coking and deactivation (Kechagiopoulos et al., 2006; Fu et al., 2012; Paasikallio et al., 2015). The fate of microbial catalysts in the presence of these compounds is of interest to determine if bioanode processes can perform better than the catalytic reforming processes. Development of a functional bioanode for utilization of the pyrolysis aqueous phase has been recently reported by our group as discussed below.

16.2.2.1 Bioanode development for pyrolysis aqueous phase

The aqueous phase associated with bio-oil generated from pyrolysis is called the bio-oil aqueous phase (BOAP). A BOAP stream generated from switchgrass via intermediate pyrolysis was investigated as a substrate for electron generation in a bioanode (Lewis et al., 2015). The goal was to produce hydrogen using a microbial electrolysis cell (MEC); however, in order to generate hydrogen, it was necessary to produce the electrons from the BOAP first. Recent work conducted by Lewis et al. demonstrated the development of an anode consortia for utilization of BOAP derived from switchgrass. The presence of a high concentration of furanic and phenolic compounds in BOAP poses a challenge for microbial growth because these molecules are known to be inhibitors and toxic to the microorganisms (Palmqvist and Hahn-Hagerdal, 2000). The concentration of these compounds in the pyrolysis streams is much higher than that generated via the bioconversion route. The method of microbial selection and enrichment was based on the patented process developed at Oak Ridge National Laboratory (Borole, 2010). Furthermore, enhancements were made by Lewis et al. (2015) to manage the toxicity of the said compounds via management of the concentrations used in the bioanode during enrichment. A biofilm transfer process was employed in addition to the optimization method outlined in Figure 16.3. This was reported

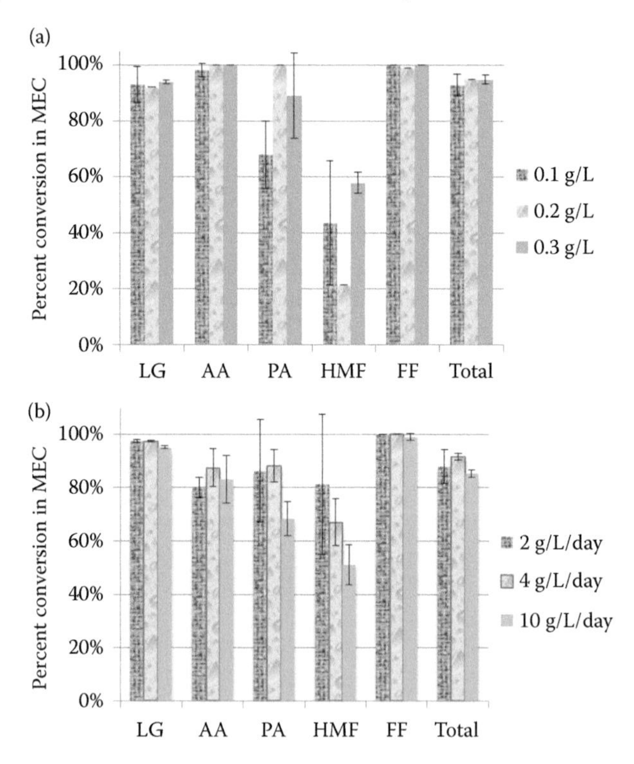

Figure 16.9 Extent of removal of key compounds from bio-oil aqueous phase in an MEC bioanode; total column includes all peaks quantified by HPLC, which contribute 33% to the COD of BOAP (LG = levoglucosan, AA = acetic acid, PA = propionic acid, FF = furfural). (From Lewis, A. J., S. Ren, X. Ye, P. Kim, N. Labbe, and A. P. Borole, *Bioresource Technology*, 195(2015) 231–241.)

to result in a community that was capable of utilization of the complete range of molecules identified in BOAP via HPLC including carboxylic acids, aldehydes, ketones, sugar derivatives, and furanic and phenolic compounds (Lewis et al., 2015). Figure 16.9 shows the percent removal of the major components of BOAP using the electroactive consortium in the anode of an MEC. A number of phenolic compounds were identified in the BOAP including guaiacol, 3-ethylphenol, and vanillic acid, which were reportedly removed.

16.2.2.2 *Bioanode performance using BOAP*

The electrochemical performance of the bioanode is the most important factor in advancing the bioelectrochemical biorefining technology. A range of process conditions have been investigated by Lewis and others including batch versus continuous feeding of BOAP into the anode, a range of concentrations and organic loading rates, flow rate of the anode recirculation, and hydraulic retention time (Lewis et al., 2015; Lewis and Borole, 2016). More recently, analysis of the charge transfer rates, particularly the proton transfer rates, were quantified (Borole and Lewis, 2017). The limitations to achieving high rates of electron and hydrogen yields were identified and the process conditions necessary to alleviate some of the limitations were overcome. This has resulted in the maximum current density reaching above 7 A/m2 using the BOAP substrate (Lewis and Borole, 2016). Considering the presence of the inhibitory compounds in the stream, this is a high current density. Figure 16.10 shows the currents resulting in batch and continuously fed MECs. The average current

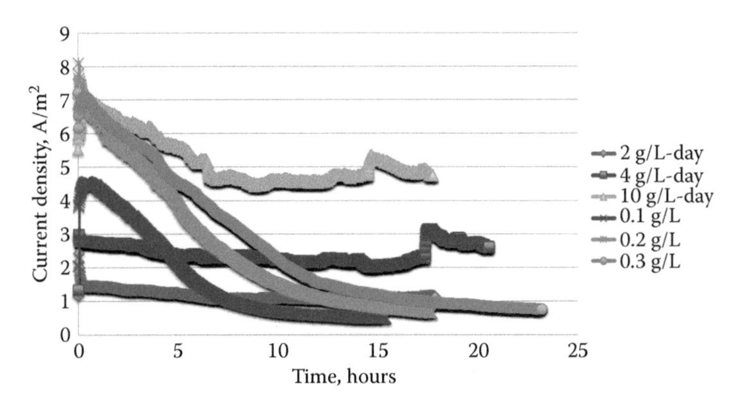

Figure 16.10 Current production profile during batch and continuous operation of MEC generating hydrogen from BOAP; the concentration of BOAP was varied from a COD of 0.1 to 0.3 g/L; in continuous experiment, the organic loading rate was increased from 2 to 10 g/L-day, resulting in a proportional increase in currentproduction. (From Lewis, A. J., S. Ren, X. Ye, P. Kim, N. Labbe, and A. P. Borole, *Bioresource Technology*, 195(2015) 231–241.)

densities over the duration of the run shown in Figure 16.10 are reduced as the substrate is depleted in batch runs; however, the current densities can be maintained fairly steadily in continuously fed MECs.

The coulombic efficiency of the anode was reported to range from 60% to 90% for batch systems (Lewis et al., 2015). The CE decreased with loading rate for the continuously fed bioanode; however, there is potential to improve this via further biocatalyst optimization. The overall energy efficiency for hydrogen production using BOAP as the substrate was reported to be between 50% and 60%. Additionally, the electrical efficiency was also reported, which is the efficiency of hydrogen production from the electrical input to the MEC and is determined as the ratio of the energy content of hydrogen produced to the electrical energy input. This was reported to be in the range of 150%–170% (Lewis et al., 2015). Because the biomass-derived electrons contribute to the hydrogen production, the electrical efficiency can exceed 100%. This is an advantage of the system compared to conventional electrolysis, which yields an electrical efficiency in the range of 70%–75%. Thus, the hydrogen yield can be doubled using MEC and waste as the resource via the bioelectrochemical route.

16.2.2.3 *Managing inhibitory effects of furanic and phenolic compounds*

Studies targeted at understanding the inhibitory effects of furanic and phenolic compounds on electroactive communities have been reported (Zeng et al., 2015, 2016a,b, 2017). A better understanding of the inhibitory effects can lead to improvements in strategies to convert lignocellulosic streams with higher concentrations of these compounds into electrons, which can lead to efficiency and economic gains for the biorefinery. One study reported the use of a mixture of two furanic and three phenolic compounds to investigate inhibition. It was reported that all five compounds were used and transformed by the mixed community used in the study via a route consisting of fermentation, followed by exoelectrogenesis (Zeng et al., 2015). Acetate was reported to be the primary intermediate between fermentation and exoelectrogenesis. Experiments conducted using 1.2 g/L of the substrate mixture were reported to cause inhibition and a decrease in current production compared to the initial concentration of 0.8 g/L. Further investigation into the inhibition mechanism revealed that the inhibition was primarily due to the parent compounds targeted at the exoelectrogenesis reactions, while fermentation was uninhibited (Zeng et al., 2016a). The inhibition occurred only when the initial substrate concentration was higher than 0.8 g/L. Sequential addition of the substrates to higher concentration did not cause inhibition. This suggests that the continuous addition of substrates or the biorefinery wastewater into bioanode at fixed intervals may not be inhibitory, if the total concentration of the mixture of phenolic and

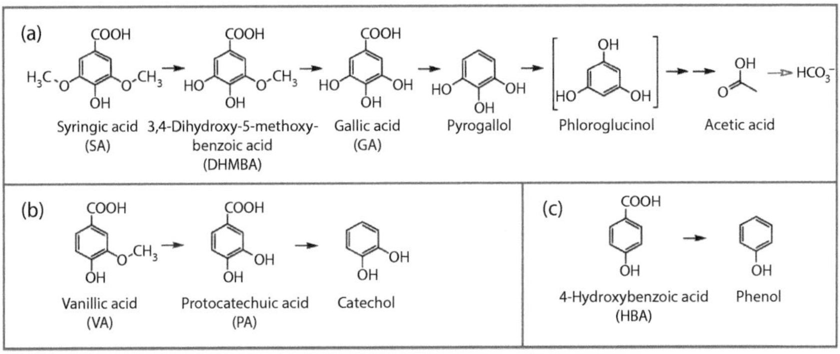

(a) Syringic acid (SA) → 3,4-Dihydroxy-5-methoxy-benzoic acid (DHMBA) → Gallic acid (GA) → Pyrogallol → [Phloroglucinol] → → Acetic acid → HCO_3^-

(b) Vanillic acid (VA) → Protocatechuic acid (PA) → Catechol

(c) 4-Hydroxybenzoic acid (HBA) → Phenol

Figure 16.11 Proposed biotransformation pathways of (a) syringic acid, (b) vanillic acid, and (c) hydroxybenzoic acid under MEC bioanode conditions based on identified metabolites; solid arrows denote fermentative steps; hollow arrow denotes exoelectrogenic step; phloroglucinol in brackets is a hypothetical metabolite. (From Zeng, X., M. A. Collins, A. P. Borole, and S. G. Pavlostathis, *Water Research*, 109(2017) 299–309.)

furanic compounds is kept below 0.8 g/L. The mechanism of biotransformation of the phenolic compounds indicated a demethylation step, followed by decarboxylation (Figure 16.11) (Zeng et al., 2017). The products of transformation were reported to be hydroxylated analogs. Ring cleavage occurred for compounds that contained three hydroxyl groups in the intermediate, resulting in formation of acetate via further degradation. This led directly into the exoelectrogenesis pathway, producing electrons, which were detected as current. Thus, complete transformation of some of the phenolic compounds is possible, yielding electrons as the end product. This implies that use of lignin as a substrate may be possible if it is degraded into monomeric phenolic compounds. This could lead to an opportunity to use the lignin fraction for much better end use as compared to its use now, which is primarily in the production of steam or heat.

Certain phenolic compounds generated during the bioelectrochemical transformation, however, resulted in end products such as phenol and catechol, which persisted in the bioanode investigated by Zeng et al. (2017). Other studies reported by Lewis et al., however, have reported conversion of these compounds. Further investigations targeting phenol and catechol are needed using pure substrates to understand their fate in the bioanode fed with whole BOAP streams versus pure substrates.

16.2.3.4 Characterization of anode consortia utilizing BOAP

Microbial communities developed in the bioanode utilizing biomass-derived substrates are expected to contain fermentative as well as exoelectrogenic microbes at a minimum. In addition, degraders of specific substrates such as furanic and phenolic compounds are likely to be present. The microbial community reported by Lewis et al. (2015) is shown in Figure 16.12 a and b. It included Proteobacteria as the dominant phylum, and Geobacteraceae, Clostridiaceae, Rhodocyclaceae, Enterococcaceae, and Sphingobacteriaceae, plus a few other unique families are in the consortia. The primary role of the Geobacteraceae can be predicted to be exoelectrogenesis; however, the function of most other members is difficult to predict without further investigations. Several other members have known exoelectrogens in the family as well, including Enterococcaceae, Rhodocyclaeceae, and Comamonadaceae; however, these members could also play fermentative or other roles in the community. The community developed using model furanic and phenolic substrates as reported by Zeng et al. (2015) is shown in Figure 16.12c. This community is slightly different than the one reported using BOAP as substrate at the phylum level; however, there were differences at the family and genus level. The 16S rRNA-based identification has limitations and deeper genomic

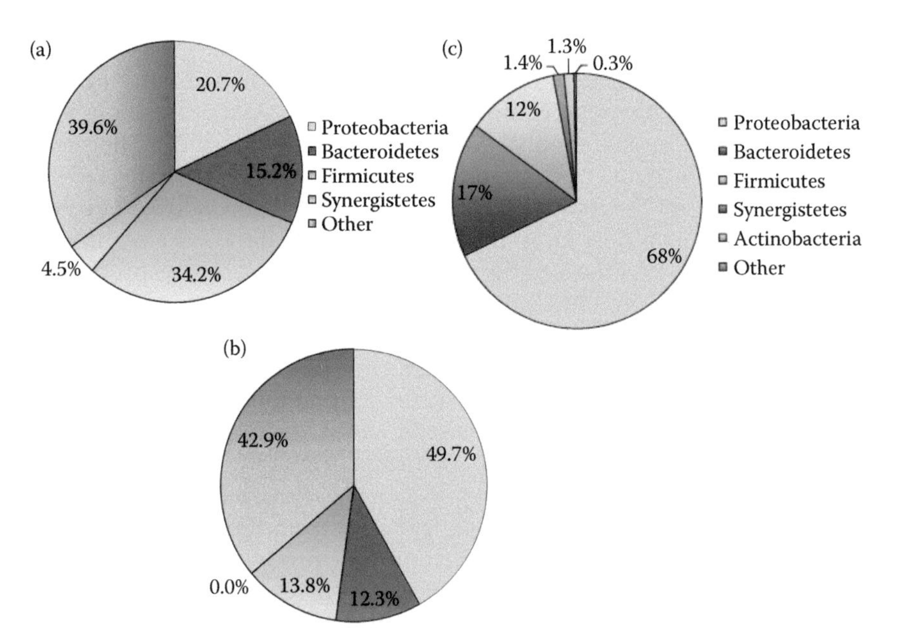

Figure 16.12 Microbial 16S rRNA-based characterization of mixed community developed in bioanode using biomass-derived substrates: (a) bio-oil aqueous phase as the substrate (Replicate 1); (b) Replicate 2 with BOAP; and (c) consortia developed using model furanic and phenolic compounds. (From Lewis, A. J., S. Ren, X. Ye, P. Kim, N. Labbe, and A. P. Borole, *Bioresource Technology*, 195(2015) 231–241; Zeng, X., M. A. Collins, A. P. Borole, and S. G. Pavlostathis, *Water Research*, 109(2017) 299–309.)

information is needed to distinguish between the functional capacities of the communities.

16.3 Cathode processes

The electrons generated at the anode in a BER cell can be used at the cathode of the cell to generate a variety of products as outlined in Figure 16.2. Products such as alcohols, acids, and diols represent high-value products which can improve economic viability of the biorefinery.

16.3.1 Production of alcohols

Butanol, isobutanol, and other higher alcohols are traditionally derived from petroleum, but recent work has shown potential to derive them from biomass via fermentation. These alcohols fetch higher market value because they can also serve as reagents in production of higher-value products. Thus, production of specialty alcohols or use of common alcohols for production of specialty chemicals have potential economic implications for the biorefinery. A method to yield alcohols from waste electrons therefore would be valuable. This is possible via use of electrofermentation, a method that uses electrical current to enhance fermentation. This was investigated by a few researchers using different *Clostridia* sp. (Choi et al., 2014; He et al., 2016; Mostafazadeh et al., 2016). Choi et al. (2014) recently demonstrated production of butanol from glucose at a 2.4-fold higher productivity in a bioelectrochemical reactor versus a typical fermenter. *C. pasteurianum* was used as the biocatalyst, which demonstrated use of reducing equivalents from the electrode in the presence of glucose. This dual use of electron source was a new discovery, which indicates that more reduced chemicals can be produced from sugars than typically possible using fermentation. Furthermore, it was reported that the organism utilized lactate as the substrate for increased butanol production, which was a by-product of the typical fermentation (Choi et al., 2014). Mostafazadeh et al. (2016) reported improvements in butanol production using a similar method and a different strain of *C. pasteurianum*. They reported batch experiments to reach higher titer of butanol and reached 13.3 g/L by applying 1.32 V external potential.

Several other biotechnological routes for production of butanol exist, including the acetone, butanol, and ethanol (ABE) route, which is a well-established method of butanol production. This method produces ABE using sugar as the substrate. Typically, the yield of butanol is low due to coproduction of acetone and ethanol; however, if these coproducts could be reduced, the economic feasibility of producing butanol via this route can be increased. He et al. (2016) reported use of electrofermentation for enhancing butanol production using *C. beijerinckii* IB4. A 1.6-fold higher productivity was reported using the electrofermentation method.

16.3.2 *Production of acids*

Organic acids have numerous applications in the chemical, food, and pharmaceutical industries. Acetic, lactic, propionic, succinic, butyric, and other acids can be produced via fermentation. Production of the acids in pure form, however, requires expensive separation steps. Therefore, production of the acids at higher concentration with less coproducts is desirable. Production of acetic acid was shown by Nevin et al. (2011), using carbon dioxide and electrons as the substrates. While generating products from carbon dioxide, however, this pathway uses a significant number of electrons. More recently, the electrofermentation alternative has been investigated for production of acids. This requires fewer electrons and uses other waste products from fermentation or other processes. Choi et al. (2012) reported production of butyrate using *C. tyrobutyricum* in a bioelectrochemical reactor. Acetate was used to generate the butyrate in the process unlike the usual fermentation in which acetate is generated as a coproduct. Production of acetate was eliminated in the electrofermentation process while increasing butyrate titer from 5 to 8.8 g/L. The yield was also increased from 0.33 to 0.44 g/g. They demonstrated that the increase in yield was due to alteration of $NADH/NAD^+$ ratio in the microbial cells, which was controlled by the external redox potential of the fermenter.

Production of another acid, lactate, was demonstrated by Kracke et al. (2016) via the electrofermentation pathway. A 35-fold improvement in titer was reported via control of external redox potential. They used *C. autoethanogenum* as the biocatalyst, which altered its metabolic pathways producing less acetate, which increased the yield of lactate.

16.3.3 *Production of diols*

Propanediol and butanediol are two common diols that are high-value products useful in production of polymers, composites, adhesives, coatings, polyesters, and solvents, among others. Production of 1,3-propanediol was demonstrated by Zhou et al. (2013) using electrical current to increase yield from glycerol. The fermentation pathways were altered from production of propionate to 1,3-propanediol, increasing the yield from 24.8% to 50.1%. They used a mixed culture and demonstrated that the electrofermentation strategy also works with consortia. Choi et al. (2014) also reported production of 1,3-propanediol using the *C. pasteurianum* biocatalyst they used for production of butanol. Apparently, the microbe can also use glycerol as a substrate and direct it to produce 1,3-propanediol. A 21% increase in its yield was reported based on NADH turnover in the electrically enhanced system.

Production of 2,3-butanediol has been demonstrated by Kracke et al. (2016) via a redox-dependent metabolic shift in the metabolism of *C. autoethanogenum*. Fermentation of glucose was reported to result in threefold higher production of 2,3-butanediol in addition to enhancement in lactate production. This shift was shown to occur via a change in intracellular NADH ratio. It was further demonstrated that the improvement was primarily due to a change in product slate and reduction in products with higher redox state, and that electron uptake from the electrode did not contribute significantly to the increase in reduced products. The results reported by Choi et al. (2014), on the other hand, reported that *C. pasteurianum* could uptake electrons from the electrode directly to produce 1,3-propanediol with glycerol as the substrate. Thus, both pathways—one in which change in NADH/NAD+ ratio enhances the reduced products, as well as the other, where direct uptake of electrons from the electrode yields reduced products—can be active during electrofermentation depending on the biocatalyst used. This demonstrates the versatility of the approach and that use of electrons provided by the cathode can be utilized directly to generate more reduced products via one of the other mechanisms.

16.4 Biorefinery integration

The demonstration of anodic processes to generate electrons from waste materials as well as the use of electrons provided by the cathode to enhance production of valuable, reduced products demonstrates that integration of these two half-reactions can enable production of value-added products using the BER concept described in Section 16.1. Availability of waste products in the biorefinery and the need for production of value-added chemicals are perfectly aligned, and therefore this is a concept that is ripe for further research and development in advancement of the biorefineries and the bioeconomy.

16.5 Future prospects

The recent advancements in demonstration of high current densities reaching more than 7 A/m^2 in the bioanode and the demonstration of production of value-added acids, alcohols, and diols in the cathode of bioelectrochemical cells show the prospects of further research in this area to be very bright. Furthermore, the advantage of this technology in converting waste into chemicals has real economic potential. The concept of bioelectrochemical biorefining can be married to metabolic engineering to enhance the yields of desirable products further toward economic feasibility. This can be a novel route to develop technologies for enabling the bioeconomy in the 21st century.

Acknowledgments

The manuscript has been authored by UT-Battelle, LLC, under Contract No. DEAC05-00OR22725 with the U.S. Department of Energy. The funding from Laboratory Director's Research and Development program at Oak Ridge National Laboratory and support from the Bioenergy Technologies Office (BETO) within the DOE Office of Energy Efficiency and Renewable Energy (EERE) is acknowledged.

References

Bond, D. R. and D. R. Lovley, Electricity production by *Geobacter sulfurreducens* attached to electrodes, *Applied and Environmental Microbiology*, 69(2003) 1548–1555.

Borole, A. P. 2010. Microbial fuel cell with improved anode. U.S. Patent 7,695,834.

Borole, A. P. 2012. Microbial fuel cell treatment of ethanol fermentation process water. U.S. Patent 8,192,854 B2.

Borole, A. P., C. Hamilton, and D. Schell, Conversion of residual organics in corn stover-derived biorefinery stream to bioenergy via microbial fuel cells, *Environmental Science & Technology*, 47(2013), 642–648.

Borole, A. P. and A. J. Lewis, Proton transfer in microbial electrolysis cells, *Sustainable Energy & Fuels*, 1(2017) 725–736.

Borole, A. P., G. Reguera, B. Ringeisen et al., Electroactive biofilms: Current status and future research needs, *Energy & Environmental Science*, 4(2011) 4813–4834.

Catal, T., Y. Z. Fan, K. C. Li, H. Bermek, and H. Liu, Effects of furan derivatives and phenolic compounds on electricity generation in microbial fuel cells, *Journal of Power Sources*, 180(2008a) 162–166.

Catal, T., K. Li, H. Bermek, and H. Liu, Electricity production from twelve monosaccharides using microbial fuel cells, *Journal of Power Sources*, 175(2008b) 196–200.

Choi, O., T. Kim, H. M. Woo, and Y. Um, Electricity-driven metabolic shift through direct electron uptake by electroactive heterotroph *Clostridium pasteurianum*, *Scientific Reports*, 4(2014) 6961.

Choi, O., Y. Um, and B. I. Sang, Butyrate production enhancement by *Clostridium tyrobutyricum* using electron mediators and a cathodic electron donor, *Biotechnology and Bioengineering*, 109(2012) 2494–2502.

Esteve-Nunez, A., M. Rothermich, M. Sharma, and D. Lovley, Growth of *Geobacter sulfurreducens* under nutrient-limiting conditions in continuous culture, *Environmental Microbiology*, 7(2005) 641–648.

Fu, J., S. H. Hakim, and B. H. Shanks, Aqueous-phase processing of bio-oil model compounds over Pt-Re supported on carbon, *Topics in Catalysis*, 55(2012) 140–147.

He, A.-Y., C.-Y. Yin, H. Xu et al., Enhanced butanol production in a microbial electrolysis cell by *Clostridium beijerinckii* IB4, *Bioprocess and Biosystems Engineering*, 39(2016) 245–254.

Huang, L. P. and B. E. Logan, Electricity production from xylose in fed-batch and continuous-flow microbial fuel cells, *Applied Microbiology and Biotechnology*, 80(2008), 655–664.

Huang, L., R. J. Zeng, and I. Angelidaki, Electricity production from xylose using a mediator-less microbial fuel cell, *Bioresource Technology*, 99(2008) 4178–4184.

Huggins, T., H. Wang, J. Kearns, P. Jenkins, and Z. J. Ren, Biochar as a sustainable electrode material for electricity production in microbial fuel cells, *Bioresource Technology*, 157(2014) 114–119.

Hussain, A., S. R. Guiot, P. Mehta, V. Raghavan, and B. Tartakovsky, Electricity generation from carbon monoxide and syngas in a microbial fuel cell, *Applied Microbiology and Biotechnology*, 90(2011) 827–836.

Kechagiopoulos, P. N., S. S. Voutetakis, A. A. Lemonidou, and I. A. Vasalos, Hydrogen production via steam reforming of the aqueous phase of bio-oil in a fixed bed reactor, *Energy & Fuels*, 20(2006) 2155–2163.

Kracke, F., B. Virdis, P. V. Bernhardt, K. Rabaey, and J. O. Kromer, Redox dependent metabolic shift in *Clostridium autoethanogenum* by extracellular electron supply, *Biotechnology for Biofuels*, 9(2016) 249–260.

Lewis, A. J. and A. P. Borole, Understanding the impact of flow rate and recycle on the conversion of a complex biorefinery stream using a flow-through microbial electrolysis cell, *Biochemical Engineering Journal*, 116(2016) 95–104.

Lewis, A. J., S. Ren, X. Ye, P. Kim, N. Labbe, and A. P. Borole, Hydrogen production from switchgrass via a hybrid pyrolysis-microbial electrolysis process, *Bioresource Technology*, 195(2015) 231–241.

Lu, L. and Z. J. Ren, Microbial electrolysis cells for waste biorefinery: A state of the art review, *Bioresource Technology*, 215(2016) 254–264.

Luo, Y., G. Liu, R. Zhang, and C. Zhang, Power generation from furfural using the microbial fuel cell, *Journal of Power Sources*, 195(2010) 190–194.

Mostafazadeh, A. K., P. Drogui, S. K. Brar et al., Enhancement of biobutanol production by electromicrobial glucose conversion in a dual chamber fermentation cell using *C. pasteurianum*, *Energy Conversion and Management*, 130(2016) 165–175.

Nevin, K. P., S. A. Hensley, A. E. Franks et al., Electrosynthesis of organic compounds from carbon dioxide is catalyzed by a diversity of acetogenic microorganisms, *Applied and Environmental Microbiology*, 77(2011) 2882–2886.

Paasikallio, V., J. Kihlman, C. A. S. Sanchez et al., Steam reforming of pyrolysis oil aqueous fraction obtained by one-step fractional condensation, *International Journal of Hydrogen Energy*, 40(2015) 3149–3157.

Palmqvist, E. and B. Hahn-Hagerdal., Fermentation of lignocellulosic hydrolysates. I: Inhibition and detoxification, *Bioresource Technology*, 74(2000) 17–24.

Pannell, T. C., R. K. Goud, D. J. Schell, and A. P. Borole, Effect of fed-batch vs. continuous mode of operation on microbial fuel cell performance treating biorefinery wastewater, *Biochemical Engineering Journal*, 116(2016) 85–94.

Pant, D., G. Van Bogaert, L. Diels, and K. Vanbroekhoven, A review of the substrates used in microbial fuel cells (MFCs) for sustainable energy production, *Bioresource Technology*, 101(2010) 1533–1543.

Park, L. K., S. Ren, S. Yiacoumi et al., Separation of switchgrass bio-oil by water/organic solvent addition and pH adjustment, *Energy Fuels*, 30(2016) 2164–2173.

Ren, Z., L. M. Steinberg, and J. M. Regan, Electricity production and microbial biofilm characterization in cellulose-fed microbial fuel cells, *Water Science and Technology*, 58(2008), 617–622.

Ren, S., X. P. Ye, and A. P. Borole, Separation of chemical groups from bio-oil water-extract via sequential organic solvent extraction, *Journal of Analytical and Applied Pyrolysis*, 123(2017), 30–39.

Ren, S., P. Ye, A. P. Borole, P. Kim, and N. Labbe, Analysis of switchgrass-derived bio-oil and associated aqueous phase generated in a pilot-scale auger pyrolyzer, *Journal of Analytical and Applied Pyrolysis*, 119(2015) 97–103.

Rezaei, F., T. L. Richard, and B. E. Logan, Enzymatic hydrolysis of cellulose coupled with electricity generation in a microbial fuel cell, *Biotechnology and Bioengineering*, 101(2008), 1163–1169.

Rezaei, F., D. F. Xing, R. Wagner et al., Simultaneous cellulose degradation and electricity production by *Enterobacter cloacae* in a microbial fuel cell, *Applied and Environmental Microbiology*, 75(2009) 3673–3678.

Rismani-Yazdi, H., A. D. Christy, B. A. Dehority et al., Electricity generation from cellulose by rumen microorganisms in microbial fuel cells, *Biotechnology and Bioengineering*, 97(2007) 1398–1407.

Tomasini, D. B., F. Cacciola, F. Rigano et al., Complementary analytical liquid chromatography methods for the characterization of aqueous phase from pyrolysis of lignocellulosic biomasses, *Analytical Chemistry*, 86(2014) 11255–11262.

Wang, X., Y. Feng, H. Wang et al., Bioaugmentation for electricity generation from corn stover biomass using microbial fuel cells, *Environmental Science & Technology*, 43(2009) 6088–6093.

Yates, M. D., P. D. Kiely, D. F. Call et al., Convergent development of anodic bacterial communities in microbial fuel cells, *ISME Journal*, 6 (2012) 2002–2013.

Zeng, X., A. P. Borole, and S. G. Pavlostathis, Biotransformation of furanic and phenolic compounds with hydrogen gas production in a microbial electrolysis cell, *Environmental Science & Technology*, 49(2015) 13667–13675.

Zeng, X., A. P. Borole, and S. G. Pavlostathis, Inhibitory effect of furanic and phenolic compounds on exoelectrogenesis in a microbial electrolysis cell bioanode, *Environmental Science & Technology*, 50(2016a) 11357–11365.

Zeng, X., A. P. Borole, and S. G. Pavlostathis, Performance evaluation of a continuous-flow bioanode microbial electrolysis cell fed with furanic and phenolic compounds, *RSC Advances*, 6(2016b) 65563–65571.

Zeng, X., M. A. Collins, A. P. Borole, and S. G. Pavlostathis, The extent of fermentative transformation of phenolic compounds in the bioanode controls exoelectrogenic activity in a microbial electrolysis cell, *Water Research*, 109(2017) 299–309.

Zhou, M., J. Chen, S. Freguia, K. Rabaey, and J. Keller, Carbon and electron fluxes during the electricity driven 1,3-propanediol biosynthesis from glycerol, *Environmental Science & Technology*, 47(2013) 11199–11205.

Index

D

E

G

O

P